PURINE AND PYRIMIDINE METABOLISM IN MAN VII

Part A: Chemotherapy, ATP Depletion, and Gout

ADVANCES IN EXPERIMENTAL MEDICINE AND BIOLOGY

Recent Volumes in this Series

Volume 303
IMMUNOBIOLOGY OF PROTEINS AND PEPTIDES VI:
Human Immunodeficiency Virus, Antibody Immunoconjugates, Bacterial Vaccines, and Immunomodulators
Edited by M. Zouhair Atassi

Volume 304
REGULATION OF SMOOTH MUSCLE CONTRACTION
Edited by Robert S. Moreland

Volume 305
CHEMOTACTIC CYTOKINES: Biology of the Inflammatory Peptide Supergene Family
Edited by J. Westwick, I. J. D. Lindley, and S. L. Kunkel

Volume 306
STRUCTURE AND FUNCTION OF THE ASPARTIC PROTEINASES: Genetics, Structures, and Mechanisms
Edited by Ben M. Dunn

Volume 307
RED BLOOD CELL AGING
Edited by Mauro Magnani and Antonio De Flora

Volume 308
CELLULAR AND MOLECULAR MECHANISMS IN HYPERTENSION
Edited by Robert H. Cox

Volume 309A
PURINE AND PYRIMIDINE METABOLISM IN MAN VII, Part A:
Chemotherapy, ATP Depletion, and Gout
Edited by R. Angus Harkness, Gertrude B. Elion, and Nepomuk Zöllner

Volume 309B
PURINE AND PYRIMIDINE METABOLISM IN MAN VII, Part B:
Structural Biochemistry, Pathogenesis, and Metabolism
Edited by R. Angus Harkness, Gertrude B. Elion, and Nepomuk Zöllner

---

A Continuation Order Plan is available for this series. A continuation order will bring delivery of each new volume immediately upon publication. Volumes are billed only upon actual shipment. For further information please contact the publisher.

PURINE AND PYRIMIDINE METABOLISM IN MAN VII

Part A: Chemotherapy, ATP Depletion, and Gout

Edited by

R. Angus Harkness
Institute of Child Health
London, United Kingdom

Gertrude B. Elion
Burroughs Wellcome Co.
Research Triangle Park, North Carolina

and

Nepomuk Zöllner
Universität München
München, Germany

PLENUM PRESS • NEW YORK AND LONDON

Library of Congress Cataloging in Publication Data

International Symposium on Purine and Pyrimidine Metabolism in Man (7th: 1991: Bournemouth, England)
Purine and pyrimidine metabolism in man VII / edited by R. Angus Harkness, Gertrude B. Elion, and Nepomuk Zöllner.
p. cm.—(Advances in experimental medicine and biology; v. 390A–B)
Proceedings of a joint meeting of the Seventh International Symposium on Purine and Pyrimidine Metabolism in Man and the Third European Symposium on Purine and Pyrimidine Metabolism in Man, held June 30–July 5, 1991, in Bournemouth, England.
Includes bibliographical references and index.
Contents: Pt. A. Chemotherapy, ATP depletion, and gout—Pt. B. Structural biochemistry, pathogenesis, and metabolism.
ISBN 0-306-44092-X (hardbound): pt. A.—ISBN 0-306-44093-8 (hardbound): pt. B
1. Purines—Metabolism—Disorders—Congresses. 2. Pyrimidines—Metabolism—Disorders—Congresses. 3. Purines—Therapeutic use—Congresses. 4. Pyrimidines—Therapeutic use—Congresses. I. Harkness, R. A. (Robert Angus) II. Elion, Gertrude B. III. Zöllner, Nepomuk. IV. European Symposium on Purine and Pyrimidine Metabolism in Man (3rd: 1991: Bournemouth, England) V. Title. VI. Series.
[DNLM: 1. Purines—metabolism—congresses. 2. Pyrimidines—metabolism—congresses. W1 AD559 v. 309 / QU 58 I609p 1991]
RC632.P87I59 1991
616.3′9—dc20
DNLM/DLC 91-39326
for Library of Congress CIP

Proceedings of the Seventh International/Third European Joint Symposium on Purine and Pyrimidine Metabolism in Man, held June 30–July 5, 1991, in Bournemouth, United Kingdom

ISBN 0-306-44092-X

Printed in the United States of America

JOSEPH FRANKLIN HENDERSON

These proceedings are dedicated to Dr. Joseph Franklin Henderson whose achievements in the biochemistry and pharmacology of purines and pyrimidines over the years have been outstanding and have markedly influenced progress in this area of science. How many of us faced with an unusual finding have turned, for example, to the book entitled 'Nucleotide Metabolism: an Introduction' and found with gratitude the explanation we were seeking there.

Frank, as he is known affectionately to his many friends throughout the world, has contributed immeasurably to our understanding of the basic pathways of purine and pyrimidine metabolism and their interaction. His most important contributions to the field of purine biochemistry have included quantitative studies of both normal and alternative pathways of purine metabolism in intact cells and the regulation of purine metabolism in a variety of tissues. Moreover, he has alerted us all to the fact that there can be pitfalls in the extrapolation of results in cell extracts to the behaviour of enzymes in intact cells and tissues. The work of his group has focussed on anticancer chemotherapy as well as inherited diseases and, together, they have published many studies on the biochemical basis of purine-related anticancer drug action and resistance. This has also resulted in the development of a large number of new methods.

Frank was educated in the University of Arizona where he gained his Bachelor's and Master's degrees in Science and subsequently the University of Wisconsin where he was awarded a Ph.D. He has been associated for more than thirty years with the Cancer Research Unit and Department of Biochemistry of the University of Alberta, Canada. He has published over two hundred papers and, in addition, has written many excellent review articles and books. He has communicated his enthusiasm and critical approach to research to a large number of young investigators now working throughout the world. During his long and distinguished career he has been editor of the Canadian Journal of Biochemistry, invited participant at many international symposia and member of scientific committees in both Canada and the USA. He has, of course, been a long-standing member of the International Scientific Committee in the 'Purine and Pyrimidine in Man' series and a consistent contributor to successive Symposia. He has accumulated many honours for his achievements, which include the Ayerst Award of the Canadian Biochemical Society and the Geigy Prize in Rheumatology.

Frank is now retiring from active research. We hope that this will give him time to update his Nucleotide Metabolism book, as well as pursue the many other interests which he shares with his wife, Ruth. Frank is a truly remarkable man. Besides his achievements in purine biochemistry, his intellectual curiosity extends to diverse areas of science and philosophy. His enthusiasm for research in the field of purine and pyrimidine metabolism has enriched basic research in the area immeasurably. His repeated exhortations that we should temper that enthusiasm with caution, and recognise the limitations of our experimental models and working assumptions, should be heeded by all young investigators.

PREFACE

These two volumes record the scientific and clinical work presented at the VIIth International and 3rd European joint symposium on purine and pyrimidine metabolism in man held at the Bournemouth International Conference Centre, Bournemouth, UK, from 30th June to 5th July 1991. The series of international meetings at three yearly intervals have previously been held initially in 1973 in Israel, then Austria, Spain, the Netherlands, USA and Japan. The European Society for the Study of Purine and Pyrimidine Metabolism in Man (ESSPPM) which has its own executive and some finance first met in Switzerland in 1987, then in Germany in 1989.

The steady evolution of the science in this series of meetings is intellectually satisfying; the subsequent clinical progress is emotionally and economically reassuring. As befits the position of purines and pyrimidines at the centre of biochemistry, there has been steady scientific development into molecular genetics and now onto developmental controls and biochemical pharmacology. The complexities of the immune system are being unravelled but an understanding of the human brain largely eludes us. Laboratory based scientists now predominate over those who work as clinical specialists in rheumatology, immunology, oncology and paediatrics.

However, there continue to be major clinical objectives since large sections are concerned with major causes of death like ATP depletion, cancer and now AIDS; the laboratory work is providing clinical solutions.

Because basic science is often linked to a clinical problem these volumes do not separate into clinical and laboratory sections. It is clear that some articles in these two books are preliminary short communications on major work which allows priority to be established and should not prevent further full publication of major new results. However, other articles are worthwhile additional records which are adequate in themselves as contributions to our accummulated experience and do not require additional recording.

Recognition has been given to this area of science. The other participants at the meeting had great pleasure in being able to express their pleasure at the award of the 1988 Nobel Prize for Medicine to Gertrude Elion and George Hitchings for their work in this area of science; the award was announced just after the previous meeting in the international series, in Japan. The European editors are honoured by being accompanied by Dr. Elion as coeditor of these two volumes.

The international scientific community is especially endebted to the Mayor and Council of the Borough of Bournemouth, representing its citizens for the use of the superb Bournemouth International Conference Centre and the services of its efficient and friendly staff. The list of generous sponsors of the meeting must be led by the Arthritis and Rheumatism Council of the UK and include Amersham International plc, Burroughs Wellcome Co (USA) Division, Cancer Research Fund, Edward Arnold, Glaxo Group Research Ltd, Hodder & Stoughton, Millipore (UK) Ltd, Napp Laboratories, National Institute of Health (USA), Parke-Davis Pharmaceutical Research, Pfizer Ltd, The Royal Society, Sandoz Pharmaceuticals, Sigma Chemical Co Ltd, Warner Lambert and The Wellcome Foundation Ltd. Our plenary lecturers deserve special thanks for providing their own support in order to reduce costs for other participants.

The success of the meeting can in part be judged from the fact that 27 countries contributed. The Europeans especially enjoyed the freer communications amongst the peoples of our small planet.

Although the meeting was the product of many teams' work, including that of the International Committee, the untiring efforts of Anne Simmonds from 1987 onwards tower over other contributions. The rest of the local committee were Tom Scott (treasurer), Dave Perrett (printing), Simon Jarvis & George Nuki (workshops), Richard Watts & Neopomuk Zollner (abstracts) and Angus Harkness (editor) with Francoise Roch-Ramel (ESSPPM finance) who was co-opted for budgeting. At the meeting itself the volunteer administrators, Catherine Potter, Faith Scott, Helen Lees, Maureen Morris and Eveline Harkness were crucial to the success of PP 91.

We look forward to the 8th International meeting to be held in the University of Indiana, USA, in 1994 and to the 4th European meeting in Nijmegen, the Netherlands, in 1993.

G B Elion
Nobel Laureate (1988)
Burroughs Wellcome, USA

*R A Harkness
Institute of Child Health
University of London, UK

N Zöllner
University of Munich, Germany

---

*Contact Editor, 15 St Thomas's Drive, Hatch End, Pinner, Middlesex HA5 4SX, UK.

CONTENTS

PART A

CHEMOTHERAPY

1. Historical Development

2. Cancer and AIDS Treatment

3. Development and Effects of New Analogues

4. Analogue Effects and Metabolism

6. Pyrimidine Analogues

7. Hypouricaemic Compounds

8. Parasitic Diseases

9. Enzyme Replacement

B. GOUT AND URIC ACID

C. ATP DEPLETION

1. Biochemistry of ATP Depletion

2.1 Xanthine Oxidase and Reperfusion Injury

2. DNA Repair

3. Nucleoside Transport

4. Adenosine Receptors

BIOCHEMICAL APPROACH TO NEW MEDICATIONS

George H. Hitchings

Wellcome Research Laboratories
Burroughs Wellcome Co.
Research Triangle Park, NC 27709

My story began with the discovery of sulfanilamide--the fourth useful chemotherapeutic agent the world had known. Morris Fishbein in the Boston Herald of December 31, 1939, described the three previous specific remedies as: "quinine for malaria, arsphenamine for syphilis and diphtheria antitoxin. Sulfanilamide entered this "hall of fame" by acting as a specific for a strain of the streptococcus germ." It was important for its therapeutic effects but perhaps more important for the development of the antimetabolite theory.

I joined Burroughs Wellcome Co. (then in Tuckahoe, N.Y.) July 10, 1942--the only member of its Biochemistry Department. I received my Ph.D. degree in 1933 from Harvard Medical School under the supervision of Cyrus Fiske, Associate Professor of Biochemistry. He, with Yellapragada Subba Row had discovered adenosine triphosphate (ATP) in tissues (among other exciting metabolites including phosphocreatine). I was assigned the development of methods of study of ATP. By the time I got my degree I was saturated with knowledge of nucleic acids. The year 1933 was the nadir of the great depression--no new positions offered and no interest in nucleic acids. I spent a somewhat frustrating nine years before I joined Burroughs Wellcome Co. which had in the preceding three months undergone some drastic changes. The result was that, when I joined, there was no president and no research director, so I was free to develop my own program. So I interested Elvira Falco, then an assistant in bacteriology in working with me. I saw a way to combine interests in anti-metabolites and in nucleic acids and Falco and I developed the *Lactobacillus casei* test (1). This organism could grow (and produce titrable acidity) with a purine and thymine (not very unusual) and also with a metabolite that later was recognized as a form of folic acid. (This was unusual--most microorganisms are dependent on internal synthesis of this important metabolite and unable to use exogenous forms). We proposed to study analogs of purines, thymine, and *L. casei* factor. This was an unusual enterprise in a pharmaceutical house--not only because its primary purpose was to gather basic information, but also because information was obtainable with 25 mg of compound where customarily 25 g of compound had to be submitted before biological work would be done.

By 1944 we had a Research Director (Erwin Nelson) and a scientist on the Wellcome Foundation Board in the United Kingdom (Charles Kellaway). By then *L. casei* was indicating interesting biological results and my superiors agreed to an addition to my staff. I chose Gertrude Elion, a graduate of Hunter College, *summa cum laude*. She soon became responsible for developments in purine analogs while Falco engineered the pyrimidine (and soon folic acid) analogs.

By 1947 we had had some additions to our staff (Peter Russell and Norman Whittaker from England and two or three others). The *L. casei* and other tests were so exciting that the word in the laboratories was: "Now we have the chemotherapeutic agents; we have only to find the diseases to go with them." Both the purine and antagonists and the antifolate pyrimidines were to lead to important new drugs. The purine analogs found utility in the field of leukemia, immunosuppression for transplantation and autoimmune disease, gout and herpes virus infections. The 2,4-diaminopyrimidines were dihydrofolate reductase inhibitors and several became valuable drugs in the treatment of malaria, bacterial infections and toxoplasmosis.

By 1948 we had established a collaboration with the Sloan-Kettering Institute for testing our compounds against solid tumors and leukemia in mice and subsequently in man. Our first real encouragement in the anticancer field came with the finding by Joseph Burchenal that 2,6-diaminopurine, which we had identified as an adenine antagonist in *L. casei*, produced several remissions in acute leukemia. One of these remissions lasted for three years (2). Unfortunately, the drug proved to be too toxic for further development but it convinced us that we were on the right track. Within three years we had synthesized and tested 6-mercaptopurine and 6-thioguanine, which soon became important antileukemic drugs (3-5). We were enthralled to watch the progress in the treatment of leukemia, from a life expectancy of 3 months in children with acute leukemia in 1948 to 1 year in 1954, after the introduction of 6-mercaptopurine, to 85% cures in 1985 with the use of multi-drug therapy. 6-Mercaptopurine continues to play an important role in consolidation and maintenance therapy in childhood leukemia to this day.

The finding by Robert Schwartz (6) that 6-mercaptopurine could suppress the antibody response in rabbits to a foreign antigen, e.g. bovine gamma globulin, began a train of events which led to making kidney transplantation successful between unrelated donor and recipient (5). Roy Calne, a young surgeon in England, intrigued by Schwartz's findings tested 6-mercaptopurine in kidney transplants in dogs and found prolonged survival of the graft (7). He then came to work with Joseph Murray at the Peter Bent Brigham Hospital in Boston. Murray had perfected the surgical techniques for kidney transplantation, successful in identical twins, but had been unable to overcome the rejection that genetic differences between host and donor induced. In a few exciting years, azathioprine (IMURAN®), the derivative which we had synthesized as a pro-drug of 6-mercaptopurine, made possible cross-genetic transplants (8). Transplantation today encompasses many organs in the body. Joseph Murray and E. Donnall Thomas shared the Nobel Prize in 1990, mainly for work on transplantation.

Metabolic studies of the thiopurines revealed a number of catabolic pathways which made the drugs less available for conversion to their nucleotides (9). This led us to search for ways of preventing such destruction. One of these attempts was azathioprine, the compound described above. It is now used not only for preventing transplant rejection but for the treatment of autoimmune diseases such as rheumatoid arthritis.

One of the few purine-metabolizing enzymes which was available for study in our laboratory in the early 1950's was xanthine oxidase, which could be prepared from unpasteurized milk. Doris Lorz's studies on the substrate and inhibitor specificities of this enzyme were very useful. We chose one of the inhibitors which she had identified, allopurinol, to inhibit the catabolism of 6-mercaptopurine *in vivo* and to inhibit the formation of uric acid in conditions of hyperuricemia (10). For the past 25 years allopurinol has been a standard treatment for gout as well as for secondary hyperuricemias arising from the treatment of malignancy. Allopurinol and allopurinol riboside have also found utility as antileishmanial and antitrypanosomal agents due to the different metabolism of these compounds in these protozoa (11,12).

I would now like to return to the other major part of our program, the 2,4-diaminopyrimidines (13,14). By 1948 we had established a collaboration with our colleagues in the Wellcome Research Laboratories in the United Kingdom, Len Goodwin and Stanley Bushby, which led to the development of the antimalarial drug, pyrimethamine (15), and the antibacterial drug, trimethoprim (16,17). It was during the examination of these 2,4-diaminopyrimidines in a variety of test systems that we came to appreciate how selective these compounds were in their inhibition of the growth of different organisms (13,14). The true understanding of this selectivity came with the isolation of the dihydrofolate reductases from different species and the recognition of the selectivity of these small molecule antifolates for the specific dihydrofolate reductases (18,19). Thus, pyrimethamine, which was a potent inhibitor of the enzyme from the malarial parasite *Plasmodium berghei*, was much less inhibitory to the enzyme from rat liver, while trimethoprim was most active against

Table 1. Inhibition of Dihydrofolate Reductases

	$IC_{50} \times 10^{8}$, μM		
	E. coli	Rat Liver	*P. berghei*
Methotrexate	0.1	0.2	0.07
Pyrimethamine	250	70	0.05
Trimethoprim	0.5	3000	7.0

the enzyme from *E. coli* (Table 1). Methotrexate, the full structural analog of folic acid, did not distinguish among the specific enzymes. These investigations of structure-activity relationships permitted us to find the best inhibitor for a particular organism.

The idea that antifolates could be so selective met with considerable skepticism at first. However, the skeptics were convinced by the evidence of synergism in combinations of our pyrimidines and sulfonamides which led

Table 2. Therapeutic Agents

Base Analogs		
Antileukemic		Immunosuppressant
2,6-Diaminopurine		
6-Thioguanine	→	
6-Mercaptopurine	→	
	→	Azathioprine
Hypouric		Antiprotozoal
Allopurinol	→	
	→	Allopurinol riboside
DHFR Inhibitors		
Pyrimethamine		Antimalarial
Trimethoprim		Antibacterial
Piritrexim		Antitumor
Nucleoside Analogs		
Acyclovir		Antiherpetic
Zidovudine		Anti-AIDS

not only to increases in potency but to the conversion of bacteriostatic to bactericidal activity (13,14,17,20). The sulfonamides, which were products of Hoffman-LaRoche, were combined with the diaminopyrimidines which most closely matched their pharmacokinetic profiles.

At present the combination of trimethoprim and sulfamethoxazole (Septra, Bactrim and other trade names) is a leading antibacterial, especially in the Third World where its low cost, stability and broad spectrum of activity make it a leading antibacterial. One writer estimated that it had saved 1,000,000 lives in the world.

The program, which dealt at first with the heterocyclic bases, was later expanded to include analogs of the nucleosides. Two important antiviral agents resulted from these studies. Howard Schaeffer, Gertrude Elion and their groups successfully developed the acyclic nucleoside analog acyclovir (ZOVIRAX®) for the treatment of herpes virus infections such as genital herpes, cold sores, mucocutaneous herpes, chicken pox, shingles and herpes encephalitis (21-23). David Barry led a team which developed zidovudine (AZT, RETROVIR®) for the treatment of AIDS (24).

The work goes on. New analogs and new uses continue to appear. It has been an exciting and productive program, now in existence for nearly 50 years. Table 2 summarizes some of its accomplishments. An extensively annotated review of our work appears in the Nobel lectures which Gertrude Elion and I delivered in 1988 (9,14).

Acknowledgment: The author is very grateful for the assistance and contributing criticism of Gertrude B. Elion.

REFERENCES

1. G.H. Hitchings, E.A. Falco, M.B. Sherwood, Science 102:251 (1945).

2. J.H. Burchenal, D.A. Karnofsky, E.M. Kingsley-Pillers, C.M. Southam, W.P.L. Meyers, G.C. Escher, L.F. Craver, H.W. Dargeon, C.P. Rhoads, Cancer 4:549 (1951).

3. G.B. Elion, G.H. Hitchings, H. VanderWerff, J. Biol. Chem. 192:505 (1951).

4. J.H. Burchenal, M.L. Murphy, R.R. Ellison, M.P. Sykes, C.T.C. Tan, L.A. Leone, D.A. Karnofsky, L.F. Craver, H.W. Dargeon, C.P. Rhoads, Blood 8:965 (1953).

5. G.H. Hitchings and G.B. Elion, Cancer Res. 45:2415 (1985).

6. R. Schwartz, J. Stack and W. Dameshek, Proc. Soc. Exp. Biol. Med. 99:164 (1958).

7. R.Y. Calne, Lancet I:417 (1960).

8. J.E. Murray, J.P. Merrill, J.H. Harrison, R.E. Wilson, G.J. Dammin, N. Engl. J. Med. 268:1315 (1963).

9. G.B. Elion, in Les Prix Nobel 1988, Norstedts Tryckeri, Stockholm, 1989, pp. 267-288; also publ. In Vitro Cellular and Developmental Biology 25:321 (1989).

10. R.W. Rundles, J.B. Wyngaarden, G.H. Hitchings, G.B. Elion, H.R. Silberman, Trans. Assoc. Am. Physicians 76:126 (1963).

11. J.J. Marr, R.L. Berens, D.J. Nelson, Science 201:1018 (1978).

12. D.J. Nelson, C.J.I. Bugge, G.B. Elion, R.L. Berens, J.G. Marr, J. Biol. Chem. 254:3959 (1979).

13. G.H. Hitchings, Cancer Res. 29:1895 (1969).

14. G.H. Hitchings in Les Prix Nobel 1988, Norstedts Tryckeri, Stockholm, 1989, pp. 296-313; also publ. In Vitro Cellular and Development Biololgy 25:303 (1989).

15. E.A. Falco, L.G. Goodwin, G.H. Hitchings, I.M. Rollo, P.B. Russell, Brit. J. Pharmacol. 6:185 (1951).

16. B. Roth, E.A. Falco, G.H. Hitchings, J. Med. Pharm. Chem. 5:1103 (1962).

17. S.R.M. Bushby, G.H. Hitchings, Brit. J. Pharmac. Chemother. 33:72 (1968).

18. J.J. Burchall, G.H. Hitchings, Molec. Pharmacol. 1:126 (1965).

19. G.H. Hitchings, J.J. Burchall, Adv. Enzymol. 27:417 (1965).

20. S.R.M. Bushby, in Trimethoprim-Sulfamethoxazole, Microbiological Pharmacological and Clinical Considerations, ed. by M. Finland and E.H. Kass, pp. 10-30 (1973).

21. G.B. Elion, P.A. Furman, J.A. Fyfe, P. de Miranda, L. Beauchamp, H.J. Schaeffer, Proc. Natl. Acad. Sci. USA 74:5716 (1977).

22. H.J. Schaeffer, L. Beauchamp, P. de Miranda, G.B. Elion, D.J. Bauer, P. Collins, Nature 272:583 (1978).

23. G.B. Elion, Cancer Res. 45:2943 (1985).

24. P.A. Furman, J.A. Fyfe, M.H. St. Clair, K. Weinhold, J.L. Rideout, G.A. Freeman, S.N. Lehrman, D.P. Bolognesi, S. Broder, H. Mitsuya, D.W. Barry, Proc. Natl. Acad. Sci. USA 83:8333 (1986).

ANTIMETABOLITES IN CANCER CHEMOTHERAPY

Stephen J. Clarke, Ann L. Jackman and Kenneth R. Harrap

Drug Development Section, The Institute of Cancer Research, Sutton Surrey, UK

INTRODUCTION

Some 43 years have elapsed since the first successful use of an anti-metabolite, methotrexate (MTX), in the treatment of malignant disease. In the intervening years much attention has been devoted to the discovery and development of other anticancer anti-metabolites: several, notably 5-fluorouracil (5FU), 6-mercaptopurine (6MP), 6-thioguanine (6-TG), cytosine arabinoside (ARA-C), together with MTX, now fulfill an established role in the chemotherapy of malignant diseases. Enthusiasm for the discovery and development of novel antimetabolites persists and there are several which are presently at various stages of development. In this discussion we will restrict our comments to those new antimetabolites listed in Table 1, which have reached the clinic. The structures of these agents are shown in Figure 1 and sites of action of some of the agents are shown in Figure 2.

Table 1. New antimetabolites

1.	**Antifolates**			
	(a)	TS inhibitors: CB3717, ICI D1694.		
	(b)	DHFR inhibitors:	(i) Classical:	10-EDAM
			(ii) Lipophilic:	Piritrexim, Trimetrexate.
	(c)	GAR transformylase inhibitor:		DDATHF
3.	**Antipyrimidines**			
		DHO dehydrogenase inhibitor:		Brequinar
		multifactorial DNA inhibitor:		Gemcitabine
4.	**Antipurines**			
		IMP dehydrogenase inhibitor:		Tiazofurin
		ADA inhibitor:		Deoxycoformycin
		multifactorial DNA inhibitors:		Fludarabine monophosphate
				2-chloro-2'-deoxyadenosine

Abbreviations: GAR - glycinamide ribonucleotide; DHO - dihydroorotate; RR - ribonucleotide reductase; 10 EDAM-10-ethyl, 10-deazaaminopterim; DDATHF-5,10-dideaza tetrahydrofolate.

Purine and Pyrimidine Metabolism in Man VII, Part A
Edited by R.A. Harkness *et al.*, Plenum Press, New York, 1991

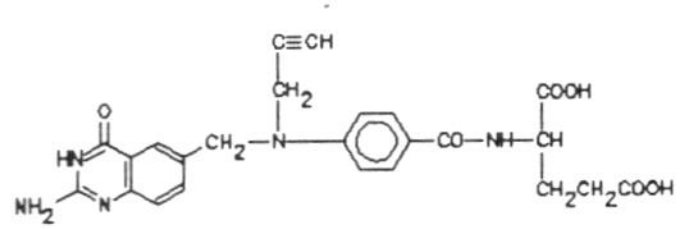

CB3717

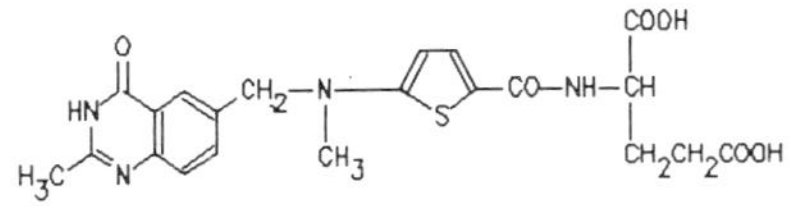

ICI D1694

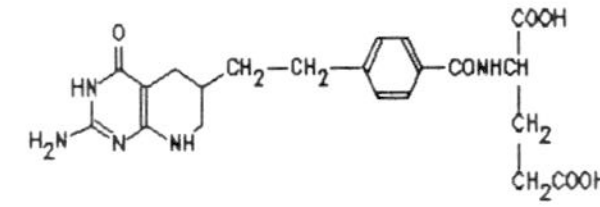

DDATHF

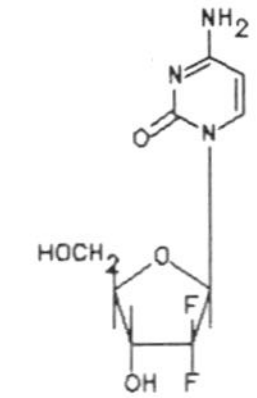

Gemcitabine

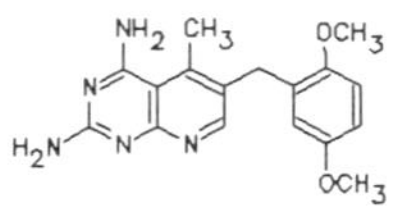

Piritrexim

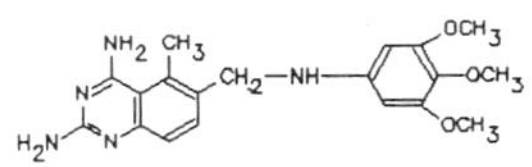

Trimetrexate

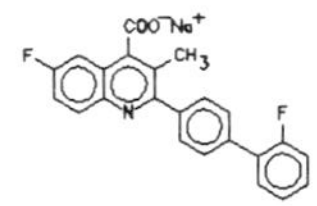

Brequinar

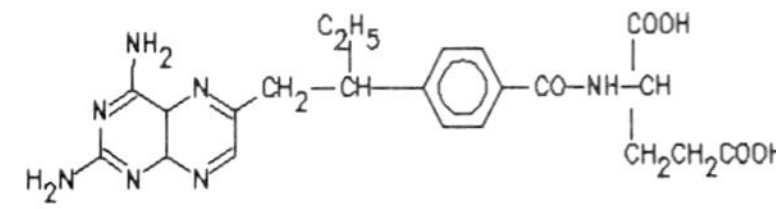

10-EDAM

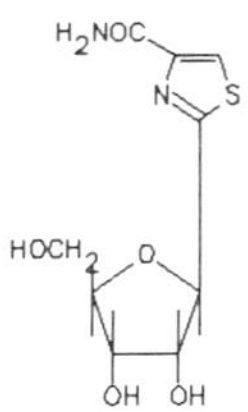

Tiazofurin

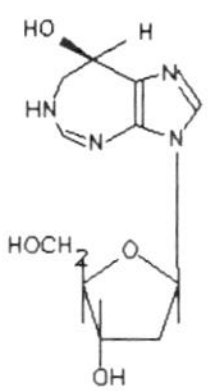

Deoxycoformycin

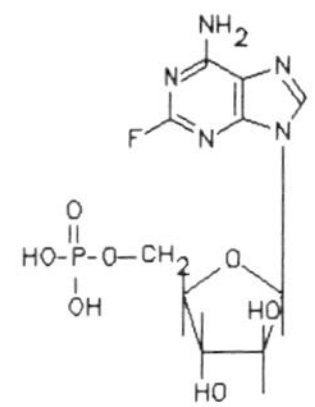

Fludarabine

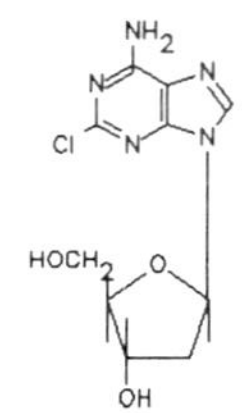

2'Chlorodeoxyadenosine

Figure 1

NEW ANTIMETABOLITES

(i) Antifolates

(a) Thymidylate Synthase (TS) inhibitors; **CB3717** (N^{10}-propargyl-5,8-dideazafolic acid; PDDF) is a potent quinazoline inhibitor of TS with a Ki of 3nM. CB3717 does not generally use the reduced folate/MTX carrier (RFC) to enter cells and is therefore active in cell lines resistant to MTX due to impaired uptake via the RFC. As TS is the locus of action of CB3717, activity is also seen against MTX resistant cell lines which overproduce dihydrofolate reductase. CB3717 is a substrate for folylpolyglutamate synthetase (FPGS) and polyglutamates can be formed intracellularly. These metabolites are up to two orders of magnitude more potent as TS inhibitors than the parent compound and are retained intracellularly. In phase I studies nephrotoxicity was found to be dose limiting and sometimes life-threatening and occurred in 70% of patients at doses >450mg/m^2. A debilitating malaise and elevations in hepatic transaminases were also seen. Activity was observed in refractory ovarian cancer, breast cancer and mesothelioma. In phase II studies at 300-400mg/m^2 every 3 weeks, activity was seen in breast cancer (8 of 48 PR - 16.6%) and hepatoma (6 of 14 PR - 43%). The toxicities seen in phase I were reproduced and were thought to be related to poor water solubility and drug precipitation and not to its locus of action. The drug was withdrawn from further development. The development of CB3717 was through a collaboration between the Institute of Cancer Research and ICI Pharmaceuticals and this collaboration continued in order to find a less toxic analogue.

ICI D1694; Stuctural refinement of CB 3717 led to the discovery of the more water-soluble and highly cytotoxic ICI D1694 (N-(5[N-(3,4 dihydro-2-methyl-4-oxoquinazolin-6-ylmethyl)-N-methylamino]-2-thenoyl)-L-glutamic acid). ICI D1694 uses the RFC and is an excellent substrate for FPGS. These factors are responsible for the rapid and extensive intracellular polyglutamation of the drug. The polyglutamate metabolites are substantially better as TS inhibitors than the parent drug (Ki ICI D1694 = 60nM; ICI D1694 tetraglutamate = 1nM) and are retained intracellularly. Against a panel of human tumour xenografts ICI D1694 had a significant degree of activity at doses below the MTD. Neither MTX, CB 3717, nor 5FU showed comparable activity. In animals the dose limiting toxicities were haematological and gastrointestinal and the toxicities seen with CB 3717 were not repeated. ICI D1694 entered phase I study in The Royal Marsden Hospital early in 1991.

Lometrexol (DDATHF); 5,10-dideaza-5,6,7,8-tetrahydrofolate, blocks de novo purine synthesis by inhibiting glycinamide ribonucleotide transformylase. DDATHF, like ICI D1694, utilises the RFC and is a good substrate for FPGS which leads to the formation of more active polyglutamate forms. DDATHF has shown activity against a range of tumour models, including colon 26, B-16 melanoma, Lewis lung carcinoma, Madison lung carcinoma and several human colon carcinoma xenografts. The drug is currently undergoing clinical evaluation, the results of which are awaited with interest.

(b) Dihydrofolate reductase (DHFR) Inhibitiors

(i). *Classical* - designed to exploit differences in reduced folate carrier between normal and tumour tissues.

10-Ethyl-10-deaza aminopterin (10-EDAM); This classical DHFR inhibitor is an analogue of methotrexate. It has greater potency in transplanted murine tumours and human tumour xenografts than methotrexate, presumably due to its greater accumulation and retention (as polyglutamates). In phase I trial the dose limiting toxicity was oral mucositis and a dose of $80mg/m^2$ weekly or $100mg/m^2$ weekly times three followed by a two week rest was recommended for further use. Activity has been seen in non-small cell lung cancer, both as a single agent and in combination with cyclophosphamide (4/19-21%PR) and cisplatinum (3CR/8PR in 32 patients - 37%) and with mitomycin C and vinblastine (60% major response rate). 10-EDAM elicits response rates similar to methotrexate in head and neck cancer.

(ii). *Lipophilic* - these compounds do not possess a glutamate residue and were designed, in part, to circumvent resistance determined by impaired transport and/or polyglutamation.

Trimetrexate (2,4-diamino-5-methyl[(3,4,5-trimethoxyanilino) methyl] quinazoline); Trimetrexate is a lipophilic DHFR inhibitor which does not enter cells via the RFC and polyglutamation is structurally precluded. These features may enable it to circumvent some of the resistance mechanisms seen with MTX, such as transport defects and failure of polyglutamation, while it also has the potential to cross the blood brain barrier. *In vitro* experiments suggested a broader range of activity than methotrexate. In phase II studies in multiple malignancies no impressive responses have been seen.

Piritrexim (2,4-diamino-6-(2,5-dimethoxybenzyl)-5-methylpyrido[2,3-d] pyrimidine); Piritrexim is another lipid soluble DHFR inhibitor that is not dependent on an active transport mechanism and cannot be polyglutamated. In phase I, dose limiting toxicity was myelosuppression (mainly thrombocytopoenia) and fatigue. Thrombophlebitis necessitates central vein cannulation, although oral administration has been favoured recently. This causes nausea and vomiting, which decreases with divided dosing. In phase II trials activity has been seen in malignant melanoma with prolonged low oral dosing (30% response rate).

The Role of Folate Metabolism in the *De Novo* Synthesis of Thymidylate and Purines

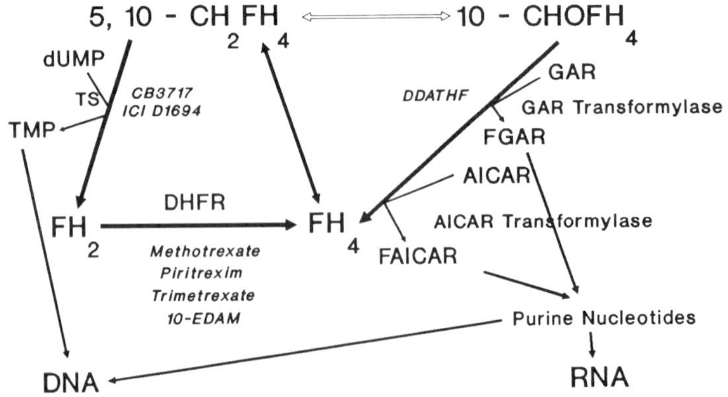

Figure 2

(2) Antipyrimidines

Brequinar (6-fluoro-2(2'-fluoro-1,1'-biphenyl-4-yl)-3-methyl-4-quinoline carboxylic acid). This quinoline carboxylic acid derivative blocks pyrimidine synthesis by inhibiting dihydroorotate dehydrogenase. It inhibits the growth of a broad spectrum of nude mouse implanted human tumour xenografts, including breast (MX-1), lung (LX-1), stomach (BL-STX-1) and various human colon carcinomas (including CX-1). Prolonged exposure to Brequinar produces greater growth inhibition than shorter exposures. In multiple phase I trials utilising differing schedules, the dose limiting toxicities were mucositis and skin rash in some and, myelosuppression in others. Other toxicities included nausea, vomiting, malaise and burning at the injection site. Phase II trials have failed to reveal significant activity in lung cancer, melanoma, head and neck cancer, gastrointestinal and breast cancers.

(3) Antipurines

Tiazofurin (2-β-D-ribofuranosylthiazole-4-carboxamide), a C nucleoside, is an analogue of ribavirin. It is active against L1210 and P388 murine leukaemias and the Lewis lung carcinoma. Tiazofurin is a prodrug for the corresponding adenine dinucleotide (TAD), which inhibits IMP dehydrogenase by competing with the NAD^+ cofactor, leading to decreased de novo guanine nucleotide synthesis and cell death. Phase I studies showed non-haematologic toxicities to be dose limiting, the most worrying being neurotoxicity, expressed by confusion, lethargy, obtundation and some occurrences of focal neurological deficits. Other side effects were nausea and vomiting, serositis, lethargy and malaise, muscle pain, increased creatinine phosphokinase, mucocutaneous changes, myelosuppression, increased transaminases and increased uric acid production. Allopurinol co-administration decreases excess uric acid and results in inhibition of the purine salvage pathway, thus enhancing the inhibitory effects on IMP dehydrogenase. In phase II trials activity has been seen in refractory leukaemias, with little activity in solid tumours. Weber and associates have shown that Tiazofurin can be used safely when given by prolonged intravenous infusion, together with biochemical and pharmacological monitoring.

2'Deoxycoformycin(2'dCF;pentostatin;R-3-(2-deoxy-β-D-erythropentofuranosyl-3, 6,7,8-tetrahydroimidazole[4,5-d][1,3] diazepin-8-ol); This is not a new drug, although interest in it has revived in the last few years. 2'dCF is a tight- binding inhibitor of adenosine deaminase. Its potential as a lympholytic agent was recognised when it was appreciated that lymphocytes have higher levels of adenosine deaminase than other mammalian cells. Early trials of 2'dCF at high doses were complicated by the occurrence of non-haematological toxicity, especially hepatotoxicity and neurotoxicity. However, once it was appreciated that adenosine deaminase is inhibited at low doses of 2'dCF, new schedules were evaluated, and, significant activity has been seen in hairy cell leukaemia, chronic lymphocytic leukaemia (CLL), and low grade non-Hodgkin's lymphomas (NHL). Septicaemia and other life threatening infections are the most significant side effects at low doses.

Gemcitabine (2',2'-difluorodeoxycytidine; dFdC); dFdC like Ara-C, is a deoxycytidine analogue. Initially synthesised as an antiviral agent, nevertheless it showed greater activity than Ara-C against murine solid tumours. It was also found to be active against human tumour xenograft models, including LX-1 lung carcinoma, MX-1 breast carcinoma, CX-1 and GC3 colon carcinomas. dFdC shows greater intracellular accumulation and retention than Ara-C, which may explain its improved anti-tumour properties. The increased accumulation is due to increased membrane permeation. dFdC has greater affinity for dC kinase and dFdC triphosphate is retained longer than is the case with Ara-CTP. Phase I trials showed that the dose limiting toxicity was myelosuppression, with thrombocytopoenia and anaemia greater than neutropoenia. Sporadic fever and hypertension have also been seen. Phase II studies at 800mg/m^2 weekly x 3 have been performed in a variety of malignancies, with activity seen in non-small cell lung cancer (7/29 PR - 24%) and pancreatic cancer (5/39 PR - 13%).

Fludarabine monophosphate (9-β-D-arabinosyl-2-fluoroadenine); This is the 2-fluoro, 5'-monophosphate derivative of ara-A. It has an advantage over ara-A in that it is relatively resistant to deamination by adenosine deaminase. The 5'-monophosphate is more soluble than 2-fluoro ara-A, which was also difficult to formulate. The mechanism of action is through incorporation of F-ara-ATP into DNA. Fludarabine phosphate demonstrated *in vivo* activity against murine tumours, including L1210 leukaemia, CD8F1 mammary carcinoma, P388 leukaemia and the human LX-1 lung tumour xenograft. In phase I trials, dose limiting toxicity was myelosuppression, regardless of schedule; lymphopoenia developed rapidly in all patients at all dose levels. At doses $\geq$100mg/m^2, delayed, fatal neurotoxicity was seen. Lactic acidosis was also seen in some patients at moderate doses. In Phase II trials, at low dose, fludarabine monophosphate is inactive against solid tumours, but is highly active against low grade lymphoproliferative disorders such as CLL, and NHL.

2'Chloro-2'deoxyadenosine (2'CdA); This analogue of deoxyadenosine is resistant to the action of adenosine deaminase. 2'CdA is phosphorylated to the triphosphate and high levels of the phosphorylated 2'CdA accumulate in cells high in deoxycytidine kinase, particularly lymphocytes and monocytes. The exact intracellular mechanism of action is not completely established, but is probably multi-factorial. Postulated mechanisms include incorporation of CldATP into DNA, inhibition of ribonucleotide reductase by CldATP, inhibition of DNA polymerases and decrease in intracellular NAD by poly ADP-ribose polymerase activation. The end result is DNA strand breakage. Dose limiting toxicity has been myelosuppression with minimal non-haematological toxicities. Activity has been seen in Hairy cell leukaemias, CLL, low grade NHL and some childhood leukaemias.

ACKNOWLEDGEMENTS

This work was supported by grants to The Institute of Cancer Research: Royal Cancer Hospital from the Cancer Research Campaign and the Medical Research Council.

REFERENCES

Full referencing is not possible due to limitation of space. We would refer readers to the following recent articles and reviews.

CB3717

Calvert A.H., Newell D.R., Jackman A.L., Gumbrell L.A., Sikora E., Grzelakowska-Sztabert B., Bishop J.A.M., Judson I.R., Harland S.J. and Harrap K.R., 1987, Recent Preclinical and Clinical Studies with the Thymidylate Synthase Inhibitor N^{10}-Propargyl-5,8-dideazafolic acid (CB 3717),NCI Monographs, 5:213-218.

ICI D1694 - See Jackman A.L. in this volume.

DDATHF - See Beardsley G.P. in this volume.

Grindey, G. B., Alati, T.,and Shih, C. Reversal of the Toxicity, but not the Antitumour Activity of Lometrexol by Folic Acid. Proc. Amer. Assoc. Cancer Res. 1991 Abs 1921.

10-EDAM
Kris M.G., Kinaham J.J., Gralla R.J., Fanucchi M.P., Wertheim M.S., O'Connell J.P., Marks L.D., Williams L., Farag F., Young C.W. and Sirotnak F.M., 1988, Phase I Trial and Clinical Pharmacological Evaluation of 10-Ethyl-10-deazaaminopterin in Adult Patients.Cancer Res., 48:5573-5579.
Piritrexim
Sigel C.W., Machlin A.W. and Woolley J.L., Johnson N.W., Collier M.A., Blum M.R., Clendeninn N.J., Everitt B.J.M. Grebe G., Mackars A., Foss R.G., Duch D.S., Bowers S.W. and Nichol C.A., 1987, Preclinical Biochemical Pharmacology and Toxicology of Piritrexim, a Lipophilic Inhibitor of Dihydrofolate Reductase, NCI Monographs, 5:111-120.
Trimetrexate
Jackson R.C., Leopold W.R. Hamelehle K.L. and Fry D.W., 1988, Preclinical Studies with Trimetrexate: A Review of Conclusions and Unanswered Questions, Semin. Oncol. 15(Suppl2):1-7.
Bertino J.R., 1988, Trimetrexate: Overall Clinical Results, Semin. Oncol., 15(Suppl2):50-51.
Brequinar
Peters G.J., Schwartsmann G., Nadel J.C. Laurensse E.J., van Groeningen C.J., van der Vijgh W.J.F. and Pinedo H.M., 1990, *In Vivo* Inhibition of the Pyrimidine *de novo* Enzyme Dihydroorotic Acid Dehydrogenase by Brequinar Sodium (DUP-785; NSC 368390) in Mice and Patients, 50:4644-4649.
Gemcitabine
Abbruzzese J.L., Grunewald R., Weeks E.A., Gravel D., Adams T., Nowak B., Mineishi S., Tarassoff P., Satterlee W., Raber M.N. and Plunkett W., 1991, A Phase I Clinical, Plasma, and Cellular Pharmacology Study of Gemcitabine, J Clin Oncol, 9:491-498.
Tiazofurin - see Weber G. in this volume.
Grem J.L., Rubinstein L., King S.A., Cheson B.D., Hawkins M.J. and Shoemaker D.D., 1990, Clinical Toxicity associated with Tiazofurin, Invest. New Drugs, 8:227-238.
Deoxycoformycin
O'Dwyer P.J., Wagner B., Leyland-Jones B., Wittes R.E., Cheson B.D. and Hoth D.F., 1988, 2'-Deoxycoformycin (Pentostatin) for Lymphoid Malignancies: Rational Development of an Active New Drug, Ann. Intern. Med., 108:733-743.
Fludarabine monophosphate
Chun H.G., Leyland-Jones B. and Cheson B.D., 1991, Fludarabine Phosphate: A Synthetic Purine Antimetabolite with Significant Activity against Lymphoid Malignancies, J. Clin. Oncol. 1:175-188.
Von Hoff D.D. 1990, Phase I Clinical Trials with Fludarabine Phosphate, Seminars in Oncol., 17:33-38.
2'-Chlorodeoxyadenosine
Piro L.D., Carrera C.J., Beutler E. and Carson D.A., 1988, 2'-Chlorodeoxyadenosine: An Effect New Agent for the Treatment of Chronic Lymphocytic Leukemia, Blood, 72:1069-1073.

2-CHLORODEOXYADENOSINE CHEMOTHERAPY TRIGGERS PROGRAMMED CELL DEATH IN NORMAL AND MALIGNANT LYMPHOCYTES

Carlos J. Carrera[1], Lawrence D. Piro[2], Alan Saven[2], Ernest Beutler[2], Chihiro Terai[3], and Dennis A. Carson[1]

[1]University of California, San Diego, La Jolla, California
[2]Research Institute of Scripps Clinic, La Jolla, California
[3]Institute of Rheumatology, Tokyo Women's Medical College Tokyo, Japan

INTRODUCTION

2-Chloro-2'-deoxyadenosine (CdA) is a congener of deoxyadenosine that is resistant to adenosine deaminase. Rationale for the development of CdA as an immunosuppressive and antineoplastic agent derives from the lymphocytotoxicity of elevated plasma deoxyadenosine in congenital deficiency of adenosine deaminase[1]. Recent clinical trials have documented promising efficacy of CdA in the treatment of chronic B-cell and T-cell lymphoid malignancies, notably chronic lymphocytic leukemia (CLL) and hairy cell leukemia[2,3]. Among nucleoside antimetabolites, CdA is uniquely toxic to non-dividing lymphocytes, as well as to most proliferating T-lymphoblasts[4]. Surprisingly, mature blood monocytes are also lysed by CdA at the nanomolar concentrations readily achieved by continuous infusion in vivo[5]. In both non-dividing lymphocytes and monocytes, CdA causes the prompt accumulation of DNA strand breaks, with subsequent NAD depletion, inhibition of RNA synthesis, and eventual cell lysis[5,6].

Lymphoid cell death after treatment with DNA-damaging drugs or with γ-irradiation occurs by apoptosis. Distinct from necrosis, apoptosis is characterized by early condensation of chromatin, cytoplasmic compaction, and budding of the nuclear membrane[7]. The hallmark of apoptosis is the cleavage of chromatin into oligonucleosomal-length fragments by an endogenous, Ca^{++}/Mg^{++}-dependent endonuclease. Apoptosis occurs during fetal development and during the involution of thymus and hormone-dependent tissues, and has thus been referred to as programmed cell death. Recently, mouse thymocytes exposed to purine deoxynucleosides in vitro were reported to undergo oligonucleosomal DNA fragmentation[8].

Because the morphology of normal or leukemic lymphocytes exposed to CdA in vitro was similar to that of irradiated cells, we examined CdA-treated cells for evidence of internucleosomal DNA fragmentation as a consequence of the early accumulation of DNA single-strand breaks.

METHODS

Normal peripheral blood lymphocytes and malignant cells from patients with CLL and HCL were isolated by centrifugation over Histopaque (Sigma, St. Louis, MO) and cultured in RPMI 1640 medium supplemented with

L-glutamine, 2-mercaptoethanol, and autologous plasma 20% v/v. Cultures were maintained in 5% CO_2 at 37°C. Following treatment with CdA or other compounds for various intervals, cells were analyzed for viability by erythrosin B dye exclusion and for DNA damage by either the alkaline unwinding assay or by agarose gel electrophoresis.

The fluorescent assay for DNA unwinding in alkaline solution[9] was used to detect the early accumulation of DNA breaks in cells exposed to CdA or to γ-irradiation from a ^{137}Cs source. In this assay, the fluorescence of ethidium bromide bound to double-stranded DNA after 1 h unwinding at pH 12.8 is expressed as a percent of total double-stranded DNA fluorescence in aliquots not exposed to alkali. The percent double-stranded DNA after unwinding is a measure of DNA strand breaks in control or treated cells.

Soluble, low molecular weight DNA fragments (**S**) were separated from high molecular weight chromatin (**P**) by Triton-X100 lysis at 4°C for 20 min, followed by centrifugation at 13,000 g [10]. Both DNA fractions were treated with proteinase K 100 µg/ml, extracted with phenol/chloroform, and then precipitated with NaCl 0.5 M and 50% isopropanol at -20°C. After dissolution in Tris/EDTA buffer, pH 8, the **P** and **S** DNA fractions were treated with RNase 50 µg/ml and then subjected to electrophoresis for 3 h in 2% agarose gels in Tris acetate/EDTA buffer, pH 8 containing ethidium bromide 5 µg/ml. DNA bands were visualized under uv illumination.

RESULTS AND DISCUSSION

Although both CLL cells and normal lymphocytes repaired radiation-induced DNA breaks rapidly, DNA damage caused by exposure to CdA in vitro was repaired slowly, if at all (Fig. 1). Incubation of CLL cells with CdA 1 µM for 12 h generated DNA strand breaks equivalent to approximately 1 Gy (100 rad). As shown in Fig. 1A, this damage persisted despite removal of the drug, and was sufficient to cause subsequent lysis after 24 h. In contrast, the initial DNA breaks induced by γ-irradiation were rapidly ligated.

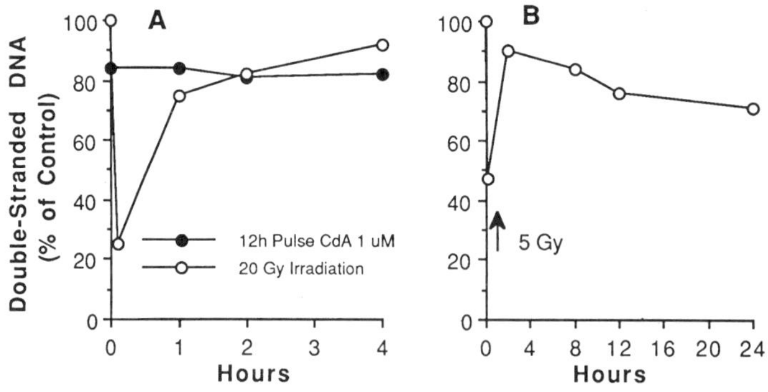

Fig. 1. DNA strand breaks and repair in CLL cells (**A**) and normal lymphocytes (**B**) treated with γ-irradiation (O), or with CdA 1 µM for 12 h followed by removal of drug (●).

Despite the repair of radiation-induced DNA breaks, lymphocytes are known to be quite sensitive to small radiation doses. By monitoring DNA strand breaks at longer intervals after irradiation, we found that DNA damage re-appeared between 8 and 24 h (Fig. 1B). Moreover, the irradiated lymphocytes developed morphologic features of apoptosis, with contraction of cytoplasmic volume and marked karyorrhexis (not shown). These findings are consistent with the delayed chromatin fragmentation by Ca^{++}/Mg^{++}-endonuclease in lymphoid organs of irradiated rodents[11].

Normal lymphocytes and monocytes treated in vitro with CdA 1 μM accumulate substantial DNA damage within the first hours[5,6]. However, loss of viability by dye exclusion is not detected until after 24 h. Gel electrophoresis of DNA from CdA-treated lymphocytes shows the appearance of a nucleosomal ladder pattern between 12 - 18 h, prior to cell lysis (Fig. 2, left panel). The temporal pattern of internucleosomal DNA cleavage in monocytes is identical to that in lymphoid cells (not shown). Although lymphocyte DNA strand breaks are readily detected by the alkaline unwinding assay at 4 h, this initial damage does not result in the release of soluble, oligonucleosomal fragments. Addition of nicotinamide to lymphocyte cultures containing CdA preserves cellular NAD levels and delays cell death (presumably by inhibition of poly[ADP-ribose] polymerase)[6]. However, Fig. 2 shows that the internucleosomal cleavage pattern in the CdA-treated cells is unaffected by nicotinamide 5 mM. Because CdA is phosphorylated by deoxycytidine kinase[4], simultaneous addition of deoxycytidine 100 μM abolishes the initial accumulation of DNA strand breaks and all subsequent manifestations of CdA toxicity (not shown).

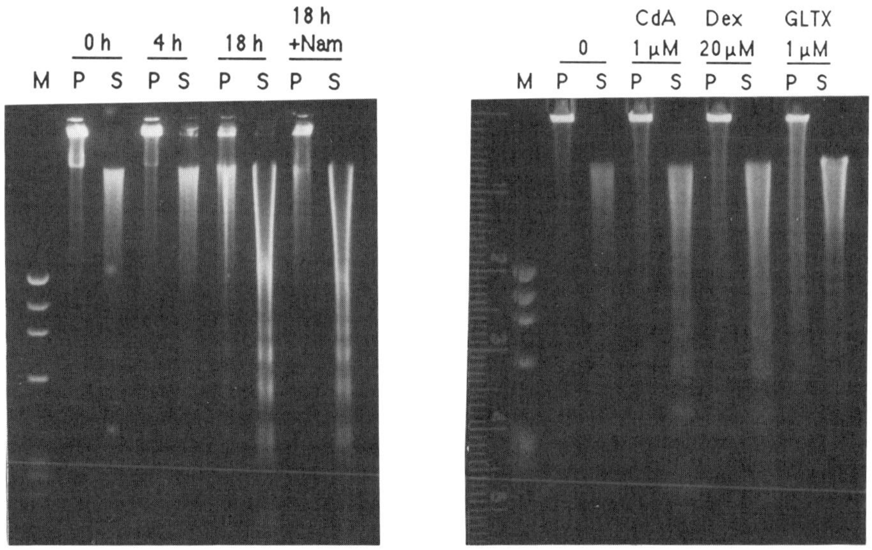

Fig. 2. Oligonucleosomal DNA fragmentation in normal and leukemic lymphocytes detected by agarose gel electrophoresis. Normal lymphocytes (left panel) were exposed to CdA 1 μM ± nicotinamide (Nam) 5 mM for 0, 4, and 18 h prior to extraction of soluble DNA fragments (**S**) from chromatin pellets (**P**) for electrophoresis. CLL cells (right panel) were treated for 18 h with CdA, dexamethasone (Dex), or gliotoxin (GLTX) prior to extraction and electrophoresis of the low and high molecular weight DNA fractions as described in Methods. **O** = control cells without drug, **M** = ΦX174 DNA size markers. The smallest nucleosomal fragment visible corresponds to 180 - 190 base pairs.

The toxicity of CdA towards CLL cells also initiates with DNA strand breaks[2], and Fig. 2, right panel shows that Ca^{++}/Mg^{++}-endonuclease is activated as a consequence. Dexamethasone 20 μM triggers apoptosis to a similar extent in the leukemic cells, confirming the previous report by McConkey et al.[12]

CONCLUSION

CdA toxicity toward both normal and leukemic lymphocytes requires phosphorylation by deoxycytidine kinase and the early accumulation of DNA strand breaks. By biochemical mechanisms still to be defined, the initial DNA damage triggers the activation of a Ca^{++}/Mg^{++}-dependent endonuclease, resulting in the oligonucleosomal fragmentation pattern and morphologic changes typical of apoptosis. Apoptosis is not generally observed in healthy organs, apart from senescent blood cells and intestinal crypts. Thus, it is highly relevant that mechanisms of programmed cell death remain operative in malignant cells. The demonstration of apoptosis in leukemic cell samples offers the intriguing prospect that the biochemical mechanisms involved might be used to advantage in developing novel anticancer strategies.

ACKNOWLEDGMENT

This work was supported by National Institutes of Health grants CA 01100, GM 23200, and RR 00833.

REFERENCES

1. D. A. Carson, D. B. Wasson, and E. Beutler, Antileukemic and immunosuppressive activity of 2-chloro-2'-deoxyadenosine, Proc. Natl. Acad. Sci. USA 81:2232 (1984).
2. L. D. Piro, C. J. Carrera, E. Beutler, and D. A. Carson, 2-Chlorodeoxyadenosine: An effective new agent for the treatment of chronic lymphocytic leukemia, Blood 72:1069 (1988).
3. L. D. Piro, C. J. Carrera, D. A. Carson, and E. Beutler, Lasting remissions in hairy-cell leukemia induced by a single infusion of 2-chlorodeoxyadenosine, N. Engl. J. Med. 322:1117 (1990).
4. D. A. Carson, D. B. Wasson, R. Taetle, and A. Yu, Specific toxicity of 2-chlorodeoxyadenosine toward resting and proliferating human lymphocytes, Blood 62:737 (1983).
5. C. J. Carrera, L. D. Piro, C. Terai, M. Lotz, E. Beutler, and D. A. Carson, Potent toxicity of 2-chlorodeoxyadenosine towards human monocytes in vitro and in vivo: A novel approach to immunosuppressive therapy, J. Clin. Invest. 86:1480 (1990).
6. S. Seto, C. J. Carrera, M. Kubota, D. B. Wasson, and D. A. Carson, Mechanism of deoxyadenosine and 2-chlorodeoxyadenosine toxicity to non-dividing human lymphocytes, J. Clin. Invest. 75:377 (1985).
7. A. H. Wyllie, J. F. R. Kerr, and A. R. Currie, Cell death: The significance of apoptosis, Int. Rev. Cytol. 68:251 (1980).
8. H. Kizaki, H. Shimada, F. Ohsaka, and T. Sakurada, Adenosine, deoxyadenosine, and deoxyguanosine induce DNA cleavage in mouse thymocytes, J. Immunol. 141:1652 (1988).
9. H. C. Birnboim and J. J. Jevcak, Fluorometric method for rapid detection of DNA strand breaks in human white blood cells produced by low doses of radiation, Cancer Res. 41:1889 (1981).
10. K. S. Sellins and J. J. Cohen, Gene induction by gamma-irradiation leads to DNA fragmentation in lymphocytes, J. Immunol. 139:3199 (1987).
11. M. Skalka, J. Matyasova, and M. Cejkova, DNA in chromatin of irradiated lymphoid tissues degrades in vivo into regular fragments, FEBS Lett. 72:271 (1976).
12. D. J. McConkey, M. Aguilar-Santelises, P. Hartzell, I. Eriksson, H. Mellstedt, S. Orrenius, and M. Jondal, Induction of DNA fragmentation in chronic B-lymphocytic leukemia cells, J. Immunol., 146:1072 (1991).

ICI D1694, AN INHIBITOR OF THYMIDYLATE SYNTHASE FOR CLINICAL STUDY

Ann L. Jackman, Duncan I. Jodrell, William Gibson and *Trevor C. Stephens

Drug Development Section, Institute of Cancer Research, Sutton, Surrey, UK and *ICI Pharmaceuticals, Alderley Park, Macclesfield, Cheshire UK

INTRODUCTION

ICI D1694[1] is a new quinazoline antifolate that acts via inhibition of thymidylate synthase (TS) and is in clinical study. TS is crucial for the de novo synthesis of thymidylate and interest in developing a TS inhibitor as an anticancer agent arose from the concept that drugs such as methotrexate (MTX) and 5-fluorouracil, which probably in part act by TS inhibition, were complicated by their multiple effects[2]. The folate analogue, 10-propargyl-5,8-dideazafolic acid (CB3717) fulfilled the biochemical criteria for selective inhibition of TS and established the usefulness of such a drug in some animal models and in man[2,3]. However renal toxicity, related to the poor aqueous solubility of CB3717, limited its therapeutic efficacy and it was therefore withdrawn from clinical study. Further chemical syntheses led to the more water-soluble analogues of CB3717 (2-desamino and 2-methyl) which were devoid of renal toxicity in rodents and were better antitumour agents both in vitro and in vivo[4,5]. Definition of the biochemical events leading to this increased potency aided the development of ICI D1694. This paper describes some features of ICI D1694 important for its in vitro and in vivo activity.

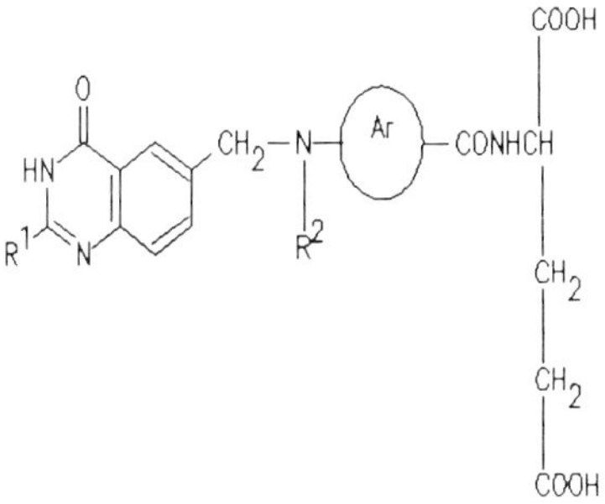

R^1	R^2	Ar	
NH_2	$CH_2C{\equiv}CH$	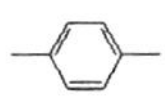	CB3717
H			desamino-CB3717
CH_3			2-methyl-CB3717
CH_3	CH_3	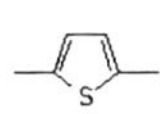	ICI D1694
NH_2			2-amino-D1694

Fig.1

Purine and Pyrimidine Metabolism in Man VII, Part A
Edited by R.A. Harkness *et al.*, Plenum Press, New York, 1991

RESULTS AND DISCUSSION

ICI D1694 is ~500-fold more cell growth inhibitory when compared with CB3717 despite being 20-fold poorer as an inhibitor of TS (Table 1). TS remains the cytotoxic locus of action of ICI D1694 since thymidine (dThd) alone prevents the activity in vitro

Table 1. In vitro activity of some quinazoline TS inhibitors

	Inhibit^n of TS Ki (nM)	Inhibition of cell growth, IC_{50} (μM)				L1210 cell growth in the presence of 10μM dThd and 10 x IC_{50} drug concentration (% control)
		MCF7	L1210	L1210:1565	L1210:1565 / L1210	
CB3717	2.7	0.4	3.5	3.6	1	96
Desamino-CB3717	27	0.1	0.36	4.1	11	100
2-methyl-CB3717	10	0.015	0.085	8.2	96	100
2-amino-ICI D1694	~40	-	0.7	0.47	0.7	-
ICI D1694	60	0.0007	0.0076	0.93	122	94

Cells were grown in suspension with drugs for 48 hrs (mouse leukemia L1210) and 72 hrs (L1210:1565) as previously described [4]. Human breast MCF-7 cells (~$5x10^4$/plate;5mls) in RPMI 1640 supplemented with 5% foetal calf serum were incubated for 5 days. Cell numbers were measured by coulter counting.

(Table 1). Reduction in cytotoxic activity of ICI D1694 was found against a variant L1210 cell line (L1210:1565) with a greatly impaired reduced-folate/MTX cell membrane carrier (RFC). Similar resistance was also seen with most other examples of 2-methyl compounds e.g. 2-methyl-CB3717 (ICI 198583; DMPDDF) (Table 1), however less resistance was seen to desamino-CB3717 and none to CB3717 or 2-amino-ICI D1694. We conclude from these and other studies[5,6] that the potent activity of 2-methyl compounds and notably ICI D1694 is at least in part due to their use of the RFC.

Another factor that influences cytotoxicity for this class of antifolates is their differential ability to act as substrates for the folate-metabolising enzyme, folylpolyglutamate synthetase (FPGS)[7]. ICI D1694 was found to be an excellent substrate for this enzyme (Table 2). Confirmation of this was obtained in L1210 cells where exposure to as little as 100nM ^{3}H ICI D1694 led to the rapid accumulation of

Table 2. Substrate activity of some quinazoline TS inhibitors for mouse liver FPGS

	Km, μM	*Vmax	*Vmax/Km
#Folic acid	140	1.0	1.0
#CB3717	40	0.88	2.3
#desamino-CB3717	48	0.9	3.2
+2-methyl-CB3717	40	1.2	3.5
+ICI D1694	1.3	1.5	220

*Relative to folic acid in the same experiment. #reference 8; +reference 7

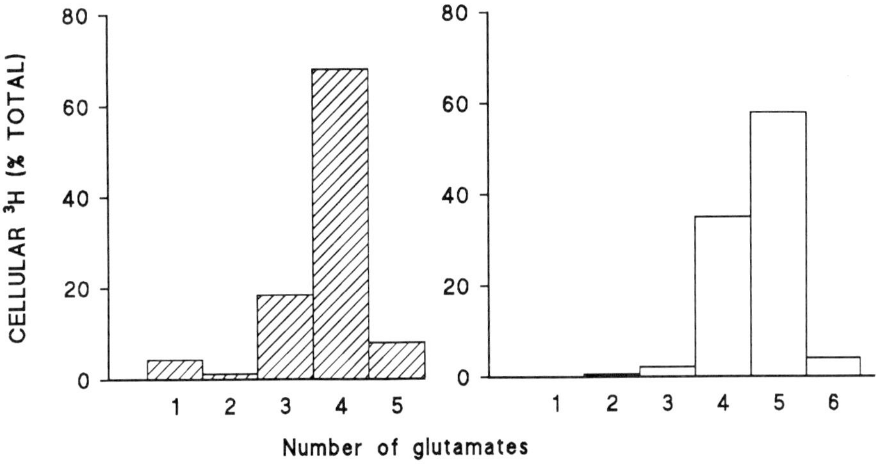

Fig 2. The intracellular polyglutamation of ICI D1694. L1210 cells were incubated with 100nM ^{3}H ICI D1694 for 4 hrs (hatched bars) plus a further 24 hrs in drug-free medium (open bars). Total cellular ^{3}H = 1.5μM and 1.1μM respectively. After extraction of ^{3}H material from the cells, ICI D1694 derived material was analysed by HPLC using an ion-pairing method. Identification of the radiochromatograhic peaks was by the co-chromatography using synthetic standards.

polyglutamates so that by 4 hrs 95% of the cellular ^{3}H was associated with these fractions (principally tetraglutamate) (Fig.2). These polyglutamates are not readily effluxed from cells (Fig 2) and are ~100-fold better as inhibitors of TS than the parent drug[7].

From the data presented so far we conclude that the increases in cytotoxic potency from CB3717 through the desamino and 2-methyl analogues correlated with their increasing use of the RFC. Although FPGS substrate activity was similar the rate of polyglutamation should be greater due to increased intracellular drug levels. A further enhancement in cytotoxic potency is seen with ICI D1694 which is attributed both to its use of the RFC and to the marked improvement in FPGS substrate activity.

Given as a single bolus i.v. injection to mice, ICI D1694 is only marginally active against most tumour models. However this is predicted from the high level of plasma dThd found in mice (~1μM) which seems to be sufficient to by-pass any TS inhibition via the dThd salvage pathway[9]. For this reason a variant L5178Y tumour line deficient in dThd salvage (TK-/-) has been used to assess the relative activity of folate-based TS inhibitors and, as predicted, sensitivity was seen[10]. ICI D1694 at the non-toxic single dose of 10mg/kg cures mice bearing this tumour while CB3717 was marginally active at its MTD of 200mg/kg[11]. At 500mg/kg ICI D1694 was devoid of renal and hepatic toxicities while CB3717 caused acute toxicity to these organs at 100mg/kg[2,12]. Renal toxicity was also absent when ICI D1694 was administered at a dose of 250mg/kg/wk for 6 weeks.

Following daily x 5 dosing of ICI D1694 toxicity was observed. At 10mg/kg x 5 approximately 15% weight loss (probably related to gut toxicity), leucopenia and thrombocytopenia was seen (Table 3). The MTD in mice varied between strains and was difficult to define because it was also dependent on the maturity of the mice and diet. However doses $\geq$ 100mg/kg daily x 5 were routinely tolerated. Generally these antiproliferative toxicities were prevented by co-administration of thymidine consistent with these toxicities being related to TS inhibitory effects[12]. This is in contrast to CB3717 where liver and kidney toxicities were related to the insolubility of the drug rather than its anti-proliferative properties[2]. Multiple dosing of ICI D1694 also produced antitumour activity against most tumours refractory to a single dose. At 6.6 mg/kg daily

Table 3. ICI D1694 induced bone marrow toxicity in C57/DBA2 F1 hybrid mice

	*Total white cells ($x\ 10^9/l$)	Neutrophils ($x\ 10^6/l$)	Platelets ($x\ 10^9/l$)
Solvent control	6.4 $\pm$ 2.4	750 $\pm$ 250	898 $\pm$ 82
10mg/kg ICI D1694 daily x 5	3.6 $\pm$ 0.4	230 $\pm$ 110	353 $\pm$ 47

*8 days after commencing treatment. Mean $\pm$ S.D; n=5

for 5 days ICI D1694 was active against the thymidine salvage competent L5178Y tumour (5 day growth delay)[11] and against the L1210:ICR tumour it was curative to >70% of the mice at 0.4mg/kg daily for 5 days[13]. Co-administration of dThd prevented activity against the L1210:ICR tumour[13]. Both these tumours were approximately two orders of magnitude less sensitive to CB3717. ICI D1694 was also found to be active when given as a daily x 15 schedule against a number of human tumour xenografts. For example 1mg/kg daily x 15 produced a 15 day growth delay in the HX62 human ovarian xenograft[11]. ICI D1694 also had superior activity to MTX and FU[14]. We speculate that the reason prolonged treatment protocols are needed for activity in mice is related to the requirement for plasma dThd levels to fall to a level that does not compromise inhibition of TS. A reduction in plasma dThd was observed after mice were treated with CB3717[9]. It is pertinent to point out that plasma dThd levels in man are lower (~0.1μM) and should not seriously limit the efficacy of ICI D1694 in humans[9].

SUMMARY

The TS inhibitor, ICI D1694, is a highly potent inhibitor of tumour growth in vitro and in vivo. Uptake via the RFC and rapid metabolism to polyglutamate forms appear to be responsible for potency. Antiproliferative toxicity in mice was evident although the dose-limiting renal toxicity experienced with CB3717 was not apparent.

REFERENCES

1. Marsham, P.R; Hughes, L.R; Jackman, A.L; Hayter, A.J; Oldfield, J;Wardleworth, J.M; Bishop, J.A; O'Connor, B.M. and Calvert, A.H. J.Med. Chem. 34: 1594-1605, 1991.
2. Harrap, K.R; Jackman, A.L; Newell, D.R; Taylor, G.A; Hughes, L.R. and Calvert, A.H. Adv. Enz. Regul. 29: 161-179, 1989.
3. Jones, T.R; Calvert, A.H; Jackman, A.L; Brown, S.J; Jones, M. and Harrap, K.R. Eur. J. Cancer 17: 11-19, 1981.
4. Jackman, A.L; Taylor, G.A; O'Connor, B.M; Bishop, J.A; Moran, R.G. and Calvert, A.H. Cancer Res. 50: 5212-5218, 1990.
5. Jackman, A.L; Newell, D.R; Jodrell, D.I; Taylor, G.A; Bishop, J.A; Hughes, L.R. and Calvert, A.H. In: Chemistry and Biology of Pteridines (eds. H.-Ch. Curtius, S.Ghisla, N.Blau) Walter de Gruyter, Berlin, pp 1023-1026, 1990.
6. Patil, S.D; Jones, C; Nair, M.G; Galivan, J, Maley, F; Kisliuk, R.L; Gaumont, Y; Duch, D. and Ferone, R. J.Med. Chem. 32: 1284-1289, 1989.
7. Jackman, A.L; Marsham, P.R; Moran, R.G; Kimbell, R; O'Connor, B.M; Hughes, L.R. and Calvert, A.H. Adv. Enz. Regul. 31: 13-27, 1991.
8. Moran, R.G; Colman, P.A. and Jones, T.T. Mol. Pharmacol. 36: 736-743, 1990.
9. Jackman, A.L; Taylor, G.A; Calvert, A.H. and Harrap, K.R. Biochem Pharmacol. 33: 3269-3275, 1984.
10. Stephens, T.C; Calvete, J.A; Janes, D; Hughes, L.R. and Jackman, A.L. Proc. Amer. Assoc. Cancer Res. 30: 477, 1989.
11. Stephens, T.C; Calvete, J.A; Janes, D; Waterman, S.E; Valcaccia, B.E; Hughes, L.R. and Calvert, A.H. Proc. Amer. Assoc. Cancer Res. 31: 342, 1990.

12. Jodrell, D.I; Newell, D.R; Calvete, J.A; Stephens, T.C. and Calvert, A.H. Proc. Amer. Assoc. Cancer Res. 31: 341, 1990.
13. Jackman, A.L; Taylor, G.A; Bishop, J.A; O'Connor, B.M; Bisset, G; Hughes, L.R; Moran, R.G. and Calvert, A.H. Proc. Amer. Assoc. Cancer Res. 31: 342, 1990
14. Stephens, T.C; Valcaccia, B.E; Sheader, M.L; Hughes, L.R. and Jackman, A.L. Proc. Amer. Assoc. Cancer Res. 32: 328, 1991.

ACTION AND SEQUENCE DEPENDENT INTERACTION OF ACIVICIN AND 6-THIOGUANINE IN HUMAN DERIVED MALIGNANT T-ALL AND CALLA+ CELL LINES

R.C. Trueworthy*, R.A. DeAbreu, L.H.J. Lambooy, J.P.M. Bökkerink, and E.H. Stet.

*University of Kansas School of Medicine; K.C. KS, USA
University Hospital of Nijmegen, Netherlands

INTRODUCTION

Acivicin is an irreversible inhibitor of L-glutamine amino transferases and therefore effects several enzymatic steps in purine de novo sythesis (PDNS) including phosphoribosyl pyrophosphate(PRPP)amidotransferase and enzymes which are specifically rate limiting for GTP, for example, GMP synthetase. This drug biochemically targets GTP concentration in leukemic blast cells (fig. 1).[1] Acivicin may also be effective as a modulator of other anti-leukemic agents such as AraC and 6-thioguanine (6TG).[2] Acivicin interferes with PDNS resulting in PRPP elevations which may allow for synergistic action with antipurines.[3] In response to inhibition of (PDNS), an increase in the availability of 5-phosphoribosyl-1-pyrophosphate (PRPP) occurs. As a result of this increased availability of PRPP potentiation of 6TG incorporation can be expected in cells with an active PDNS.

6TG is a likely drug for synergy. This drug interaction may mimic the synergy of Methotrexate and 6MP. PDNS is the basis of the synergism of Methotrexate and 6MP in human malignant lymphoblasts of different lineages.[4] This mechanism also probably underlies the synergy of Acivicin with 6TG in L1210.[5] This synergy is based upon high intracellular PRPP levels, is schedule dependent and is maximized by pre-treatment with Acivicin. Pre-treatment with Acivicin in L1210 cells potentiates the action of subsequently administered 6TG. The mechanism may involve both biochemical as well as cytokinetic factors. Combination regimens of Acivicin with other agents other than AraC and 6TG have also been investigated for synergistic anti-tumor activity.[6] Acivicin demonstrates more activity in human tumor cloning systems as a continuous exposure than as a one hour exposure.[7]

Observations of cell growth and cell viability in both KM3 and MOLT4 cell lines were used to describe the cytotoxicity of Acivicin and 6TG and the enhanced cytotoxicity of 6TG following pretreatment with Acivicin.

METHODS

Cell Culture

The experiments were performed in MOLT-F4, a T-ALL cell line and KM3, a non-T non-B ALL cell line. The cells were passaged in RPMI 1640DM

Purine and Pyrimidine Metabolism in Man VII, Part A
Edited by R.A. Harkness *et al.*, Plenum Press, New York, 1991

(Gibco, Netherlands) supplemented with 10% non-dialyzed fetal calf serum v/v (Gibco, Netherlands), 2mM L-Glutamine (Sigma Chemical, USA) 2 mM Sodium Pyruvate (BDH Chemicals limited, UK) and Gentamicin/Streptomycin three times a week. The cell cultures were maintained in a 5% CO_2 humidified atmosphere at 37° C. The cells cultured were tested for the absence of mycoplasma contamination and the presence of marker antigens regularly.

Cell Growth and Viability Assays

Cell growth and cell viability experiments were performed in logarithmically growing cells which were resuspended in fresh medium in a concentration of 0.3 X 10^6 cells per ml 24 hours prior to the start of each experiment. At the start the initial drug was added as a single dose. In certain experiments a second drug was added 24 hours later. During the experiment, 2 mM glutamine was supplemented every 24 hours in order to prevent glutamine exhaustion of the medium. After incubation, the cells were harvested and counted in a coulter counter. Cell viability was determined by means of trypan blue exclusion.

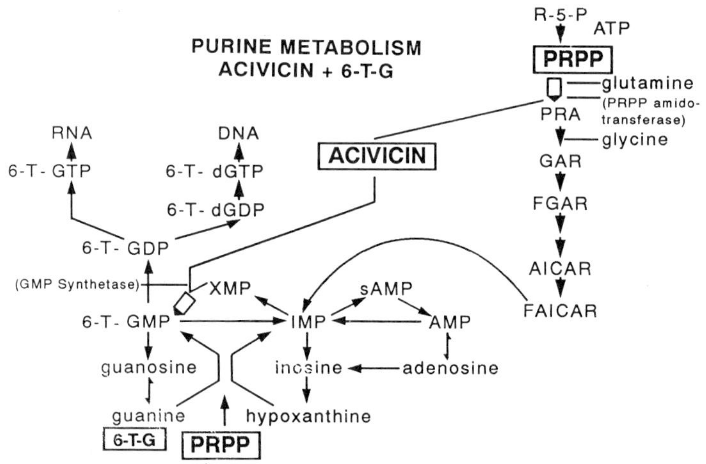

Figure 1

RESULTS

Analysis of the growth of viable cells reveals a dose and time dependent relationship in both MOLT4 and KM3 cells (fig. 2, table 1). Viable cell growth of MOLT4 cells was 89% of control following a 48 hour exposure to 1 μm Acivicin and decreased to 60% after a 48 hour exposure to 2 μm Acivicin and 19% after a 72 hour exposure to 2 μm Acivicin. In KM3 cells, viable cell growth was 84% of control after a 48 hour exposure to 1 μm Acivicin and decreased in a time and dose dependent manner to 57% after a 72 hour exposure to 2 μm Acivicin. These variations in both cell growth and cell viability between MOLT4 and KM3 cells may reflect the more active PDNS in MOLT4 cells.

In the MOLT4 cells exposure for 48 hours to 6TG concentrations between 0.5 - 2 μm resulted in a dramatic decrease in viable cell growth from 63% to 14%. Viable cell growth decreased from 85% to 35% of control with increasing concentrations of 6TG between 0.5 - 2 mM at 48 hours in KM3 cells. Only at the highest concentration did this effect seem to be time dependent

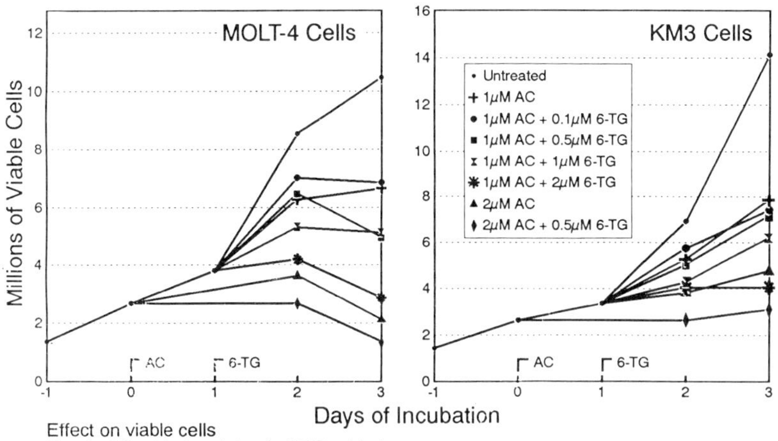

Effect on viable cells
Day 0: AC added and day 1: 6-TG added
AC= acivicin, 6-TG= 6-thioguanine

Figure 2

Table 1

um	MOLT-4 cells, AC, TG, AC + TG % of control 1d%	2d%	3d%	KM3 cells 1d%	2d%	3d%
AC				AC		
1	100	89	64	100	84	78
2	86	60	19	100	77	57
5	59	16	8	100	69	47
TG				TG		
0.1	77	87	93	100	100	100
0.5	67	63	75	85	85	85
1.0	46	39	58	77	55	55
2.0	43	14	15	55	35	19
AC + TG				AC + TG		
1 + 0.1	100	85	67	100	86	54
1 + 0.5	100	90	50	100	71	52
1 + 1	100	65	50	100	60	43
1 + 2	100	74	30	100	59	28
2 + 0.5	89	50	14	-	40	22
TG + AC						
2 + 1	38	11	-			

with only 19% of cells still viable after 72 hour exposure to 2 μM 6TG.

6TG concentrations 0.1 - 2 μm may be ineffective or somewhat effective alone, but appear more effective when used following the administration of Acivicin. Apparent sequence dependent additive effects upon viable cell growth were seen in MOLT4 cells when 6TG followed Acivicin administration. Viable cell growth dropped from 85% of control to 14% with increasing Acivicin (1 - 2 μm), increasing 6TG (0.1 - 2 μm), and with time of total exposure to drugs (48 - 72 hours). The reverse sequence i.e. adding 2 μm 6TG prior to the addition of 1 μm Acivicin showed no additive affect. The result of 11% viable cell growth being similar to that observed with the use of 2 μm 6TG alone, 14%.

Apparent synergistic affects were seen in KM3 cells in both cell growth and cell viability when 6TG followed Acivicin administration. Viable cell growth dropped from 86% of control to 22% with increasing Acivicin (1 - 2 μm), increasing 6TG (0.1 - 2 μm), and with time of total exposure to drugs (48 - 72 hours). These changes were greater than expected from single agent data.

DISCUSSION

Acivicin is an inhibitor of PDNS. As the PDNS becomes inhibited, the intracellular PRPP concentration rises. PRPP is a cofactor needed for the conversion of 6TG into 6TGMP; so, as a result of the increase of the PRPP concentration more 6TG can be utilized by the cells. In this way, stimulation of 6TG incorporation may occur.

The failure of several powerful inhibitors of PDNS to have a significant impact in the clinic may be due to the replenishment of intracellular nucleotides through the purine salvage pathways. A synergistic combination should prevent the rescue of cells from the effects of interrupted PDNS by blocking nucleotide salvage. Acivicin and 6TG may represent such a combination. Alternatively, Acivicin and 6TG cytotoxicity may result only from depletion of guanine nucleotides and inhibition of DNA and RNA synthesis.

Studies involving the sequential administration of these agents, in vivo, seem warranted.

REFERENCES

1. Y. Natsumeda, T. Ikegami, E. Olah, G. Weber. Significance of purine salvage in circumventing the action of antimetabolites in rat hepatoma cells. Cancer Res., 49, 88-92 (1989).
2. B.L. Powell, J.B. Craig, R.L. Capizzi, F. Richard 2nd. Phase I-II trial of Acivicin in adult acute leukemia. Invest New Drugs, 6, 41-43, (1988).
3. C.S. Gordon, R.C. Sawyer, D.S. Martin, Et Al. Effect of L-glutamine antagonists on 5-phosphoribosyl 1-pyrophoshate levels in P388 leukemia and in muring colon adenocarcinoma in vivo. Biochem Pharmacol, 31, 1509-1513, (1982).
4. J.P.M. Bökkerink, M.A.H. Bakker, T.W. Hulscher, R.A. DeAbreu, D.A.M. Schretlen. Purine De Novo Synthesis as the basis of Synergism of Methotrexate and 6-Merrcaptopurine in human malignant lymphoblasts of different lineages. Biochemical Pharmacology, 37, 2321-2327, (1988).
5. K. Bhalla, R. Nayak, A. Dehch, S. Grant. Synergistic interation between Acivicin (AT-125) and 6-thioguanine in the murine leukemia L1210. Biochemical Pharmacology, 33, 247-254, (1984).
6. R.L. Stolfi. Combination chemotherapy with acivicin, PALA, and Ara C against CD8F1 breast tumors. Proc. Am. Assoc. Can. Res., 26, 241, (1985).
7. B. Ethridge, D. Von Hoff. Activity of one hour and continuous exposure of Acivicin in a human tumor cloning system. Proc. ASCO, 4, C-111, (1985).

CONVERSION OF ACYCLIC NUCLEOSIDE PHOSPHONATES TO THEIR DIPHOSPHATE DERIVATIVES BY 5-PHOSPHORIBOSYL-1-PYROPHOSPHATE (PRPP) SYNTHETASE

J. BALZARINI and E. DE CLERCQ

Rega Institute for Medical Research
K.U.Leuven
B-3000 Leuven. Belgium

INTRODUCTION

Recently, we discovered a group of acyclic nucleoside phosphonate derivatives with potent and selective activity against a broad spectrum of DNA viruses (1,2). The prototype compound of this new class of compounds is (*S*)-9-(3-hydroxy-2-phosphonylmethoxypropyl)adenine, designated HPMPA (Fig. 1). In these nucleoside phosphonate derivatives, the phosphorus atom is attached to the alkyl side chain of the purine base via a P-C bond. The resulting phosphonate linkage is not susceptible to hydrolysis by esterases and is chemically very stable. While HPMPA itself has poor, if any, activity against retroviruses, several other phosphonylmethoxyalkylpurine derivatives, i.e. 9-(2-phosphonylmethoxyethyl)adenine (PMEA), show marked antiretroviral activity. PMEA (Fig. 1) is active against viruses that belong to the herpes family and also against human immunodeficiency virus type 1 (HIV-1) and type 2 (HIV-2), simian immunodeficiency virus (SIV), feline immunodeficiency virus (FIV) and Moloney murine sarcoma virus (MSV) (2-9). Also, recently, a new subclass of acyclic nucleoside phosphonate derivatives, with (*R,S*)-9-(3-fluoro-2-phosphonylmethoxypropyl)adenine (FPMPA) as the prototype (Fig. 1), was shown to be highly selective inhibitors of retroviruses (i.e. HIV-1, HIV-2, SIV, MSV) (10,11).

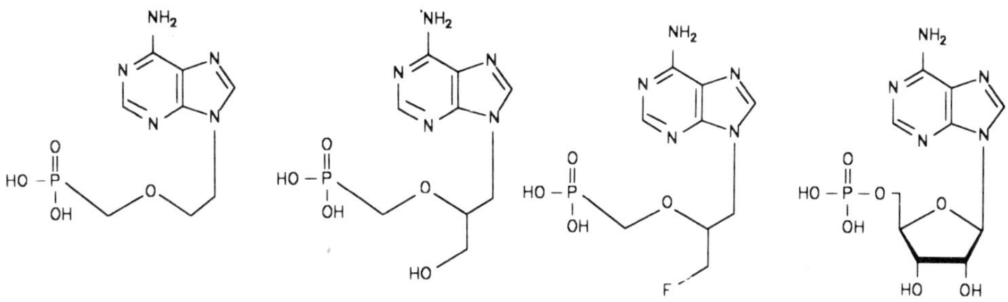

(PMEA) (HPMPA) (FPMPA) (AMP)

Fig. 1. Structural formulae of HPMPA, PMEA, FPMPA and AMP.

Purine and Pyrimidine Metabolism in Man VII, Part A
Edited by R.A. Harkness *et al.*, Plenum Press, New York, 1991

The acyclic nucleoside phosphonates can be considered as analogues of AMP (Fig. 1). The diphosphorylated derivatives of PMEA, HPMPA and FPMPA (designated PMEApp, HPMPApp and FPMPApp, respectively) correspond to the antivirally active metabolites of these compounds and should be considered as ATP analogues. PMEApp and FPMPApp have been shown to be potent and selective inhibitors of HIV-1 reverse transcriptase (K_i/K_m values: 0.01 and 0.002, respectively). They act as DNA chain terminators in the HIV-1 reverse transcriptase reaction, and this property may obviously explain their antiretroviral activity (10-13).

Cell culture experiments revealed that incubation of human T4-lymphocyte MT-4 cells with radiolabeled PMEA results in the formation of the mono- and diphosphorylated derivatives as the predominant metabolites (12). Adenylate kinase could not be identified as the enzyme responsible for the phosphorylation of PMEA, when ATP was used as the phosphate donor. However, we found that the acyclic nucleoside phosphonates were converted in a one-step reaction to their diphosphorylated derivatives by 5-phosphoribosyl-1-pyrophosphate (PRPP) synthetase purified from *E.coli* (10-12,14).

The rate of the PRPP synthetase reaction proceeds 20-fold higher in the direction of PMEA than of PMEApp synthesis. This equilibrium is comparable with that observed for the natural substrates AMP and ATP. PMEA has a ~ 8-fold lower affinity for PRPP synthetase than AMP (Table 1). The V_{max} of PMEA conversion to PMEApp is 150-fold lower than the V_{max} of AMP conversion to ATP. HPMPA has similar kinetic properties as PMEA, whereas for FPMPA both the K_m and V_{max} are ~ 3-fold lower than for HPMPA. In the "forward" reaction, PMEApp and HPMPApp have similar substrate affinities for PRPP synthetase as ATP, although their V_{max} is 50- to 200-fold lower than that of ATP (Table 1).

Table 1. Kinetics of PRPP synthetase with acyclic nucleoside phosphonate derivatives

Substrate	K_m[a] (mM)	V_{max}[a] (μmole/unit/hr)	K_m/V_{max}
PMEA	1.47	0.087	16,890
HPMPA	1.13	0.097	11,651
FPMPA	0.435	0.024	18,125
AMP	0.178	13	11
PMEApp	0.040	0.082	488
HPMPApp	0.050	0.362	138
ATP	0.020	188	0.106

[a]Data taken from ref. 14.

In its substrate affinity for the acyclic nucleoside phosphonates PRPP synthetase is very stringent with respect to the nature of the purine base but not the acyclic phosphonate side chain. In fact, only the derivatives of adenine, 2,6-diaminopurine and the N^6-hydrazine and N^6-hydroxylamine derivatives of adenine, but not hypoxanthine, guanine, 2-aminopurine, 6-thiopurine, N^6-methylaminopurine, N^6-dimethylaminopurine and 1,6-ethenoadenine derivatives are recognised as substrate by PRPP synthetase. These data point to the role of the NH_2 group at C-6 of the purine ring in the affinity of the acyclic nucleoside phosphonates for the enzyme. Much broader structural variations are allowed for the acyclic part of the molecules. In fact, the phosphonylmethoxy moiety can be translocated from the C-2 of the alkyl side chain to the C-3 position (as in HPMPA). The 3-hydroxyl group of HPMPA can be replaced by an hydrogen or fluorine without much loss of affinity for the enzyme. Also, PMEA derivatives in which the ethyl of the side chain has been extended to pentyl, and phosphonylmethoxyethyl, phosphonylethoxyethyl, phosphonylmethoxymethyl and phosphonylethoxymethyl derivatives of adenine are prominent substrates for PRPP synthetase.

Among the acyclic purine nucleoside phosphonate derivatives, we also identified several congeners that did not act as substrate but proved to be efficient inhibitors of PRPP synthetase. PMEHx, the hypoxanthine derivative of PMEA was found to be a more potent inhibitor of the enzyme than the 5'-monophosphate of MRPP (4-methoxy-8-β-D-ribofuranosylamino)pyrimido[5,4-d]pyrimidine) when AMP (or ATP) was used as substrate (15). The IC_{50} (50% inhibitory concentration) of PMEHx was 140 μM (Fig. 2) and its K_i and K_i/K_m for PRPP synthetase were 41 μM and 0.26, respectively. FPMPHx, the hypoxanthine counterpart of FPMPA was 6- to 7-fold less inhibitory than PMEHx (IC_{50}: 900 μM), while HPMPHx was devoid of any marked inhibitory potential at 2 mM (Fig. 2).

Another interesting feature of PRPP synthetase is its lack of stereospecificity with regard to the substrate recognition of the acyclic nucleoside phosphonates. The (*S*) and (*R*) enantiomers of HPMPA and FPMPA are equally effective as substrates for the enzyme. In this respect, PRPP synthetase differs from many other enzymes in nucleotide metabolism, i.e. reverse transcriptase, adenosine kinase, and adenosine deaminase, which are quite stereospecific in their substrate recognition.

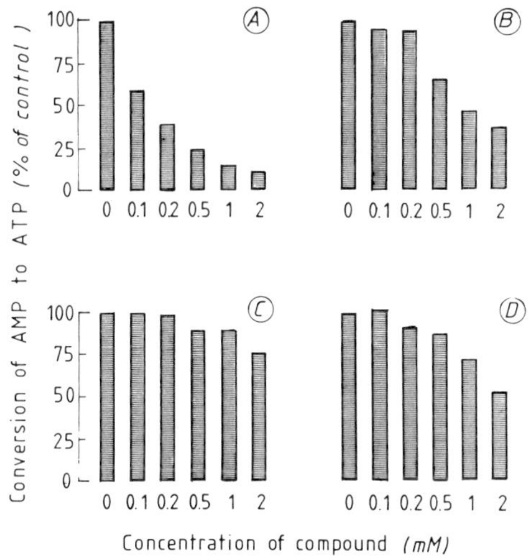

Fig. 2. Inhibitory effect of PMEHx (Panel A), FPMPHx (Panel B), HPMPHx (Panel C) and IMP (Panel D) on the conversion of AMP to ATP by PRPP synthetase. The AMP concentration was invariably 200 μM.

It is not clear to what extent PRPP synthetase plays a role in the conversion of the acyclic nucleoside phosphonate derivatives to their active metabolites in intact cells. Attempts at reversing the antiretroviral activity of PMEA in HIV-1-infected MT-4 cells and MSV-infected C3H/3T3 cells by PRPP synthetase inhibitors (i.e. 4-amino-8-(β-D-ribofuranosylamino)pyrimido[5,4-d]pyrimidine (ARPP) and MRPP have failed. It is not excluded, however, that at their subtoxic concentrations ARPP and MRPP are not efficient inhibitors of their target enzyme. As has also been reported by Holy and coworkers (13,16) for crude extracts of murine leukemia L1210 cells, we found that partially purified murine kidney homogenates are able to convert radiolabeled PMEA to its corresponding monophosphate (PMEAp) and diphosphate (PMEApp) in the presence of ATP as the phosphate donor. In our experiments ATP was superior over GTP and dATP, whereas dTTP, CTP, UTP, dCTP and dGTP were not efficient as phosphate donors to convert PMEA to its phosphorylated derivatives. These observations thus indicate that alternatively or additionally to PRPP synthetase other enzymes may be involved in the metabolism of PMEA (and related analogues).

In conclusion, several acyclic purine nucleoside phosphonates have been identified as novel substrates and/or inhibitors of PRPP synthetase. Much has yet to be learned about the role of PRPP synthetase in the intracellular metabolism of the compounds, and the kinetics of their interaction with the enzyme. The substrate affinity of several acyclic nucleoside phosphonates for PRPP synthetase provides an interesting approach toward the synthesis of the diphosphorylated (antivirally active) derivatives through a one-step conversion assay using PRPP synthetase as the catalyzing enzyme.

REFERENCES

1. E. De Clercq, A. Holy, I. Rosenberg, T. Sakuma, J. Balzarini and P.C. Maudgal. Nature 323:464 (1986).
2. E. De Clercq, T. Sakuma, M. Baba, R. Pauwels, J. Balzarini, I. Rosenberg and A. Holy. Antiviral Res. 8:261 (1987).
3. R. Pauwels, J. Balzarini, D. Schols, M. Baba, J. Desmyter, I. Rosenberg, A. Holy and E. De Clercq. Antimicrob. Agents Chemother. 32:1025 (1988).
4. J. Balzarini, L. Naesens, P. Herdewijn, I. Rosenberg, A. Holy, R. Pauwels, M. Baba, D.G. Johns and E. De Clercq, Proc. Natl. Acad. Sci. USA 86:332 (1989).
5. J. Balzarini, L. Naesens and E. De Clercq. Int. J. Cancer 46:337 (1990).
6. J. Balzarini, L. Naesens, J. Slachmuylders, H. Niphuis, I. Rosenberg, A. Holy, H. Schellekens and E. De Clercq. AIDS 5:21 (1991).
7. H. Egberinck, M. Borst, H. Niphuis, J. Balzarini, H. Neu, H. Schellekens, E. De Clercq, M. Horzinek and M. Koolen. Proc. Natl. Acad. Sci. USA 87:3087 (1990).
8. J.D. Gangemi, R.M. Cozens, E. De Clercq, J. Balzarini and H.-K. Hochkeppel. Antimicrob. Agents Chemother. 33:1864 (1989).
9. L. Naesens, J. Balzarini, I. Rosenberg, A. Holy and E. De Clercq. Eur. J. Clin. Microbiol. Infect. Dis. 8:1043 (1989).
10. J. Balzarini, A. Holy, J. Jindrich, H. Dvorakova, Z. Hao, R. Snoeck, P. Herdewijn, D.G. Johns and E. De Clercq. Proc. Natl. Acad. Sci. 88: 4961 (1991).
11. J. Balzarini, 1991, in: Advances in Chemotherapy of AIDS. R.B. Diazio & J.-P. Sommadossi, eds., in press (1991).
12. J. Balzarini, Z. Hao, P. Herdewijn, D.G. Johns and E. De Clercq. Proc. Natl. Acad. Sci. USA 88:1499 (1991).
13. A. Holy, I. Votruba, A. Merta, J. Cerny, J. Vesely, J. Vlach, K. Sediva, I. Rosenberg, M. Otmar, H. Hrebabecky, M. Travnicek, V. Vonka, R. Snoeck and E. De Clercq. Antiviral Res. 13:295 (1990).
14. J. Balzarini and E. De Clercq. J. Biol. Chem. 266:8686 (1991).
15. R.C. Willis, L.D. Nord, J.M. Fujitaki and R.K. Robins, in: "Advances in Enzyme Regulation", vol. 28, G. Weber, ed., Pergamon Press, Oxford, New York (1989).
16. A. Merta, J. Vesely, I. Votruba, I. Rosenberg and A. Holy. Neoplasma 37:111 (1990).

DERIVATION OF NEW DRUGS FOR THERAPY OF AIDS

Kenneth L Powell, Head of Cell Biology and
Dorothy J Purifoy, Head of Antiviral Research

The Wellcome Research Laboratories
Langley Court, Beckenham, Kent BR3 3BS, UK

Introduction:

It is remarkable given the very short period since AIDS was discovered to have a viral cause that there are about 30 potential anti-HIV drugs in clinical development. The reasons for this rapid progress are fairly obvious, and include the advanced knowledge of retroviral structure and function accumulated over the previous thirty years, the exciting progress made in antiviral research just prior to the emergence of the AIDS epidemic and particularly the success of acyclovir. This latter event meant that the nucleosides which had previously been regarded as a rather non-selective group of compounds would now be seen as potent and selective agents able to distinguish between virus and host cell enzymes. Thus when the AIDS epidemic emerged we were prepared as never before to deal with the threat.

The Current Therapy - AZT (Retrovir$^{(R)}$):

The initial discovery of AZT activity was made in the laboratories of the Burroughs Wellcome Company where the compound was shown to have potent activity. This was followed by the well known and in itself remarkable story of the drugs development (1-3) which lead to demonstration of clinical benefit of AZT treatment and marketing of the drug within three years of discovery of its activity. Of course, like any other drug, AZT is not the simple solution to the AIDS problem. High doses of the drug are known to cause toxicity and more recently there has been much speculation about the emergence of AZT resistant virus. This latter phenomenon was first reported by Larder *et al.*, (4) who showed that when patients treated for relatively long periods with AZT were sampled and laboratory tests done on their virus, this showed the virus was relatively resistant to AZT. Clearly this did not mean that these patients did not benefit from AZT treatment but resistance might be a

Purine and Pyrimidine Metabolism in Man VII, Part A
Edited by R.A. Harkness *et al.*, Plenum Press, New York, 1991

contributing factor to the patients continuing decline despite AZT therapy. There are many questions relating to the clinical significance of AZT resistance which currently remain unanswered however it can be clearly be surmised that if resistance were clinically significant then combination therapy utilising AZT and another antiviral agent might be more beneficial to the patient than treatment with either drug alone. Thus the discovery of new antiviral agents and their use in combination with AZT is a major goal of current research.

Drugs in Late Clinical Development - ddI and ddC

The dideoxynucleosides ddI and ddC were discovered by an intense screening programme of nucleoside compounds though likely to have anti-HIV activity (5) ddC remains one of the most active compounds known in terms of *in vitro* anti-HIV activity. The drug appeared to have a good therapeutic index *in vitro* and to be relatively non-toxic in standard preclinical toxicology testing. On testing in clinical trials the drug was shown to cause peripheral neuropathy, fever and mouth sores. These toxic effects could be titrated out at very low doses of the drug (<0.01 mg/kg) and even at those doses the drug had some clinical effect (6). ddC is this unlikely to be used in monotherapy but may have a role in combination therapy.

Dideoxyinosine (ddI) has much lower antiviral activity *in vitro* than either, ddC or AZT, however it also is less toxic to cultured cells (5). The drug appeared to have no overt problems in preclinical testing but once again on entry to clinical trials was found to cause peripheral neuropathy and sporadic pancreatitis at high doses. On titration these toxic effects could be largely avoided at doses where the compound has signficant antiviral activity (7). These results mean that once again the most appropriate use of the compound is probably in combination therapy.

Potential Targets for Antiviral Therapy for HIV

The surprising difference between the human retroviruses and the majority of those isolated from animals is the presence of a considerable group of novel functions encoded by the virus genome. These proteins as well as the other well known retrovirus proteins ie. gag, env and pol gene products are all potential targets for chemotherapy. Since there is insufficient space here to consider all these possibilities we will concentrate on what we consider to be the four prime targets for attack. These targets are the following:

(1) The interaction between the virus glycoprotein (gp120) and its cell surface receptor (CD4).
(2) The transcriptional control protein tat.
(3) The reverse transcriptase of the virus.
(4) The virus encoded protease.

CD4/gp120

Although there are some clear indications that the CD4/gp120 interaction is not the only route by which HIV can enter cells it is clear that inhibition of this interaction can prevent virus growth in some cases. The evidence for this has come from studies of the soluble form of the CD4 molecule. This protein produced excellent *in vitro* results causing the inhibition of infection of uninfected cells by laboratory strains of HIV (8). Treatment of patients with CD4 has been much less exciting. There has been no clear cut evidence so far of therapeutic effect of the recombinant protein in patients (9). One reason for this lack of success may be that strains of virus differ in susceptibility to CD4 from the laboratory strains (10). It is not clear that small molecules with the ability to inhibit CD4/gp120 interactions will not succeed in inhibiting virus replication. The problem is to design or discover such molecules and efficient tests for CD4/gp120 interaction have been slow to develop but are now emerging.

HIV tat

This protein has been shown to upregulate the expression of HIV genes. The exact mechanism is still the basis of some speculation and its elucidation will probably require a detailed knowledge of the structure and function of tat. Nevertheless, it is comparatively simple to assay tat activity (as was originally described by Pavlakis *et al.* (11)) using a marker gene linked to an HIV promoter stably expressed in a suitable cell line. The addition of purified tat protein to such cells leads to a signal from transactivation of the resident marker gene (12). This type of cell line can thus be readily used as an assay for anti-tat drugs and leads to screening of potentially active compounds. Several companies are examining their compounds for such activity and Roche are reported to have a compound in development. Design of anti-tat agents will require a much more detailed knowledge of this fascinating protein.

Reverse Transcriptase (RT)

This protein is unique amongst those of the HIV virion in that it has been shown to be a suitable target for chemotherapy by the success of AZT in the clinic. There are now many other nucleoside analogues which also inhibit the virus by this route, including the series of dideoxy nucleosides mentioned above. Unfortunately many of these compounds share the unfortunate side effect of peripheral neuropathy and they appear not to offer marked advantage over the AZT molecule. It remains to be seen if nucleoside analogues with real advantage over AZT can be derived; certainly many are in the early stages of development but in the meantime it would clearly be worthwhile to attempt the understanding of RT structure in detail to enable the design of non-nucleosides. To this end, we at Wellcome began a study some four years ago to obtain the structure of this enzyme. We expressed the enzyme by recombinant methods to high levels firstly, in bacteria (13) and subsequently, in yeast and baculovirus. The protein has been purified by conventional methods and used to prepare monoclonal antibodies. Subsequently these antibodies were used to prepare immunoaffinity columns for the large scale preparation of RT (14). We discovered that RT forms as a heterodimer either by cleavage by the HIV protease in virus infected cells or by exposure to a variety of proteases *in vitro* (15). The enzyme has been crystallized both alone and in combination with a variety of ligands including oligonucleotides. The crystals diffract to 5-6Å and should ultimately provide a low resolution structure of the enzyme.

The design of new reverse transcriptase inhibitors has continued. The most interesting series to date were discovered by very distinct routes. Thus TIBO and its analogues were discovered by the Janssen group using a rational approach to drug screening (16). Utilising a wide variety of structural classes of chemical compounds they discovered this series of benzodiazopines with excellent antiviral activity. Refining this lead gave them a compound with remarkable anti-HIV-1 activity. This compound has no activity against HIV-2 which may by analogy to AZT indicate a potential problem when the question of mutation to resistant phenotype is examined. Two other groups at Boehringer Ingelheim and Merck have since produced compounds with related structures and similar activity (17, 18). De Clerq and his collaborators on the other hand synthesised a series of compounds beginning from an acyclic nucleoside design HEPT (19). Again this series was refined to yield highly selective compounds with no HIV-2 activity. It will be fascinating to learn how these compounds interact with RT which *in vitro* tests demonstrate is their target.

HIV Protease

The HIV protease was seen to be an excellent target for antiviral chemotherapy as soon as its primary sequence was determined. From that data it could be deduced (20) that the protease was related to the aspartic protease family of mammalian proteases. As such medicinal chemists were familiar with the routes that might be used to design peptide based inhibitors of the enzyme and from these to design non-peptidic second generation molecules. With the HIV protease there was a bonus in that the enzyme could be made both by chemical methods and through recombinant technique in amounts sufficient to provide material for protein crystallography. Further, the enzyme so obtained was found to crystallise easily producing crystals which diffracted to high resolution enabling determination of the structure of the enzyme to about 2Å resolution (21).

Thus in the case of the HIV protease we are in a unique situation. The enzyme is available in reasonable amounts for structural studies and enzymic studies. The structure of the enzyme is well known and thanks to the generousity of the crystallographers involved the enzyme can be studied at high resolution for modelling experiments. This work has enabled the design of several series of protease inhibitors from for example Upjohn (22), Roche (23) and Abbot (24) which are also inhibitors of virus production in tissue culture. It remains to be seen how effective such inhibitors will be in obtaining the desired therapeutic effects.

Combination Therapy using Novel Compounds

Thus it can be seen that there are already available a considerable number of potential therapies which are currently being tested clinically in attempts to demonstrate their utility. How should these new compounds be used as they emerge from the research laboratory to the clinic? If it proves that AZT resistance is clinically significant then there are major implications for the derivation and use of new anti-HIV drugs. It clearly would imply that resistance would be a problem with many other potential therapies. We should therefore consider the mechanism of AZT resistance and how this might influence future strategy.

Mechanism of AZT Resistance

Larder and his colleagues have considered the role of the RT gene in AZT resistance and have shown by examining virus that various mutatations in the gene will confer resistance. These mutations occur at a restricted number of sites in the gene but multiple mutations are needed to obtain highly resistant virus (25). Genetically engineered virus with mutations in the RT gene only is similar to virus isolated from patients in its resistance to AZT. Thus it is clear that the virus has to accumulate multiple mutations in the RT to acquire AZT resistance. Such multiple mutations must be difficult to acquire in a highly conserved structure like RT and must mitigate against the acquisition of resistance. Since AZT resistance is produced by multiple mutations in the RT gene this implies that further selective pressure on the RT gene eg. by introducing another inhibitor of RT would make the emergence of resistant virus less likely. This would suggest that the best combinations to be used to prevent or slow the emergence of resistant virus would be one of two RT inhibitors. Clearly if it proves that clinical resistance to AZT is not a problem then a combination of inhibitors of RT and of another virus function particularly at a different stage of replication might make the most sense.

Future of anti-HIV in the Therapy of AIDS

It is likely that antiviral agents will remain one of the major routes to controlling the disease. Given the success of AZT and the reduction in the toxicity observed with the drug now that we are learning how to use it correctly it means that future anti-viral drugs can build on this success. The attitude of the Regulatory authorities may change to the introduction of new antiviral agents as monotherapy given the potential resistance problem. The use of combinations of new antiviral drugs with AZT, or of biological agents like interferon with AZT, will become the focus of anti-HIV research.

References

1. P.A. Furman, J. A. Fyfe, M.H. St.Clair, K. Weinhold, J.L. Rideout, G.A. Freeman, S.N. Lehrman, D.P. Bolognesi, S. Broder, H. Mitsuya and D. W. Barry. Phosphorylation of 3'-azido-3'-deoxythymidine and selective interaction of 5'-triphosphate with human immunodeficiency virus reverse transcriptase. Proc. Natl. Acad. Sci. USA 83: 8333 (1986).

2. H. Mitsuya, K.J. Weinhold, P.A. Furman, M. St.Clair, S.N. Lehrman, R.C. Gallo, D. Bolognesi, D.W. Barry and S. Broder. 3'-azido 3' deoxythymidine (BW509U): an antiviral agent that inhibits the infectivity and cytopathic effect of human T-lymphotrophic virus type III/lympho-adenopathy associated virus in vitro. Proc. Natl. Acad. Sci. USA 82: 7096 (1985).

3. M.A. Fischl, D.D. Richman, M.H. Grieco, M.S. Gottlieb, P.A. Volberding, O.S. Laskin, J.M. Leedom, J.E. Groupman, D. Mildvan, R.T. Schooley, G.G. Jackson, D.T. Durack, P.O. King, The AZT Collaborative Working Group. The efficacy of Azidothymidine (AZT) in the treatment of patients with AIDS and AIDS-related complex. A double-blind placebo controlled trial. N. Engl. J. Med. 317: 185 (1987).

4. B.A. Larder, G. Darby and D.D. Richman. HIV with reduced sensitivity to zidovudine (AZT) isolated during prolonged therapy. Science 243: 1731 (1989).

5. H. Mitsuya and S. Broder. Inhibition of the *in vitro* infectivity and cytopathic effect of human T-lymphotropic virus type III/lymphoadenopathy-associated virus (HTLV III/LAV) by 2', 3' dideoxynucleosides. Proc. Natl. Acad. Sci. 83: 1911 (1986).

6. T.C. Merigan and G. Skowron. Safety and tolerance of dideoxycytidine as a single agent. Results of early phase studies in patients with acquired immunodeficiency syndrome (AIDS) or advanced AIDS-related complex. Study group of the AIDS clinical trials group of the National Institute of Allergy and Infectious Diseases. Am. J. Med. 88: 11S (1990).

7. R. Yarchoan, J.M. Pluda, C.F. Perno, H. Mitsuya, R.V. Thomas, K.M. Wyvill and S. Broder. Initial Clinical Experience with dideoxynucleosides as single agents and in combination therapy. Ann. N.Y. Acad. Sci. 616: 328 (1990).

8. D.H. Smith, R.A. Byrn, S.A. Marsters, T. Gregory, J.E. Groopman and D.J. Capon. Blocking of HIV-1 infectivity by a soluble, secreted form of the CD4 antigen. Science 238: 1704 (1987).

9. J.O. Kahn, J.D. Allan, T.L. Hodges, L.D. Kaplon, C.J. Arri, H.F. Fitch, A.E. Izu, J. Mordenti, S.A. Sherwin, J.E. Groopman and P.A. Volberding. The safety and pharmacokinetics of recombinant soluble CD4 (rCD4) in subjects with the Acquired Immunodeficiency Syndrome (AIDS) and AIDS-related Complex. Annals of Int. Med. 112: 254 (1990).

10. D.J. Looney, S. Hayashi, M. Nicklas, R.R. Redfield, S. Broder, F. Wong-Staal and H. Mitsuya. Differences in the interaction of HIV-1 and HIV-2 with CD4. J. Acquired Immune Deficiency Syndromes 3: 649 (1990).

11. B.K. Felber and G.N. Pavlakis. A quantitative bioassay for HIV-1 based on transactivation. Science 239: 184 (1988).

12. C.M. Roberts and E.D. Blair. Exploitation of a rapid and sensitive assay to analyse transactivation of the human immunodeficiency virus type 1 (HIV-1) long terminal repeat. Antiviral Chemistry and Chemotherapy 1: 139 (1990).

13. B.A. Larder, D.J.M. Purifoy, K.L. Powell and G. Darby. AIDS virus reverse transcriptase defined by high level expression in E.coli. EMBO J. 6: 3133 (1987).

14. M. Tisdale, B.A. Larder, D.M. Lowe, D.K. Stammers, D.J.M. Purifoy, P.F. Ertl, C. Bradley, S. Kemp, G.K. Darby and K.L. Powell. Characterization of HIV Reverse Transcriptase using monoclonal antibodies: The role of the C-terminus in antibody reactivity and enzyme function. J. Virol. 62: 3662 (1988).

15. D.M. Lowe, A. Aitken, C. Bradley, G. Darby, B.A. Larder, K.L. Powell, D.J.M. Purifoy, M. Tisdale and D.K. Stammers. HIV reverse transcriptase: crystallization and analysis of domain structure by limited proteolysis Biochemistry 27: 8884 (1988).

16. R. Pauwels, K. Andries, J. Desnyter, D. Schols, M.J. Kukla and H.J. Breslin. Potent and selective inhibition of HIV-1 replication *in vitro* by a novel series of TIBO derivatives. Nature 343:470 (1990).

17. D. Richman, A.S. Rosenthal, M. Skoog, R.J. Eckner, T-C Chou, J.P. Sabo and V.J. Merluzzi. BI-RG-587 is active against zidovudine-resistant human immunodeficiency virus type 1 and synergistic with zidovudine. Antimic. Agents Chemother. 35: 305 (1991).

18. W.S. Saari, J.S. Wai, M.E. Goldman, J.A. O'Brien, J.H. Nunberg, J.C. Quintero, E.A. Emini, T.E. Fisher, J.M. Hoffman, C.S. Rooney, A.M. Smith, C.M. Thomas and P.S. Anderson. HIV-1 specific pyridinone RT inhibitors: II synthesis and structure activity relationships WA 1061, 7th International Meeting on AIDS, Florence (1991).

19. T. Miyasaka, H. Tanaka, M. Baba, H. Hayakawa, R.T. Walker and J. Balzarini. A novel lead for specific Anti-HIV-1 Agents: 1-(2-Hydroxyethoxy) methoyl) 6-(phenylthio) thymine. J. Med. Chem. 32: 2507 (1989).

20. L.H. Pearl and W.R. Taylor. Sequence specificity of retroviral proteases. Nature 328: 482 (1987).

21. A. Wlodawer, M. Miller, M. Jaskolski, B.K. Sathyanrayana, E. Baldwin, I.T. Weber, L.M. Selk, L. Clawson, J. Schneider and S.B.H. Kent. Conserved folding in retroviral proteases: crystal structure of a synthetic HIV-1 protease. Science 245: 616 (1989).

22. T.J. McQuade, A.G. Tomasselli, L. Liu, V. Karacostas, B. Moss and T.K. Sawyer. A synthetic HIV-1 protease inhibitor with antiviral activity arrests HIV-like particle maturation. Science 247: 454 (1990).

23. N.A. Roberts, J.A. Martin, D. Kinchington, A.V. Broadhurst, J.C. Craig and I.B. Duncan. Rational design of peptide based HIV proteinase inhibitors. Science 248: 358 (1990).

24. D.J. Kempf, D.W. Norbeck, L. Codacovi, X.C. Wang, W.E. Kohlbrenner, N.E. Widebury, D.A. Paul, M.F. Knigge, S. Vasavanonda, A. Craig-Kennard, A. Saldivar, W. Rosenbrook Jr, J.J. Clement, J.J. Plattner and J. Erickson. Structure-based, C2 symmetric inhibitors of HIV protease. J. Med. Chem. 33: 26887 (1990).

25. B.A. Larder and S.D. Kemp. Multiple mutations in HIV-1 reverse transcriptase confer high-level resistance to zidovudine (AZT). Science 246: 1155 (1989).

BIOCHEMICAL AND PHARMACOLOGICAL PROPERTIES OF CI-972, A NOVEL 9-DEAZAGUANINE ANALOG PURINE NUCLEOSIDE PHOSPHORYLASE (PNP) INHIBITOR

Richard B. Gilbertsen[†], Mi K. Dong[†], David J. Wilburn[†], Lynn M. Kossarek[†], Jagadish C. Sircar[§], Catherine R. Kostlan[§], and Mary Carol Conroy[†]

Departments of Immunopathology[†] and Chemistry[§]
Parke-Davis Pharmaceutical Research Division
Warner-Lambert Company, Ann Arbor, MI, 48105

ABSTRACT

CI-972 (2,6-diamino-3,5-dihydro-7-(3-thienylmethyl)-4H-pyrrolo[3,2-d]pyrimidin-4-one monohydrochloride, monohydrate) is a novel inhibitor of PNP ($K_i = 0.83\ \mu M$) under development as a T cell-selective immunosuppressive agent. CI-972 inhibited proliferation (^3H-thymidine uptake) of human MOLT-4 (T cell) but not MGL-8 (B cell) lymphoblasts with respective IC_{50}s of 3.0 and >50 μM when tested with 10 μM 2'-deoxyguanosine. Without addition of exogenous 2'-deoxyguanosine, CI-972 was not inhibitory to any human T or B lymphoblastoid cell line tested. 2'-Deoxycytidine (10 μM), but not hypoxanthine or adenine, restored MOLT-4 cell growth. Inhibition of ^3H-thymidine uptake in MOLT-4 cells correlated with accumulation of dGTP, while alterations in guanine nucleotides were not observed. 2'-Deoxycytidine (10 μM) also blocked dGTP accumulation in MOLT-4 cells. CI-972 showed activity in vivo over a broad dose range: At 5-150 mg/kg p.o., CI-972 produced dose-dependent elevation of plasma inosine one hr after administration to rats (mean maximum of 2.62 vs. 0.06 μM in controls). Guanosine was also significantly elevated in a concentration-dependent manner, although the effect was not as impressive. Plasma nucleosides remained statistically-significantly elevated for up to four hr following a single oral dose of CI-972.

INTRODUCTION

Patients with homozygous deficiency of PNP have a T cell-selective form of immune deficiency and also exhibit several biochemical features consistent with the absence or impairment of this enzyme[1,2]. The biochemical changes observed belong to two major processes: those resulting from accumulation and subsequent metabolism of the substrates of PNP, and those resulting from a

Purine and Pyrimidine Metabolism in Man VII, Part A
Edited by R.A. Harkness *et al.*, Plenum Press, New York, 1991

depletion of the products of PNP-mediated catalysis. From an immunological perspective, the accumulation of deoxyguanosine and its phosphorylation to 2'-deoxyguanosine triphosphate (dGTP) is felt to be most important[1,2]. Of the four major physiologic PNP substrates, only deoxyguanosine is readily phosphorylated in mammalian tissues. The dGTP which accumulates is a potent inhibitor of ribonucleotide reductase, an enzyme that catalyzes formation of the deoxyribonucleoside diphosphates required for DNA synthesis. Of particular importance, phosphorylation of deoxyguanosine and accumulation of dGTP have generally been found to occur more extensively in thymus-dependent lymphocytes (T cells) or T lymphoblasts than in B cells or B lymphoblasts. Inhibition of PNP in replicating T cells therefore causes dGTP to accumulate and stops DNA synthesis, which results in cytostasis or cell death. While dGTP formation is the most widely-held hypothesis to explain the T cell deficiency in homozygous PNP deficiency, conflicting data do exist, and conclusive evidence that this pathway is sufficient to explain T cell-selective immune deficiency is lacking[3-5].

All of the major substrates of PNP are present in elevated concentrations in urine and/or plasma from homozygous PNP deficient patients[1,2]. However, inosine and guanosine are most readily-detected, while deoxyinosine and deoxyguanosine are minimally elevated. For that reason, models designed to monitor PNP inhibitors in vivo through alterations in purine metabolism have utilized measurement of plasma inosine and guanosine rather than deoxyribonucleosides[6-8]. This report extends observations reported elsewhere on CI-972, a compound under study in the clinic as a potential T cell-selective immunosuppressive agent[9,10].

MATERIALS AND METHODS

Human Lymphoblast Cultures, Thymidine Uptake, and Nucleotide Extractions. Details of these methods have been described elsewhere[8]. Briefly, ^{3}H-thymidine (^{3}H-TdR) was added during the final 16 hours of a 64 hour culture. The effects of CI-972, alone or with 2'-deoxyguanosine (GdR, 10 μM), 2'-deoxycytidine (CdR, 10 μM), hypoxanthine (Hx, 40 μM), or adenine (Ad, 40 μM) on ^{3}H-TdR uptake were determined. Nucleotides were extracted with cold 60% methanol and analyzed by HPLC as described previously[7,8].

Effects of CI-972 on Plasma Nucleosides. Oral dose-response studies were performed using 10 rats per dose group in each study. CI-972 was dissolved in propylene glycol and administered by gavage to male Wistar rats at doses of 5 to 150 mg/kg. Rats were exsanguinated under ether anesthesia one hour after dosing, the plasmas were processed, and plasma nucleosides were quantitated by HPLC[6]. Time-course studies were performed using both the i.v. and the p.o. routes in a vehicle of 5% dextrose in water.

RESULTS

CI-972 failed to inhibit ^{3}H-TdR uptake in MOLT-4 cells even at the highest concentration tested (50 μM), but in the presence of 10 μM GdR, had an IC_{50} of 3.0 $\pm$ 0.7 μM (SEM) for MOLT-4 but remained non-inhibitory for MGL-8 (data not shown). Similar effects were observed using another human T cell line, CEM

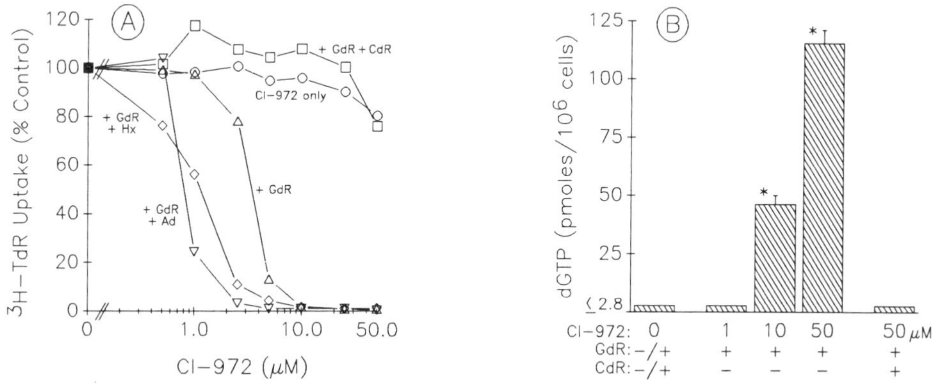

Figure 1. Effects of CI-972 on MOLT-4 Cells. **(A)** ^{3}H-TdR uptake. Data are representative of two studies. **(B)** Accumulation dGTP. Results are the means ± SEM of three experiments. Values marked with an asterisk are statistically-significantly different from control (Student's t-test, $p < 0.05$).

(data not shown). Addition of CdR (10 μM) to MOLT-4 cultures treated with CI-972 and GdR restored ^{3}H-TdR uptake, while adenine and hypoxanthine tended to enhance inhibition, showing that GTP accumulation was not responsible for toxicity (Fig. 1A). No dGTP was detected in MOLT-4 or MGL-8 lymphoblasts treated with CI-972 alone. Coaddition of 10 μM GdR with CI-972 resulted in CI-972 concentration-dependent elevation of dGTP in MOLT-4 cells (Fig. 1B) but not in MGL-8 cells (data not shown). Levels of the other nucleotides examined, ATP and GTP, were not significantly affected in MOLT-4 or MGL-8 cultures treated with any combination of test reagents (data not shown).

CI-972 produced statistically-significant elevation of inosine in rats at all doses tested (Fig. 2A). Elevation of inosine was dose-dependent, reaching a maximal concentration of 2.62 μM at 150 mg/kg in the first study, and 1.98 μM in the

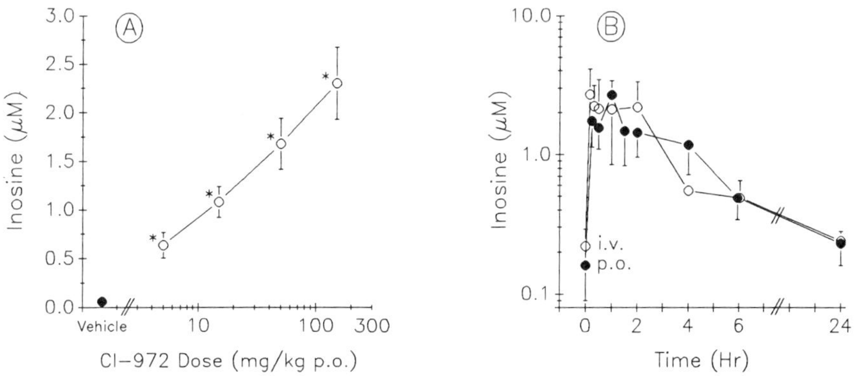

Figure 2. Elevation of Plasma Nucleosides in Rats Treated with CI-972. **(A)** Oral dose-response studies. Each point represents the mean ± SEM of two studies (n = 10 rats per dose per study). *$p < 0.05$ versus vehicle control, Student's t-test. **(B)** Time-course study. CI-972 (15 mg/kg) was administered p.o. or i.v. Values are mean ± SEM (n = 8-10 rats per group).

second. Plasma guanosine was also significantly elevated, although not as profoundly (data not shown). Inosine was highly elevated by both the i.v. and p.o. routes at the first time points assessed (15 min for i.v., 10 min for p.o.) (Fig. 2B). The biochemical duration of action of CI-972, as determined by plasma nucleoside elevation, was approximately four hours by either route of administration. These data show that CI-972 is an orally-active PNP inhibitor with sufficient activity and duration of biologic action in rats to warrant further study.

REFERENCES

1. Martin, D. W., Jr., and Gelfand, E. W. Biochemistry of diseases of immunodevelopment. Ann. Rev. Biochem. **50**:845-877, 1981.
2. Kredich, N. M., and Hershfield, M. S. Immunodeficiency diseases caused by adenosine deaminase deficiency and purine nucleoside phosphorylase deficiency. In The Metabolic Basis of Inherited Disease (J. A. Stanbury, J. B. Wyngaarden, and D. S. Fredrickson, consulting eds.), 6th edition, pp. 1045-1075, McGraw-Hill Information Services Co., 1989.
3. Goday, A., Simmonds, H. A., Morris, G. S., and Fairbanks, L.D. B cells as well as T cells form deoxynucleotides from either deoxyadenosine or deoxyguanosine. Clin. exp. Immunol. **56**: 39-48, 1984.
4. Sidi, Y., and Mitchell, B. S. 2'-Deoxyguanosine toxicity for B and mature T lymphoid lines is mediated by guanine ribonucleotide accumulation. J. Clin. Invest. **74**:1640-1648, 1984.
5. Simmonds, H. A., Goday, A., Morris, G. S., and Brolsma, M. F. J. Metabolism of deoxynucleosides by lymphocytes in long-term culture deficient in different purine enzymes. Biochem. Pharm. **33**: 763-670, 1984.
6. Gilbertsen, R. B., Dong, M. K., and Kossarek, L. M. Aspects of the purine nucleoside phosphorylase (PNP) deficient state produced in normal rats following oral administration of 8-amino-9-(2-thienylmethyl)guanine (PD 119,229), a novel inhibitor of PNP. Agents and Actions **22**:379-384, 1987.
7. Sircar, J. C., and Gilbertsen, R. B. Purine nucleoside phosphorylase (PNP) inhibitors: Potentially selective immunosuppressive agents. Drugs of the Future **13**:653-668, 1988.
8. Gilbertsen, R. B., and Sircar, J. C. Enzyme cascades: Purine metabolism and immunosuppression. In Comprehensive Medicinal Chemistry (C. Hansch, P. G. Sammes, and J. B. Taylor, Eds.), Vol. 2, pp. 443-480, Pergamon Press, Oxford, 1990.
9. Gilbertsen, R. B., Dong, M. K., Sircar, J. C., Kossarek, L. M., Shaw, M. K., Wilburn, D. J., Kostlan, C. R., and Conroy, M. C. Reduced thymidine uptake and augmented dGTP accumulation in MOLT-4 T cells and stimulated normal human lymphocytes treated with deoxyguanosine and CI-972, a new inhibitor of purine nucleoside phosphorylase. J. Cell. Biochem. **15E**: 181, 1991.
10. Sircar, J. C., Kostlan, C. R., Gilbertsen, R. B., Dong, M. K., and Cetenko, W. A. Pyrrolo[3,2-d]pyrimidines, a new class of purine nucleoside phosphorylase (PNP) inhibitors as potential T-cell selective immunosuppressive agents. In Purine and Pyrimidine Metabolism in Man, in press.

PYRROLO[3,2-D]PYRIMIDINES, A NEW CLASS OF PURINE NUCLEOSIDE PHOSPHORYLASE (PNP) INHIBITORS AS POTENTIAL T-CELL SELECTIVE IMMUNOSUPPRESSIVE AGENTS

Sircar, J. C.*, Kostlan, C. R., Gilbertsen, R. B., Dong, M. K., and Cetenko, W. A.

Parke Davis Pharmaceutical Research Division
Warner-LambertCompany, Ann Arbor, Michigan 48105, USA

ABSTRACT

Previously, we have described the synthesis and biological activity of 2,8-diamino-1,9-dihydro-9-(2-thienylmethyl)-6H-purin-6-one (PD 119229; CI-950) as a potent and competitive PNP inhibitor. As a part of our continuing efforts to develop a PNP inhibitor for autoimmune diseases, we have synthesized a series of pyrrolo[3,2-d]-pyrimidines as PNP inhibitors. In this series, 2,6-diamino-3,5-dihydro-7-(3-thienylmethyl)-4H-pyrrolo- [3,2-d]pyrimidin-4-one (CI-972) was found to be a potent, competitive inhibitor of PNP with K_i of 0.83 μM. It was also found to be selectively cytotoxic to human MOLT-4 (T cell) (IC_{50} = 3.0 μM) but non-toxic to MGL-8 (B cell) lymphoblasts. CI-972 is under development as a potential T-cell selective immunosuppressive agent. Synthesis and biological activities of the series are discussed.

INTRODUCTION

Purine nucleoside phosphorylase (PNP) is an enzyme in the purine cascade which reversibly catalyzes the degradation of inosine, deoxyinosine, guanosine, and deoxyguanosine to the corresponding purines hypoxanthine and guanine (equation 1). Because of its importance in the purine cascade, it has been a target for drug design for sometime[1-4].

(Deoxy)inosine/(deoxy)guanosine		Hypoxanthine/Guanine	
+	<----->	+	(equation-1)
Inorganic phosphate		Ribose-1-phosphate	

Inhibitors of other enzymes in the purine cascade, such as the adenosine deaminase inhibitor 2′-deoxycoformycin (pentostatin; DCF)[5] and the xanthine oxidase inhibitor allopurinol, have been developed, but so far no PNP inhibitor has been developed. PNP-deficient children show profound impairment in their T cell function with minimal or no effect on their B cell function[6]. Thus it has been hypothesized that a potent PNP inhibitor will mimic nature and will cause T cell cytotoxicity without having any effect on B cell function and thus be a T cell selective immunosuppressive agent.

Purine and Pyrimidine Metabolism in Man VII, Part A
Edited by R.A. Harkness *et al.*, Plenum Press, New York, 1991

CHEMISTRY

Previously, we have reported the synthesis and PNP inhibitory activity of 2,8-diamino-1,9-dihydro-9-(2-thienylmethyl)-6H-purin-6-one (**1**; PD 119229; CI-950)[2,3]. Since 9-deazainosine[7] (**2**), 9-deazaguanosine[7] (**3**), 5′deoxy-5′-chloro/5′deoxy-5′-iodo-9-deazainosines[7] (**4,5**) are known potent PNP inhibitors, we have synthesized the 9-deaza-analogs or pyrrolopyrimidine analogs of CI-950 (**6**) to investigate the contribution of the arylmethyl group at C_9 in the 9-deaza guanine series. The compounds (**6**) were synthesized[8] and assayed for their PNP inhibitory activity.

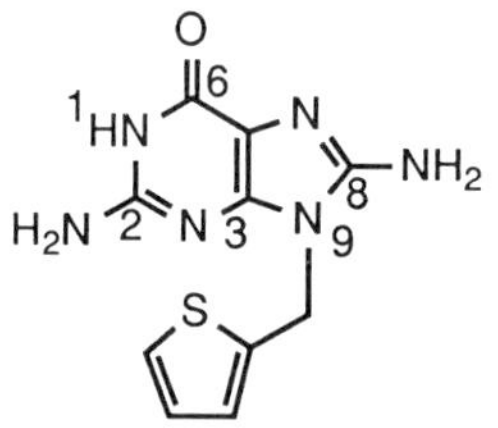

1 (PD 119229, CI-950)

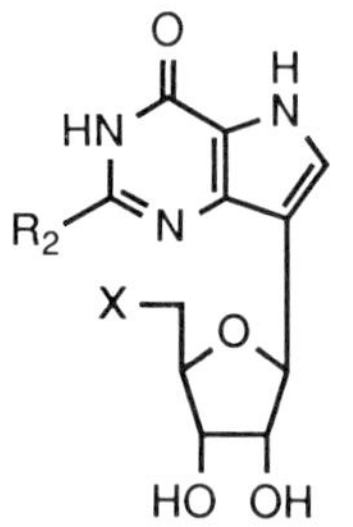

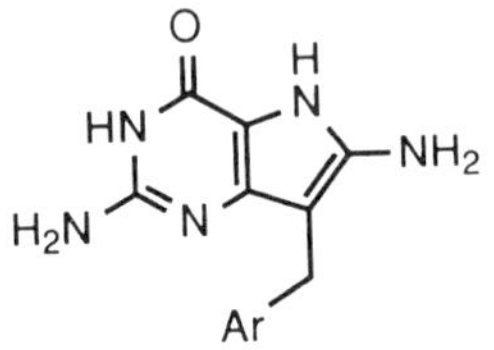

6

2, R_2 = H;	X = OH;	Ki = 2.0 μM
3, R_2 = NH_2;	X = OH;	Ki = 2.9 μM
4, R_2 = H;	X = Cl;	Ki = 0.20 μM
5, R_2 = H;	X = I;	Ki = 0.176 μM

PURINE NUCLEOSIDE ASSAY

The radiochemical assay was performed as described[9] before. In brief, 20 μL of PNP substrate (^{14}C-inosine, 20 to 100 μM) was added to each reaction vessel, followed by the addition of inhibitor of varying concentrations and buffer totalling 60 μL. The reaction was started by the addition of 20 μL of human PNP enzyme diluted 1:5000 with potassium phosphate dibasic buffer. The reaction was allowed to proceed for 10 min at 37°C and was stopped by the addition of 20 μL 8 M formic acid. Twenty μL of each sample was applied to TLC sheets, which were developed with 0.8 M LiCl. Spots containing inosine and hypoxanthine, the enzymatic product, were cut out and radioactivity was determined by liquid scintillation methodology.

BIOCHEMISTRY

The biochemistry of PNP and the abnormalities associated with the PNP deficiency in children have been thoroughly reviewed elsewhere[1]. It has been demonstrated that the absence of PNP leads to the accumulation of the PNP substrates, (deoxy-)inosines and (deoxy-)guanosines. However, only the accumulation of deoxyguanosine (GdR) affects T-lymphocytes. Intracellular GdR is phosphorylated to dGTP by deoxycytidine (CdR) kinase within T-lymphocytes. The accumulated dGTP within the cells allosterically inhibits the enzyme ribonucleoside diphosphate reductase (ribonucleotide reductase), thereby preventing conversion of pyrimidine ribonucleoside diphosphates to the corresponding deoxyribonucleoside diphosphates (e.g. CDP to dCDP and UDP to dUDP). The depletion of dCTP prevents DNA synthesis thereby causing cell death.

RESULTS AND DISCUSSION

The compounds (**7-10**) were evaluated for their inhibitory activity in human PNP enzyme and they were found to be potent inhibitors (Table-1). As was found in the CI-950 series, substitution on the benzene ring reduced activity, compare compounds **7** and **10**. The 3-thienyl compound (**9**, CI-972; 2,6-diamino-3,5-dihydro-7-(3-thienylmethyl)-4H-pyrrolo[3,2-d]pyrimidin-4-one) was found to be most potent and was evaluated further. CI-972 was found to be a competitive inhibitor of human PNP with a K_i of 0.83 μM (fig 1). As discussed in the biochemistry section, CI-972 (**9**) is expected to show selective cytotoxicity towards T cells but not to B cells. Thus it was evaluated for its selectivity towards human MOLT-4 (T-cell) lymphoblasts and human MGL-8 (B-cell) lymphoblasts and was found to be cytotoxic to MOLT-4 (IC_{50} = 3.0 μM) and not to MGL-8 (IC_{50} >50 μM) lymphoblasts. CI-972 has been evaluated further in pharmacology and toxicology and is presently under development clinically[10].

Table 1. 9-Deazaguanines (6)

Compound Number	Ar[a]	PNP-4 (IC_{50} (μM))
7	C_6H_5	1.6
8	2-Th	1.0
9	3-Th	0.9
10	2-CH_3O-C_6H_4	31.4

[a] Th = Thiophene

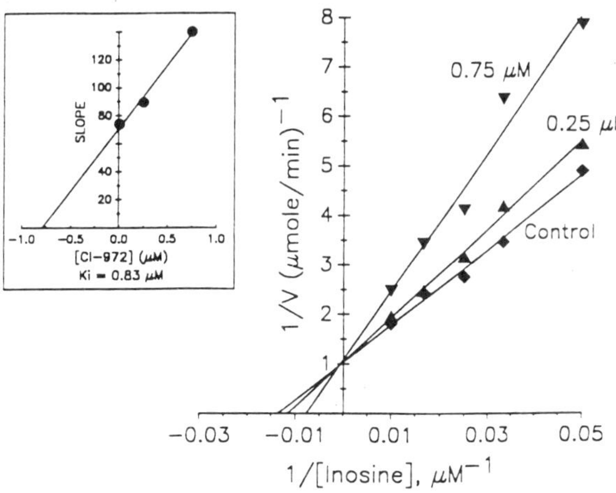

FIGURE 1. Kinetics of the Inhibition of Human Erythrocyte PNP by CI-972. The PNP enzyme assay consisted of a 2-min preincubation of CI-972 with a 1:5000 dilution of enzyme, followed by the addition of ^{14}C-inosine and an additional 10-min incubation. Inosine concentrations of 20 to 100 μM were used in the kinetics assay. Conversion of inosine to hypoxanthine was used as the endpoint. Hypoxanthine and inosine were separated using thin layer chromatography and spots containing these materials were cut out and counted in a liquid scintillation counter. Data are graphed in a Lineweaver-Burke plot (double reciprocal of inosine concentration versus velocity) and the K_i calculated from a replot of these data (inset).

REFERENCES

1. Gilbertsen, R. B.; Sircar, J. C.; 8.3 Enzyme Cascades: Purine Metabolism and Immunosuppression. In Comprehensive Medicinal Chemistry, Vol.-2.; Hansch, C.; Sammes, P. G.; Taylor, J. B., Eds.; Pergamon Press: Oxford, **1990**; pp 443-480.

2. Sircar, J. C.; Kostlan, C. R.; Pinter, G. W.; Suto, M. J.; Bobovski, T. P.; Capiris, T.; Schwender, C. F.; Dong, M.K.; Scott, M. E.; Bennett, M. K.; Kossarek, L.M.; Gilbertsen, R. B.; 8-Amino-9-substituted guanines: Potent purine nucleoside phosphorylase (PNP) inhibitors. Agents and Actions, **1987**, 21, 253-256.

3. Gilbertsen, R. B.; Scott, M. E.; Dong, M. K.; Kossarek; L. M.; Bennett, M. K.; Schrier, D. J.; Sircar, J. C.; Preliminary report on 8-amino-9-(2-thienylmethyl)guanine (PD 119229), a novel and potent purine nucleoside phosphorylase inhibitor. Agents and Actions, **1987**, 21, 272-274.

4. Sircar, J. C.; Gilbertsen, R. B.; Purine nucleoside phosphorylase (PNP) inhibitors: Potentially selective immunosuppressive agents. Drugs of the Future, **1988**, 13, 653-668.

5. Cheson, B. D.; Martin, A.; Clinical Trials in Hairy Cell Leukemia. Current Status and Future Directions; Ann. Intern. Med., **1987**, 106, 871-878.

6. Giblett, E. R.; Ammann, A. J.; Wara, D. W.; Sandman, R; Diamond, L.K.; Nucleoside-phosphorylase deficiency in a child with severely defective T cell immunity and normal B cell immunity. Lancet, **1975**, 1, 1010-1013.

7. Stoeckler, J. D.; Ryden, J. B.; Parks, Jr., R. E.; Chu, M. Y.; Lim, M.I.; Ren, W. Y.; Klein, R. S.; Inhibitors of purine nucleoside phosphorylase: Effects of 9-deazapurine ribonucleosides and synthesis of 5′-deoxy-5′-iodo-9-deazainosine. Cancer Res., **1986**, 46, 1774-1778.

8. Sircar, J. C. et. al, a detailed article will be published elsewhere. Analyses of the elements of new compounds were within ±0.4% of theoretical value.

9. Sircar, J. C.; Suto, M. J.; Scott, M. E.; Dong, M. K.; Gilbertsen, R. B.; Inhibitors of Human Purine Nucleoside Phosphorylase. Syntheis, Purine Nucleoside Phosphorylase Inhibition, and T-cell Cytotoxicity of 2,5-Diaminothiazolo[5,4-d]pyrimidin-7(6H)-one and 2,5-Diaminothiazolo-[4,5-d]pyrimidin-7(6H)-one. Two Thio Isosteres of 8-Aminoguanine. J. Med. Chem., **1986**, 29, 1804-1806.

10. Gilbertsen, R. B.; Dong, M. K.; Wilburn, D. J.; Kossarek, L. M.; Sircar, J. C.; Kostlan, C. R.; Conroy, M. C.; Biochemical and pharmacological properties of CI-972, a novel 9-deazaguanine analog purine nucleoside phosphorylase (PNP) inhibitor. In Purine and pyrimidine metabolism in man VII, in press.

* Author for correspondence

MODULATION OF METABOLISM AND ANTI-HIV-1 ACTIVITY OF PURINE 2′,3′-DIDEOXYNUCLEOSIDES BY IMP DEHYDROGENASE INHIBITORS

Larry L. Bondoc, Jr., Brian L. Robbins, Gurpreet S. Ahluwalia, Hiroaki Mitsuya, David G. Johns and Arnold Fridland

Division of Biochemical and Clinical Pharmacology, St. Jude Children's Research Hospital, Memphis, Tennessee 38101 (L.L.B., B.R., A.F.); Laboratory of Medicinal Chemistry, Developmental Therapeutics Program (G.S.A., D.G.J.), and Clinical Oncology Program (H.M.), Division of Cancer Treatment, National Cancer Institute, National Institutes of Health, Bethesda, Maryland 20892

1. INTRODUCTION

The dideoxynucleosides (ddNs) are having a substantial impact on the treatment of the human immunodeficiency syndrome (AIDS). The purine 2′,3′-dideoxyadenosine and its deaminated product 2′,3′-dideoxyinosine are potent inhibitors of human immunodeficiency virus (HIV) in human T lymphocytes and monocytes/macrophages (1,2). The anti-HIV effects of these drugs are mediated by conversion to the 5′-triphosphate ddATP. This triphosphate has been shown to act as both an inhibitor and substrate of HIV reverse transcriptase (3). The basis for the selectivity of ddI against HIV replication is related to the finding that its active metabolite ddATP binds more efficiently to reverse transcriptase than any of the cellular DNA polymerases (4). Although ddNs differ only slightly from their natural counterparts, their metabolism in cells differ substantially from that of the normal nucleosides. This diversity has manifested itself as differences in both antiviral potency and toxicity. Thus, these results imply a critical role for the anabolic pathways in determining the antiviral effectiveness of nucleoside analogs in anti-HIV therapy.

2. COMBINATION CHEMOTHERAPY WITH INHIBITORS OF INOSINATE DEHYDROGENASE

Our interest in IMP dehydrogenase as a potential target for combination chemotherapy with the ddNs started with the observation of Baba et al. (5) who showed that ribavirin could significantly enhance the anti-HIV activity of dideoxyadenosine and its deoxyguanosine derivative ddG in the MT-4 cell system. The combination of ddA with ribavirin resulted not only in increased antiviral activity of these compounds, but also in an increase in therapeutic activity; i.e., the increase in anti-HIV activity was greater than the cytotoxicity. In addition, Balzarini et al. (6) utilizing the Moloney sarcoma retrovirus

Purine and Pyrimidine Metabolism in Man VII, Part A
Edited by R.A. Harkness *et al.*, Plenum Press, New York, 1991

found that ribavirin interacted synergistically with ddA or its analog 2',3'-dideoxy-2,6-diaminopurine *in vivo*. Despite the therapeutic effect of the combinations, the basis for the synergistic interaction between these drugs had not been elucidated.

Ribavirin synthesized by Robins and colleagues (7) is a broad spectrum antiviral agent which has demonstrated anti-HIV activity in some *in vitro* assays, but at relatively high concentrations. Biochemically, ribavirin exerts a number of effects prominent among which is the inhibition of the enzyme IMP dehydrogenase, a rate-limiting enzyme in guanylate biosynthesis, by its 5'-monophosphate. We have shown (8,9) that the phosphorylation of ddI in human T cells to the active triphosphate occurs through the action of a cytosolic 5'-nucleotidase and that IMP is probably the primary intracellular phosphate donor for this activation of the drug. Thus, ribavirin by inhibiting IMPD and by increasing intracellular level of IMP could in principle cause an enhanced formation of phosphorylated ddNs in infected cells through modulation of the nucleotidase pathway. These considerations led to the studies of antiviral activity and mechanism of action to be described below.

Mechanism of action

In a comparative study, we examined the effect of two inhibitors of IMPD, mycophenolic acid and ribavirin on ddN phosphorylation in the HIV infectable human T cells, CCRF-CEM. As shown in Table 1, low concentration of both IMPD inhibitors which are noncytotoxic stimulated the accumulation of the anabolites from ddA, ddG and its carbocyclic analog 2',3'-dideoxy-2',3'-didehydroguanosine (carbovir) in CEM cells. Of the three ddNs evaluated, the enhancement was most pronounced for

Table 1. Effect of IMPD inhibitors on 5'-phosphorylation of ddNs in human lymphoid cells

ddN (μM)	IMPD inhibitor	ddNDP	ddNTP	ddNDP	ddNTP
		pmol/10^9 cells		fold-increase	
ddA (5)	none	127	190	1.0	1.0
	MPA	383	551	3.0	2.9
	RV	235	352	1.8	1.8
ddG (10)	none	250	90	1.0	1.0
	MPA	4000	310	16.0	3.4
	RV	2950	430	12.0	4.7
CARBOVIR	none	54	27	1.0	1.0
(10)	MPA	7330	282	135	10.4
	RV	1130	360	21.0	13.3

CCRF-CEM cells growing in log phase (10^6 cells/ml at time zero) were incubated with 5 μCi/ml [^{3}H]ddN in the presence or absence of 1 μM mycophenolic acid (MPA) or 5 μM ribavirin (RV). Cells were incubated with the respective IMPD inhibitors for 30 minutes prior to the addition of the radiolabeled ddN and incubated for a further 5 h. At the end of the incubation, cells were extracted with 60% cold methanol (-20° C) and extracts analyzed by ion exchange HPLC as described (8). Each value represents the mean value of duplicate samples. ddAMP and ddGMP eluted at 18 and 14-18 min, respectively, but could not be accurately quantitated by this method.

carbovir where nucleotide analog levels increased remarkably up to 125-fold followed by ddG and ddA. Of interest also, is the observation that the diphosphate of ddG and carbovir levels were increased to much greater degree than their triphosphates, indicating a rate-limiting step at the diphosphate step. This differential effect was not observed with ddADP where its activation was almost identical to that of ddATP.

Next we asked how IMPD inhibitors influence the level of IMP in these human cells. For each ddN, the concentration of IMPD inhibitor was increased to noncytotoxic concentrations (0-2 μM for mycophenolic acid and 0-40 μM for ribavirin) and the level of intracellular IMP produced in cells quantitated by HPLC and plotted versus the corresponding ddNTP level accumulated. Fig. 1 shows that there was a good correlation between ddNTP levels formed from the corresponding ddN and IMP levels formed in the presence of ribavirin and mycophenolic acid. GTP levels fell in parallel with the increase in IMP indicating inhibition of IMPD activity in the cells. A similar correlation was also observed for the ddDNP and IMP levels.

Enhancement of antiviral activity

In order to determine whether the increase in ddNTP formation seen in the presence of IMPD inhibitors was accompanied by an increase in anti-HIV formation, we examined the effect of ribavirin or its analog tiazofurin on the activity of ddI or ddG in the ATH-8 and Molt-4 cells. The assays were done over a 7 day period. Neither ribavirin nor

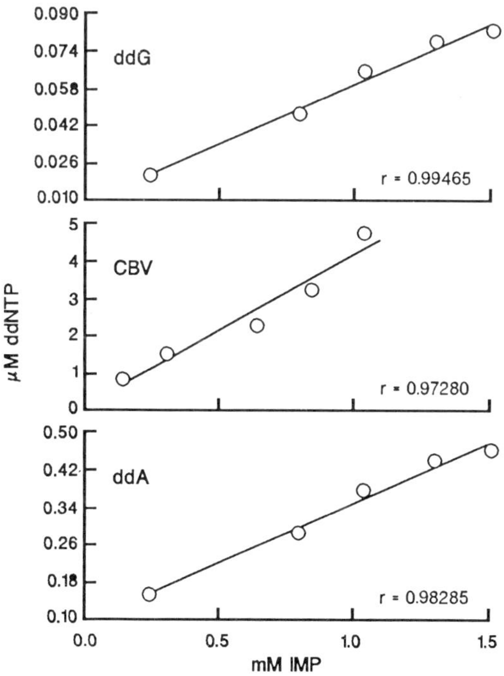

Fig. 1. Relationship between dideoxynucleotide formation and intracellular IMP concentration. CEM cells growing in log phase were incubated with ddN as in Table 1 and varying concentrations of IMPD inhibitors. Intracellular IMP and ddNTPs concentrations were determined by HPLC as described.

tiazofurin by itself appeared to have any anti-HIV effect at the concentrations used in this study (1-5 μM). However, both these compounds potentiated the anti-HIV activity of ddI and ddG. As shown in Fig. 2, a significant decrease in the concentration of ddI needed to inhibit HIV replication was seen in ATH-8 cells in the presence of 5 μM ribavirin. A similar effect of the combination of purine dideoxynucleotides:ribavirin was noted by previous investigators (4,5).

3. DISCUSSION

In summary, ribavirin and mycophenolic acid, two potent inhibitors of the enzyme IMP dehydrogenase and guanylate biosynthesis, potentiate the activity of a number of purine dideoxynucleoside analogs against HIV by as much as 5-fold in T cells considered as principal target for the virus. In some studies, but not all, ribavirin appears to have intrinsic antiviral activity. By demonstrating the two protypical inhibitors of IMPD dehydrogenase ribavirin and mycophenolic acid share the same properties of augmenting IMP levels and the anabolites of ddNs, it warrants the speculation that the former of these effects is causative of the potentiation seen with these combinations. The effect of IMPD inhibitors of the activation of the dideoxypurine nucleosides is in contrast to that seen with the analog 3'-azido-3'-thymidine (AZT). In the case of ribavirin:AZT combination, a strong antagonism was observed in the effect on HIV infectivity which was probably related to

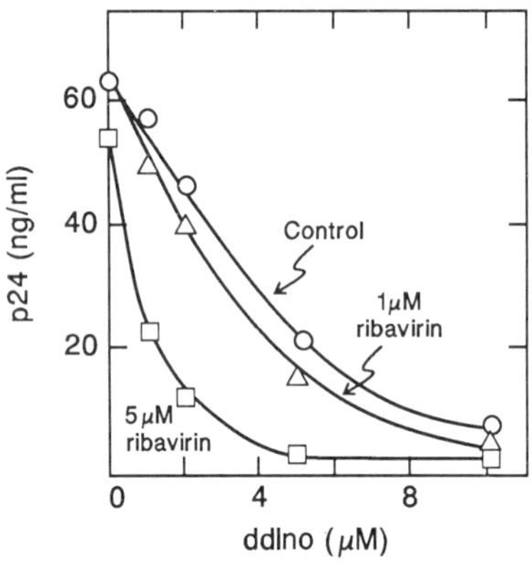

Fig. 2. Effect of ribavirin on anti-HIV-1 activity of ddIno in ATH-8 cells. Cells ($2x10^4$) were exposed to HIV-1/III_b for 40 min and incubated with indicated concentrations of drugs in 5 ml culture medium. On day 7 in culture, supernatants were collected and amount of p24 gag protein was quantitated.

feedback inhibition on thymidine kinase and phosphorylation of AZT by an increase in thymidine triphosphate pool (10).

In view of these differences between different drugs when used in combination, i.e., synergy when combined with purine 2',3'-dideoxypurine analog ddA, ddG and carbovir and antagonism when combined with pyrimidine analog such as AZT, the studies emphasize the importance of elucidating the metabolic pathways responsible for metabolism of these agents. Further explorations of the modulation of the anti-HIV activity of the purine ddNs by inhibitors of nucleotide metabolism appear to be warranted, particularly since many of these agents have previously been approved for investigational study.

Acknowledgements - This work was supported by USPHS Grant 1 R01 AI 27652, 1 R01 AI 31145, CORE Grant P30 CA 21765 from the National Institutes of Health and by the American Lebanese Syrian Associated Charities.

REFERENCES

1. H. Mitsuya and S. Broder, Inhibition of the in vitro infectivity and cytopathic effect of human T-lymphotropic virus type III/lymphoadenopathy associated virus (HTLV-III/LAV) by 2',3'-dideoxynucleosides. Proc. Natl. Acad. Sci. USA 83:1911 (1986).
2. C. F. Perno, R. Yarchoan, D. A. Cooney, N. R. Hartman, D. S. Webb, Z. Hao, H. Mitsuya, D. G. Johns, and S. Broder, Replication of HIV in monocytes:macrophage colony stimulating factor potentiates viral production yet enhances the antiviral effects mediated by AZT and other dideoxynucleosides. J. Exp. Med. 169:933 (1989).
3. R. Yarchoan, H. Mitsuya, C. E. Myers, and S. Broder, Clinical pharmacology of 3'-azido-2',3'-dideoxythymidine (zidovudine) and related dideoxynucleosides. N. Engl. J. Med. 321:726 (1989).
4. H. Mitsuya, R. Yarchoan, and S. Broder, Molecular targets for AIDS therapy. Science 249:1533, 1990.
5. M. Baba, R. Pauwels, J. Balzarini, P. Herderwijn, E., De Clercq, J. Desmyter, Ribavirin antagonizes inhibitory effects of pyrimidine 2',3'-dideoxynucleosides but enhances inhibitory effects of purine 2',3'-dideoxynucleosides on replication of human immunodeficiency virus in vitro. Antimicrob. Agents and Chemother. 31:1613 (1987).
6. J. Balzarini, L. Naesens, M. J. Robins, and E. De Clercq, Potentiating effect of ribavirin on the in vitro and in vivo antiretrovirus activities of 2',3'-dideoxyinosine and 2',3'-dideoxy-2,6-diaminopurine riboside. J. Acq. Immune Defi. Synd. 3:1140 (1990).
7. J. T. Wikowski, R. K. Robins, and R. W. Sidwell. The design and synthesis and broad-spectrum antiviral activity of 1-β-D-ribofuranosyl-1,2,4-triazole-3-carboxamide and related nucleosides. J. Med. Chem. 15:1150 (1972).
8. M. A. Johnson, G. Ahluwalia, M. C. Connelly, D. A. Cooney, S. Broder, D. G. Johns, and A. Fridland, Metabolic pathways for the activation of the antiretroviral agent 2',3'-dideoxyadenosine in human lymphoid cells. J. Biol. Chem. 263:15354 (1988).
9. M. A. Johnson and A. Fridland, Phosphorylation of 2',3'-dideoxyinosine by cytosolic 5'-nucleotidase of human lymphoid cells. Mol. Pharmacol. 36:291 (1989).
10. M. W. Vogt, K. L. Hartshorn, P. A. Furman, T-C. Chou, J. A. Fyfe, L. A. Coleman, C. Crumpacker, R. T. Schooley, and M. S. Hirsch, Ribavirin antagonizes the effect of azidothymidine on HIV replication. Science 235:1376 (1987).

ACADESINE (AICA RIBOSIDE) INHIBITS PLATELET AGGREGATION IN HUMAN WHOLE BLOOD

Bullough D. A., Zhang C., and Mullane K. M.

Gensia Pharmaceuticals, Inc.
San Diego, CA 92121, USA

Interactions between platelets and erythrocytes can influence platelet aggregation and hemostasis. Adenosine inhibits platelet aggregation and the levels of adenosine can be modified by the presence of erythrocytes. Acadesine (AICA riboside; 5-amino-1-β-D-ribofuranosylimidazole-4-carboxamide) is a nucleoside which enters cells and is phosphorylated by adenosine kinase to ZMP, an intermediate in de novo purine biosynthesis. Some of the pharmacological properties of acadesine are attributed to it's ability to increase local adenosine concentrations. Consequently we investigated whether acadesine inhibits platelet aggregation in whole blood using an impedance aggregometer. Acadesine induced a time- and concentration-dependent inhibition of aggregation elicited by ADP, collagen or arachadonic acid. The IC_{50} for acadesine against ADP-induced aggregation decreased from greater than 5mM after 5 min incubation to 0.47 ±0.05 mM (n=5) after 60 minutes at 37°C. Dipyridamole showed an IC_{50} of 0.20 ± 0.05 mM (n=6) independent of the incubation period. Inhibition of adenosine kinase by 20μM 5'-deoxy 5-iodotubercidin suppressed the antithrombotic activity of acadesine by 74 ± 7%. Similarly, addition of adenosine deaminase (3.6 U/ml) reversed the antiaggregatory effects by 78 ± 5%. A background of exogenous adenosine (10μM), insufficient to modify platelet behavior directly, decreased the concentration of acadesine required to inhibit aggregation 5-fold. Thus, acadesine inhibits human platelet aggregation in whole blood. The response is dependent upon the metabolism of acadesine and is mediated by adenosine.

CONCENTRATION AND TIME DEPENDENT GROWTH INHIBITION AND METABOLISM IN VITRO BY 2',2'-DIFLUORO-DEOXYCYTIDINE (GEMCITABINE)

Veronique W.T. Ruiz van Haperen, Gijsbert Veerman, Paul Noordhuis, Jan B. Vermorken and Godefridus J. Peters

Dept. of Oncology, Free University Hospital, PO Box 7057, 1007 MB Amsterdam, The Netherlands

INTRODUCTION

2',2'-Difluoro-deoxycytidine (dFdC) is a new deoxycytidine analogue. 1-ß-D-arabinofuranosylcytosine (cytosine arabinoside, ara-C), also a deoxycytidine analogue, is one of the most effective agents against acute myelogenous leukemia (AML). However, activity against solid tumours both in preclinical systems and patients is limited (1). As opposed to this, dFdC has shown excellent antitumour activity in various preclinical solid tumour model systems (2,3). Several phase I trials using different schedules have been completed (4,5) and the drug is now being evaluated in some clinical phase II trials. Some activity has been observed in patients with non-small-cell lung cancer, ovarian cancer and pancreas cancer. Ara-C exhibits its antitumour effect through accumulation of its triphosphate ara-CTP, which is subsequently incorporated into DNA, leading to growth inhibition (1,6). The antitumour effect of ara-C is dose and time dependent (7). The phosphorylation of ara-C to ara-CMP, catalysed by deoxycytidine kinase (dCK), is the crucial step. The drug can be inactivated by deamination to ara-U, catalysed by deoxycytidine deaminase (dCDA) (1). Like ara-C, dFdC must be converted into dF-dCMP, catalyzed by dCK, and subsequently to dF-dCTP. dFdC can be inactivated by deamination to dFdU (difluorodeoxyuridine), catalysed by dCDA (8,9). Until now, most studies on dFdC have been limited to leukemic cell lines. Because of the promising preclinical results in head & neck, colon and ovarian cancer, we used these solid tumour cell lines as a model for solid tumours. We determined in vitro whether sensitivity to the drug and accumulation of dF-dCTP were time and concentration dependent.

MATERIALS AND METHODS

Cell culture. The following cell lines were used: C26-10 and WiDr, a murine and a human colon carcinoma, respectively; A2780 and OVCAR-3, both human ovarian carcinomas; UM-SCC-22B and UM-SCC-14C (22B and 14C), both human head and neck squamous cell carcinomas. The cells were cultured in Dulbecco's Modified Eagle's Medium (Gibco), supplemented with 10% v/v heat inactivated Foetal Calf Serum (FCS) (Gibco), 1 mM L-glutamine (Sigma) and 250 ng/ml gentamycin, at 37°C, at 5% CO_2. For growth inhibition studies the culture medium was supplemented with 5% v/v FCS.

Chemosensitivity testing. For assessment of chemosensitivity, on day 0 cells were seeded in triplo in flat bottom 96-well plates. After 24 h dFdC was added in a concentration range from 10^{-4} to 10^{-11} M. After exposure for 0.5, 1, 2, 4, 24, 48, and 72 h, the cells were washed twice and allowed to grow till day 4. At day 1 and 4 the sulforhodamine B (SRB) protein assay was performed as described earlier (10,11), to quantify the number of cells. Growth inhibition was expressed as IC50: the concentration of dFdC, which caused 50% growth inhibition.

dF-dCTP accumulation. dF-dCTP accumulation was measured in exponentially growing cells, after exposure for 1, 4, or 24 h to 1, 10 and 100 μM dFdC. Cells were harvested and counted. Nucleotides were extracted using trichloro-acetic acid as described earlier (12). The HPLC analysis for the nucleotides was performed with a Partisphere SAX anion exchange column, using isocratic elution (13). The HPLC analysis for dF-dCTP was performed essentially as described by Heinemann et al. (9).

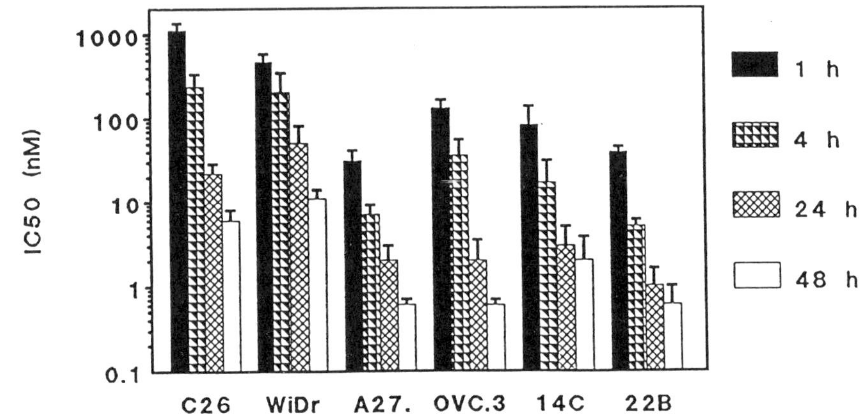

Fig. 1 IC50's of dFdC at different exposure times, in 6 tumour cell lines. Values are means ± SD of 4-10 separate experiments.

RESULTS

Growth inhibition. In all cell lines the growth inhibitory effect of dFdC was time and concentration dependent (Fig. 1). The effect of 30 min exposure was not significantly different from that of 1 h exposure. Similar results were obtained after 2 and 4 h and 48 and 72 h exposure, respectively. Therefore only data of the 1, 4, 24 and 48 h exposure are presented. Although in all cell lines a comparable time dependence was observed, the absolute IC50's varied considerably between the different cell lines (Fig. 1). For example for the 24 h exposure, the IC50 in the colon tumour cell line WiDr was 3x higher than in the squamous cell carcinoma 14C, and 26x higher than in the ovarian carcinoma A2780.

dF-dCTP accumulation. The accumulation of dF-dCTP, as determined in C26-10 and WiDr, was concentration dependent (Fig. 2). For example in C26-10, 24 h exposure to 100 μM dFdC yielded a 16x higher dF-dCTP concentration than 24 h exposure to 1 μM. More pronounced, however, is the time dependence of dF-dCTP accumulation. 24 h Exposure of C26-10 cells to 100 μM of dFdC yielded 43x more dF-dCTP than 1 h exposure. The normal nucleotide pattern changed by dFdC exposure. The major disturbances were observed in the UTP and CTP pools of WiDr. In this cell line CTP decreased to undetectable levels after 24 h exposure to dFdC, and that of UTP increased considerably. In C26-10 cells UTP accumulation was also seen, but the CTP pool did not change significantly.

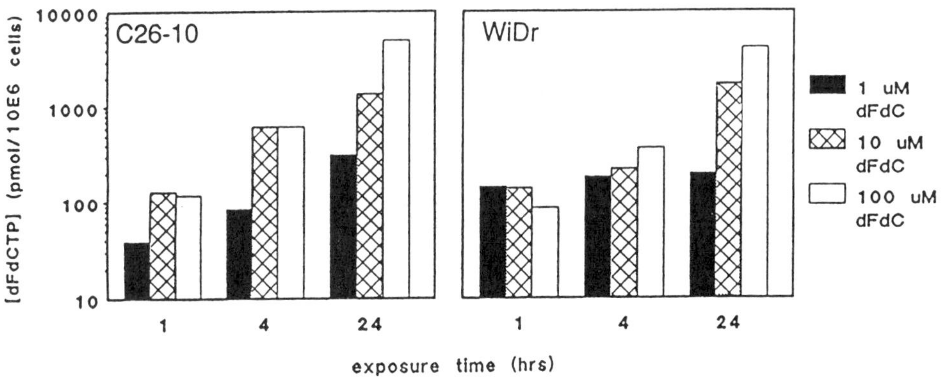

Fig. 2 dF-dCTP accumulation in C26-10 and WiDr cells, after continuous exposure to 1, 10 and 100 μM of dFdC for 1, 4 and 24 h. Data are from one experiment, performed in triplo.

DISCUSSION

The growth inhibitory effect of dFdC in solid tumour cell lines is concentration and time dependent. This also holds for the accumulation of the active triphosphate dF-dCTP. It appears that the antitumour effect of dFdC is related to the dF-dCTP accumulation, as it is for ara-C and ara-CTP. It is known that dF-dCDP can inhibit ribonucleotide reductase, resulting in a depletion of dCTP (14,15). As dCTP is the most important feedback inhibitor of dCK, a depletion of dCTP would result in a higher dCK activity. The concentration dependence of dF-dCTP accumulation becomes more pronounced with longer exposure (Fig. 2). Possibly at 1 and 4 h exposure dCK is still feedback regulated by dCTP. After 24 h, however, enough dF-dCDP could be formed to deplete dCTP, resulting in a higher dCK activity. Under these circumstances dFdC may be phosphorylated more efficiently at low concentrations. The observed increase in UTP at longer exposure may also enhance dCK activity, since UTP is a better substrate than ATP (16). Interestingly, the difference in dF-dCTP concentration after 24 h exposure between 10 μM and 100 μM dFdC is much smaller than the difference between 1 and 10 μM, after 24 h exposure. This could be due to saturation of the phosphorylating enzymes. Grünewald et al. (5) recently published similar findings in patients treated with dFdC. Longer infusion times with the same absolute dose resulted in higher concentrations of dF-dCTP in the mononuclear cells. But there was a limit to the dose dependency; accumulation of dF-dCTP did not increase at doses >900 mg/m^2, probably due to saturation of the phosphorylating enzymes (5,17).
From our in vitro results and literature data it seems therefore that dFdC would exhibit a better antitumour effect when applied by continuous infusion. It is not clear how

dFdC causes CTP depletion in WiDr cells, when exposed for 24 h. From this CTP depletion in WiDr and the UTP accumulation in both WiDr and C26-10 it seems that dFdC or one of its metabolites may inhibit CTP synthetase, an enzyme which catalyzes the synthesis of CTP from UTP. We conclude that the growth inhibitory effect of dFdC is concentration and time dependent and is related to the accumulation of its triphosphate derivative dF-dCTP.

ACKNOWLEDGEMENTS

This work was supported by the Dutch Cancer Society (grant IKA 90-19). Dr G.J. Peters is the recipient of a senior research fellowship of the Royal Netherlands Academy of Arts and Sciences.

REFERENCES

1. Momparler RL, Onetto-Pothier N: Drug resistance to cytosine arabinoside. In: Resistance to antineoplastic drugs, pp 353-367. D. Kessel, ed. CRC Press, Inc., 1989.
2. Braakhuis BJM, Van Dongen GAMS, Vermorken JB et al.: Preclinical in vivo activity of 2',2'-difluorodeoxycytidine (gemcitabine) against human head and neck cancer. Cancer Res. 51, 211-214, 1991.
3. Hertel LW, Broder GB, Kroin JS et al.: Evaluation of the antitumor activity of gemcitabine (2',2'-difluoro-2'-deoxycytidine). Cancer Res. 50, 4417-4422, 1990.
4. Tanis B, Clavel M, Guastalla J et al.: Phase I study of gemcitabine (difluorodeoxycytidine; dFdC; LY 188011) administered in a two-weekly schedule. Proc. Am. Assoc. Cancer Res. 31, 207, 1990.
5. Grunewald R, Abbruzzese JL, Tarassoff P et al.: Saturation of 2',2'-difluorodeoxycytidine 5'triphosphate accumulation by mononuclear cells during a phase I trial of gemcitabine. Cancer Chemother. Pharmacol. 27, 258-262, 1991.
6. Kufe D, Spriggs D, Egan EM et al.: Relationships among ara-CTP pools, formation of (ara-C)DNA, and cytotoxicity of human leukemic cells. Blood, 64, 54-58, 1984.
7. Heinemann V, Estey E, Keating MJ et al.: Patient-specific dose rate for continuous infusion high-dose cytarabine in relapsed acute myelogenous leukemia. J. Clin. Oncol. 7, 622-628, 1989.
8. Gandhi V, Plunkett W: Modulatory activity of 2',2'-difluorodeoxycytidine on the phosphorylation and cytotoxicity of arabinosyl nucleosides. Cancer Res. 50, 3675-3680, 1990.
9. Heinemann V, Hertel LW, Grindey GB, et al.: Comparison of the cellular pharmacokinetics and toxicity of 2',2'- difluorodeoxycytidine and 1-ß-D-arabinofuranosylcytosine. Cancer Res. 48, 4024-4031, 1988.
10. Keepers YPAM, Pizao PE, Peters GJ et al.: Comparison of the sulforhodamine B protein and tetrazolium assays for in vitro chemosenstivity testing. Eur. J. Cancer, 27, 897-899, 1991.
11. Skehan P, Storeng R, Scudiero D et al.: New colorimetric cytotoxicity assay for anticancer-drug screening. J. Natl. Cancer Inst. 82, 107-112, 1990.
12. Peters GJ, Laurensse E, Leyva A, et al.: Sensitivity of human, murine and rat cells to 5-fluorouracil and 5'-deoxy-5-fluorouridine in relation to drug-metabolizing enzymes. Cancer Res. 46, 20-28, 1986.
13. Peters GJ, Nadal JC, Laurensse EJ, et al.: Retention of in vivo antipyrimidine effects of brequinar sodium (DUP-785; NSC 368390) in murine liver, bone marrow and colon cancer. Biochem. Pharmacol. 39, 135-144, 1990.
14. Plunkett W, Gandhi V, Chubb S et al.: 2',2'-Difluorodeoxycytidine metabolism and mechanism of action in human leukemia cells. Nucleosides and Nucleotides, 8, 775-785, 1989.
15. Heinemann V, Xu Y-Z, Chubb S et al.: Inhibition of ribonucleotide reduction in CCRF-CEM cells by 2',2'-difluorodeoxycytidine. Mol. Pharmacol. 38, 567-572, 1990.
16. White JC, Capizzi RL: A critical role for uridine nucleotides in the regulation of deoxycytidine kinase and the concentration dependence of 1-ß-D-arabinofuranosylcytosine phosphorylation in human leukemia cells. Cancer Res. 51, 2559-2565, 1991.
17. Peters G, Tanis B, Clavel M et al.: Pharmacokinetics of gemcitabine (LY188011); difluorodeoxycytidine) administered every two weeks in a phase I study. Proc. Am. Assoc. Cancer Res. 31, 180, 1990.

THIOPURINE METHYLTRANSFERASE ACTIVITY AND EFFICACY OF AZATHIOPRINE IMMUNOSUPPRESSION IN TRANSPLANT RECIPIENTS

Pedro R. Chocair,* John A. Duley,** H. Anne Simmonds,** and J. Stewart Cameron**

* Hospital das Clinicas, Unidade de Transplante Renal (UTR), Sao Paulo, Brazil, ** Purine Research Laboratory and Renal Unit, Clinical Science Laboratories, UMDS Guy's Hospital, London, UK

INTRODUCTION

Azathioprine (AZA) has been given to transplanted patients since 1963 (1), but despite extensive use its metabolism is still not completely understood. After oral administration, it is converted rapidly to 6-mercaptopurine (6MP), mainly in the liver and the gut (2). 6MP is metabolised by three competing pathways (Figure 1): A) conversion to 6-thioinosinic acid by hypoxanthine guanine phosphoribosyltransferase (HPRT) and thence thioguanine nucleotides which exert their cytotoxicity by incorporation into DNA and RNA (2,3); B) catabolism by xanthine oxidase (XOD) to 6-thiouric acid; C) conversion to 6-methyl mercaptopurine (6MeMP) by the enzyme thiopurine methyltransferase (TPMT), which has a wide range of activity in the normal population and is inherited in a autosomal codominant fashion (3,4).

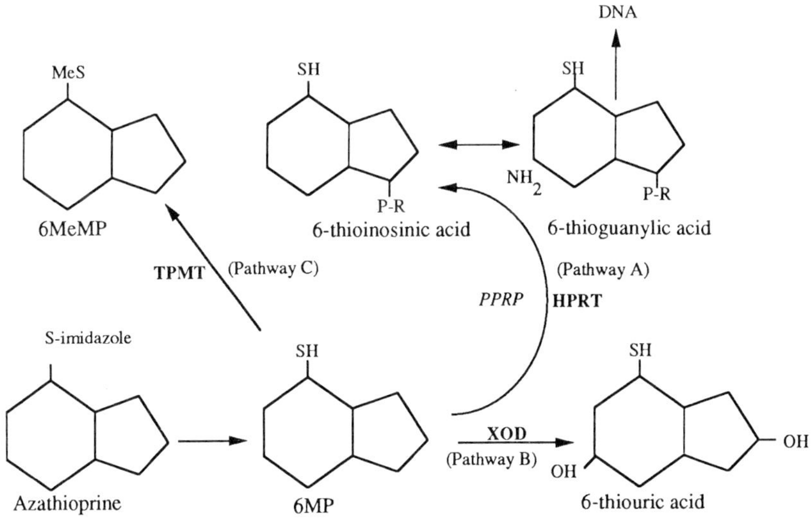

Fig 1. Schematic representation of azathioprine metabolism.

Purine and Pyrimidine Metabolism in Man VII, Part A
Edited by R.A. Harkness *et al.*, Plenum Press, New York, 1991

One study suggested that about 1 in 300 individuals have a complete deficiency of TPMT, while 11% show intermediate activity (3). Very low activity would obviously increase the availability of 6MP for the anabolic pathway and put subjects at special risk for myelotoxicity from AZA therapy. Conversely, high TPMT levels could limit the availability of 6MP for the anabolic pathway and reduce its effectiveness. Published studies in patients given AZA for other disorders have shown no clear correlation between myelotoxicity (as indicated by leucopenia or thrombocytopenia), TPMT activity and the subsequent clinical course. The purpose of this study was to evaluate TPMT activity and clinical parameters in renal transplant recipients receiving AZA.

PATIENTS AND METHODS

Patients and controls: TPMT activity was analysed in red blood cell (RBC) lysates of 25 normals (14 males, 11 females, mean 35 years, range 6-57), 25 uremic patients maintained on dialysis (15 males, 10 females, mean 56 years, range 23-76), and 68 renal transplant out-patients (37 males, 31 females, mean 42 years, range 20-73), who were cadaver-donor recipients. Nine of the cadaver-transplant recipients were on cyclosporin A and prednisolone, but not AZA. Of the remainder, 29 were receiving prednisolone together with AZA in a single dose, adjusted to the white blood cell (WBC) count and 30 were on triple therapy comprising AZA, prednisolone and cyclosporin. The mean post-transplant follow-up at investigation was 83 months (range 3-170). The WBC count was the mean of 10 previous assessments, while plasma creatinine was the most recent result.

Blood samples: Freshly collected heparinised venous blood was centrifuged for 6 min at 420g, the plasma, buffy coat and top fifth layer containing platelets, reticulocytes and WBC (which have high TPMT activity), was removed and the RBC washed twice with saline.

TPMT assay: 100 ul of washed packed RBC was added to 500 ul of distilled water, and freeze-thawed twice in dry ice, then the stroma was removed by centrifugation at 12,000g for 15 min. 100ul aliquots of supernatant were used for the assay method, employing radiolabelled S-adenosylmethionine, reported by Weinshilboum et al.(5), with slight modifications.

Statistical analysis: Because TPMT activity is not normally distributed, non-parametric statistical tests were used for analysis.

RESULTS

Weinshilboum (5) and others reported TPMT activity as nmol/h/mlRBC, but in this study, we have expressed activity as pmol/h/mgHb. Activity was calculated by both methods in a small group of patients: comparison with Weinshilboum's results could be achieved by dividing our results by 2.8.

TPMT activity in 25 healthy controls (median: 31 pmol/h/mgHb; range: 16.2-43.0) was not significantly different from that in 9 transplanted patients taking cyclosporin but not AZA (median 31.7; 12.7-43.5). However, TPMT was significantly higher ($p < 0.01$) in the 25 uremic patients (median 35.5; 18.6-62.6), and also in all 59 transplant recipients whose treatment included AZA (median 36.1; 16.1-71.3). No significant difference was found in the TPMT activity of the transplanted group between males and females.

TPMT activity was categorised into 3 groups: very low (less than 15), intermediate (15 to 26) and high (greater than 26 pmol/h/mgHb). Two transplant patients, showed undetectable activity, and only 12.7 pmol/h/mg Hb respectively, and both exhibited absolute intolerance to AZA (Fig 2). Two other patients (TPMT 16.1 and 21.7 pmol/h/mg Hb respectively) had thrombocytopenia but not leucopenia, controlled by lowering the AZA dose. All other patients in this intermediate group showed no evidence of myelotoxicity, as indicated by leucopenia and/or thrombocytopenia. One patient with TPMT of 29.8 pmol/h/mgHb exhibited persistent leucopenia while taking cotrimoxazole as well as AZA. Correlation between WBC and TPMT activity in all AZA treated patients was only weakly positive (rs=0.22, $p < 0.05$).

Analysis was made of the results in 29 transplant patients taking only AZA plus prednisolone: 24 had creatinines lower and 5 greater than 200 uM/l. This showed that although mean age, time of transplant, WBC count, drug dose, and number of previous blood transfusions did not differ, TPMT activity was significantly higher ($p < 0.04$) in the 5 patients with significant renal impairment (median 42.5 pmol/h/mgHb; range 33.8-51.5) than in the other 24 subjects (33.1 pmol/h/mg Hb; range 16.1-46.10).

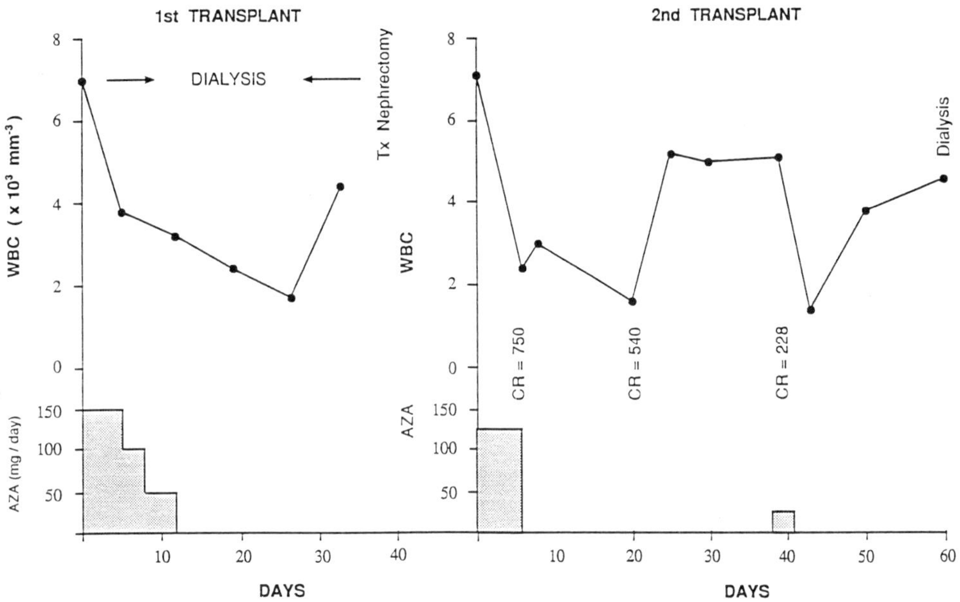

Fig 2. Azathioprine-induced leucopenia in a patient with undetectable TPMT activity after 2 cadaver kidney transplants. A 3rd transplant was successful and the patient is now well, receiving only cyclosporin and prednisolone.

DISCUSSION

Azathioprine (AZA) together with prednisolone, and more recently with cyclosporin as 'triple' therapy, has been used extensively for the prevention of rejection of transplanted organs and for the treatment of other immune-related diseases. This study compared activity of TPMT, an enzyme which may have significance for the therapeutic efficacy of AZA, in erythrocyte lysates of controls, uraemic patients on dialysis, and transplanted patients of whom the majority were receiving an immunosuppressive regime including AZA. Median activity was significantly lower in controls and transplanted patients given only cyclosporin and prednisolone, than in the AZA treated group or the uremic group on dialysis. Enzymic induction is a common phenomenon during drug therapy and would explain the raised TPMT in patients on AZA. The increased activity observed in uremic patients has been noted by others and may be due to a metabolite accumulating in uraemia. The similarity of TPMT activities in cyclosporin treated patients and controls would exclude such an effect for cyclosporin alone.

Lennard et al.(4) have observed severe bone marrow depression in patients taking AZA with very low TPMT activity (less than 15 pmol/h/mgHb) and a consequent excess of cytotoxic 6-thioguanine nucleotide. In our study, only 3 patients demonstrated total intolerance to AZA, 2 of whom also had very low TPMT activity. The intolerance of the third patient, despite high TPMT activity, was attributed to concomitant cotrimoxazole therapy (5). Patients with intermediate activity (15-26 pmol/h/mgHb) tolerated AZA well.

Of 29 cadaver recipients given only AZA plus prednisolone, 5 with reduced allograft function had a significantly higher median activity than 24 with a better clinical outcome. TPMT activity in these 24 patients was also significantly lower than in the general group of AZA-treated patients. It appears that the immunosuppressive efficacy of AZA is inversely related to its catabolism by TPMT, and that self selection of the best recipients for AZA immunosuppression occurred, based on low catabolism of the drug.

It has been shown that the correlation between 6-thioguanine nucleotide levels and cytotoxicity, as indicated by neutropenia, is improved when potentially myelosuppressive influences or concurrent cotrimoxazole and folate were excluded (6). Probably, because of such variable factors, the correlation between TPMT activity and WBC was poor in this group of transplanted patients in our study. This finding confirms that the WBC count alone is not a reliable index of drug efficacy.

We conclude that total intolerance to AZA is rare, and usually appears in patients with very low TPMT activities. Our results also suggest that the wide range of TPMT activity may be an important factor in determining long-term graft survival in AZA treated patients; those with high activity might benefit from doses at the upper limit of that generally recommended.

Acknowledgements: We thank the Special Trustees of Guy's Hospital, the Arthritis and Rheumatism Council, and Fundacao de Amparo a Pesquisa do Estado de Sao Paulo for their support.

REFERENCES

1. J.E. Murray, J.P Merrill, J.H. Harrison, R.E. Wilson and G.J. Dammin. Prolonged survival of human-kidney homografts by immunosuppressive drug therapy. N Engl J Med 268:1315 (1963).
2. D.M. Tidd and A.R.P. Paterson. A biochemical mechanism for the delayed cytotoxic reaction of 6-mercaptopurine. Cancer Res 34: 738 (1974).
3. R.M. Weinshilboum and S.L.Sladek. Mercaptopurine pharmacogenetics: monogenic inheritance of erythrocyte thiopurine methyltransferase activity. Hum Genet 32:651 (1980).
4. L.Lennard, J.A. Van Loon and R.M. Weinshilboum. Pharmacogenetics of acute azathioprine toxicity: relationship to thiopurine methyltransferase genetic polymorphism. Clin Pharmacol Ther 46:149 (1989).
5. R.M. Weinshilboum, F.A. Raymond and P.A. Pazmino. Human erythrocyte thiopurine methyltransferase: radiochemical microassay and biochemical properties. Clin Chim Acta 85:323 (1978).
6. C.A. Rees, L. Lennard, J.S. Lilleyman and J.L. Maddocks. Disturbance of 6-mercaptopurine metabolism by cotrimoxazole in childhood lymphoblastic leukaemia. Cancer Chemother Pharmacol 12:87 (1984).
7. G.L.C. Chan, G.R. Erdmann, S.A. Gruber, A.J. Matas and D.M. Canafax. Azathioprine metabolism: Pharmacokinetics of 6-mercaptopurine, 6-thiouric and 6-thioguanine nucleotides in renal transplant patients. J Clin Pharmacol 30: 358 (1990).

INCREASE IN 2',5'-OLIGOADENYLATE SYNTHETASE CAUSED BY DEOXYCOFORMYCIN IN HAIRY CELL LEUKAEMIA

K. GANESHAGURU, W.C.P. de MEL, G. SISSOLAK, D. CATOVSKY*, C.E. DEARDEN*, A.B. MEHTA, A.V. HOFFBRAND

Department of Haematology, The Royal Free Hospital School of Medicine
Rowland Hill Street, London NW3 2PF

* Academic Department of Haematology
Royal Marsden Hospital, Fulham Road, London SW7

Introduction: Deoxycoformycin (dCF) is a potent inhibitor of the enzyme adenosine deaminase (ADA). Although used initially to treat T-cell leukaemias, dCF has also been found to be effective in the treatment of hairy cell leukaemia (HCL) (Spiers et al, 1984). Inhibition of ribonucleotide reductase by accumulation of dATP is thought to be the dominant mechanism of toxicity of dCF in rapidly proliferating cells as in T-acute lymphoblastic leukaemia (T-ALL)(Mitchell et al, 1983) but how dCF kills slowly proliferating cells such as in HCL and other chronic B or T-cell tumours is unclear. In these diseases no definite correlation has been found between clinical response and the results of in vivo or in vitro studies of some postulated mechanisms of cytotoxicity including depletion of nicotinamide adenine dinucleotide (NAD) or ATP, accumulation of DNA strand breaks or of dATP. However, a correlation of clinical response with decreased S-adenosyl-homocysteine hydrolase (SAHH) levels has been shown (Ganeshaguru et al, 1987, Ho et al, 1988).

Patients with HCL were found to respond to alpha-Interferon (α-IFN) (Quesada et al, 1984). However, the variant form of HCL (Catovsky et al, 1984) does not respond to α-IFN therapy. α-IFN has been shown to induce the enzyme 2',5' oligoadenylate synthetase (2-5OAS) which converts ATP to 2',5' oligoadenylates (2-5A) (Lengyel, 1982). 2-5A activates a latent endonuclease (RNAse) which cleaves RNA. Degradation of mRNA for the cytokines tumour necrosis factor (TNF), interleukin 1 (IL-1) and interleukin 6 (IL-6) following the induction of 2-5OAS mRNA by α-IFN in HCL and B-chronic lymphocytic leukaemia (B-CLL) in vitro has been demonstrated (Heslop et al, 1990). Clinical response in patients with haematological malignancies treated with α-IFN has been related to the degree of induction of 2-5OAS mRNA (de Mel et al, 1990). Since HCL also reponds to therapy with dCF (Spiers et al, 1984) we postulated that induction of 2-5OAS might also be a mechanism by which dCF kills HCL cells. In this study we have assessed 2-5OAS mRNA in peripheral blood mononuclear cells (PBMC) from HCL patients treated with dCF or α-IFN.

Patients and methods: All chemicals used were from BDH and Sigma Chemicals Co. Tissue culture medium was from GIBCO and the isotopes and probe labelling kit from Amersham International. dCF was obtained from the National Cancer Institute, Bethesda, USA and α-IFN from Hoffman La Roche.
Peripheral blood was obtained from 6 patients with typical HCL and 1 patient with a

variant form of HCL all treated with dCF 4mg/m^2 weekly intravenously. Two patients with HCL and treated with α-IFN at 2 MU/m^2 three times weekly were also studied. The initial number of circulating hairy cells in the 9 patients studied varied from < 5% to > 90%. The blood samples in EDTA or preservative free heparin were studied immediately before and 4 and 24h after infusion of dCF or subcutaneous α-IFN. The response to therapy was assessed at 9 weeks from the first injection. Also studied were 15 other patients with circulating tumour cells treated with dCF which included 2 T-CLL, 3 B-CLL, 2 T-ALL, 3 adult T-cell leukaemia lymphoma (ATLL), 2 Sezary, 1 non-Hodgkin's lymphoma (NHL), 1 B-prolymphocytic leukaemia (PLL) and 1 T-PLL.

For patients treated with dCF or α-IFN, complete remission (CR) was defined as: (i) complete disappearance of all clinical evidence of disease, (ii) haemoglobin > 12 g/dl, (iii) platelets > 100 x 10^9/l, (iv) neutrophils > 1.5 x 10^9/l and (v) the absence of leukaemic cells from the peripheral blood and bone marrow. A partial remission (PR) was defined as (i) > 50% reduction of the leukaemic cells in peripheral blood and bone marrow, with Hb, platelets and neutrophils as in CR. Any response less than PR was considered as no response or stable disease (SD).

PBMC were separated from blood samples on a lymphoprep (Nycomed, UK) density gradient. The rapid dot-blot 2-5OAS mRNA assay was similar to that described by Cheley and Anderson (1984). Separated PBMC were washed with PBS by centrifugation and 1 x 10^7 cells lysed in 750 ul 7.6 M guanidine_HCL, homogenised by aspiration through a needle and RNA precipitated with 450 ul 95% ethanol at -20°C overnight. The pelleted RNA was reconstituted in 15% formaldehyde, denatured and serial dilutions applied onto nitrocellulose sheets on a minifold filtration apparatus (Biorad). The sheets were baked, prehybridised and hybridised with a cDNA probe to 2-5OAS labelled with ^{32}P as described by Feinberg and Vogelstein (1983). The probe used was the 1397bp cDNA 6-2, as described by Saunders et al (1985). All blots were also hybridised to the K alpha 1 tubulin probe to verify RNA loading.

TABLE 2-5OAS mRNA LEVELS IN VIVO IN CLINICAL SAMPLES AFTER dCF OR α-IFN TREATMENT AND CLINICAL RESPONSE - EXPRESSED AS A FACTOR OF PRETREATMENT VALUES OF RESPECTIVE PATIENTS.

Patient No.	2-5OAS mRNA (fold increase) week	4h	24h	Clinical Response
dCF treated				
1	1	8	16	CR
	5	2	-	
2	1	8	16	CR
	6	8	8	
3	1	4	-	CR
4	1	-	8	CR
	5	-	4	
5	1	6	6	PR
6	1	8	-	CR
	2	-	4	
7 (variant)	1	1	1	SD
α-IFN treated				
8	1	-	4	PR
9	1	-	1	SD

CR=complete remission, PR=partial remission, SD=stable disease

Results

Six of the 7 HCL patients treated with dCF and studied for 2-5OAS mRNA in PBMC in vivo showed an increase ranging from 2 to 8 times of pre-treatment values at 4 h and 4 to 16 times 24 h post treatment. All 6 responded well to dCF (table 1). The highest increase in 2-5OAS mRNA were in patients with high circulating hairy cells. The HCL variant, however did not demonstrate an increase in 2-5OAS mRNA and did not respond clinically to dCF therapy. Of the two HCL patients treated with α-IFN, one showed increase in 2-5OAS mRNA level corresponding to a clinical response. The other showed no increase in the 2-5OAS mRNA levels and did not respond to therapy (table 1). Of the other 15 patients treated with dCF, none showed an in vivo increase in the 2-5OAS mRNA. However, of these 15 patients, one achieved a CR (T-PLL) and three a PR (1 ATLL, 1 Sezary and 1 T-CLL).

Discussion: The results show that in the B-cell tumours a relation between clinical response to dCF therapy and induction of the enzyme 2-5OAS in vivo. Six patients with HCL demonstrated an in vivo increase of 2-5OAS mRNA in PBMC when treated with dCF and these patients also responded clinically. On the other hand in the patient with HCL variant who did not respond to dCF therapy and and in 5 other patients with B-cell malignancies (3 CLL, 1 PLL, 1NHL) there was no increase of 2-5OAS mRNA. In the two HCL patients treated with α-IFN, one showed an increase in 2-5OAS and corresponding clinical improvement and the other showed no induction and no clinical response.

The present results suggest therefore that one mechanism of anti tumour activity of dCF in vivo may be mediated by the 2-5OAS system in patients with hairy cell leukaemia. The increase in the 2-5OAS mRNA in vivo is not necessarily in the hairy cell themselves however, since of the patients treated with dCF who showed increased mRNA levels, not all had high circulating hairy cells. The mononuclear cell present in some of these patients are largely lymphocytes and monocytes. However, there seem to be a relationship between the proportion of circulating hairy cells and the degree of 2-5OAS induction. Any difference that may exist in the circulating mononuclear cells between those that show 2-5OAS induction and those that do not is yet to be evaluated.

The mechanism by which dCF causes an increase of 2-5OAS mRNA in HCL remains unclear. It is possible that changes in ATP concentrations caused by dCF are responsible for changes in the 2-5OAS mRNA levels. ATP is the substrate for 2-5OAS. Some studies have shown a decrease in ATP with dCF therapy (Siaw et al 1980) and others have shown an increase in rat kidney tissue (Stromski et al 1988). On the other hand, dCF is a known inhibitor of the enzyme S-adenosyl homocysteine hydrolase (SAHH)(Russell te al 1981), thus affecting methylation reactions. Demethylation or non-methylation of DNA is associated with the activation of silent genes (Cedar 1988). It may be that dCF by inhibiting methylation activates the 2-5OAS gene. Our earlier studies have shown no correlation between clinical response and cell ATP levels but correlation with decreased SAHH levels (Ho et al 1988).

Recent results from our laboratory on patients with haematological malignancies undergoing therapy with α-IFN show that clinical response may be related to the degree of induction of 2-5OAS mRNA in vivo (De Mel et al 1990). This rise was clearly demonstrated here in one HCL patient treated with α-IFN who showed a clinical response. The observations reported here support our hypothesis that increase of 2-5OAS enzyme activity may also be a mechanism of action of dCF in HCL. How a rise in 2-5OAS kills these cells is uncertain, but recent studies (Cordingley et al 1988, Bianchi et al 1988, Heslop et al 1990) suggest that 2-5OAS stimulated degradation of the mRNA for tumour necrosis factor and possibly for other autocrine growth factors for hairy cells and other B-cell malignancies may be one mechanism of α-IFN action. Further studies are now needed to determine the exact cell type in which the rise in 2-5OAS in vivo occurs after dCF therapy.

References

Bianchi ACM, Heslop HE, Drexler HG, Cordingley FT, Turner M, De Mel WCP, Hoffbrand AV, Brenner MK (1988) Effects of tumour necrosis factor and alpha interferon on chronic B-cell malignancies. Nouvelle Revue Francaise d'haematologie,30,317-319.

Catovsky D, O'Brien M, Meto JV, Wardle J, Brozovic M (1984) Hairy cell leukaemia (HCL) variant: An intermediate disease between HCL and B prolymphocytic leukaemia. Seminars in Oncology,11,362-369.

Cedar H (1988) DNA methylation and gene activity, Cell,53,3-4.

Cheley S, Anderson RA (1984) A reproducible microanalytical method for the detection of specific RNA sequences by dot blot hybridisation. Analytical Biochem,137,15-19.

Cordingley FT, Bianchi A, Hoffbrand AV, Reittie, Heslop HE, Vyakarnam A, Turner M, Meager A, Brenner MK (1988) Tumour necrosis factor as an autocrine tumour growth factor for chronic B-cell malignancies, Lancet,1,969-971.

De Mel WCP, Hoffbrand AV, Giles FJ, Goldstone AH, Mehta AB, Ganeshaguru K (1990) Alpha interferon therapy for haematological malignancies: correlation between in vivo induction of the 2',5' oligoadenylate system and clinical response. British Journal Haematol.74,452-456.

Feinberg AP, Vogelstein B (1983) A technique for radiolabelling DNA restriction endonuclease fragments to high specific activity. Analytical Biochem.132,6-13.

Ganeshaguru K, Ho AD, Piga A, Catovsky D, Hoffbrand AV (1987) Biochemical mechanisms of deoxycoformycin toxicity in chronic leukaemias. Leukaemia Research,11,941-945.

Heslop HE, Bianchi ACM, Cordingley FT, Turner M, De Mel WCP, Hoffbrand AV Brenner MK (1990) Effects of interferon α on autocrine growth loops in B-lymphoproliferative disorders. J Expt Med,172,1729-1734.

Ho AD, Ganeshaguru K, Knauf WU, Dietz G, Trede I, Hunstein W, Hoffbrand AV (1988) Clinical response to deoxycoformycin in chronic lymphoid neoplasms and biochemical changes in circulating malignant cells in vivo. Blood,72,1884-1890.

Lengyel P (1982) Biochemistry of interferons and their actions. Annual Review of Biochem,51,251-282.

Mitchell BS, Edwards NL, Koller CA (1983) Deoxyribonucleoside triphosphate accumulation by leukaemic cells, Blood,62,419-424.

Quesada JR, Reuben J, Manning JT, Hersh EM, Gutterman JU (1984) Alpha interferon for induction of remission in hairy cell leukaemia. N Engl J Med,310,15-18.

Russel NH, Prentice HG, Lee N, Piga A, Ganeshaguru K, Smyth JF, Hoffbrand AV (1981) Studies in the biochemical sequelae of therapy in Thy-acute lymphoblastic leukaemia with adenosine deaminase inhibitor 2'-deoxycoformycin. Brit J Haematol,49,1-9.

Saunders ME, Gewert DR, Tugwell ME, McMahon M, Williams BRG (1985) Human 2-5A synthetase: characterization of a novel cDNA and corresponding gene structure. EMBO,4,1761-1768.

Siaw MFE, Mitchell BS, Koller CA, Coleman MS, Hutton JJ (1980) ATP depletion as a consequence of adenosine deaminase inhibition in man. PNAS,USA,77,6157-6161.

Spiers ASD, Parekh SJ, Bishop MB (1984) Hairy cell leukaemia: induction of complete remission with pentostatin (2'-deoxycoformycin).J Clin Oncol,2,1336-1342.

Stromski ME, van Waarde A, Avison JM, Thulin G, Gaudio KM, Kashgarian M, Shulman RG, Siegel NJ (1988) Metabolic and functional consequences of inhibiting adenosine deaminase during renal ischaemia in rats. J Clin Invest,82,1694-1699.

A DUAL ANTI-TUMOR EFFECT OF A COMBINATION OF INTERFERON-α AND 5-FLUROURACIL OR 2-CHLORODEOXYADENOSINE ON NATURAL KILLER (NK) CELL MEDIATED CYTOTOXICITY

Zvi Reiter, Osman N. Ozes, Sue Tomson, Lawrence M. Blatt* and Milton W. Taylor

Institute for Molecular and Cellular Biology, Department of Biology, Indiana University, Bloomington, IN 47405, and *Amgen Co., Thousand Oaks, CA 91320, USA.

INTRODUCTION

Interferon (IFN)-α is now widely used in the treatment of a number of specific neoplasms, such as hairy cell leukemia and Kaposi's sarcoma (1,2). However the therapeutic effects of IFNs are still rather limited and the success in the treatment of other cancers has not been great. One possible approach to improving the efficacy of IFN treatment is to combine it with the use of chemotherapeutic agents, such as purine and pyrimidine analogs. Preliminary data indicates that a combination of 5-flurouracil (5-FU) and IFN-α is clinically relevant (3-9), and this combination has been used in the treatment of colon cancer and renal carcinoma with some success (10-12). 2-chlorodeoxyadenosine (2-CdA) was reported recently to be an effective agent in the treatment of hairy cell leukemia, non-Hodgkin lymphoma and chronic lymphocytic leukemia patients (13,14). However, there have been no reports of combining this drug with IFN.

IFN has been shown to stimulate the immune response, including the activation of natural killer (NK) cells against tumor target cells. It has been postulated that NK cells are the first line of defense against tumor cells and their metastasis. IFN has been shown to increase the lytic effect of NK cells against sensitive target cells and it induces cytotoxic activity against normally resistant target cells. Pre-treatment of target cells, on the other hand, with IFN, reduces their sensitivity to NK-CMC (15,16).

In this study, we examined the effect of pre-treating both NK cells and target cells with the pyrimidine analog, 5-FU, and the purine analog, 2-CdA, in combination with IFN-α. We report that this combination has additive antiproliferative effect on tumor cells, sensitized the target cells to NK cytotoxic effects and abolished the protection of target cells by IFN.

RESULTS

The effect of IFN and 5-FU or 2-CdA on cell growth. When a series of tumor cells, including a hairy cell leukemia-like cell line (Eskol, 17) (Fig. 1), ME-180 (a cervical carcinoma cell line), K-562 (a chronic myelogenous leukemia cell line) and AIDS-related Kaposi's sarcoma (18) (not shown) were treated in combination with IFN-α (recombinant consensus IFN-α, specific activity $2x10^9$ units/mg; Amgen, Inc., Thousand Oaks, CA) and 5-FU (Sigma) (Fig. 1a) or IFN-α and 2-CdA (2-CdA was provided by Dr. D.A. Carson, Scripps Clinic, La Jolla, CA) (Fig. 1b), a significant ($P<0.05$) additive effect on cell growth was found. This effect was both time- (Fig. 1) and dose-dependent (not shown). No synergy was noted.

Measurement of cell proliferation was as described (19).

5-FU and 2-CdA do not affect NK cell activity. Purification of NK cells, ^{51}Cr-release cytotoxicity assay for NK activity, conjugate formation assay, cold target competition assay, natural killer cytotoxic factor (NKCF) preparation and assay were performed as described (16). Neither 5-FU nor 2-CdA had any effect on the induction of NK cell mediated cytotoxicity in the presence or absence of IFN-α. Various doses of 5-FU and 2-CdA (0.01 ng-10 μg) were used in these experiments (not shown).. All the results are expressed as an arithmetic mean ± a standard error of the mean (SEM). Comparisons were carried out by a *student t-test.*

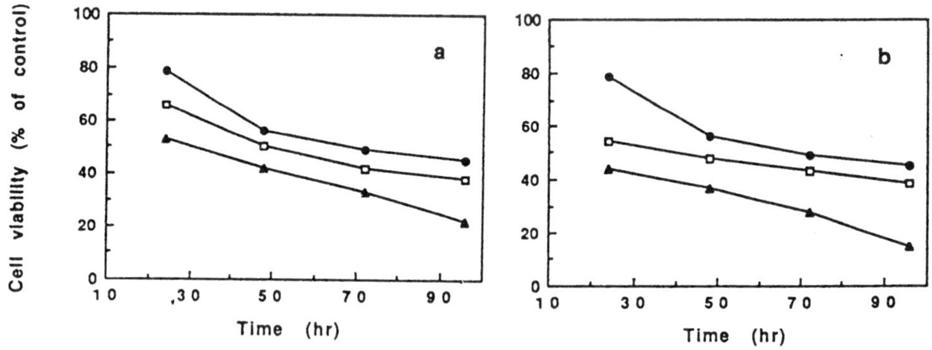

Fig. 1. **Growth inhibition with combined treatment of 5-FU + IFN and 2-CdA + IFN.** Eskol cells at concentration of $1x10^5$ were cultured for varying periods of time in the presence of IFN (300 units/mL, ●); 5-FU (a; 1 μg/mL, □); 2-CdA (b; 100 ng/mL, □) or combination of IFN and either 5-FU or 2-CdA (▲) . The growth inhibition assays expressed as the percentage increase in the number of cells present in the treated cultures relative to the increase in the number of cells in un-treated control cultures. The curves represent the data of one of 2 or 3 experiments.

5-FU and 2-CdA increased the sensitivity of Eskol cells to NK activity. Eskol cells were found to be almost completely resistant to NK-cell mediated cytotoxicity. The susceptibility of these cells was increased significantly ($P<0.01$-$P<0.05$) following pre-incubation with either 5-FU (Fig. 2a) or 2-CdA (Fig. 2b). The effect

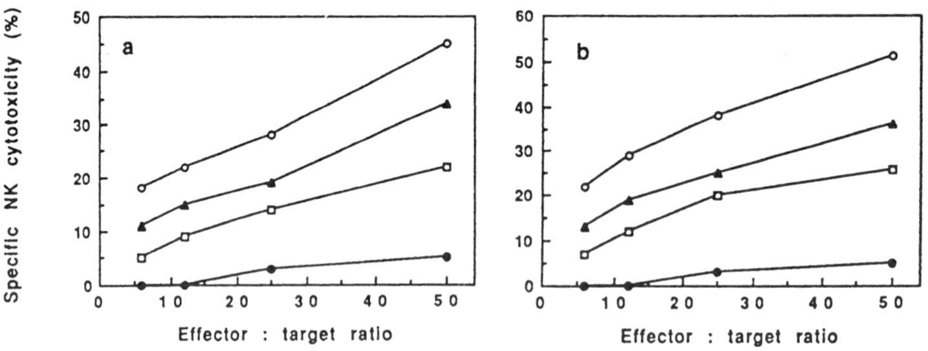

Fig. 2. **5-FU and 2-CdA increase NK-susceptibility.** Eskol cells were cultured for 18 hr in absence (●) or presence of 5-FU (a) or 2-CdA (b) (1 ng/mL, □; 100 ng/mL, ▲; 1 μg/mL, o). Various amounts of NK effector cells were added for 4 hr of incubation. The curves represent the % specific ^{51}Cr release from target cells. Spontaneous ^{51}Cr release was 5-17% and SEM was 1.3-4.8%. Fig. 2 represents the data of one of three experiments using effector cells from three different donors.

was found to be dose-dependent. Similarly, the sensitivity of these tumor cells to either IFN-primed NK activity (not shown) was increased when the target cells were pre-treated with the analogs. A similar effect was found when the other tumor cell lines were pretreated with the purine or pyrimidine analogs (not shown).

5-FU and 2-CdA abolished the protective effect against NK-CMC induced by IFN.

Treatment of target cells with IFN reduces their susceptibility to NK-cell mediated cytotoxicity. Combination treatment of IFN-α and 5-FU (Fig. 3a) or IFN-α and 2-CdA (Fig. 3b) abolished this protective effect. A dose dependent relationships was found, low doses of the analogs increased the sensitivity of Eskol cells to NK activity, while high dose of each analog resulted in a significant ($P<0.05$ and $P<0.01$) loss of the protective effect. Similar results were found with the other target cells (not shown).

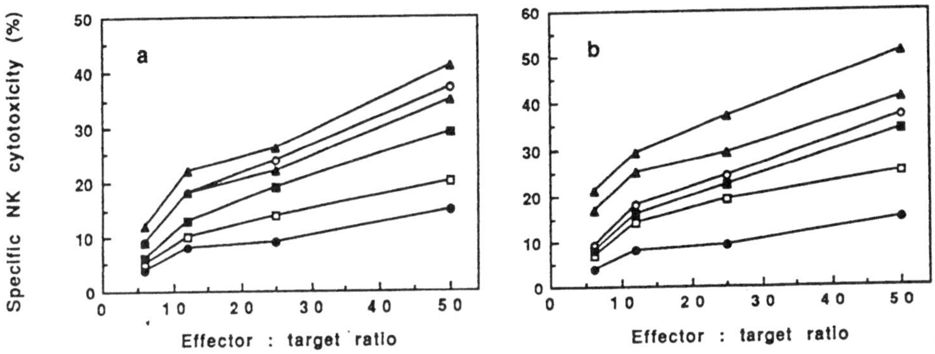

Fig. 3. **5-FU and 2-CdA abolished the IFN-induced protection against NK activity.** Eskol cells were cultured in the presence of medium alone (o), IFN (300 units/mL, ●) or with combination of IFN and either 5-FU (a) or 2-CdA (β) (1 ng/mL, □; 10 ng/mL, ■; 100 ng/mL, Δ; 1 μg/mL, ▲) for 18 hr. NK effector cells were added for 4 hr of incubation. The curves represent the % specific ^{51}Cr release from Eskol cells. Spontaneous ^{51}Cr release was 7-16% and SEM was 1.8-5.4%. Fig. 3 represents the data of one of three experiments using effector cells from three different donors.

5-FU and 2-CdA increased the sensitivity of tumor cells to NK cytotoxic factor (NKCF). We investigated the mechanism of action of 5-FU and 2-CdA at the cellular level by measuring the capacity of treated tumor cells to form conjugates with NK cells. None of the treatments changed the levels of conjugate formation (Table 1). However, when conjugated dead cells were counted there was an increase in percentage of dead cells following treatment with either base analog. This was particularly pronounced following incubation of the NK cells with IFN-α (Table 1). The cold target competition assay supported this finding. Neither IFN nor 5-FU or 2-CdA pre-treatment of competitor cells (Eskol or K-562) changed their ability to compete for ^{51}Cr-labeled K-562 cells (not shown). NKCF activity can be separated from the effector cells following incubation of NK cells with their targets. This factor when added to target cells resulted in lysis. Pretreatment of tumor cells with the analogs sensitized the cell to this cytotoxic activity. IFN reduces the ability of tumor cells to induce activation of effector cells and NKCF release following effector-target cell conjugate formation. 5-FU and 2-CdA did not show a significant effect in this step of the killing process (not shown).

DISCUSSION

Clinical trials using a combinations of IFN-α and 5-FU in treating colon cancer and renal carcinoma have been reported (9-12). The results are of clinical significance. We have examined whether this combined treatment has

Table 1. **Conjugate formation between Eskol and effector cells.**

Effector cells	Target cells Treatment	Conjugate formation (%)	Conjugated dead cells (%)
NK cells	-	4.5	2.4
NK cells	IFN-α(1)	3.9	1.6
NK cells	5-FU(2)	4.7	6.3
NK cells	5-FU+IFN	5.0	4.3
NK cells	2-CdA(3)	4.6	5.8
NK cells	2-CdA+IFN	4.9	4.6
IFN-α(1)-primed NK cells	-	5.1	22.3
IFN-α-primed NK cells	IFN-α	4.9	11.5
IFN-α-primed NK cells	5-FU	5.6	34.5
IFN-α-primed NK cells	5-FU+IFN	4.7	24.0
IFN-α-primed NK cells	2-CdA	5.8	38.0
IFN-α-primed NK cells	2-CdA+IFN	4.9	27.5

(1) Target cells or effector were treated with IFN-α (300 units/mL) for 18 hr. (2) 1 μg/mL. (3) 100 ng/mL. The conjugate formation assay was done in duplicate using effector cells from three different donors. SEM was ± 0.6-3.4%.

any effect on one arm of the immune system, namely NK cell activity. We have shown that neither analog affects the behavior of NK cells directly. However, pretreatment for a short period of time (18 hours) with either 5-FU or 2-CdA results in sensitization of the target cell to NK cell activity and the loss of IFN protection. Whether this sensitization is the result of a perturbation of cellular metabolism in general, or an effect on specifically RNA or protein metabolism awaits further analysis. The fact that we find this phenomenon with both NK resistant-cells and NK-sensitive cells, and with both analogs would suggest that this is a very general mechanism in which a population or subpopulation of tumor cells is made more sensitive to the cytotoxic effects of factors (NKCF) released by NK cells. Similar finding have been reported by Ujie in the treatment of murine EL-4 cells with 5-FU (20). On the other hand, we have found that both 5-FU and 2-CdA neither affected the activity of NK effector cells nor was involved in the priming of NK cells by IFN. Charamella et al (21) have reported similar results at a 5-FU concentrations used in clinical treatments.

Despite the lack of knowledge of the mechanism involved, this work indicates the clinical importance of combination therapy between lymphokine and a chemotherapeutic agent. A major problem with both immunotherapy and chemotherapy is high toxicity. By using a combination of IFN and 5-FU or 2-CdA, the total dose of both agents can be lowered, without losing efficacy.

REFERENCES

1. Quesada J.R., E.M. Hersh, J. Manning, J. Reuben, M. Keating, E. Schnipper, L. Itri and J.U. Gatterman. 1986. *Blood* 68:493. **2**. Volberding P.A., R.T. Mitsuyasu, J.P. Golando and R.J. Spiegel. 1987. *Cancer* 59:620. **3**. Namba, M., T. Miyoshi, T. Kanamori, M. Nobuhara, T. Kimoto and S. Ogawa. 1982. *Gann*. 73:819. **4**. Miyoshi, T., S. Ogawa, T. Kanamori, M. Nobuhara and M. Namba. 1983. *Cancer Lett.* 17:239. **5**. Yamamoto, S., H. Tanaka, T. Kanamori, M. Nobuhara and M. Namba. 1983. *Cancer Lett.* 20:131. **6**. Stolfi, R.L. and D.S. Martin. 1985. *J. Biol. Response Modif.* 4:634. **7**. Kimoto, Y. 1986. *Gan. to Kagaku. Ryoho.* 13:293. **8**. Elias, L. and H.A. Crissman. 1988. *Cancer Res.* 48:4868. **9**.Walder, S., E.L. Schwartz and M. Goldman. 1989. *J. Clin. Oncol.* 7:1769. **10**. Pazdur, R., J. Abbruzzese and J. Faintuch. 1990. *Proc. ASCO* 9:455. **11**. Elias, L. and J.M. Sandoval. 1989. *Biochem. Biophys. Res. Commun.* 163:867. **12**. Logothetis, C.J., E. Hossan, A. Sella, F.H. Dexeus and R.J. Amato 1991. *J. Nat. Cancer Inst.* 83:285. **13**. Piro, L.D., C.J. Carrera, E. Beutler and D.A. Carson. 1988. *Blood* 72:1069. **14**. Piro, L.D., C.J. Carrera, D.A. Carson and E. Beutler.

1990. *N. Engl. J. Med.* 322:1117. **15**. Trinchieri, G. and D. Santoli. 1978. *J. Exp. Med.* 147:1314. **16**. Reiter, Z., D.G. Fischer and M. Rubinstein. 1988. *Immunol. Lett.* 17:232. **17**. Harvey, W., E.F. Srour, R. Turner, R. Carey, R. Maze, B. Starrett, R. Kanagala, D. Pereira, P. Merchant, M. Taylor and J. Jansen. 1991. *Leukemia Res.* In press. **18**. Roth, W.K., S. Werner, W. Risau, K. Remberger and P.H. Hofschneider. 1988. *Int. J. Cancer* 42:767. **19**. Ozes, O.N., Z. Reiter, M.L. Blatt and M.W. Taylor. Submitted. **20**. Ujiie, T. 1989. *Japan. J. Exp. Med.* 59:17. **21**. Charamella, L.J., C. Meyer, G.E. Thompson and N.V. Dimitrov. 1985. *J. Immunopharm.* 7:53.

THE EFFECT OF METHOTREXATE ON THE FOLATE COENZYME POOLS IN HUMAN HEPATOMA CELLS IN CULTURE

John Galivan[1], Myung Rhee[1], David G. Priest[2], Marlene Bunni[2], James H. Freisheim[3], and John M. Whiteley[4]

[1]Wadsworth Center for Laboratories and Research
New York State Department of Health
Albany, NY 12201

[2]Department of Biochemistry and Molecular Biology
Medical University of South Carolina
Charleston, South Carolina 29405

[3]Department of Biochemistry and Molecular Biology
Medical College of Ohio
Toledo, Ohio 43699

[4]Department of Molecular and Experimental Medicine
Research Institute of Scripps Clinic
La Jolla, California 92037

INTRODUCTION

Recent studies on the activity of antifolates have been directed at the perturbations in folate pools caused by inhibitors of dihydrofolate reductase. For the most part these studies have been restricted to the effects of methotrexate, but the effects of trimetrexate have also been examined. A constant feature of these results is the elevation of dihydrofolate which accompanies inhibition by both these agents (1-7). A variable result is the extent to which the reduced folate coenzymes are depleted following exposure to the antifolates. In several cell lines the reduction in tetrahydrofolate is quite modest (less than 40%) with the 10-formyltetrahydrofolate being maintained (1-5). In other cases such as L1210 cells *in vivo* there is a marked depletion of the reduced folate coenzyme pools (6). Preliminary studies with a rodent hepatoma cell line indicated that the reduced folate coenzymes are depleted by 80% or greater and this is compensated for by an increase in H_2PteGlu$_n$ and 10-HCO-H_2PteGlu$_n$ (7). The current study extends these investigations to the effect of methotrexate on the folate coenzymes in human HEPG2 hepatoma cells *in vitro*.

METHODS

Cell culture. Human HepG2 cells were grown in monolayer culture in Dulbecco's MEM containing 10% FBS for 72 hr for all experiments. The cells were pulsed with MTX for 2 hr, washed twice and placed in fresh medium for 2 hr or 16 hr prior to analysis.

Thymidylate and purine biosynthesis. [6-^{3}H]Deoxyuridine incorporation into DNA was used as a measure of de novo thymidylate biosynthesis (8). De novo purine biosynthesis was measured by the incorporation of [^{14}C] glycine into DNA according to the method of Cadman et al. (9). The results were expressed as dpm/mg cell protein.

Cellular reduced folates. 5,10-CH_2H_4PteGlu$_n$, H_4PteGlu$_n$, and H_2PteGlu$_n$ in the cell extracts were measured by the ternary complex formation of [^{3}H]FdUMP-TS-5,10-CH_2H_4PteGlu$_n$ as previously described (10, 11). 10-HCO-

$H_4PteGlu_n$ and 10-$HCOH_2PteGlu_n$ were measured following conversion to $H_4PteGlu_n$ in the presence of glycineamideribonucleotide (GAR) and GAR transformylase (GARTF), or in the presence of DHFR and GARTF, respectively (12). 5-$CH_3H_4PteGlu_n$ was reacted with methylenetetrahydrofolate reductase to form 5,10-$CH_2H_4PteGlu_n$ (13). 5-$HCOH_4PteGlu_n$ was reacted with 5,10-methenyltetrahydrofolate synthetase in the presence of ATP (0.5 mM) (14). The product, 10-$HCOH_4PteGlu$ (formed at neutral pH from 5,10-CH=$H_4PteGlu_n$) is converted to $H_4PteGlu_n$ with GAR and GARTF as above. Individual reduced folate species were calculated by the appropriate subtractions based on each enzymatic conversion.

RESULTS AND DISCUSSION

The effects of methotrexate on human HEPG2 cells were examined over a 72 hr culture period. Cultures exposed to methotrexate continuously over this period result in 50% growth inhibition at a concentration of 8.1 nM. In order to determine the effects of methotrexate on the cellular folate coenzymes, shorter incubation periods are necessary. The concentration chosen for these studies is based upon the fact that 25 μM methotrexate causes not less than 90% inhibition of glycine and deoxyuridine incorporation after a 4 and 18 hr incubation (Table 1). In these studies methotrexate is present for only the first two hours of the incubation. It is then removed and the incubation continued for 2 or 16 hr. The intracellular concentration of total methotrexate is 9.4 and 7.0 μM respectively at the 4 and 18 hr time points. The composition of methotrexate is greater than 80% polyglutamates and is predominately tri- and tetraglutamates.

The folate pools in control cultures and those exposed to methotrexate as described in Table 1 were evaluated. Control cultures contained primarily 10-$HCOH_4PteGlu_n$ with lesser amounts of 5-CH_3-,5,10-CH_2-, and $H_4PteGlu_n$ (Table 2). Following methotrexate exposure for 4 hr there was a near complete depletion of all reduced folates with an increase in $H_2PteGlu_n$ and 10-$HCOH_2PteGlu_n$ to approximately 20 pmol/mg. The folate profile after a 16 hr exposure to methotrexate is essentially the same as that at the 4 hr point. Complete recovery of the folate pool was not observed in these experiments. However, the amount of folate recovered after methotrexate exposure has to be corrected upward by 25% to account for the loss of total cellular folate caused by methotrexate. This results in the recovery of total folates of approximately 60% the methotrexate treated cells are compared with untreated cultures.

The results of these studies demonstrate that treatment of HEPG2 cells with methotrexate causes an extensive and nearly complete loss of reduced folate coenzymes which is accompanied by an increase in the dihydro- species. The latter does not completely account for the loss in the tetrahydro-derivatives, and the reason for this is under investigation. Analysis of similar studies with other cell lines reveals

Table 1. Effect of methotrexate on glycine and deoxyuridine incorporation by HepG2 cells.[a]

MTX (μM)	Time of exposure (hr)	[6-^{3}H] UdR x10^3 dpm/mg	(%)	[^{14}C] glycine x10^3 dpm/mg	(%)
-[b]	-	121.6 ± 33.5	(100)	23.6 ± 2.8	(100)
25[c]	4	1.15	(0.95)	2.3	(9.9)
25[b]	18	0.83 ± 0.19	(0.71)	0.89 ± 0.40	(4.76)

a. HepG2 cells are cultured in Dulbecco's MEM containing 10% FBS for 72 hr. The cells were exposed to MTX for 2 hr and removed for 2 (4 hr exposure) or 16 hr (18 hr exposure), respectively. Glycine and deoxyuridine incorporation was measured as described in "Methods".
b. n = 3 ± SD.
c. average of 2 experiments.

Table 2. Effect of methotrexate on folate coenzyme pools in HepG2 cells.[a]

MTX μM	$5,10\text{-}CH_2FH_4$	FH_4	$10fFH_4$	$5mFH_4$	FH_2	$10fFH_2$
	pmol / mg cell protein					
–[b]	4.9±0.26	4.4±1.0	35.2±8.0	9.9	2.4±0.65	n.d.
25[c]	0.1	0.1	0.2	n.d.	16.2	6.5
25[d]	0.07±0.01	0.04±0.02	0.11±0.03	n.d.	18.4±6.4	2.3±1.3

a. The cultures are treated as in Table 1 and then extracted with 50 mM tris/1 mM EDTA/50 mM ascorbate buffer (pH 7.4) and folate coenzyme pools were assayed as described in Methods.
b. N = 3 ± SD
c. 4 hr exposure, average of 2 experiments
d. 18 hr exposure, N = 3 ± SD
n.d. not detected

a wide variation in reduced folate coenzyme depletion. Some cell lines such as MCF-7 cells (1), and L1210 cells in vitro (2-4) show a limited depletion of $10\text{-}HCOH_4PteGlu_n$ and $5,10\text{-}CH_2H_4PteGlu_n$ whereas other cell lines such as L1210 in vivo (6) and H35 hepatoma cell lines in vitro (11) show a marked reduction in both of these coenzymes.

Further evaluation is needed to determine the reasons for the differences in the extent of reduced folate coenzyme depletion in various cell lines. The HepG2 line appears to be particularly interesting in that it exhibits the most extensive loss of reduced folate coenzymes following inhibition of dihydrofolate reductase by methotrexate. In light of the varied results with different cell lines it is not yet clear whether methotrexate inhibits all cell types by a similar mechanism or if different mechanisms operate in different cell systems. The possibility remains that the apparent perturbation in folate pools in itself may not be adequate to infer the mechanism of inhibition of purine and thymidylate synthesis by dihydrofolate reductase inhibitors.

ACKNOWLEDGEMENTS

This work was generously supported by NIH grants CA11778, CA22754, CA25933, CA41461.

REFERENCES

1. C. J., Allegra, K., Huang, G. C., Yeh, J. C. Drake, and J. Baram, Evidence for direct inhibition of de novo purine synthesis in human MCF-7 breast cells as a principal mode of metabolic inhibition by methotrexate. J.Biol.Chem. 262:13520-13526 (1987).
2. L. H. Matherly, C. K. Barlowe, V. M. Phillips, and I. D. Goldman, The effects on 4-aminoantifolates on 5-formyltetrahydrofolate metabolism in L1210 cells. J. Biol. Chem. 262:710-717 (1987).
3. M. Bunni, M. T. Doig, H. V. Donato, V. Kesevan, and D. G. Priest, Role of methylene tetrahydrofolate depletion in methotrexate mediated intracellular thymidylate synthesis inhibition in cultured L1210 cells. Cancer Research 48:3398-3404 (1988).
4. R. L. Seither, D. T. Trent, D. C. Mikalecky, T. J. Rape, I. D. Goldman, Folate-pool interconversion and inhibition of biosynthetic processes after exposure of L1210 leukemia cells to antifolates. J. Biol. Chem. 264:17016-17023 (1989).
5. P. R. Morrison, and C. J. Allegra, Folate cycle kinetics in human breast cancer cells. J. Biol. Chem. 264:10552-10566 (1989).
6. D. G. Priest, M. Bunni, and F. M. Sirotnak, Relationship of reduced folate changes to inhibition of DNA synthesis by methotrexate in L1210 cells in vivo, Cancer Research 49:4204-4209 (1989).

7. M. S. Rhee, J. K. Coward, and J. Galivan, Depletion of reduced folate coenzymes by methotrexate in cultured hepatoma cells. Proceed. Amer. Assoc. Cancer Res. 32:1909 (abs) (1991).
8. D. S. Duch, M. P. Edelstein, S. S. W. Bowers, and C. A. Nichol, Biochemical and chemotherapeutic studies on 2,4-diamino-6-(2,5-dimethoxybenzyl)-5-methylpyrido-[2,3-d] pyrimidine (BW301U), a novel lipid-soluble inhibitor of dihydrofolate reductase. Cancer Research 42:39867-3994 (1982).
9. E. Cadman, C. Benz, R. Heimer and J. O'Shaughnessy, Effect of de novo purine synthesis inhibitors on 5-fluorouracil methabolism and toxicity. Biochem. Pharmacol. 30:2469-2472 (1981).
10. D. G. Priest, and M. T. Doig, Tissue folylpolyglutamate chain length determination by electrophoresis or thymidylate synthase-fluorodeoxyuridylate ternary complexes. Methods Enzymol. 122:313 (1986).
11. M. S. Rhee, M. Balinska, M. Bunni, D. G. Priest, G. F. Maley, F. Maley, and J. Galivan, Role of substrate depletion in the inhibition of thymidylate biosynthesis by the dihydrofolate reductase inhibitor trimetrexate in cultured hepatoma cells. Cancer Research 50:3979-3984 (1990).
12. M. Balinska, M. Rhee, J. Whiteley, D. G. Priest, and J. Galivan, Inhibition of mammalian thymidylate synthase by 10-formyltetrahydropteroylpolyglutamate. Arch. Biochem. Biophys., 284:219-222 (1991).
13. M. T. Doig, J. R. Peters, P. Sur, M. Dang, and D. G. Priest, Determination of mouse liver 5-methyltetrahydrofolate concentration and polyglutamate forms. J. Biochem. Biophys. Methods 10:287-294 (1985).
14. D. G. Priest, M. A. Bunni, R. Romero-Fedes, J. C. Schmitz and J. M. Whiteley, A sensitive radioenzymatic assay for [S]5-formyltetrahydrofolate. Anal. Biochem., in press.

PURINE NUCLEOTIDE BIOSYNTHESIS IN LEUKEMIC PROMYELOCYTES TREATED WITH RETINOIDS

Hiroshi Tsutani*, Daizaburo Kagawa, Michihiko Uchida, Teruo Yoshimura, Tsuneo Tanaka, Takanori Ueda and Toru Nakamura

*First Department of Internal Medicine, Fukui Medical School
23-3 Shimoaizuki, Matsuoka, Fukui 910-11, Japan

INTRODUCTION

Recent progress in chemotherapy for acute myeloid leukemia (AML) has achieved a favorable outcome; however, treatment for patients who do not respond to conventional chemotherapy and who develop multiple drug resistance has yet to be established [1]. Differentiation induction therapy with retinoid, a family of molecules comprising both natural and synthetic analogues of retinol, seems to open the way for an alternative approach to the treatment of AML, especially acute promyelocytic leukemia (APL)[2,3]. Because karyotypic hallmark of APL, a reciprocal translocation between the long arms of chromosomes 15 and 17, includes the breakpoint region of the retinoic acid receptor α gene [4]. In order to clarify the appropriate combination therapy between conventional anti-metabolites and cell-differentiation agents, we describe here the dynamics of the activation pathway of 6-mercaptopurine (6-MP) in fresh leukemic promyelocytes and promyelocytic cell line HL-60 cells during retinoid-induced cell differentiation *in vivo* and *in vitro*.

PATIENT AND METHODS

Case report

A 67-year-old housewife was admitted to our hospital because of pancytopenia. Laboratory examination revealed a leukocyte count of 3.5 x 10^9/l with 92% leukemic cells. The marrow nucleated cell count was 190 x 10^9/l and included 79.6% leukemic cells which were rich in azure granules and stained positive for peroxidase. A marrow cytogenetic study revealed 46, XX, t(15;17). The patient was diagnosed as having APL. The peripheral leukocyte count increased to 19.5 x 10^9/l on day three of the patient's hospital stay. Following initiation of treatment with retinol palmitate at a dosage of 210 x 10^6 IU daily, a marked decrease in the number of leukemic promyelocytes was observed with a concomitant initial transient neutrophilia. After eighth days of treatment, the leukemic cell count was 120 x 10^6/l in blood and 2.6 x 10^9/l in bone marrow [2].

Cells

Mononuclear cells were separated from the patient's bone marrow fluid by Ficoll-Hypaque density gradient centrifugation. Cultures of HL-60 cells

Purine and Pyrimidine Metabolism in Man VII, Part A
Edited by R.A. Harkness *et al.*, Plenum Press, New York, 1991

were kept under incubation at 37℃ in RPMI 1640 medium containing 20% heat-inactivated fetal calf serum (FCS). These cells suspended in the above RPMI-1640 medium supplemented with FCS, were plated at 5 x 10^5 cells/ml and incubated at 37℃ in a humid atmosphere of 5% CO_2 in air for up to 96 hours (h) with or without all-*trans* retinol.

Morphological and functional studies

Cultured cells were prepared by cytospin for staining with May-Gruenwald-Giemsa and alkaline phosphatase anti-alkaline phosphatase (APAAP) to obtain differential count and immunophenotypic characteristics. Monoclonal antibodies of Ki-67 and C3bi receptor (CD11b) were used to detect proliferating cell fraction and differentiating cell fraction, respectively. To determine the individual cells' ability to generate superoxide after activation by phorbol myristate acetate, a slide test using nitro blue tetrazolium (NBT) dye was performed.

Incorporation of labeled precursors of purine nucleotides

The cell suspension (5 x 10^6cells/ml) was incubated at 37℃ for 20 min in the presence of 0.2 mM [^{14}C]-hypoxanthine (Hx) or [^{14}C]-glycine (Gly) in serum-free HB101 medium. The cell suspension was put onto a silicone oil layer (specific gravity: 1.034) and centrifuged at 12,000 *g* for 10 sec to stop the reaction. Cold 0.5 M perchloric acid was added to the cell pellets. Following a 20-sec vortex, keeping at 4℃ for 1 h and centrifugation at 12,000 *g* for 10 sec to collect the acid soluble fraction, the supernate was neutralized with freon solution as described previously [5].

Assay of enzyme activity

Activities of hypoxanthine-guanine phosphoribosyl transferase (HPRT) and phosphoribosyl pyrophosphate (PRPP) synthetase were determined in homogenate treated with activated charcoal as described previously [5]. The homogenate was centrifuged at 1.0 x 10^5*g* for 30 min, and was used as the source of crude enzyme. Each reaction was initiated by adding an enzyme solution and stopped by adding 40 μ mol of EDTA after incubation at 37℃ for 60 min. HPRT activities were measured based on the determination of the conversion of [^{14}C]-labeled Hx to inosine monophosphate (IMP). PRPP synthetase activities were measured by a single step procedure. PRPP, which was formed by the interaction of ribose-5-phosphate, ATP, and enzyme solution, was measured by the HPRT assay system.

Chromatography

Acid soluble fractions were subjected to DEAE cellulose thin layer chromatograpy to separate nucleotides from bases and nucleosides, whereas the enzyme reaction mixture was applied to Toyo No.51B chromatography paper. A 50-100 μ l aliquot of each solution was spotted and eluded with pure water or buthanol/acetic acid/water (4:1:1.7). The radioactivities of spots placed into 10 ml of toluene-based scintillation fluid were determined by counting in an Aloka liquid scintillation counter.

Table 1. Functional, Morphological and Phenotypic Changes

Cells	Treatment	Percent of Positive Cells: Control	Retinol
APL Cells	Mature Cells	4.7 ± 2.8	65.4 ± 4.9*
	NBT dye test	8.2 ± 4.3⁺	58.2 ± 10.3*
HL-60 Cells	Mature Cells	6.2 ± 3.8	58.2 ± 14.7*
	NBT dye test	8.3 ± 1.7⁺	44.0 ± 6.1*
	C3bi-R	11.6 ± 5.5	36.4 ± 10.9*
	Ki-67	70.7 ± 12.2	30.5 ± 9.6*

Control, retinol not added to the culture. ⁺mean ± S.E.(n=6); *p<0.01.

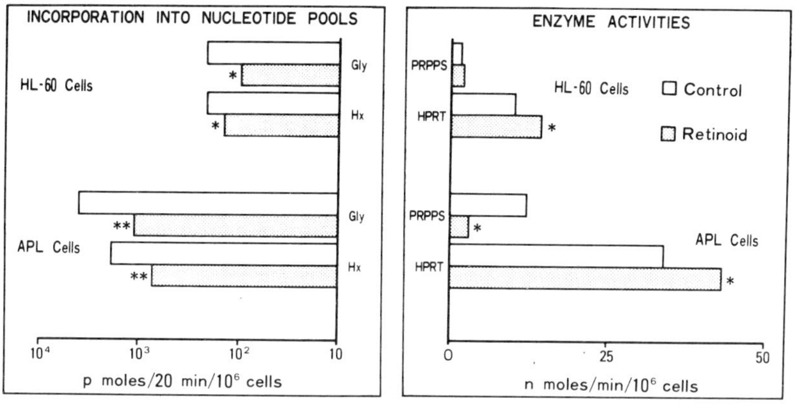

Fig. 1. PURINE NUCLEOTIDE BIOSYNTHESIS UNDER TREATMENT WITH RETINOIDS

Control, before therapy in the patient and the culture of HL-60 cells without retinol; , $^{**}p<0.01$, $^{*}p<0.05$.

RESULTS

In the patient with APL, the marrow aspirate picture showed a decrease of promyelocytes along with an increase in the number of mature myeloid cells during the treatment with retinol palmitate. The appearance of leukemic cells on the eighth day of treatment was bizarre since the lobulated nucleus contained condensed chromatin in spite of the purplish color in the cytoplasm. Fresh leukemic promyelocytes and HL-60 cells incubated for 96 h in suspension culture with or without retinol, showed little change in cell density and viability. Both types of cells showed morphological and functional maturation when cultured with 10 μM retinol. Immunohistochemically, the expression of Ki-67 decreased in retinol-treated HL-60 cells (Table 1).

Fresh leukemic promyelocytes taken on the eighth day of treatment with retinol palmitate and HL-60 cells treated with retinol for 48 h revealed significant decreases in the incorporation of Hx and Gly into the intracellular nucleotide pools compared to those leukemic promyelocytes taken before the therapy or HL-60 cells cultured without retinol. Both types of cells treated with retinol palmitate or retinol revealed a significant increase in HPRT activity and no significant change in PRPP synthetase activity (Fig. 1).

DISCUSSION

In the present study, an increase in mature myeloid cells in the bone marrow during retinol palmitate therapy and increases in morphologically mature cells and NBT-positive cells in the culture study with retinol, which is a main metabolite of retinol palmitate detected in human plasma [3], suggests that leukemic promyelocytes under treatment with retinol palmitate are induced to differentiate. Also in HL-60 cells, like the APL cases, morphological and functional maturation was presumed by increases in NBT and C3bi-R positive cells. In addition, a marked reduction in leukemic promyelocytes of APL patients during the treatment with retinol palmitate and a reduction in HL-60 cells stained with Ki-67 monoclonal antibody recognizing growth phase cells suggest that DNA synthesis might be reduced with a concomitant decrease in S phase cells as a result of induction of differentiation.

Previous reports as respect to synthesis of purine nucleotide [5-7] describes that differentiated leukemic or cancer cells show a reduction in nucleotide synthesis, whereas proliferating cells show promotion of IMP synthesis by acceleration of PRPP production or *de novo* synthesis. The present study also revealed a reduction of nucleotide synthesis as proven by a reduced incorporation of Hx and Gly into nucleotide pools during induction of differentiation. On the other hand, regarding changes in enzyme activity of fresh leukemic promyelocytes and HL-60 cells, PRPP synthetase activity was unchanged, whereas HPRT activity was increased. Such changes in enzyme activity are the ones in differentiation and maturation common to those of normal granulocytes as mentioned previously [5]. The maintained salvage system and reduced or disappeared *de novo* synthesis in the final maturation stage of granulocytes suggests that a reduction in *de novo* synthesis might actively influence the changes in nucleotide synthesis in the process of differentiation and maturation rather than the salvage system. However, increased HPRT activity seems to be a paradoxical change against reduced incorporation of Hx into the nucleotide fraction. Therefore, it is possible that reduced nucleotide synthesis depending on the metabolic control system rather than enzyme activity was deduced as changes in purine metabolism during induction of differentiation.

Treatment by differentiation of induction is considered a new therapy and progress is expected in this method. When a combination of such treatment and chemotherapy is planned, the timing for efficient administration is important. From the present study, a negative effect on activation of 6-MP is assumed by the reduced incorporation of Hx into nucleotide pool in retinoid-treated leukemic promyelocytes and HL-60 cells. Moreover, combined with previous report in retinoid-treated HL-60 cells showing same negative metabolic change on activation of cytarabine [8], prevailing or concomitant treatment with retinoid is presumed to inhibit the activation of antimetabolites in leukemic promyelocytes *in vivo* and to reduce the effect of chemotherapy. Regarding the problems of optimal treatment combinations, *etc.*, further study of the changes in metabolism during induction of differentiation should be made to establish more effective treatment.

REFERENCES

1) Nakamura T. Recent progress in treatment of haematological malignancies. *J Jpn Soc Intern Med* 78: 751, 1989.
2) Tsutani H, *et al*. Reduction of leukemia cell growth in a patient with acute promyelocytic leukemia treated by retinol palmitate. *Leukemia Res* 14: 595, 1990.
3) Tsutani H, *et al*. Pharmacological studies of retinol palmitate and its clinical effect in patients with acute non-lymphocytic leukemia. *Leukemia Res* (in press).
4) Borrow J, *et al*. Molecular analysis of acute promyelocytic leukemia breakpoint cluster region on chromosome 17. *Science* 249: 1577, 1990.
5) Tsutani H, *et al*. Purine nucleotide synthesis during terminal differentiation. *Adv Exp Med Biol* 253A: 423, 1989.
6) Korte DD, *et al*. Imbalance in the nucleotide pools of myeloid leukemia cells and HL-60 cells: correlation with cell-cycle phase, proliferation, differentiation and transformation. *Cancer Res* 47: 1841, 1987.
7) Wright DG. A role for guanine ribonucleotides in the regulation of myeloid cell maturation. *Blood* 69: 334, 1987.
8) Chiba P, *et al*. Effect of cell growth and cell differentiation on 1-β-D-arabinofuranosylcytosine metabolism in myeloid cells. *Br J Haematol* 71: 451, 1989.

6-MERCAPTOPURINE METABOLISM IN TWO LEUKEMIC CELL LINES.

E. H. Stet, R. A. De Abreu, Y. P. G. Janssen, J. P. M. Bökkerink, J. M. F. Trijbels
Center for pediatric Oncology South-east Netherlands, department of Pediatrics, University Hospital Nijmegen, Geert Grooteplein zuid 20, The Netherlands

INTRODUCTION

6-Mercaptopurine (6MP), an analogue of the purine base hypoxanthine, is commonly used in the maintenance therapy of acute lymphoblastic leukemia (ALL) in children [1].

The first step of 6MP cytotoxicity is intracellular conversion into thio-IMP (tIMP) by the enzyme hypoxanthine-guanine phosphoribosyl transferase (HGPRT). tIMP can be further metabolized by two pathways. Firstly, it can be converted into thioguanine nucleotides by the purine interconversion route [2, 3]. Subsequently, these thioguanine nucleotides are incorporated into RNA, and especially into DNA [4, 5]. Incorporation of thioguanine nucleotides into DNA results in delayed cytotoxicity, probably as a result of induction of DNA damage [6, 7]. The rate limiting enzyme in this metabolic route is IMP- dehydrogenase (IMPDH) [8], and as such its activity is of importance for the efficacy of 6MP cytotoxicity. Secondly, tIMP can be methylated by thiopurine methyltransferase (TPMT) [9, 10]. Methyl-thio IMP (Me-tIMP) formed by this reaction is a strong inhibitor of the purine de novo synthesis (PDNS) [11] and this may also contribute to the cytotoxicity of 6MP by induction of depletion of normal nucleotides for DNA and RNA synthesis.

To elucidate the role of both metabolic routes in the effectiveness of 6MP therapy, it is of importance to obtain insight into the activity of the two routes. In this study mycophenolic acid (MPA), a specific inhibitor of IMPDH was used to inhibit the conversion of 6MP into thioguanine nucleotides. A microassay for IMPDH was developed, based on the assay described by Uchida et al [3]. IMPDH activity was determined in two leukemic cell lines, Molt F4 (a T-ALL cell line) and KM_3 (a non-T-non-B cell line).
The effects of MPA on cell growth, cell viability, IMPDH activity and cellular nucleotide concentration were determined.

MATERIALS AND METHODS

Chemicals and reagents

6MP was purchased from Wellcome bv, Netherlands, MPA from Sigma Chemicals, USA, and ^{14}C-IMP from Amersham international, UK.

Cell culture

The experiments were performed with Molt F4, a T-ALL cell line, and with KM_3, a non-T-non-B cell line. The cells were cultured in RPMI 1640 DM (Gibco, Netherlands), supplemented with 10 % non-dialyzed fetal calf serum (Gibco, Netherlands), 2 mM L-glutamine (Sigma Chemicals, USA), 2 mM sodium pyruvate (BDH Chemicals Ltd, UK) and gentamycin/streptomycin, thrice a week. The cell cultures were maintained in a 5 % CO_2 humified atmosphere at 37°C. The absence of mycoplasma contamination and the presence of marker antigens were tested regularly.

IMPDH assay

For determination of the IMPDH activity a microassay was developed, based on the assay described by Uchida et al [3]. Cells were lyophilized overnight and were incubated with 1.61 μmol Tris/HCl (Sigma Chemicals, USA) pH 8.0, 12.9 nmol reduced glutathion (GSH, sigma Chemicals, USA), 3.23 nmol EDTA (Siegfried AG, BDR), 3.23 nmol NAD^+ (Sigma Chemicals, USA), 0.16 μmol KCl (Merck, BDR) and 2.9 nmol [8-^{14}C]IMP, for 60 minutes at 37°C. The reaction volume was 10 μl. Substrate and product formed were separated by means of thin layer chromatography. The radioactivity was counted in a liquid scintillation counter (Delta, Netherlands). The specific activity (SA) of the enzyme was expressed as nmol product.10^6 viable cells^{-1}.hour^{-1}.
The effect of inhibition of IMPDH activity by MPA was determined in cell lysates.

Determination of endogeneous nucleotides

After cell culture during 4 hours in the presence of various concentrations of MPA, the cells were harvested, counted in a coulter counter, and cell viability was determined. Nucleotides were extracted from 3 x 10^6 viable cells and stored at -20°C until analysis. Nucleotide concentrations were determined by means of HPLC [12].

Cell growth and viability assays

Cell growth and viability experiments were performed in logarithmically growing cells, which were seeded 24 hours prior to the start of the experiment. At this time 6MP, MPA or a combination of both were added as a single dose. During the experiment 2 mM glutamine was supplemented every 24 hour, in order to prevent glutamine exhaustion of the medium. After exposure to the drugs, cells were harvested and counted in a coulter counter. Cell viability was determined by means of the trypan blue exclusion.

RESULTS

The specific activity of IMPDH was higher in Molt F4 cells as compared to KM_3 cells (1.9 ± 0.2 nmoles product.hr^{-1}.10^6 viable cells^{-1}, n = 60 versus 1.1 ± 0.1, n = 11).

MPA appeared to be a strong inhibitor of IMPDH (Fig. 1). The inhibition was concentration dependent. A fifty percent inhibition of the enzyme activity was observed in the presence of 0.2 μM MPA. With 5 μM MPA virtually no IMPDH activity could be detected.

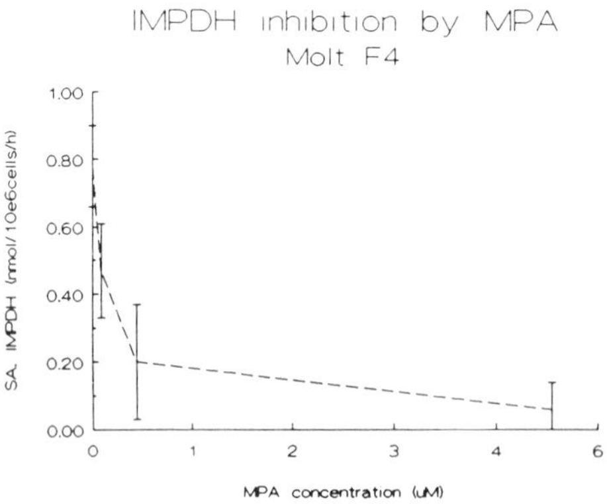

Fig. 1. Inhibition of IMPDH activity (mean ± SD, n = 15) from Molt F4 cells by MPA.
The IMPDH activity was determined in lysates of 25,000 Molt F4 cells in the presence of various concentrations of MPA.

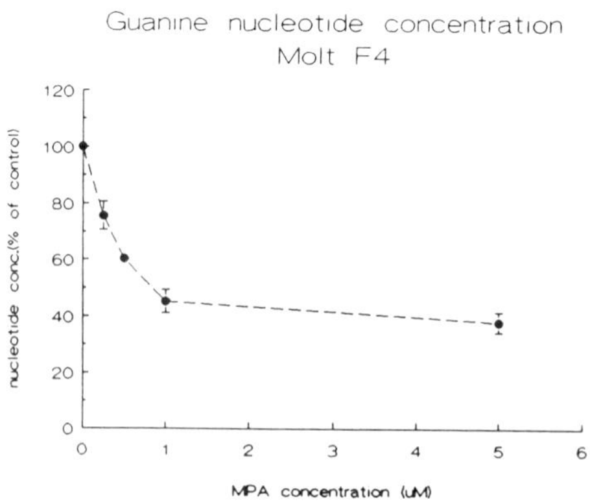

Fig. 2. Effect of MPA on the cellular guanine nucleotide concentration of Molt F4 cells.
The cells were incubated in various concentrations of MPA for 4 hours (mean ± SD, n = 3).

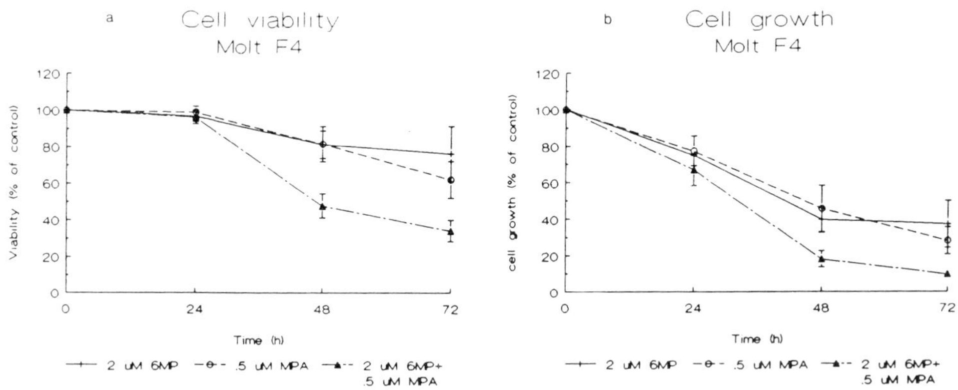

Fig. 3. Effects of 2 μM 6MP, 0.5 μM MPA and the combination of both drugs on cell viability (a) and cell growth (b) of Molt F4 cells.
Viability and cell growth were expressed as percentage of untreated cells (mean ± SD, n = 5).

The inhibition of IMPDH by MPA was reflected by a decrease of intracellular guanine nucleotide levels (Fig. 2). When cells were incubated during 4 hours with 0.8 μM MPA, a fifty percent reduction of the guanine nucleotide level was observed.

Both 6MP and MPA were cytotoxic for Molt F4 cells (Fig. 3a, 3b). The effects of the drugs were more pronounced on cell growth (fig. 3a) as compared to cell viability. The remaining viability after an incubation with 2 μM 6MP during 48 hours was 80 % of control. Cell growth, however, was decreased to 45 % of control. The combination of 6MP and MPA resulted in a synergistic decrease of cell growth and cell viability (fig. 3a, 3b).

DISCUSSION

The results of this study show that IMPDH activity is higher in Molt F4 cells compared to KM_3 cells. Earlier studies in our laboratory demonstrated that Molt F4 cells had also a more active PDNS and salvage pathway activity [13]. These observations indicate that the purine flow to guanine nucleotides in Molt F4 is higher than in KM_3.

MPA inhibits IMPDH in a concentration dependent manner (Fig. 1). The inhibition of IMPDH is reflected in a decrease of intracellular guanine nucleotide levels (Fig. 2). The guanine nucleotide concentration decreases to 40 % of control when cells are incubated during 4 hours with 5 μM MPA. These results are in accordance with data from earlier studies [14]. The inhibition of IMPDH, however, in the presence of 5 μM MPA, is much more pronounced than indicated by the guanine nucleotide concentration; hardly any IMPDH activity is detected (fig. 1). The presence of measurable guanine nucleotide levels after 4 hours may reflect the production of guanine nucleotides by the salvage pathway.

MPA is cytotoxic for Molt F4 cells. Cytotoxicity may be the result of induction of depletion of guanine nucleotides. Firstly, depletion of guanine nucleotides may inhibit DNA and RNA synthesis. Secondly, this depletion could hamper intracellular signal transduction, as guanine nucleotide binding proteins (G-proteins) are often involved in this proces [15, 16]. Alternatively, the observed cytotoxicity can be the result of increased IMP, which is an inhibitor of PDNS [11]. Inhibition of purine synthesis may lead to decreased DNA and RNA synthesis, resulting in inhibition of cell growth and cell viability, as Molt F4 cells depend on a high PDNS activity [13].

6MP cytotoxicity can be caused by incorporation of thioguanine nucleotides into DNA and RNA, and by formation of Me-tIMP, an inhibitor of PDNS. This study shows that methylation of tIMP is an important route for 6MP cytotoxicity. This can be concluded from the observed synergistic effect of the combination of 6MP and MPA on the cell viability. MPA inhibits IMPDH, and thereby conversion of 6MP into guanine nucleotides. The methylation of tIMP into Me-tIMP by TPMT remains the only route available. Also, a substantial quantity of Me-tIMP could be detected when cells were grown in the presence of 6MP (results not shown). Me-tIMP is a strong inhibitor of the PDNS [11]. As the PDNS is inhibited, the intracellular PRPP concentration rises [13]. PRPP is a co-substrate needed for the conversion of 6MP into tIMP. So, as a result of the increase of the

PRPP concentration more 6MP can be converted into tIMP. In this way, a self enhancement of 6MP incorporation may occur.

Since 6MP and MPA exhibit a severe inhibition of cell growth when added separately, no conclusion can be drawn about a synergistic effect of the combination of both drugs on cell growth.

In these experiments, the effect of 6MP and MPA on cell growth can be observed at an earlier stage, and is more pronounced in comparison with the effect on cell viability. First, inhibition of cell growth is induced by the inhibition of intracellular metabolic activity caused by 6MP and MPA. As a result of this, cell death occurs. So, biochemical changes occur before cell death can be detected.

ACKNOWLEDGMENTS

This work is supported by a grant from the Dutch Cancer Society (NUKC project 89-13).

REFERENCES

1. L. Lennard, J. S. Lilleyman, Variable mercaptopurine metabolism and treatment outcome in childhood lymphoblasic leukemia, J. Clin. Oncol. 7:1816 (1989).
2. L. Lennard, D. Keen, J. S. Lilleyman, Oral 6-mercaptopurine in childhood leukemia: parent drug pharmacokinetics and active metabolite concentrations, Clin. Pharmacol. Ther. 40:287 (1986).
3. M. Uchida, T. Nakamura, H. Uchino, Conversion of 6-mercaptopurine to 6-thioguanylic acid in L1210 cells and human leukemia cells, Jpn. J. Cancer Res. (GANN) 76:124 (1985).
4. J. P. M. Bökkerink, F. J. Damen, T. W. Hulscher, M. A. H. Bakker, J. A. van Baal, R. A. De Abreu, 6-mercaptopurine cytotoxicity and biochemical pharmacology in human malignant lymphoblasts of different lineages, submitted
5. J. A. Nelson, J. W. Carpenter, L. M. Rose, D. J. Adamson, Mechanisms of action of 6-thioguanine, 6-mercaptopurine and 8-azaguanine, Cancer Res. 35:2872 (1975).
6. J. Maybaum, H. G. Mandel, Unilateral chromatid damage: a new basis for 6-thioguanine cytotoxicity, Cancer Res. 43:3852 (1983).
7. N.T. Cristie, S. Drake, R.E. Meyn, J.A. Nelson, 6-Thioguanine-induced DNA-damage as determinant of cytotoxicity in cultured hamster ovary cells, Cancer Res. 44:3665 (1984).
8. F.R. Collart, E. Huberman, Expression of IMP dehydrogenase in differentiating HL60 cells, Blood 75:570 (1990).
9. L. L. Bennet, P. W. Allan, Formation and significance of 6-methylthiopurine ribonucleotide as a metabolite of 6-mercaptopurine, Cancer Res. 31:152 (1971).
10. C. N. Remy, Metabolism of thiopyrimidines and thiopurines; S-methylation with S-adenosyl methionine transmethylase and catabolism in mammalian tissue, J. Biol. Chem. 238:1078 (1963).
11. A. R. P. Paterson, D. M. Tidd, 6-thiopurines, in: "Antineoplastic and immunosuppressive agents II", A. C. Sartorelli, ed, Plenum Berlin, Heidelberg, New York (1975).
12. R. A. De Abreu, J. M. van Baal, J. A. J. M. Bakkeren, C. H. M. M. de Bruyn, E. D. A. M. Schretlen, High performance liquid chromatographic assay for identification and quantitation of nucleotides in lymphocytes and malignant lymphoblasts, J. Chromatogr. 227:45 (1982).
13. J. P. M. Bökkerink, M. A. H. Bakker, T. W. Hulscher, R. A. De Abreu, E. D. A. M. Schretlen, Purine de novo synthesis as the basis of synergism of methotrexate and 6-mercaptopurine in human malignant lymphoblasts of different lineages, Biochem. Pharmacol. 37:2321 (1988).
14. H. J. Lee, K. Pawlak, B. T. Nguyen, R. K. Robbins, W. Sadée, Biochemical differences among four inosinate dehydrogenase inhibitors, mycophenolic acid, ribavirin, tiazofurin, and selenazofurin, studied in mouse lymphoma cell culture, Cancer Res. 45:5512 (1985).
15. S. M. Kharbanda, M. L. Sherman, D. W. Kufe, Effects of tiazofurin on guanine nucleotide binding regulatory proteins in HL60 cells, Blood 75:583 (1990).
16. M.T. Rizzo, G.T. Tricot, R. Hoffman, H.N. Nayaram, G. Weber, J.G.N. Garcia, D. English, IMPDH inhibitors. Probes for investigation of the functions of guanine binding proteis in intact cells, Cell. Signal. 2:509 (1990).

SYNERGISTIC INTERACTION OF METHOTREXATE AND 6-MERCAPTOPURINE IN HUMAN DERIVED MALIGNANT T-ALL AND CALLA$^+$ CELL LINES

Ronney A. De Abreu, Frank van Strien, Lambert H.J. Lambooy and J.P.M.Bökkerink

Centre for Pediatric Oncology SE Netherlands, Department of Pediatrics
University Hospital Nijmegen, The Netherlands

INTRODUCTION

Methotrexate (MTX and 6-mercaptopurine (6MP) are antimetabolites which are commonly used in the treatment of acute lymphoblastic leukemia (ALL). Synergistic cytotoxic effects can be expected from combinations of MTX and 6MP, based on the biochemical pharmacology of both drugs.

In this report biochemical mechanisms are addressed which are involved in the effects of prolonged exposure of MTX and 6MP on two human derived acute lymphoblastic leukemia cell lines: KM3 (common-ALL$^+$) and MOLT F4 (T-ALL) cell lines. Effects of MTX on dihydrofolate dehydrogenase activity, thymidylate synthesis, and purine de novo synthesis have been measured and double labeling experiments have been performed, studying the incorporation of [8-^{14}C]6MP and [^{32}P]phosphate into DNA and RNA.

EXPERIMENTAL PROCEDURE

Cell cultures

MOLT F4 and KM3 cells were allowed to grow in a water-saturated atmosphere containing 5.0% CO_2 at 37 °C in RPMI medium 1640 Dutch modification (Gibco), supplemented with 10% non-dialyzed fetal calf serum (Integro), 2 mM L-glutamine, 2 mM pyruvate, penicillin (100,000 U/l) and streptomycin (100,000 μg/l). The medium was refreshed three times a week. The cells were tested for the absence of mycoplasma contamination and the presence of marker antigens regularly. All experiments were performed on logarithmically growing cells. During the experiments, glutamine was added every 24 hrs to a final concentration of 2 mM, in order to prevent glutamine exhaustion of the medium. Samples were collected at indicated points of time, to detect the various effects of 6MP and MTX. Untreated cells were used as controls. The number of viable cells was determined by means of trypan blue exclusion.

Dihydrofolate dehydrogenase activity, purine de novo synthesis and thymidylate synthesis

Dihydrofolate dehydrogenase activity and [U-^{14}C]glycine incorporation into purine metabolites were measured in samples from MOLT F4 and KM3 cells which were exposed to MTX. After incubation the cell suspension was centrifugated and the cell pellet was freeze dried overnight before it was assayed. Dihydrofolate dehydrogenase activity was measured according to Raemakers-Franken et al.[1] [U-^{14}C]glycine incorporation into purine metabolites was used as a measure of purine de novo synthesis and was assayed as described earlier.[2]

The effect of MTX on thymidylate synthesis was followed, measuring deoxy-TTP concentrations by means of HPLC.[3]

Incorporation of [8-^{14}C]6MP and [^{32}P]phosphate into DNA and RNA

MOLT F4 cells were used as a model to study the effects of 6MP and MTX on DNA and RNA synthesis, the cells were exposed to 1 μCi [^{32}P]phosphate and either 0.02 μM MTX or 2.0 μM [8-^{14}C]6MP or both. Quintuple samples were taken after 24 hrs and after 48 hrs of exposure. Incorporation of radioactive label into DNA and RNA was measured in quadruplicate. The fifth sample was used to measure the total intracellular DNA content, using a fluorimetric assay with 4',6-diamidino-2-phenylindole (Boehringer, Mannheim Germany).[4]

DNA and RNA were isolated according to Kufe et al.[5] The isolated DNA and RNA fraction was washed twice with 100% ethanol at 0 °C. The ethanol was evaporated at 60 °C and RNA was hydrolysed during incubation for 1 hr at 37 °C with 0.4 M KOH solution, containing 0.5 mg/ml calf-thymus DNA. Calf-thymus DNA was used as a carrier during DNA precipitation. DNA was precipitated at 0 °C with 5 N ice-cooled $HClO_4$. After centrifugation, the DNA pellet was resuspended in a solution containing 0.4 M KOH and 0.01 M K_2HPO_4. Label incorporated into DNA and RNA was counted by a liquid scintillation analyzer.

RESULTS

Cell growth of both, KM3 and MOLT F4, was inhibited by MTX in a concentration as well as a time dependent fashion (Figs. 1a and 1b).

Dihydrofolate dehydrogenase activity of KM3 and MOLT F4 was inhibited significantly after exposure to MTX, with 0.02 μM MTX causing a partial, and 0.2 μM MTX an almost complete inhibition after 48 hrs of incubation (Figs. 2a and 2b). The dihydrofolate dehydrogenase of KM3 cells, with a lower Vmax (Vmax = 9.1 ± 0.8 nmoles/hr/10^6 cells, Km-value = 1.42 ± 0.23 μM), appeared to be more sensitive for MTX than that of MOLT F4 cells (Vmax = 20.6 ± 0.9 nmoles/hr/10^6 cells, Km-value = 2.20 ± 0.30 μM). After exposure to 0.02 μM MTX, dihydrofolate dehydrogenase activity of KM3 cells decreased initially to 45% after 24 hrs, followed by a slow recovery to 60% after 48 hrs (Fig. 2a). Dihydrofolate dehydrogenase activity of MOLT F4 cells decreased to 60% after 24 hrs of exposure to 0.02 μM MTX and the activity seemed to recover completely within the next 24 hrs (Fig. 2b). Exposure to 0.2 μM MTX, caused an almost complete inhibition of dihydrofolate dehydrogenase activity in both cell lines, after 24 hrs. No recovery of activity was observed the next 24 hrs.

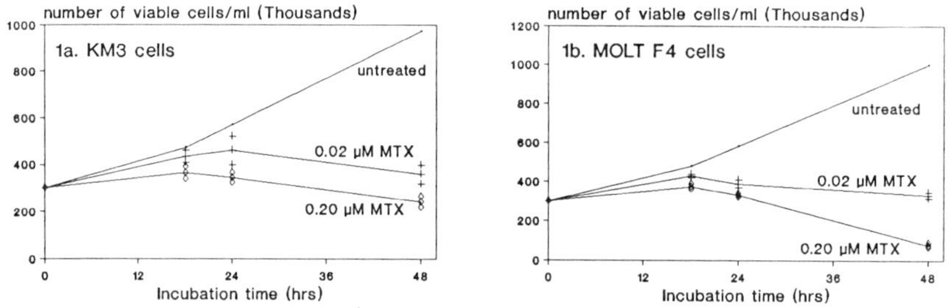

Figure 1. Inhibition of cellgrowth by MTX

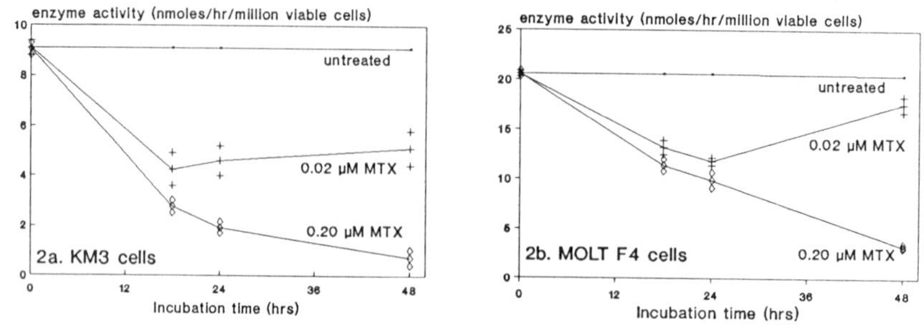

Figure 2. Effects of MTX on DHFDH activity

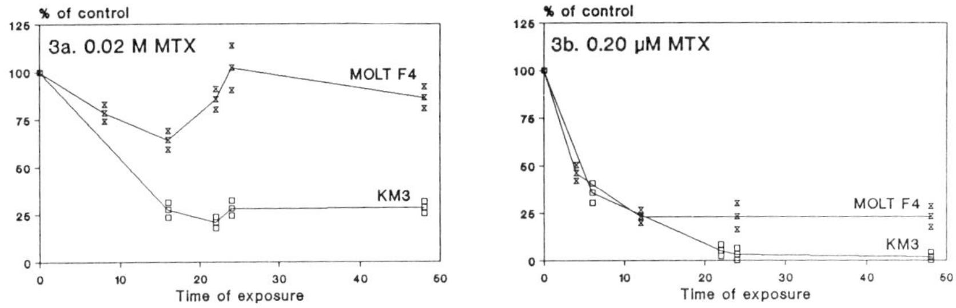

Figure 3. Effects of MTX on deoxy-TTP

TABLE I. [U-^{14}C]Glycine incorporation into purine metabolites (pmoles/hr/10^6 cells)

MTX conc (μM)	After 24 hrs KM3 cells	MOLT F4 cells	After 48 hrs KM3 cells	MOLT F4 cells
0	183 ± 28	214 ± 38	179 ± 45	231 ± 137
0.02	2.1 ± 2.1	189 ± 100	0 ± 0	219 ± 71
0.20	0 ± 0	0 ± 0		0 ± 0

number of experiments = 5

The effects of MTX on the intracellular deoxy-TTP concentration and [U-^{14}C]glycine incorporation into purine metabolites were similar to the observed effects on dihydrofolate dehydrogenase. In these experiments, a difference in the sensitivity of the two cell lines to MTX was also observed (Figs. 3a, 3b and Table I). Intracellular deoxy-TTP concentration as well as [U-^{14}C]glycine incorporation decreased significantly after exposure of the more sensitive KM3 cell line to 0.02 μM MTX. In contrast, only a initial decrease was observed on intracellular deoxy-TTP concentration and hardly any on [U-^{14}C]glycine incorporation, when MOLT F4 cells were exposed to 0.02 μM MTX. Intracellular deoxy-TTP and [U-^{14}C]glycine incorporation of both cell lines were significantly decreased by 0.2 μM MTX.

TABLE II. [^{32}P]phosphate incorporation into DNA and RNA after exposure to MTX and [8-^{14}C]6MP

exposure period (hrs)	0.02μM MTX DNA	0.02μM MTX RNA	2μM 6MP DNA	2μM 6MP RNA	0.02μM MTX + 2μM 6MP DNA	0.02μM MTX + 2μM 6MP RNA
			mean (SD)			
0 - 24	75 (8)	83 (7)	49 (7)	40 (6)	28 (5)	29 (4)
24 - 48	58 (7)	90 (7)	43 (6)	52 (7)	24 (4)	30 (5)

values are expressed as % of incorporation in untreated cells

Ratio [8-^{14}C]6MP/^{32}P

exposure period (hrs)		2 μM 6MP	0.02 μM MTX + 2 μM 6MP
0 - 24	DNA	$(3.72 \pm 1.04)10^{-2}$	$(6.74 \pm 1.42)10^{-2}$
	RNA	$(1.20 \pm 0.74)10^{-2}$	$(1.62 \pm 0.94)10^{-2}$
24 -48	DNA	$(3.81 \pm 1.35)10^{-2}$	$(5.34 \pm 1.52)10^{-2}$
	RNA	$(1.57 \pm 0.45)10^{-2}$	$(1.65 \pm 0.63)10^{-2}$

number of experiments = 4

The effects of MTX and/or 6MP on DNA and RNA synthesis of MOLT F4 cells are shown in Table II. The inhibitory effects of these antimetabolites on DNA and RNA synthesis are reflected by the decreased incorporation of ^{32}P into newly formed DNA and RNA (Table II). An at least additive inhibition of DNA and RNA synthesis was observed by a combination of 0.02 μM MTX + 2 μM 6MP, as compared with 0.02 μM MTX or 2 μM 6MP alone. Moreover, a relatively higher thiopurine nucleotide incorporation into DNA was noted with the combination of 0.02 μM MTX + 2 μM [8-^{14}C]6MP, since the ratio [8-^{14}C]/[^{32}P] in DNA was higher when the cells were exposed to 0.02 μM MTX + 2 μM [8-^{14}C]6MP than to 2 μM [8-^{14}C]6MP alone, despite the increased inhibition of total DNA synthesis by the combination of these drugs.

DISCUSSION

After exposure to MTX, a good correlation can be observed between the inhibition of the dihydrofolate dehydrogenase activity, the depletion of deoxy-TTP concentration and the decrease of [U-^{14}C]glycine incorporation in purine metabolites. The exposure of the cells to methotrexate, which is a tight-binding inhibitor of dihydrofolate dehydrogenase, results in a reduction in the generation of deoxy-TMP from deoxy-UMP. The inhibition of the methyl transfer to deoxy-UMP to form deoxy-TMP, which is catalysed by thymidylate synthetase, occurs only when the methyl-donor 5-10 methylene tetrahydrofolate, has been reduced below levels which are necessary for this reaction to proceed.[6,7]

After exposure of MOLT F4 cells to 0.02 μM MTX, only an initial inhibition of dihydrofolate dehydrogenase activity and decrease of deoxy-TTP is observed after 24 hrs, followed by a complete recovery after 48 hrs. So, it seems that the inhibition of dihydrofolate dehydrogenase does not lead to a significant reduction of 5-10 methylene tetrahydrofolate, for long enough to obtain long term biochemical effects from 0.02 μM MTX in MOLT F4 cells. The results from the [U-^{14}C]glycine incorporation study confirm these findings: very little effect of 0.02 μM MTX is observed on [U-^{14}]glycine incorporation into intracellular purine metabolites of MOLT F4 cells. Reduced folates are required in two steps of purine de novo synthesis: the formation of formylglycinamide ribonucleotide (FGAR) from glycinamide ribonucleotide (GAR) requires 5,10-methylenetetrahydrofolate and the synthesis of 5-fomamidoaminoimidazole-4-carboxamide ribonucleotide (FAICAR) from aminoimidazole-4-carboxamide ribonucleotide (AICAR) requires 10-formyltetrahydrofolate.[8,9] Therefore, the inhibitory effect of MTX on purine de novo synthesis is indirect and the consequence of decreased tetrahydrofolate pools.

KM3 cells with a lower Vmax of dihydrofolate dehydrogenase, are contrary to MOLT F4 cells, more sensitive to 0.02 μM MTX. After 24 hrs, the dihydrofolate reductase activity has decreased to 45% of control, followed by a slow recovery to 60% of control after 48 hrs. As a result, deoxy-TTP is dramatically depleted and [U-^{14}C]glycine incorporation is reduced to 2% of control after 24 hrs and completely inhibited after 48 hrs. So, the partial inhibition of dihydrofolate dehydrogenase of KM3 cells by 0.02 μM MTX appeares to be sufficient for long term effects on thymidylate synthetase and purine de novo synthesis.

Exposure to 0.2 μM MTX, corroborates the effects of MTX on dihydrofolate dehydrogenase, thymidylate synthesis and purine de novo synthesis in both cell lines. An almost complete inhibition of dihydrofolate dehydrogenase activity is observed at 24 hrs (KM3 and MOLT F4 8% and 15% of control, respectively). Deoxy-TTP concentration of MOLT F4 cells decreased to 25% of control and in KM3 cells no measurable amount could be detected after 24 hrs. In both cells [U-^{14}C]glycine incorporation was blocked completely after 24 hrs.

Previously we observed increase of intracellular 5-phospho-1-pyrophosphate (PRPP) and AICAR concentrations of MOLT F4 and KM3 cells after exposure to 0.02 μM MTX.[2,10] Intracellular PRPP and AICAR were elevated, as a consequence of inhibition of purine de novo synthesis by MTX, within 24 hours from 100 to 1500 and from <5 to 90 pmoles/10^6 cells, respectively. The enhanced PRPP and AICAR concentrations, measured in MOLT F4 after 0.02 1M MTX, seem in contrast with the results from the [U-^{14}]glycine incorporation study. There is, however, a possible explanation for this. [U-^{14}C]glycine incorporation reflects the maximal capacity of purine de novo synthesis after exposure to 0.02 μM MTX, whereas PRPP and AICAR concentrations reflect the actual capacity of purine de novo synthesis. So, it seems that after 0.02 μM MTX the actual purine de novo synthesis is partly inhibited. When MOLT F4 and KM3 cells were exposed to 0.2 μM MTX, only the PRPP concentration was elevated but not the AICAR concentration, hence, the purine de novo synthesis must be completely inhibited at the GAR-formyltransferase level under these conditions. GAR formyltransferase catalyzis the conversion of GAR to FGAR.

The results from the ^{32}P and [8-^{14}C]6MP incorporation into DNA and RNA, show that there is at least an additive inhibition of [^{32}P]phosphate incorporation into DNA and RNA when the MOLT F4 cells are pretreated with 0.02 μM MTX for 24 hrs, followed by treatment with 2.0 μM 6MP, as compared with 0.02 μM MTX or 2.0 μM 6MP alone (Table II), this means that DNA and RNA synthesis are inhibited in a more than additive fashion by the combination of these drugs. Besides, the ratio of [8-^{14}C]6MP incoporation/[^{32}P]phosphate incorporation into DNA of cells which are exposed to the combination of the drugs is significantly higher than in the DNA of cells exposed to 6MP, which means that relatively more 6MP per nucleotide is incorporated into newly formed DNA by the combination, despite a greater inhibition of DNA synthesis. Hence, the results indicate that MTX acts synergistically upon the incorporation of 6MP into DNA. Indeed, increased intracellular conversion of 6MP into thionucleotides has been reported at time points at which methotrexate pretreatment causes an increase of PRPP levels.[2,10] PRPP is needed as a cosubstrate for ribophosphorylation of 6MP into thio-IMP by HGPRT. Enhanced conversion by MTX of 6MP to thio-IMP and ultimately via thio-guanine ribonucleotides and thio-guanine deoxyribonucleotides, into RNA and DNA respectively, may lead to observed synergy.

REFERENCES

1. P.C. Raemakers-Franken, R.A. De Abreu, J.G. Willems, C. van der Drift and G.D. Vogels, Biochem. Pharmac. 41: 561 (1991)

2. J.P.M. Bökkerink, M.A.H. Bakker, T.W. Hulscher, R.A. De Abreu, E.D.A.M. Schretlen, J.P.R.M. van Laarhoven and C.H.M.M. de Bruyn, Biochem. Pharmac., 35: 3549 (1986).

3. J.P.M. Bökkerink, R.A. De Abreu, M.A.H. Bakker, T.W. Hulscher, J.M. van Baal and G.A.M. de Vaan, Biochem. Pharmac. 35: 3557 (1986)

4. J. Kapuacinski and B. Skocylas, Anal. Biochem: 252 (1977)

5. D. Kufe, D. Herrick, C Crumpacker and L. Schnipper, Cancer Res. 44: 69 (1984)

6. R.C. Moran, M. Mulkins and C. Heidelberger, Proc. Natl. Acad.Sci. U.S.A. 11: 5924 (1979)

7. J.R. Bertino, Cancer Chemother., III: 311 (1981)

8. E. Cadman, C. Benz, R. Heimer and J. O'Shaughnessy, Biochem. Pharmac. 30: 2469 (1981b)

9. E. Chapman in Developments in Cancer Chemotherapy, R.I. Glazer ed., CRC Press Inc., Boca Raton, Florida, 61 (1984)

10. J.P.M. Bökkerink, M.A.H. Bakker, T.W. Hulscher, R.A. De Abreu,E.D.A.M. Schretlen and C.H.M.M. De Bruyn, Biochem Pharmacol. 37: 2329 (1988)

6-THIOGUANINE METABOLISM IN HUMAN ERYTHROCYTES

A. Lucano, C. Crifò, and C. Salerno

Departments of Biochemical Sciences and Human Biopathology, University of Rome La Sapienza, Italy

INTRODUCTION

Thiopurines must undergo enzymatic conversion into a variety of nucleotide derivatives in order to display cytotoxic activity [1]; the major ones in human erythrocytes are the 6-thioguanine nucleotides [2]. These compounds derive from the immunosuppressive agent azathioprine as well as from the cytotoxic drugs 6-thioguanine and 6-mercaptopurine [3]. Increased concentrations of erythrocyte 6-thioguanine nucleotides have been related to neutropenia in leukaemic children receiving chronic 6-mercaptopurine chemotherapy [4]. Moreover, 6-thioguanine nucleotides are an index of bone marrow failure in adults subjected to azathioprine immunosuppression [5].

We report here preliminary results concerning the mechanisms of interconversion of 6-thioguanine nucleotides in human red blood cells. This work is part of a project directed to engineer human erythrocytes as a vehicle for carrying purine antimetabolites in a slow-release form [6].

MATERIALS AND METHODS

The experiments were performed in isotonic 50-mM Tris/NaCl buffer, pH 7.4, unless otherwise specified. Human erythrocytes were prepared from blood freshly drawn in heparin. Cellular ghosts were obtained as reported in reference [7]. Glutaraldehyde cross-linked cells and ghosts were prepared according to Deloach et al. [8] with minor modifications. Because of the binding of thiopurines to acid denatured erythrocyte proteins [9], deproteinization was achieved by heating the cellular extracts for 10 min in a boiling bath. Control experiments showed that this procedure caused only slight losses in the analytes.

6-Thioguanine metabolites were separated by reversed-phase HPLC using a 0.1 M potassium phosphate, pH 6.0 / methanol linear gradient and a 340-nm photometric detector. The peaks were identified by means of commercial standards (6-thioguanine and 6-

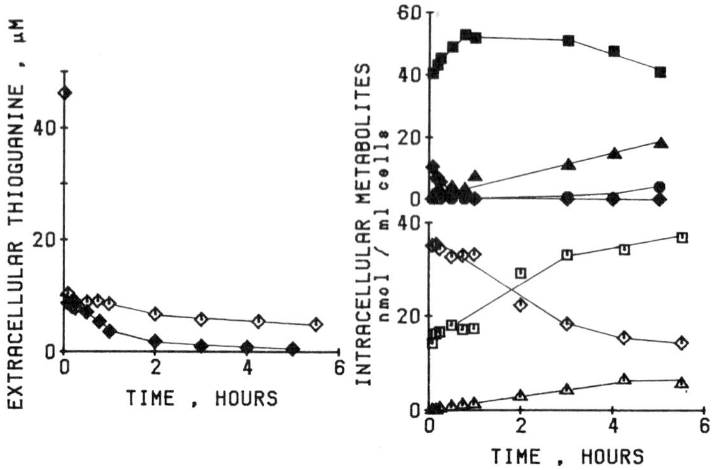

Fig. 1. Time course of 6-thioguanine uptake by human erythrocytes. *Left:* extracellular compartment. *Right:* intracellular compartment. ◇ : 6-thioguanine; ▣ : t-GMP; ▲ : t-GDP; ● : t-GTP. Closed symbols refer to experiments in the presence of 50 mM inorganic phosphate.

Table I. Intracellular 6-thioguanine derivatives after 150 min incubation at 37°C

	6-thioguanine	t-GMP	t-GDP	t-GTP
	% total thiopurine			
50 mM P_i	0	71	27	2
50 mM P_i + 10 mM glu	0	6	70	24
50 mM P_i + 10 mM ino	20	42	29	9

thioguanosine; Sigma Inc.) and/or by treating the samples with alkaline phosphatase, nucleosidediphosphate kinase and guanylate kinase. Adenine nucleotides were separated by reversed-phase HPLC as reported in reference [10].

RESULTS AND DISCUSSION

6-Thioguanine (50 μM) was taken up by human erythrocytes in isotonic Tris-buffered media through a biphasic process. At first, the purine analog was rapidly incorporated and partially utilized for the synthesis of 6-thioguanosine 5′-monophosphate (t-GMP). The subsequent slower phase of the process was characterized by intracellular synthesis of di- and triphosphate

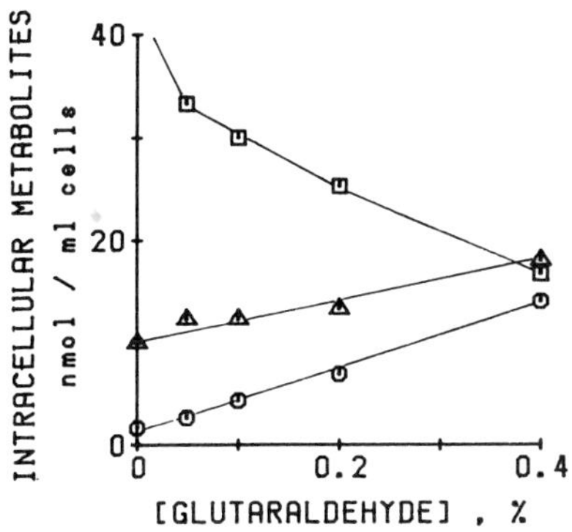

Fig. 2. Effect of glutaraldehyde treatment on the intracellular metabolism of 6-thioguanine nucleotides after 150 min incubation at 37°C. ◘ : t-GMP; ▲ : t-GDP; ⊖ : t-GTP.

Table II. 6-Thioguanine metabolism in cell-free systems

	t-GMP	t-GDP	t-GTP
	% total	thiopurine	
Blank	12	67	21
Untreated ghosts	28	72	0
+ 10 mM ATP	11	70	19
+ 10 mM ATP + 10 μM GTP[S]	10	50	40
+ 10 mM ATP + 100 μM GTP[S]	13	43	44
Glutarald.-treated ghosts	33	67	0
+ 10 mM ATP	13	50	37
+ 10 mM ATP + 10 μM GTP[S]	12	45	43
+ 10 mM ATP + 100 μM GTP[S]	11	44	45

Ghosts were treated with 0.2% glutaraldehyde. Ghost concentration was 1.6 mg protein/ml.

derivatives (t-GDP, t-GTP), as shown in Fig. 1. Intracellular concentration of 6-thioguanosine was always negligible.

Addition of inorganic phosphate in the suspending medium increased 6-thioguanine incorporation, probably by promoting phosphoribosylation of the purine base (Fig. 1). In the presence of inorganic phosphate and glucose, synthesis of t-GDP and t-GTP was further favoured (Table I). By contrast, addition of inosine in the medium strongly inhibited nucleotide synthesis from the base analog. The t-GDP/t-GMP ratio was also affected by conditions employed in erythrocyte preparation. If the cells were repeatedly washed in isotonic saline before addition of inorganic phosphate and 6-thioguanine, t-GMP was prevalent. Synthesis and accumulation of t-GDP was predominant in cells firstly washed in isotonic saline and then resuspended in plasma or in cells not washed at all (data not shown).

Nucleotide degradation and release of diffusible thioderivatives from inorganic phosphate-treated cells appeared to be limited even after exhaustive washing in isotonic saline or addition of heparinized autologous plasma. On the contrary, intracellularly unmetabolized 6-thioguanine was rapidly lost by washed cells. We studied the effect of cross-linking agents on 6-thioguanine permeability through the erythrocyte membrane. In phosphate-containing media, glutaraldehyde-treated erythrocytes incorporated 6-thioguanine as well as native cells. However, we found that treatment with this bifunctional agent markedly promoted t-GTP synthesis (and, to a less extent, t-GDP synthesis) by decreasing the intracellular t-GMP content (Fig. 2) but did not appreciably affect adenine nucleotide levels (data not shown).

The effect of glutaraldehyde on 6-thioguanine metabolism was studied in a cell-free system by adding native or cross-linked membranes to solutions of 6-thioguanine nucleotides (obtained from deproteinized cellular extracts). As reported in Table II, membranes caused a sharp decrease of t-GTP level in ATP-free mixtures. Addition of ATP (10 mM) prevented t-GTP loss or induced a slight increase of t-GTP concentration above the initial value (data not shown). When glutaraldehyde-treated membranes were used instead of native membrane in the ATP-enriched suspension, t-GTP accumulation was more pronounced. Similar results were obtained by adding guanosine 5′-[γ-thio] triphosphate (GTP[S]), a specific inhibitor of GTPase enzymes [11], to mixtures of native ghosts and 6-thioguanine nucleotides. The effects elicited by glutaraldehyde and GTP[S] were not additive, thus suggesting common target(s) in the thiopurine metabolism.

REFERENCES

1. G.Elion, Fed. Proc. 26:898 (1967)
2. L.E.Lavi and J.S.Holcenberg, Anal. Biochem. 144:514 (1985)
3. L.Lennard and J.L.Maddocks, J. Pharm. Pharmacol. 35:15 (1983)
4. L.Lennard, D.Keen and J.S.Lilleyman, Clin. Pharmacol. Ther. 40:287 (1986)
5. J.L.Maddocks, L.Lennard, J.Amess, R.Amos and R.M.Thomas, Lancet, i:156 (1986)
6. R.Green and K.J. Widder, Meth. Enzymol. 149:217 (1987)
7. S.L.Marchesi, E.Steers, V.T.Marchesi and T.W.Tillack, Biochemistry 9:50 (1970)
8. J.Deloach, S.Peters, O,Pinkard, R.Glew and G.Ihler, Biochim. Biophys. Acta 496:507 (1977)
9. M.R.Sheen, B.K.Kim and R.E.Parks, Molec. Pharmacol. 4:293 (1968)
10. P.D.Schweinsberg and T.L.Loo, J. Chromatogr. 181:13 (1980)
11. B.D.Gomperts, Nature 306:64 (1983)

THE IMPACT OF SELECTED NUCLEOSIDES ON THE CYTOTOXICITY OF ARA-C IN HL60 CELLS

Johannes Mejer

The Finsen Institute
The Laboratory of the Medical Department
Copenhagen, Denmark

INTRODUCTION

1-ß-D-arabinofuranosylcytosine, Ara-C, is one of the main drugs in the treatment of acute myeloid leukemia, AML. The Ara-C metabolism is similar to that of deoxycytidine. Thus Ara-C is either degradated to 1-ß-D-arabinofuranosyluracil, Ara-U, by a process catalyzed by the enzyme cytidine deaminase or activated through phosphorylation initially catalyzed by the enzyme deoxycytidine kinase. Further phosphorylation to 1-ß-D-arabinofuranosylcytosine 5′-triphosphate, Ara-CTP, is required for the cytotoxicity. This effect is counteracted by deoxycytidine 5′-triphosphate, dCTP. Ultimately, the concentration of Ara-CTP and dCTP are of decisive importance for the effect of Ara-C (1).

It is possible to interfere in the process of Ara-CTP and dCTP formation. Thus deoxycytidine kinase can be inhibited by deoxycytidine, cytidine, deoxyadenosine, deoxyguanosine and dCTP (2,3). Furthermore, the concentration of dCTP can be increased by the addition of either deoxycytidine or cytidine and decreased by the addition of deoxyadenosine and deoxyguanosine (4). The present study examines the impact of these four nucleosides on the Ara-C growth inhibitory effect in the leukemic derived HL60 cells and in normal myeloid cells from bone marrow.

METHODS

The HL60 cells were cultured in RPMI 1640 (with 10% fetal calf serum, glutamine (0.06 mg/ml), penicillin (0.05 mg/ml), and streptomycin (0.05 mg/ml)) in a Heraeus incubator with humidified air 4.5 % CO_2 at 37° C. The cell line was regularly tested for mycoplasma infection and discarded if positive. The normal human bone marrow myeloid cells were cultivated as described above with the addition of 20% placenta conditioned medium.

Cell counts were performed using a coulter counter (Coulter Electronic Ltd., Dunstable, Beds U. K.) with a leucocyte count tube (100 μm aperture). The threshold of the counter was adjusted to match the counts by microscope in a Bürker-Túrk cell chamber.

The cells were incubated in multiwell plates (2 x 10^{5}/ml/well). The cells were incubated for 72 h after the addition of Ara-C and the modulators. The effect was evaluated as the changes in growth compared to untreated controls.

In all experiments the viability of the cells evaluated with nigrosin was higher than 95 % independent of the treatments.

Purine and Pyrimidine Metabolism in Man VII, Part A
Edited by R.A. Harkness *et al.*, Plenum Press, New York, 1991

Statistic: Quadratic polynominals were fitted to each treatment groups by least square regression. The coefficients were compared in order to study differences between curves. The first order coefficients were used to compare differences between growth in the controls and the treated cells.

RESULTS AND DISCUSSION

Although deoxyadenosine and deoxyguanosine inhibite deoxycytidine kinase, which catalyzes the initial phosphorylation of Ara-C, both nucleosides increased the growth inhibitory effect of Ara-C additionally (Fig. 1). The reason may be that the Km value for the enzyme inhibition is high (2) and the decreasing effect on the Ara-C growth inhibition of dCTP is more pronounced. Cytidine and deoxycytidine had no inhibitory effect on the growth of HL60 cells if added alone, as seen at zero Ara-C concentration (Fig. 2). However, cytidine could antagonize the Ara-C growth inhibition ($p<0.05$). This reduction of the Ara-C effect of cytidine was abolished, if also uridine was added. The effect of cytidine is therefore exerted by the formation of dCTP and not by the inhibition of deoxycytidine kinase.

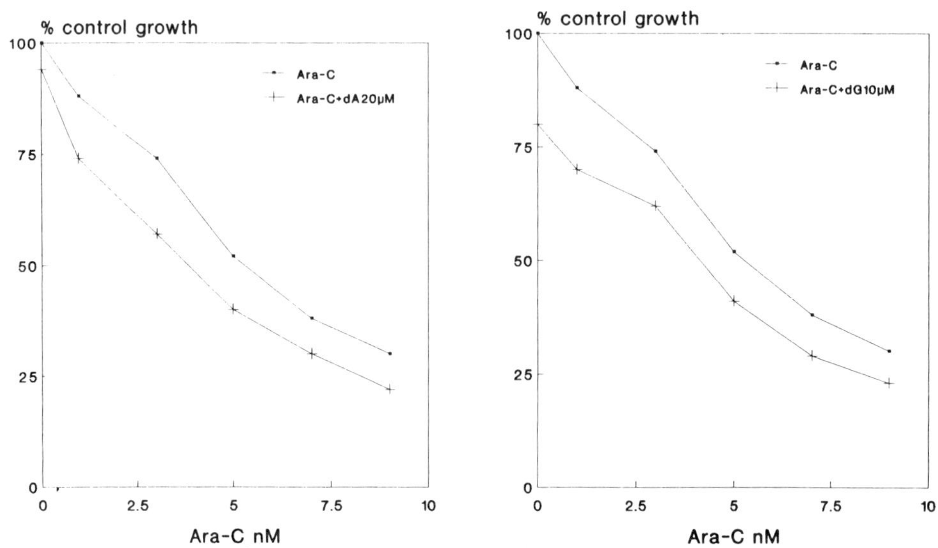

Fig. 1 The effect of Ara-C +/- deoxyadenosine (dA) or +/deoxyguanosine (dG) on HL60 cells. Mean of 5 independent observations. 2SD < 10.

At concentrations 100 times less than that of cytidine deoxycytidine could also antagonize the Ara-C effect ($p<0.05$), and this effect could be reversed if also teprahydrouridine, THU, an cytidine deaminase inhibitor, was present ($p<0.05$). The reason for this reversal effect of THU could be, that the activity of deoxycytidine kinase was higher with Ara-C as substrate than with deoxycytidine as substrate as previously described (5).

In myeloid cells from normal bone marrow no difference between the activities with the two substrates has been observed (5). This could be the explanation of the finding, that THU in normal bone marrow myeloid cells did not increase but decreased the growth inhibitory effect of Ara-C, when deoxycytidine was added concomitantly (Fig. 3). It must be notified, that the dosages of the different modulators in the normal cell were 100 times higher than that used in the HL60 cells.

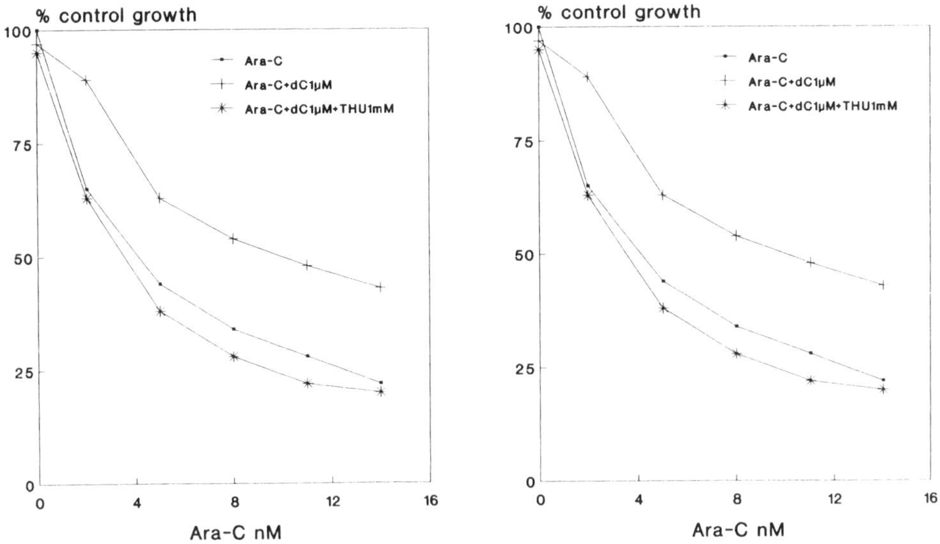

Fig. 2. Ara-C +/- deoxycytidine (dC) +/-tetrahydrouridine (THU), or +/-Cytidine (C) +/- uridine (U), the effect on HL60 cells. Mean of 5 indepenent observations. 2SD < 10.

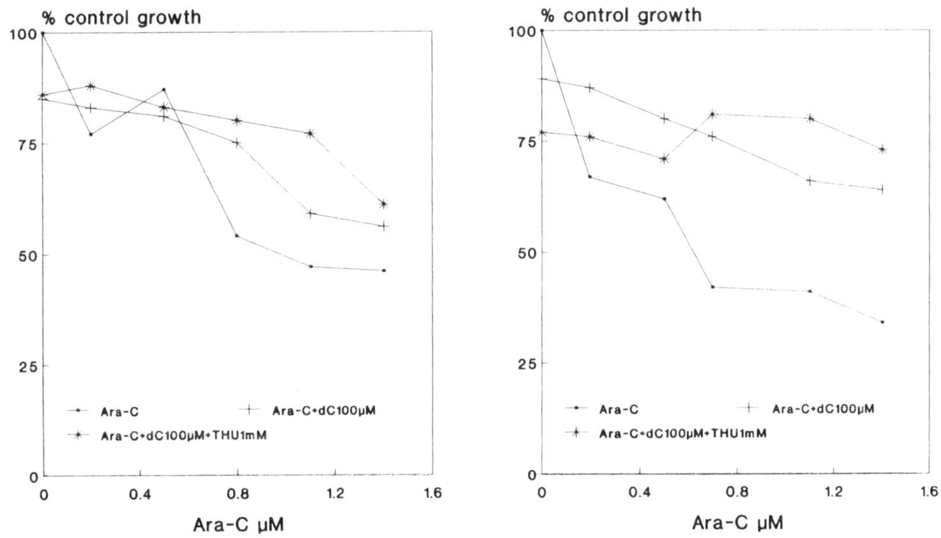

Fig. 3. Ara-C +/- deoxycytidine (dC) +/- tetrahydrouridine (THU) normal bone marrow 2 cases.

Km values with the two substrates in leukemic cells and normal cells are not available. The addition of deoxycytidine, which competetively inhibits the Ara-C phosphorylation, will increase the Km value, but the Vmax value will remain unchanged. In this way it can be deducted, that at high Ara-C concentrations the production of Ara-CTP compared with that of dCTP is much higher in the leukemic cells compared with the normal cells, in accordance with the results of others (6).

Deoxytidine kinase catalyses the phosphorylation of both deoxytidine and Ara-C (2). Recently it has been described, that thymidine kinase 2 catalyses the phosphorylation of deoxycytidine but not the phosphorylation of Ara-C (7). Differences in the distribution of deoxycytidine kinase and thymidine kinase 2 can be the explanation of the described difference in the phosphorylations capacity of deoxycytidine and Ara-C in normal and leukemic cells, and also the precented difference of the effect of THU on the combination of deocycytidene and Ara-C. This may have clinical implications in the treatments with drugs phosphorylated by deoxycytidine kinase but not by thymidine kinase 2 for instance Ara-C.

REFERENCES

1. Liliemark, J. O. and Plunket, W.. Regulation of 1-ß-D-arabinofuranosylcytosine 5′-triphosphate accumulation in human leukemia cells by deoxycytidine 5′-triphosphate. Cancer Res. 46, 1079-83, 1986.

2. Kierdaszuk, B. and Eriksson, S.. Selective inactivation of the deoxyadenosine phosphorylating activity of pure human deoxycytidine kinase: stabilization of different forms of the enzyme by substrates and biological detergents. Biochem. 29, 4109-14, 1990.

3. Harris, A. L. and Graham-Smith, D. G.. Cytosine arabinoside triphosphate production in human leukemic myeloblasts: Interactions with deoxycytidine. Cancer Chemother. Pharmacol. 5, 185-92, 1981.

4. Ross, D. D., Akman, S. A., Schreker, A. W. and Bachur, N. R.. Effect of deoxynucleosides on cultured human leukemia cell growth and deoxynucleotide pools. Cancer Res., 41, 4493-6, 1981.

5. Mejer, J., and Nygaard, P.. Cytosine arabinoside phosphorylation and deamination in acute myeloblastic leukemia cells. Leukemia Res., 2, 127-31, 1978.

6. Bhalla, K., MacLaughlin, W., Cole, J., Arlin, Z., Baker, M., Graham, G., and Grant, S.. Deoxycytidine preferentially protects normal versus leukemic myeloid cells from cytosine arabinoside-mediated cytotoxicity. Blood 70, 568-571, 1987.

7. Munch-Petersen, B., Cloos, L., Tyrsted, G., Eriksson, S.. Diverging substrate specificity of pure human thymidine kinases 1 and 2 aginst antiviral dideoxynucleosides. J. Biol. Chem. 266, 9032-8, 1991.

EFFECT OF CHLOROADENOSINE ON PHOSPHOINOSITIDE TURNOVER IN HUMAN T LYMPHOBLASTS ACTIVATED THROUGH THE TcR/CD3 COMPLEX

Buc HA, Hamet M, Moncion A and Pérignon JL

Laboratoire de Biochimie, INSERM U75
Faculté de Médecine Necker
156 rue de Vaugirard 75730 Paris, France

INTRODUCTION

The inherited deficiency of adenosine deaminase activity in humans is associated with a severe combined immunodeficiency disease with marked lymphopenia and loss of the proliferative response of lymphocytes to phytohaemagglutinin. The inhibition of ribonucleoside diphosphate reductase by dATP, a mechanism initially proposed to explain the immune dysfunction , cannot account for the earliest effects of dAdo on the proliferative response of the cells (reviewed in 1). This prompted us to study the consequences of adenosine deaminase deficiency on the early biochemical steps of lymphocyte activation. Indeed, we and others have recently demonstrated that the mitogen-induced hydrolysis of membrane phosphoinositides is markedly impaired in conditions of adenosine deaminase deficiency (2,3). In particular, we could demonstrate a significant inverse correlation between the cellular level of dATP and the hydrolysis of phosphoinositol bisphosphate (PIP2) estimated by [^{32}P]phosphatidic acid formation, [^{3}H]inositol phosphate generation and assay of inositol trisphosphate by radiocompetition. This suggested that phosphorylated derivatives of adenosine analogs might affect the activity of phospholipase C. In the present work, we used the same model of T-cell activation to study the effect of chloroadenosine (ClAdo), an adenosine analog known as an A2 receptor agonist (linked to the adenylate cyclase pathway), but which has recently been shown to be phosphorylated into ClATP (4).

MATERIALS AND METHODS

A CD2/CD3-positive subclone of Jurkat cells, a human T cell leukemia, was kept in continuous culture as described (2). 2-Chloroadenosine (ClAdo) was from Sigma and the anti-CD3 monoclonal antibody OKT3 from Ortho Pharmaceutical. The source of other reagents has been already described (2). Membrane phosphoinositides were labeled with either myo-2[^{3}H]inositol or [^{32}P]orthophosphate (Amersham) and the hydrolysis of phosphoinositides was estimated either by the analysis of

Table I. Effect of Chloroadenosine on the generation of cAMP and of IP3

3-hours preincubation	5-min stimulation	Second messenger pmol/10^7 cells cAMP	IP3
buffer	buffer	1.72	6.05
	ClAdo (500 μM)	5.22	4.85
	OKT3 (3 μg/ml)	1.68	78.20
	Forskolin (100 μM)	77.20	4.75
ClAdo (500 μM)	buffer	2.28	5.92
	ClAdo (500 μM)	1.65	5.50
	OKT3 (3 μg/ml)	3.22	28.40
	Forskolin (100 μM)	62.30	7.55

data are means of determinations from two independent experiments.

membrane phospholipids or by the production of inositol phosphates as described previously (2). Alternatively, inositol trisphosphate (IP3) was quantitated by a radiocompetition assay (Amersham). The cellular nucleotides were analysed by HPLC on a Partisil 10 SAX column (Whatman), and cAMP by a cAMP assay kit (Amersham), by previously described methods (2).

RESULTS

Effect of chloroadenosine on phosphoinositide hydrolysis

The ligation of the monoclonal antibody OKT3 to the membrane antigen CD3, a molecular complex tightly linked to the T-cell antigen receptor, induced a rapid hydrolysis of membrane phosphoinositides; this could be estimated either by the liberation of intracellular inositol trisphosphate or by the generation of phosphatidic acid (2). Table I shows that phosphoinositide hydrolysis was reduced by 64% in the cells that had been incubated for 3 hours with ClAdo. This inhibition could also be demonstrated by the incorporation of ^{32}P into phosphatidic acid (45% inhibition, not shown). We checked that ClAdo did not inhibit the basal synthesis of phosphoinositides, as estimated by the incorporation of ^{32}P into phosphoinositol, and phosphoinositol mono and bisphosphate. As ClAdo is a known agonist of the adenylate cyclase pathway, the possibility exists that the observed effects could be mediated by an increase of intracellular cAMP. However, several lines of evidences indicate that the inhibition of phosphoinositide hydrolysis was not due to the intracellular accumulation of cAMP: 1) the cAMP level at the end of the 3-hours incubation with ClAdo was no longer increased (2.28 vs 1.72 pmol/10^7 cells) 2) OKT3 alone, and the simultaneous addition of ClAdo and OKT3, induced similar increases of intracellular IP3 3) as seen on table I forskolin, a direct activator of adenylate cyclase, induced a much higher increase of intracellular cAMP with a modest inhibition of OKT3-induced liberation of IP3 (51.15 vs 78.20 pmol/10^7 cells).

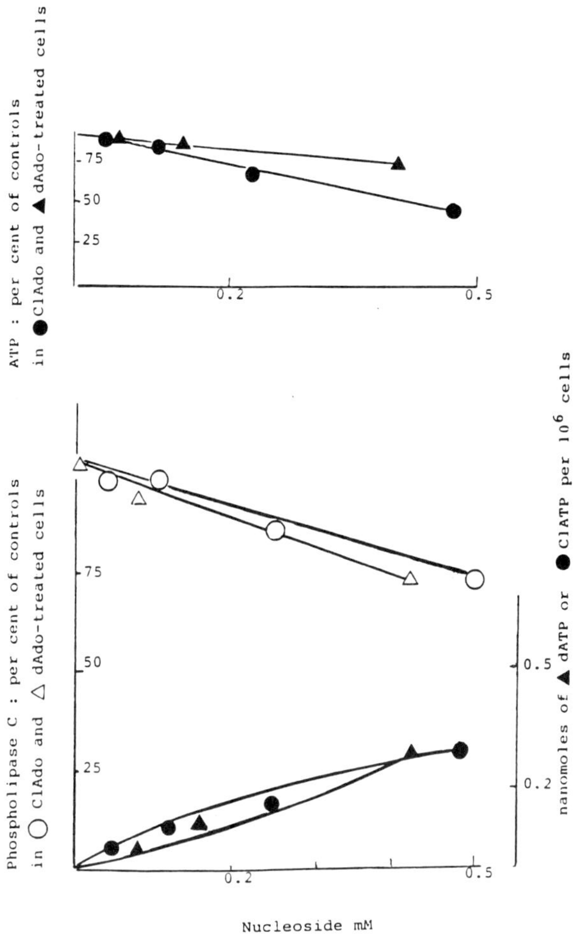

Figure 1. Intracellular content of ATP, dATP and ClATP, and inhibition of OKT3-induced phosphoinositide hydrolysis, in Jurkat cells incubated with dAdo (2) or ClAdo.

ATP, ClATP, and inhibition of phosphoinositide hydrolysis

The fact, mentioned above, that ClAdo exerted its inhibitory effect only after a 3-hours preincubation (we have not tested shorter periods of preincubation) strongly suggested that this effect was due to a metabolite of ClAdo, rather than to the nucleoside itself. Indeed, we have recently demonstrated that ClAdo could be phosphorylated into ClATP (4). This prompted us to perform an HPLC analysis of intracellular nucleotides. As described in Fig.1, the 3-hours incubation with ClAdo induced both a dose-dependent increase of ClATP and decrease of ATP. The inhibition of OKT3-induced hydrolysis of phosphoinositides was similarly dose-dependent.

Comparison of the effects of deoxyadenosine and chloroadenosine on phosphoinositide hydrolysis and cAMP concentration

The effects of ClAdo are thus similar to those previoulsy observed with dAdo (2): both nucleosides inhibit the ligand-induced hydrolysis of phosphoinositides, and in both cases the inhibition depends on the phosphorylation of the nucleoside. We took advantage of the fact that ClAdo is an agonist of the adenylate cyclase pathway to compare the effects of ClAdo and dAdo on the ClAdo-induced increase of intracellular cAMP. As shown on Table I, the preincubation of Jurkat cells with ClAdo made the cells insensitive to a subsequent stimulation by ClAdo. This was not due to a direct effect on adenylate cyclase itself, because in the same conditions there was no inhibition of the rise of cAMP induced by forskolin, a direct activator of adenylate cyclase. Similarly, after a 3-hours incubation with 1 mM dAdo (in the presence of the adenosine deaminase inhibitor deoxycoformycin) (2), there was a 50 to 65 % inhibition of the ClAdo-induced increase of cAMP (not shown).

DISCUSSION

This report, and our previous work (2), demonstrate that dAdo and ClAdo affect the signal transduction pathways in the human T lymphoblast cell line Jurkat, inhibiting both the phosphoinositide pathway and the adenylate cyclase pathway. In both cases, the inhibition appears to be dependent on the metabolization of the nucleoside, and is correlated with the accumulation of the corresponding nucleotide triphosphate. In both cases also, the apparition of the corresponding nucleotide triphosphate is accompanied by a decrease of intracellular ATP. However, we have previously shown that Ado could also inhibit the phosphoinositide pathway, although in this case ATP was moderately increased (2). Taken together, these data suggests that the effects of ClAdo and dAdo are mediated by ClATP and dATP rather than by the decrease of ATP.

The fact that ClAdo and dAdo inhibit both the posphoinositide and the adenylate cyclase pathways suggests that the inhibition may take place at a step similar in the two pathways. Good candidates could be G proteins. It is well established that adenylate cyclase is coupled to its receptors by well characterized GTP-binding proteins (5). There are also evidences that in Jurkat cells, the hydrolysis of phosphoinositides can be affected by G proteins (6), that have not yet been identified. Whether dATP or ClATP (and other nucleotides of purine analogs) affect the fonction of G proteins remains to be demonstrated.

REFERENCES

1) Kredich NM, Hershfield MS. The metabolic basis of inherited disease, Edited by Scriver CR, Beaudet AL, Sly WS, Valle D, 6th edition, 1989, McGraw-Hill, NY, pp. 1045-1075.
2) Buc HA, Moncion A, Hamet M, Houllier AM, Thuillier L, Pérignon JL. Eur J Immunol 1990; 20:611-615.
3) Scharenberg JGM, Rijkers GT, Akkerman J-W, Staal GES, Zegers BJM. Int J Immunopharmacol 1990; 12:113-120.
4) Claessens S, Hamet M, Chedeville A, Basuyan JP, Lavoinne A. FEBS Lett 1988; 232:317-322.
5) Casey PJ, Gilman AG. J Biol Chem 1988; 263:2577-2580.
6) Augustine JA, Secrist JP, Daniels JK, Leibson PJ, Abraham RT. J Immunol 1991; 146:2889-2897.

ISOLATED LIVER PERFUSION VERSUS HEPATIC ARTERY INFUSION WITH 5-FLUOROURACIL IN A RAT MODEL; EFFECTS ON THYMIDYLATE SYNTHASE

Clasina L. Van der Wilt, Andreas Marinelli[1], Jacqueline Cloos, Kees Smid, Herbert M. Pinedo, Godefridus J. Peters

Dept. Oncology, Free University Hospital, PO BOX 7057, 1007 MB Amsterdam [1] Dept. Surgery, University Hospital, PO BOX 9600, 2333 RL Leiden, The Netherlands

INTRODUCTION

5-Fluorouracil (5FU) is frequently used for locoregional chemotherapy of non-resectable hepatic metastases of advanced colorectal cancer [1,2]. This drug is metabolized to FdUMP, a potent inhibitor of thymidylate synthase (TS), an essential enzyme in the pyrimidine de novo synthesis. Inhibition of TS leads to a decreased DNA synthesis [3].

Regional 5FU administration into the liver allows the use of a higher 5FU dose than systemic therapy. Most of the 5FU is metabolized in the liver and the concentration in the systemic circulation is lower compared to peripheral intravenous infusion. This circumnvents toxic side effects of the high dose 5FU in other tissues [1,2]. The route, usually applied for this therapy, is hepatic artery infusion (HAI) with an implanted pump, which delivers the drug into the hepatic artery [4]. Isolated liver perfusion (ILP) technique has been developed to selectively expose the liver to higher doses of chemotherapy, while systemic toxicity is minimal. This method is still experimental and with 5FU clinical succes is limited [5].

We studied the biochemical effects of both ways of regional 5FU administration in the rat by evaluating the extent and retention of TS inhibition in tumor and normal tissues, in which toxic side effects of 5FU might occur.

METHODS

Hepatic metastases of the rat colon tumor line CC531 were evoked in WAG/Rij rats by subcapsular injection of viable tumor cells into the liver. Two weeks after inocculation of the tumor cells, the animals received a locoregional treatment with 5FU. 5FU was administrered at its maximally tolerated dose (MTD) either by ILP (150 mg/kg) or by HAI (50 mg/kg). For ILP, the pyloric vein, the gastroduodenal artery and the intrahepatic caval vein were cannulated.

Purine and Pyrimidine Metabolism in Man VII, Part A
Edited by R.A. Harkness *et al.*, Plenum Press, New York, 1991

During perfusion the caval vein was clamped above and below the liver, the aorta proximal to the coeliac axis and the portal vein and common hepatic artery were clamped just below the tips of the two cannulas. Two low-flow roller pumps and an adapted oxygenator were used to recirculate the perfusate, containing 150 mg/kg 5FU, during 20 min [6]. For HAI, the gastroduodenal artery side arm of the common hepatic artery was cannulated and 50 mg/kg 5FU was administered [6].

Tumors, liver tissue and bone marrow were removed 3, 24 or 48 hr after treatment and immediately frozen in liquid nitrogen. Samples from untreated rats served as controls. TS inhibition was evaluated with the FdUMP binding assay, a ligand binding assay with [6-^{3}H]-FdUMP, used to determine the free binding sites of the enzyme for FdUMP, performed essentially according to published methods [7-9]. Cytosols of samples from treated rats were washed with neutral charcoal before the assay in order to remove endogenous nucleotides. Assays were performed with saturating folate concentrations to allow maximal binding of FdUMP [10].

RESULTS

The activity of TS in tissues from untreated rats was higher in tumors than in normal livers (34 vs 9 pmol/g wet weight). Also other normal tissues (intestinal mucosa and bone marrow) had a lower TS activity than tumor tissue. Treatment with 5FU resulted in a marked inhibition of TS in tumor tissue after 3 hr (Fig 1). In tumors from animals treated by HAI, recovery of TS activity was observed after 24 hr, while in tumors from rats treated by ILP, TS inhibition was prolonged. A gradual recovery of TS activity was observed for ILP tumors after 48 hr.

In liver tissue no significant inhibition of TS was observed for either route of administration (Fig 1). After 3 hr FdUMP binding to TS was comparable to control, but after 24 and 48 hr an increase of FdUMP binding was detected. This consequent increase of TS activity was most pronounced in liver tissue of rats treated by ILP. In bone marrow we observed a strong inhibition of TS 3 hr after HAI (Fig 1). Retention of TS inhibition in bone marrow was at least 48 hr in rats treated by HAI and the inhibition of TS was higher compared to bone marrow from rats treated by ILP. The latter showed a recovery of TS inhibition after 48 hr.

DISCUSSION

Biochemical analysis of tumors after ILP or HAI delivery of 5FU showed a difference in the effect on TS activity. This might have consequences for the antitumor effect of both treatments. TS activity was higher in tumor tissues compared to other highly proliferative tissues such as intestinal mucosa and bone marrow. It has been shown that TS activity is high in another proliferative tissue such as regenerating liver [11]. The difference in TS activity between hepatic metastases and normal liver tissues was about 3.5 fold. This was comparable to results of Berne et al. [12] in another rat colon tumor model.

Evaluation of TS inhibition in tumors after treatment with 5FU showed a more pronounced effect for ILP. This can partially be explained by the higher MTD of

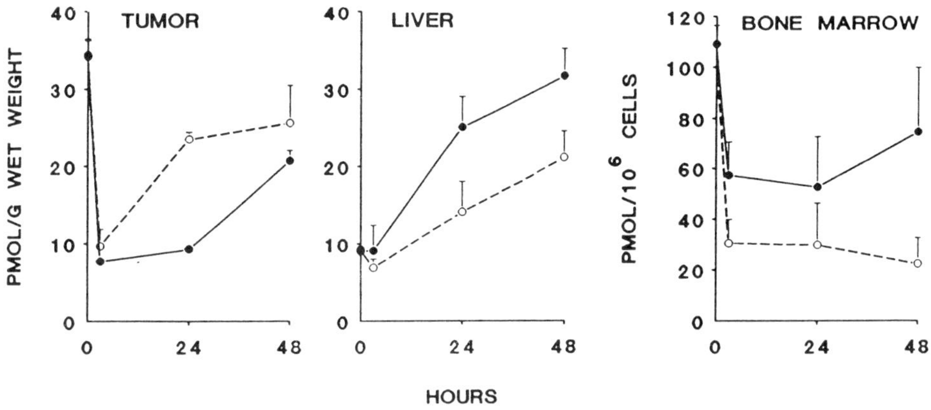

Figure 1. Comparison of the residual binding sites for [^{6}H]-FdUMP of TS after ILP (150 mg/kg 5FU) [-●-] and HAI (50 mg/kg 5FU) [-o-] in CC531 tumors, liver tissue and bone marrow of WAG/Rij rats. Values are means ± SEM, n=3-5.

5FU, that can be used for ILP (150 vs 50) and the exposure of the metastases to a higher 5FU concentration during the perfusion [13]. A slightly better antitumor effect, although not significantly different, was observed for ILP. This might correlate with the greater TS inhibition of ILP. Retention of TS inhibition in tumors was longer for ILP than for HAI. Measurements of FdUMP binding to TS in subcutaneous implanted rat colon tumors after i.p. administration of 100 mg 5FU/kg performed by others [7,12], showed that TS was completely inhibited up to 4 hr. This might be related to a higher sensitivity of the tumors to TS inhibition or a lower sensitivity of the assay.

ILP did not cause liver toxicity at this dose, this was supported by the absence of TS inhibition in the liver. For liver tissue we observed an increase of TS, probably in response to cytotoxic stress caused by 5FU exposure which can synchronize cells into S-fase [14]. It has been shown that TS activity is cell cycle dependent with a higher TS activity in the S-phase [14,15]. New enzyme synthesis may also be responsible for the increase of TS [15]. An increase of TS has been observed in rat and murine tumors shortly after i.p. 5FU treatment [7,8,12]. This phenomenom might be a resistance mechanism to 5FU. HAI resulted in a stronger TS inhibition in bone marrow compared to ILP. This more severe toxicity might be related to higher systemic exposure of normal tissues to 5FU, since plasma 5FU levels were significantly higher for HAI compared to ILP [13]. Nevertheless dose limiting toxicity for ILP is still systemic, and not liver toxicity.

These data show that ILP resulted in a better retention of TS inhibition in tumors than HAI, while HAI resulted in stronger TS inhibition in normal tissues. This points to a better therapeutic efficacy of ILP with selective exposure of the tumor and low systemic toxicity. However, ILP is a rather complicated surgical technique with several drawbacks. The initial extent of TS inhibition does not seem to be dose dependent, but TS inhibition may be prolonged by a longer administration period. So both HAI and ILP may be improved by using a longer infusion period of the drug or by combination of drugs such as leucovorin, which can prolong TS inhibition [16,17]. Systemic toxicity, such as leukopenia may be regulated by the use of a protecting agent such as uridine [18].

REFERENCES

1. De Bruijn EA, Van Oosterom AT, Tjaden UR. Site-specific delivery of 5-fluorouracil with 5'-deoxy-5-fluorouridine. Reg Cancer Treat 2: 61-76 (1989).
2. Diasio RB, Harris BE. Clinical Pharmacology of 5-Fluorouracil. Clin Pharmacokin 16: 215-237 (1989).
3. Pinedo HM, Peters GJ. 5-Fluorouracil: Biochemistry and pharmacology. J Clin Oncol 6: 1653-1664 (1988).
4. Van Groeningen CJ. New technical developments in antineoplastic drug delivery and their role in cancer treatment. In: Drug Delivery in Cancer Treatment. L Domellof (ed.), Springer Verlag, Berlin, pp 39-52 (1987)
5. Aigner KR, Tonn JC, Walter H et al. The isolated liver perfusion technique for high-dose chemotherapy of metatases of colorectal cancer: two years clinical experience. In: Liver Metastasis, CJH Van de Velde, PH Sugarbaker (eds.), Martinus Nijhoff Publishers, The Hague, pp 346-357 (1984).
6. De Brauw LM, Marinelli A, Van de Velde CJH et al. Pharmacological evaluation of experimental isolated liver perfusion and hepatic artery infusion with 5FU. Cancer Res 51: 1694-1700 (1991).
7. Berne MHO, Gustavsson BG, Almersjö O et al. Sequential methotrexate/5FU: FdUMP formation and TS inhibition in a transplantable rodent colon adenocarcinoma. Cancer Chemother Pharmacol 16: 237-242 (1986).
8. Spears CP, Shahinian AH, Moran RG et al. In vivo kinetics of thymidylate synthetase inhibition in 5-fluorouracil-sensitive and -resistant murine colon adenocarcinomas. Cancer Res. 42: 450-456 (1982)
9. Peters GJ, Van Groeningen CJ, Laurensse EJ, Pinedo HM. Thymidylate synthase from untreated human colorectal cancer and colonic mucosa: enzyme activity and inhibition by 5-flouro-2'-deoxyuridine-5'-monophosphate. Eur J Cancer 27: 263-267 (1991).
10. Cisneros R, Dunlap RB. Characterization of the parameters affecting covalent binding stoichiometry in binary and ternary complexes of thymidylate synthase. Biochim Biophys Acta 1039: 149-156 (1990).
11. Hashimoto Y. Shiotani T, Eble JN, Glover JL, Weber G. Increased thymidylate synthase (EC 2.1.1.45) activity in normal and neoplastic proliferation. Cancer Biochem Biophys 10: 1-10 (1988).
12. Berne MHO, Gustavsson BG, Almersjö O et al. Concurrent allopurinol and 5-fluorouracil: 5-fluoro-2'deoxyuridylate formation and thymidylate synthase inhibition in rat colon carcinoma and in regenerating rat liver. Cancer Chemother. Pharmacol. 20: 193-197 (1987).
13. Marinelli A, Pons DHA, Kuppen PJK et al. Isolated liver perfusion (ILP) vs hepatic artery infusion (HAI) with 5-Fluorouracil (5FU) and mitomycin C (MMC) in a rat model. Proc Am Ass Cancer Res 31: 429 abstr 2547 (1990).
14. Cadman E, Heimer R. Levels of thymidylate synthetase during normal culture growth of L1210 cells. Cancer Res 46: 1195-1198 (1986).
15. Keyomarsi K, Moran RG. Mechanism of the cytotoxic synergism of fluoropyrimidines and folinic acid in mouse leukemic cells. J Biol Chem 263: 14402-14409 (1988).
16. Van der Wilt CL, Pinedo HM, Smid K et al. Effect of folinic acid on fluorouracil activity and expression of thymidylate synthase. Sem Oncol 18, suppl 8, (1991), in press
17. Peters GJ, Van Groeningen CJ. Clinical relevance of biochemical modulation of 5-Fluorouracil. Annals Oncol 2 (1991), in press.
18. Van Groeningen CJ, Peters GJ, Leyva A, Laurensse EJ, Pinedo HM. Reversal of 5-Fluorouracil-induced myelosuppression by prolonged administration of high-dose uridine. J Natl Cancer Instit 81: 157-162 (1989).

CHANGES OF PURINE NUCLEOTIDE METABOLISM OF EHRLICH ASCITES CELLS DURING TRANSITION OF TUMOUR GROWTH

Tilman Grune, Werner Siems, Rolf Uhlig,
Peter Langen, and Gerhard Gerber

Institute of Biochemistry, Humboldt University
of Berlin, Hessische Str. 3/4, D(O)-1040 Berlin,
and Institute of Molecular Biology Berlin-Buch,
Robert-Roessle-Str. 10, D(O)-1115 Berlin, F.R.G.

INTRODUCTION

During the in vivo growth of Ehrlich ascites tumour there exists a proliferating phase, in which the number of cells increases quasi-exponentially, followed by a resting phase, in which the number of cells stays practically constant (Benndorf et al., 1988). The transition of growth phases is accompanied by structural deterioration and decreased number of mitochondria (Siems et al., 1989), and a multitude of metabolic changes, e.g. substrate utilization, energy production (Schmidt et al., 1991), synthesis of specific proteins (Benndorf et al., 1988) and uptake of radioactive labeled adenine, hypoxanthine and inosine (Siems et al., 1989).

However, the molecular mechanisms of the transition from one growth phase to the another are poorly understood. This paper deals with the investigation of purine metabolism as well as in the so called in vitro "rejuvenation" model of Ehrlich mouse ascites cells (EMAC) and in tracerkinetic experiments with suspensions of EMAC of both growth phases.

MATERIAL AND METHODS

Female mice of ICR strain, weighing approximately 15-20 g were used. The Ehrlich ascites tumour cells were obtained from animals at the 5th and at the 12th days after inoculation. Cell handling was performed as described by Schmidt et al. (1991). For the determination of nucleotide content in the "rejuvenation" process cells were suspended in cultivation medium as described by Boehmer et al. (1984). For the tracerkinetic experiments the tumour cells were incubated in presence of radioactive labeled adenine (10.3 GBq/mmol) in an Eagle / Borsook medium containing glucose (10 mM) and the proteinogenic amino acids (each at 0.2 mM). The cytocrite was adjusted at 2 % .

Table 1. Concentrations of purine compounds of Ehrlich ascites tumour cells during their different growth phases (mean ± S.D., n=4). Dimensions: mmol/l cells (nucleotides), µmol/l cells (nucleosides and bases).

Purine compound	Exponential phase	Stationary phase
ATP	4.61 ± 0.23	2.68 ± 0.19
ADP	1.55 ± 0.17	1.09 ± 0.16
AMP	0.65 ± 0.12	0.51 ± 0.10
GTP	0.95 ± 0.23	0.35 ± 0.21
GDP	0.42 ± 0.31	0.22 ± 0.19
IMP+GMP	0.31 ± 0.07	0.15 ± 0.06
ATP:ADP	2.97 ± 0.15	2.46 ± 0.17
ATP:AMP	7.09 ± 0.35	5.25 ± 0.37
GTP:GDP	2.26 ± 0.55	1.59 ± 0.95
GTP:GMP	3.06 ± 0.74	2.33 ± 1.40
ATP+ADP+AMP	6.81 ± 0.52	4.28 ± 0.45
GTP+GDP+GMP	1.68 ± 0.71	0.72 ± 0.56
Sum of purine nucleotides	8.49 ± 1.23	5.00 ± 1.01
Ado	0.45 ± 0.11	0.59 ± 0.17
Ade	0.98 ± 0.27	1.44 ± 0.22
Hyp	1.55 ± 0.26	1.44 ± 0.33
Uric acid	3.37 ± 0.29	5.47 ± 0.35

All procedures for tumour cell handling, cell incubation techniques and nucleotide analysis were described in Schmidt et al. (1991); Grune et al. (1991a) and Grune et al. (1991b).

RESULTS AND DISCUSSION

The purine nucleotide content of ascites cells was markedly decreased during the transition from exponential into stationary phase of tumour growth (Table 1). The concentrations of the nucleotides were generally higher in cells during the exponential phase in comparison with cells of the resting phase. In cells of the resting phase the ATP level amounted to 58 % and the GTP level to 37 % of the corresponding values in tumour cells of proliferating phase. The sum of adenine nucleotides amounted to about 60 % in the resting compared with the proliferating period. In the case of purine nucleosides and bases the concentration differences between suspensions of cells in proliferating phase and cells in the stationary phase were smaller than differences of nucleotide concentrations. Furthermore, the higher values of nucleosides and bases were found in suspensions of Ehrlich mouse ascites cells of the stationary phase (12th day after transplantation). Such increase was significant for the concentrations of uric acid and adenine. During the first hours of incubation of EMAC of the stationary phase of tumour growth in a new complete cultivation medium the nucleotide pattern radidly changed.

Table 2. Changes of nucleotide concentrations during the transition of Ehrlich mouse ascites cells from the stationary to the exponential phase of tumour growth (mean ± S.D., n=6). Dimension: mmol/l cells. n.d.= not determined.

Time (min)	ATP	GTP	ATP + p18
0	2.68 ± 0.19	0.35 ± 0.21	2.68 ± 0.19
10	3.36 ± 0.33	0.46 ± 0.12	2.78 ± 0.23
30	4.82 ± 1.15	0.47 ± 0.14	4.00 ± 0.95
60	5.99 ± 0.60	0.54 ± 0.05	4.61 ± 0.46
120	5.53 ± 0.95	0.54 ± 0.08	6.03 ± 1.00
240	4.27 ± 0.84	0.52 ± 0.07	n.d.

The variation of ATP and GTP during the first hours of incubation are shown in Table 2. The ATP level of resting cells increased during the first hour of their rejuvenation followed by a slow decrease reaching a value in the range of that of EMAC of proliferation phase of growth. The GTP content of EMAC showed an analoguos increase in the first hour of rejuvenation, too. In contrast to ATP there was no overshoot of GTP during the incubation. The ATP/ADP ratio rised 3fold in the first hour of incubation and than decreased to the 2fold in comparison with the value of the EMAC of the stationary growth phase. The mammary-derived growth inhibitor (MDGI) -sequence p18 delayed the ATP restoration of resting phase cells during the rejuvenation experiments (Table 2). The changes of purine nucleotide concentrations of EMAC of exponential and stationary growth phases are due to differences of enzyme activities and of the flux rates thruogh metabolic pathways in these cells. To enlight these circumstances the metabolism of radioactive labeled adenine in the tumour cells of different growth phases was investigated. The specific radioactivities in the pools of purine compounds after 30 min of incubation are presented in Table 3. Except of the precursor adenine all pools showed higher radioactivities in suspensions of cells of the proliferating phase in comparison with those from resting phase. The specific radioactivities of ATP, ADP and AMP pools were 2.1, 1.7 and 1.1fold higher in the cells of the exponential phase. Taking into account the increased pool sizes in cells of the proliferating phase one can understand that the differences of flux rates between cells of the proliferating and of the resting phases are much higher than expressed by specific radioactivities. The increased flux rates in proliferating cells are in correspondance with their increased ATP turnover (Schmidt et al., 1991) and with data on changed enzyme activities at different growth phases of tumour cells (Weber, 1983). From the different specific radioactivities of hypoxanthine and uric acid one may conclude a considerable influx rate into the xanthine pool which is realized by the guanase reaction. All data of tracerkinetic experiments are used for the mathematical modelling of purine metabolism of EMAC of different growth phases.

Table 3. Incorporation of radioactive labeled adenine into the purine compounds of EMAC from different phases of tumour growth after 30 min of incubation.

	Specific radioactivities		Share of total radioactivity	
	Exponential phase (dpm/nmol)	Resting phase (dpm/nmol)	Exponential phase (%)	Resting phase (%)
ATP	1545	725	17.1	2.9
ADP	915	545	3.4	0.9
AMP	2365	2410	3.7	1.8
GTP	995	n.d.	7.5	n.d.
IMP+GMP	2325	4030	1.7	0.9
Ado	45509	24044	2.5	1.2
Ade	485985	842145	57.9	90.6
Hyp	18949	6900	3.6	0.7
Uric acid	6289	2522	2.6	1.0

The investigations seem to be of theoretical as well as ofmedical importance because of the low knowledge on the regulation of transition of tumour cells from proliferating into resting phase or from resting into the proliferating phase of growth.

REFERENCES

Benndorf, R., Nuernberg, P., and Bielka, H., 1988, Growth phase-dependend proteins of the Ehrlich ascites tumour analysed by one- and two-dimensional electrophoresis, Exp. Cell. Res., 174:130.

Boehmer, F. D., Lehmann, W., Schmidt, H. E., Langen, P., and Grosse, R., 1984, Purification of a growth inhibitor for Ehrlich ascites mammary carcinoma cells from bovine mammary gland, Exp. Cell Res., 150:466.

Grune, T., Siems, W., Gerber, G., Tikhonov, Y. V., Pimenov, A. M., and Toguzov, R. T., 1991a, Changes of nucleotide patterns in liver, muscle and blood during the growth of Ehrlich mouse ascites cells, J. Chromatogr. Biomed. Appl., 563:53.

Grune, T., Siems, W., Gerber, G., and Uhlig, R., 1991b, Determination of the ultraviolet absorbance and radioactivity of purine compounds separated by high-performance liquid chromatography, J. Chromatogr., in press.

Schmidt, H., Siems, W., Mueller, M., Dumdey, R., and Rapoport,S. M., ATP-producing and consuming processes of Ehrlich mouse ascites tumour cells in proliferating and resting phases, Exp. Cell Res., 194:122.

Siems, W., Schmidt, H., Werner, A., Uerlings, I., David, H., and Gerber, G., 1989, Changes in the nucleotide metabolism of the Ehrlich ascites tumour cells during their growth in vivo, Cell. Molec. Biol., 35:255.

Weber, G., 1983, Biochemical strategy of cancer cells and the design of chemotherapy, Cancer Res., 43:3466.

PURINE NUCLEOTIDES, NUCLEOSIDES AND NUCLEOBASES OF LIVER, SKELETAL MUSCLE, BLOOD AND TUMOR CELLS DURING THE GROWTH OF EHRLICH ASCITES TUMOR IN MICE

W. G. Siems, T. Grune, H. Schmidt, R. Uhlig, G. Gerber, Y. V. Tikhonov, A. M. Pimenov and R. T. Toguzov

Institute of Biochemistry, Medical Faculty, Humboldt University, O-1040 Berlin, F. R. G. and Department of Biochemistry, 2nd Medical Institute, 117437 Moscow, USSR

INTRODUCTION

Previous results showed that the transition of Ehrlich ascites tumor from the proliferating into the resting phase of growth is closely connected with a tremendeous loss of cellular purine nucleotides (Siems et al., 1989). That correlates with a decline of the incorporation rates of adenine, hypoxanthine and adenosine and a decreased ATP turnover (Schmidt et al., 1991). The reasons for these changes are still unknown. The question which should to be answered was: are the drastic changes of purine pattern in Ehrlich ascites cells accompanied by deviations of nucleotide concentrations in the so-called normal tissues of the host? We selected liver and muscle as organs with the highest production of nucleotide precursors for the organism (van Waeg, 1987), erythrocytes and ascites fluid as transport media between precursor supplying organs and purine consuming tumor for studies on host-tumor interrelationships.

MATERIALS AND METHODS

Sample Preparation and Extraction Procedure

Female ICR mice (15-20 g body weight) were used. The transplantation of the ascites cells, their harvesting, incubation and preparation for analytical procedures were described before (Müller et al., 1986; Siems et al., 1989). On the 5th day (proliferation phase) and 12th day after inoculation of the cells (resting phase) liver, femoral muscle, blood and ascites fluid were taken from anesthetized animals. Liver and muscle were rapidly frozen in fluid nitrogen. Plasma and cell-free ascites fluid were rapidly separated from the cells and extracted. Red blood cells were resuspended in saline solution before extraction procedure. The samples were deproteinized with ice-cold 6% perchloric

Table 1. Concentrations of ATP, GTP, sum of adenine nucleotides (AdN), sum of guanine nucleotides (GN), ATP:ADP and ATP:AMP ratios in control animals and on the 5th and 12th days of Ehrlich ascites tumor growth. Values are given as nmoles/g w.w. and in parentheses as % of the control value, mean ± S.D.; n = 5.

		Liver	Sceletal muscle	Erythrocytes
ATP	control	1512±172 (100)	2401±224 (100)	1702±443 (100)
	5th day	1904±122 (126)	2839±367 (118)	2017±163 (118)
	12th day	1546±185 (102)	1508±308 (63)	2020±110 (119)
GTP	control	497±154 (100)	143± 9 (100)	159± 33 (100)
	5th day	399± 35 (80)	187± 16 (131)	170± 26 (107)
	12th day	312± 35 (63)	140± 33 (98)	485± 85 (305)
AdN	control	2311±204 (100)	2837±266 (100)	1913±485 (100)
	5th day	2874±267 (124)	3360±431 (118)	2293±235 (120)
	12th day	2334±295 (101)	2143±471 (76)	2775±297 (145)
GN	control	633±192 (100)	206± 25 (100)	190± 39 (100)
	5th day	519± 53 (82)	243± 35 (118)	219± 37 (115)
	12th day	418± 51 (66)	213± 45 (103)	570± 99 (300)
ATP:ADP				
	control	2.85± 0.32	6.40± 0.60	9.78± 2.55
	5th day	2.92± 0.19	6.08± 0.79	10.19± 0.82
	12th day	2.94± 0.35	2.91± 0.59	3.94± 0.21
ATP:AMP				
	control	5.62± 0.64	39.36± 3.67	46.00±11.97
	5th day	5.97± 0.38	52.57± 6.80	36.72± 5.11
	12th day	5.90± 0.71	13.00± 2.66	8.35± 0.45

acid, centrifuged for 10 min at 1200 g and neutralized. After filtration 10 or 20 µl were analysed by HPLC.

HPLC Instrument and Determination of Purine Compounds

An equipment of Waters Assoc. Milford, USA was used. The determination of nucleotides was performed with an isocratic ion-pair reversed-phase HPLC method; for the determination of nucleosides and nucleobases different reversed-phase HPLC separations were used (Grune et al., 1991).

RESULTS AND DISCUSSION

There are reports on activities of enzymes of nucleotide metabolism and on nucleotide levels of tumor cells at different growth periods (Weber et al., 1981). The influence of interrelationships between tumor and normal tissues on changes of nucleotide metabolism during the tumor growth is poorly understood. It is shown in Table 1 that ATP levels of host organs increased during the logarithmic phase (5th day after inoculation) in comparison with control values.

Table 2. Concentrations of nucleosides and nucleobases in liver, sceletal muscle, erythrocytes, cell-free ascites fluid and blood plasma of Ehrlich ascites tumor-bearing mice at 5th and 12th days after inoculation of tumor cells in comparison with control values. Values are given as nmoles/g w.w., mean ± S.D. 5 independent experiments.

	Liver	Sceletal muscle	Erythro-cytes	Ascites fluid	Plasma
Adenosine					
control	1.9±0.1	2.6±0.7	1.0±0.1		1.2±0.3
5th day	1.7±0.3	2.5±0.7	0.9±0.2	0.01±0.00	1.3±0.2
12th day	5.1±1.1	5.3±1.7	6.2±1.9	0.03±0.00	2.9±0.6
Adenine					
control	0.9±0.6	1.3±0.1	1.6±0.1		3.2±0.9
5th day	1.5±0.3	1.6±0.7	0.8±0.1	1.2±0.3	2.7±0.2
12th day	0.7±0.1	3.3±1.8	2.6±0.3	1.3±0.2	2.8±0.5
Inosine					
control	4.7±0.6	2.4±1.0	2.6±0.4		2.1±0.4
5th day	7.2±1.1	2.5±1.1	0.7±0.2	1.2±0.1	1.7±0.4
12th day	5.4±1.0	4.0±1.8	2.6±1.4	1.3±0.1	3.3±0.6
Guanosine					
control	6.9±1.1	4.5±1.1	1.7±0.4		0.8±0.2
5th day	5.6±0.9	5.2±2.2	2.0±0.2	0.8±0.1	0.7±0.1
12th day	3.6±0.6	2.0±0.6	2.3±0.4	0.5±0.1	0.9±0.2

That is true for the three tissues which were investigated - liver, sceletal muscle and erythrocytes. During the resting phase of tumor growth the ATP concentration of liver and muscle decreased again. Table 1 demonstrates the drastic decrease of muscle ATP concentration during this growth period of the Ehrlich ascites tumor leading to a value of about 60% of the control value. The values "sum of adenine nucleotides" reflect in an analogous manner the "activation" period of host organs of the tumor-bearing animals during the proliferation period of tumor growth. This activation period measured at the 5th day after inoculation of the tumor includes both the increase of the ATP level and that of the ADP and AMP level. Therefore, the ATP:ADP and the ATP:AMP ratios are not increased at the 5th day in comparison with the control values. But there was a drastic decrease of these ratios in muscle and erythrocytes at the 12th day, i.e. during the resting phase. Table 2 documents the changes of nucleoside or nucleobase levels in host tissues and in plasma or cell-free ascites. In all tissues and body fluids the adenosine level was decreased at the 5th day in comparison with the control value. At 12th day the adenosine concentrations in the tissues, in plasma and ascites fluid were markedly increased. Table 3 documents the differences of nucleotide concentrations of Ehrlich mouse ascites cells at the 5th and the 12th days with the tremendeous decrease of those concentrations at the resting period of the tumor growth in comparison with the proliferation period.

Table 3. Nucleotide concentrations and ratios between nucleotide concentrations of Ehrlich mouse ascites tumor cells at the 5th day after inoculation of the tumor (proliferating phase of tumor growth) and at the 12th day (resting phase). Values for nucleotide concentrations in nmoles/g w.w., mean ± S.D.; n=14.

	Proliferating phase	Resting phase
ATP	3047 ± 176	1136 ± 151
ADP	629 ± 60	310 ± 37
AMP	227 ± 22	150 ± 25
GTP	799 ± 75	291 ± 47
GDP	244 ± 12	70 ± 9
UTP	180 ± 29	20 ± 2
UDP	70 ± 9	12 ± 2
ATP : ADP	4.84 ± 0.28	3.66 ± 0.49
ATP : AMP	13.42 ± 0.78	7.57 ± 1.01
GTP : GDP	3.27 ± 0.31	4.16 ± 0.23
UTP : UDP	2.57 ± 0.41	1.67 ± 0.17

It is suggested that new insights into the variations of nucleotide metabolism of tumor cells in different growth phases may open new diagnostic and therapeutic approaches because of the medical importance of the transition of tumor cells from one growth period to the another.

REFERENCES

Grune, T., Siems, W., Gerber, G., Tikhonov, Y. V., Pimenov, A. M., and Toguzov, R. T., 1991, Changes of nucleotide patterns in liver, muscle and blood during the growth of Ehrlich ascites cells - Application of reversed-phase and ion-pair reversed-phase HPLC procedures with radial compression column, J. Chromatogr. Biomed. Appl., 563: 53.

Müller, M., Siems, W., Buttgereit, F., Dumdey, R., and Rapoport, S. M., 1986, Quantification of ATP-producing and consuming processes of Ehrlich ascites tumour cells, Eur. J. Biochem., 161: 701.

Schmidt, H., Siems, W., Müller, M., Dumdey, R., and Rapoport, S. M., 1991, ATP-producing and consuming processes of Ehrlich mouse ascites tumor cells in proliferating and resting phases, Exp. Cell Res., 194: 122.

Siems, W., Schmidt, H., Werner, A., Uerlings, I., David, H., and Gerber, G., 1989, Changes in the nucleotide metabolism of Ehrlich ascites tumour cells during their growth in vivo, Cell. Molecul. Biol., 35: 255.

van Waeg, G., 1987, "The use of allopurinol and the biodynamic modelling in the quantitative assessment of purine metabolism in man", Acta universitatis upsalaensis, Upsala.

Weber, G., Olah, E., Denton, J. E., Lui, M. S., Takeda, E., Tzeng, D. T., and Ban, J., 1981, Dynamics of modulation of biochemical programs in cancer cells, Adv. Enzyme Regul., 19: 87.

URIDINE FLUXES IN HEALTHY PROLIFERATING T-LYMPHOCYTES, MOLT-3 T-ALL CELL-LINE CELLS AND DIFFERENTIATED MOLT-3 CELLS

A. André van den Berg[@], Henk van Lenthe[@], Dirk de Korte[#], Dirk Roos[#] and Albert H. van Gennip[@]

@, Depts. of Ped. and Clin. Chem. o.t. Univ. of Amsterdam, Meibergdreef 9, 1105 AZ ; # Centr. Lab. o.t. Neth. Red Cross Bloodtr. Serv., Plesmanlaan 125, 1066 CX, Amsterdam, T. Neth.

INTRODUCTION

Knowledge of transformation associated differences in pyrimidine nucleotide metabolism of human leukemic cells, compared to healthy counterparts can be a rational basis for development of new antileukemic drugs. Peripheral blood cells of patients with acute lymphoblastic leukemia (ALL) contain increased amounts of guanine-, uracil- and cytosine- nucleotides and uridinediphosphate bound sugars with a changed composition[1]. Moreover, the same deviations have also been found in the human ALL cell-line of T-lymphoblastic origin MOLT-3[2]. Comparison of pyrimidine nucleotide synthesis from labeled precursors by MOLT-3 cells and growth-stimulated T-lymphocytes (TL's), showed that CTP-synthetase 'overactivity' could cause the increase in cytosine nucleotides found in the malignant cell-line cells[2]. Uridine fluxes by pulse-chase experiments in living cells with emphasis on CTP-synthetase activity and differences herein between MOLT-3 cells, TL's and differentiated resting MOLT-3 cells, were performed to get a better insight in the fluxes and the role of CTP-synthetase.

MATERIALS AND METHODS

Chemicals Nucleotides used as chromatographic standards were obtained from Sigma Chemical Corp. (St. Louis, MO, U.S.A.). All radiochemicals were obtained from Amersham International (Amersham, U.K.). All other chemicals were of analytical grade and obtained from Merck (Darmstadt, Germany). Fetal calf serum and RPMI-1640 medium with 20 mM HEPES were obtained from Flow Laboratories (Ayrshire, U.K.). Anti TdT antibodies were obtained from Dakopatt (Denmark).

Cells T-lymphoblastic cell line MOLT-3 was maintained in humidified air with 5% CO_2 at 37° C in RPMI-1640 medium supplemented with 10% complement heat-inactivated fetal calf serum. Cells were kept in exponential growth by dilution to $3*10^5$ cells/ml every 2-3 days. MOLT-3 cells were induced to differentiate with 16.2 nM PMA five days in the same medium. Changes in reactivity with MoAb's directed against cell surface structures (CD markers), characteristic for cells of T-lineage and the cytoplasmatic enzyme TdT were measured as described originally[3]. TL's were isolated as described before[2] and suspended at 10^6/ml in anti-CD3 coated flasks leading to autologous IL-2 production and exponential growth for five days[4]. Cell viabilities were measured with vital stains as described before[5] and were over 90% throughout the experiments evaluated in this report.

Pulse-chase studies MOLT-3 cells and TL's in exponentional growth and differentiated MOLT-3 cells were labeled at a density of $1-1.5*10^6$ cells/ml with 2μM ^{14}C-uridine (^{14}C-Urd). Each experiment was started with 10 ml cell suspension. To stop the pulse, cells were spun down after one hour, an extract of 2 ml was made and the remaining cells resuspended in fresh medium. At 2, 4, and 8 hours further extracts were made.

Extraction and analysis of nucleotides Nucleotides were extracted with perchloric acid and analysed by the HPLC method with on-line radioactiviyt measurement as described elsewhere[5]. Pellets were analysed for ^{14}-Ccontents of nucleic acids on a betacounter device.

RESULTS AND DISCUSSION

Growth, growth arrest and differentiation MOLT-3 cells ceased to grow 24 hours after the addition of 16.2 nM PMA. After four to five days TdT-reactivity was restricted to 5 % of the cells. Changes registered in CD marker reactivity were essentially similar to those described by others[3], indicating development to a maturer T-lymphocyte phenotype.

Nucleotide contents, uridine salvage and NTD fluxes MOLT-3 cells, even differentiated, contained more NTD than resting periferal blood lymphocytes from healthy donors, presumably because the former maintain a high cytoplasm to nucleus ratio. Moreover, the imbalances found in MOLT-3 cells compared to peripheral blood lymphocytes, were found too, in differentiated MOLT-3 cells (not shown). MOLT-3 cells incorporated more ^{14}C-Urdinto nucleic acids than TL's and the latter more than differentiated MOLT-3 cells (table 1.)

Table 1. Nucleic acids synthesized from 2 μM ^{14}-C-uridine.

picomoles per million cells in time				
time(hours) -->	1	2	4	8
MOLT-3	292	296	436	475
T-lymphocytes	68	129	145	199
MOLT-3 PMA	40	46	51	65

Incorporation of ^{14}C-Urd into pyrimidine NTD was also greater in MOLT-3 cells and TL's as in differentiated MOLT-3 cells. The total amount of label incorporated into NTD was somewhat larger in MOLT-3 cells, probably because of the greater percentage of cells in S fase (not shown). The fractions of extractable UTP and CTP pools newly synthesized, calculated from the specific activities of the labels are shown in figure 1. As can be seen the flux of label from UTP to CTP, is higher in MOLT-3. At 3 hours until 7 hours after termination of the pulse, UTP and CTP specific activities are near equal in MOLT-3 (figure 1a), whereas these are not in TL's (figure 1b). Rises and declinations in UTP and CTP ^{14}C contents of differentiated MOLT-3 cells retain the profile of proliferating cells (compare figures 1a and 1c) at a lower level.

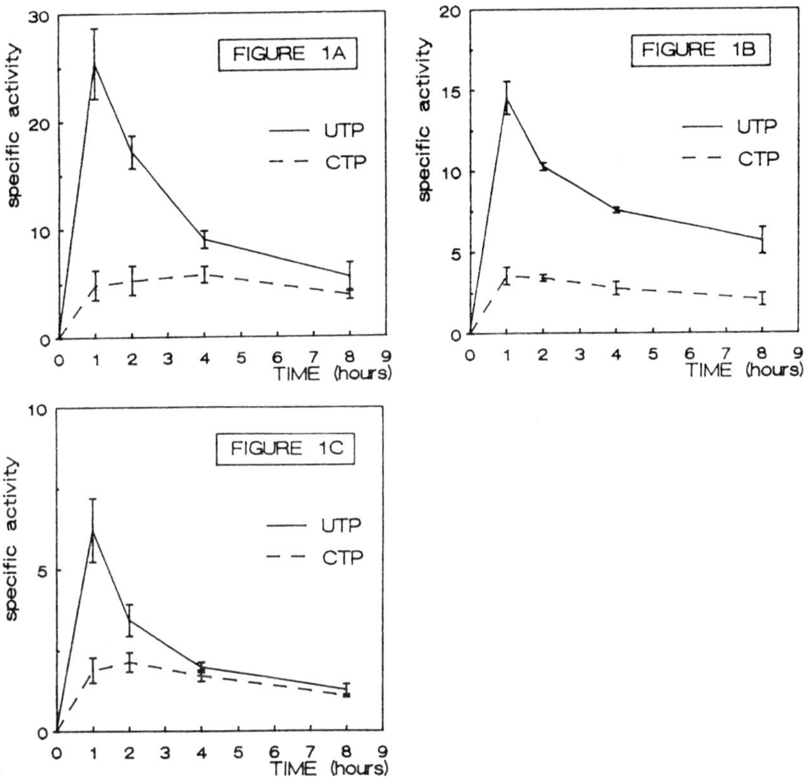

Figure 1. Changes in specific activities of UTP and CTP pools after labeling with 2 μM ^{14}C-uridine. 1A : MOLT-3 cells; 1B : T-lymphocytes; 1C : differentiated MOLT-3 cells. N.B.: mark differences in ordinates!

From these results it can be concluded that CTP-synthetase is 'overactive' and that this 'overactivity' is not related to growth nor polymer synthesis in MOLT-3 cells. The reason for this 'overactivity', however, need not be the kinetic properties of the CTP-synthetase itself. It might as well result from a less efficient synthesis of CTP in MOLT-3 cells from cytidine. Uridine and cytidine are phosphorylated by the same kinase and relative affinities might play a crucial role. To investigate this possibility we will measure CTP synthesis and fluxes from ^{14}C-cytidine, as well as the influence of cytidine on uridine fluxes. We regard similarities in UTP and CTP synthesis in fastly growing and growth-arrested MOLT-3 cells described above and the differences herein with proliferating healthy T-lymphocytes to be strong indications but no evidence of a transformation associated CTP-synthetase overactivity in leukemic cells. CTP-synthetase is shown to likely be an attractive site for anti-leukemic therapy for malignancies of T-lymphoid origin, irrespective of cell growth capacities. In this our present study confirms the expectation from earlier studies[1,2].

SUMMARY

Incorporation of ^{14}C-uridine into UTP and CTP and fluxes of label through these nucleotide pools to RNA and DNA were greater in MOLT-3 cells compared to T-lymphocytes. In growth-arrested, differentiated MOLT-3 cells overall incorporation of radiolabel into nucleotides and nucleic acids was lowered compared to exponentially growing cells. Turnover of UTP and CTP however, retained the profile of exponentionally growing MOLT-3 cells, implicating the characteristically higher conversion of UTP to CTP is independent of the MOLT-3 cells proliferative capacities. We conclude that drugs interfering with CTP-synthetase activity are good candidates to be used as selective substances in the battle against T-ALL.

REFERENCES

1. De Korte, D., Haverkort, W.A., Roos, D., Behrendt, H., Van Gennip, A.H. Imbalance in the ribonucleotide pools of lymphoid cells from acute lymphoblastic leukemia patients. Leukemia Research 10 pp. 389-396 (1986).

2. Marijnen, Y.M.T., De Korte, D., Haverkort, W.A., Den Breejen, E.J.S., Van Gennip, A.H. and Roos, D. Studies on the fluxes through 'de novo' and 'salvage' pathways of purine and pyrimidine metabolism in normal lymphocytes and lymphoblastic cell-line cells. Biochimica et Biophysica Acta 1012 pp 148-155 (1989).

3. Nagasawa, K. and Mak., T.W., Induction of differentiation in human T-lymphoblastic leukemia cell lines by 12-O-tetradecanoyl phorbol 13-acetate (TPA): studies with monoclonal antibodies to T-cells, Cellular Immmunology 71 pp 396-400 (1982).

4. Van Lier, R.A.W., Boot, J.H.A., De Groot, E.R. and Aarden, L.A. European Journal of Immunology 17 pp 1599-1604 (1987).

5. De Korte, D., Marijnen, Y.M.T., Haverkort, W.A., Van Gennip, A.H. and Roos, D. Sensitive on-line radioactivity measurement with a heterogeneous flow cell : application to HPLC-separated ribonucleotides in lymphoid cells. Journal of Chromatography 415 pp 383-387 (1987).

SOME ASPECTS OF PURINE NUCLEOTIDE METABOLISM IN HUMAN LYMPHOCYTES: NUCLEOTIDE CONTENT IN HUMAN LYMPHOBLASTOID LINES TRANSFECTED WITH HIV-1

Antonella Tabucchi, Filippo Carlucci, Roberto Pagani, Franca Ciccomascolo*, Lucia Lopalco* and Antonio Siccardi*

Institute of Biochemistry and Enzymology - University of Siena; *Department of Biology and Genetics for Medical Sciences, University of Milano

INTRODUCTION

It is well known that virus replicate in the cells by different mechanisms: a) they may possess "messenger RNA", with positive polarity which promotes the synthesis of functional proteins, such as replicase (picornavirus); b) "antimessenger RNA", with negative polarity, is transcribed by a viral transcriptase, into complementary messenger RNA, with positive polarity (paromixovirus); c) single-stranded RNA, as in b), consists of different fragments, which are separately transcribed by a viral transcriptase (orthomixovirus); d) in rheovirus, several fragments of double-stranded RNA are separately transcribed by a viral transcriptase; e) retrovirus is equipped with two identical strands of single-stranded RNA, independently translated by a reverse trancriptase: functional messenger RNA is transcribed from this DNA (1,2).

Nevertheless, all mechanisms of viral replication are obviously based on the synthesis of new RNA molecules. As a consequence, the penetration of a virus into a cell and its replication should involve variations in the purine nucleotide metabolism of the host cell. These changes could affect cell content of purine nucleotides and the incorporation of labelled precursors, such as ^{14}C-glycine or ^{14}C-formate into purines, as well as the kinetics of purine metabolism and all the enzymes related to the different pathways, namely de novo synthesis, the salvage pathway and catabolism.

Here we report some data on the intracellular content of purine nucleotides in human lymphoblastoid lines infected with HIV-1.

MATERIALS AND METHODS

Lymphocytes were prepared from heparinized peripheral blood of healthy donors as previously reported (3).

Preparation of cultures of A3Ø1 and 8E51 cells

A3.Ø1 cells, a Hat-sensitive derivative of the CEM line, were cultured in RPMI 164Ø medium containing 1Ø% fetal calf serum until needed and were routinely checked for surface Leu-3 positivity prior to infection.

Abbreviations used: RT, reverse transcriptase; IUdR, 5-iodo-2'-deoxyuridine; RV, retrovirus; PBL, peripheral blood lymphocytes.

Purine and Pyrimidine Metabolism in Man VII, Part A
Edited by R.A. Harkness *et al*., Plenum Press, New York, 1991

2x10^6 A3.01 cells were infected with a 10^3 dilution of lymphadenopathy- associated virus (LAV) stocks as described by Folks et al (4). After absorption for 1 h at 37°C the cells were washed and kept (10^8 cells/ml) in RPMI medium supplemented with 10% FCS. When the RT activity of Leu-3$^-$ cultures surviving infection became undetectable (7-10 days after peak of RT), the cells (10^8/ml) were exposed to IUdR (100 µg/ml) for 24 h and then cultivated with Leu-3$^+$ A3.01 cells, as reported by Folks et al. (4).

The induction of the virus was heralded by a second wave of RT activity in the cocultured cells. When RT was no longer detectable, the surviving cells were subjected to a second cycle of IUdR induction and cocultivation. The RT Leu-3$^-$ cells surviving these treatments were cloned by limiting dilution in 96-well microliter plates, 111 single cell clones were obtained, pooled into groups of 10, expanded to a total of 10^6 cells exposed to IUdR for 24 h, cocultivated with Leu-3$^+$ A3.01 cells, and examined daily for expression of the AIDS RV by monitoring syncytia formation. A single clone (8E5) was obtained. Both A3.01 and 8E51 cell cultures underwent a specific treatment for mycoplasma removal (Mycoplasma Removal Agent from Flow Company).

Determination of purine nucleotides

The extracts from normal PBL and lymphoblastoid cells were prepared as previously reported (3). The purine nucleotide determination was performed substantially according to De Korte (5).

The most important ratios between the different nucleotides were calculated:

a) ATP/ADP ratio
b) GTP/GDP ratio
c) energy charge for adenylate
d) energy charge for guanylate
e) total adenine nucleotides/total guanine nucleotides.

These parameters are all indices of cell energy and vaiability (6,7).

RESULTS AND DISCUSSION

A typical chromatogram of standard nucleotides is reported in Figure 1, which shows that good separation of all the most important nucleotides (NAD, AMP, XMP, IMP, GMP, ADP, GDP, ATP, GTP) can be achieved by the procedure adopted.

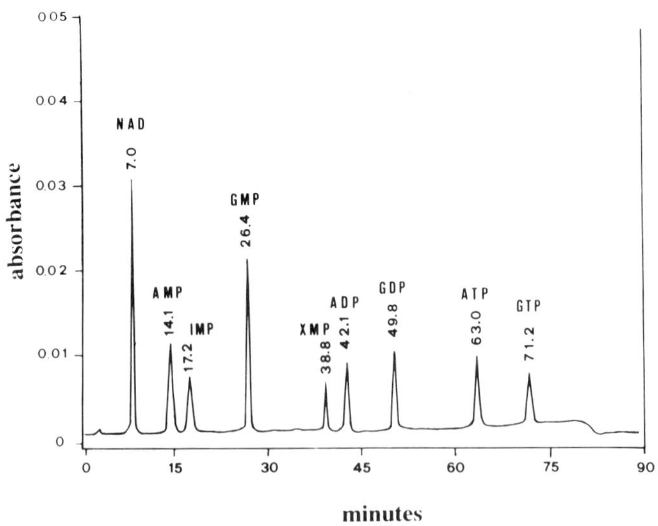

Fig. 1. Typical chromatogram of standard nucleotides.

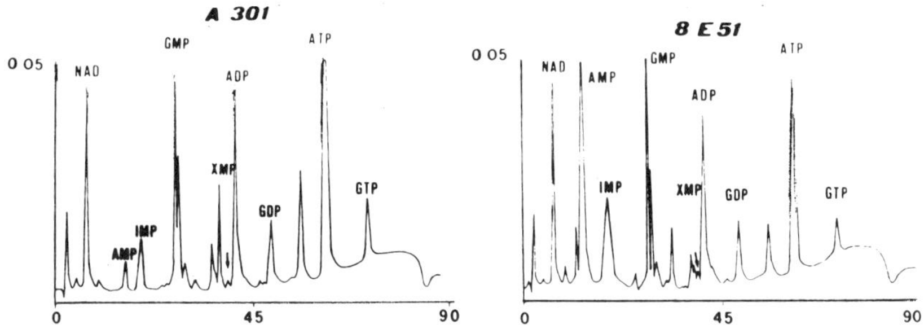

Fig. 2. Typical chromatogram from A3.01 and 8E51 cells.

Figure 2 shows a typical chromathogram from A3.01 cells and 8E51 cells.

Table 1 shows the values of purine nucleotides found in PBL, A3.01 and 8E51 cells.

From our results we can draw the following conclusions:

1) The determination of purine nucleotides can be comparatively carried out in normal PBL and normal or HIV-1 infected lymphoblastoid cell lines;
2) the values found in all types of cells are of the same order of magnitude, even if it seems that monophosphate content is higher in the cell lines,in comparison to PBL;
3) the cells infected with HIV-1 (RT^- virus) show a higher content of AMP, IMP, GMP and a lower content of ATP and GTP;
4) all the ratios were lower in infected cells, specifically ATP/ADP, GTP/GDP ratio and the energy charge, which seems to indicate that the infection causes a decrease in the energy and viability of the cells;
5) although these results need to be extended to a larger population, they give a preliminary indication that these cells can be a useful model for studying HIV-1 infection of cells and that infection induces alterations in purine nucleotide content.

Table 1. Comparation of the values in purine nucleotides found in PBL, A3.01 cells and 8E51 cells.

	NAD	AMP	IMP	GMP	XMP	ADP	GDP	ATP	GTP
PBL (17)	179 ±66	48 ±14	49 ±13	12 ±2	15 ±8	441 ±89	61 ±12	948 ±120	177 ±18
A3.01 (3)	261 ±17	85 ±37	176 ±15	388 ±15	17 ±2	454 ±27	129 ± 7	867 ±112	209 ±15
8E51 (3)	260 ± 8	562* ±57	342* ±61	438 ±16	16 ±4	447 ±64	140 ± 9	417* ±61	104* ±13

Values given as mean ± S.E. in pmoles/10^6 cells.
Number of cases in brackets.
* $p < 0.05$ with respect to A3.01 line.

Acknowledgements

This work was financied by a contribution of the Ministero della Sanità, Istituto Superiore di Sanità, AIDS Project 1990, Rome - Italy.

REFERENCES

1. H.L. Bacharach, Comparative strategies of animal virus replication, Ad. Vir. Res., 22: 163 (1978)
2. B.D. Davis, R. Dulbecco, H.N. Ersen and H.S. Giusbery, In: "Trattato di Microbiologia" II ed., Piccin Ed., Padova, 1136 (1986)
3. R. Pagani, A. Tabucchi, F. Carlucci, R. Leoncini, E. Consolmagno, M. Molinelli and P. Valerio, Some aspects of purine metabolism in human lymphocytes before and after infection with HIV-1 virus: nucleotide content, see this volume communication n71
4. T.M. Folks, S. Benn, A. Robson, T. Theodore, M.D. Hoggan, M. Martin, M.Lightfoot and K.W. Sell, Characterization of a continuous T cell line susceptible to the cytopatic effects of the acquired immunodeficiency syndrome AIDS-associated retrovirus, Proc. Natl. Acad. Sci. , 82: 4539 (1985)
5. D. De Korte, W.A. Haverkort, A.H. van Gennip and D. Roos, Nucleotide profile of normal human blood cells determination by high performance liquid chromatography, Anal. Biochem., 147: 197 (1985)
6. Y.M.T. Marijnen, D.De Korte, W.A. Haverkort, E.J.S. den Breejen, A.H. van Gennip and D. Roos, Studies on the incorporation of precursors into purine and pyrimidine nucleotides via "de novo" and salvage pathways in normal lymphocytes and lymphoblastic cell lines. Bioch. Bioph. Acta, 1012: 148 (1989)
7. G.J. Peters, R.A. De Abreu, A. Oosterhof and J.H. Veerkampf, Concentration of nucleotides and deoxynucleotides in peripheral and phytohemagglutinin-stimulated mammalian lymphocytes, Bioch. Bioph. Acta, 759: 7, (1983)

METABOLISM AND ACTION OF 2',2'-DIFLUORODEOXYCYTIDINE:

SELF-POTENTIATION OF CYTOTOXICITY

Varsha Gandhi, Peng Huang, Yi-Zheng Xu,
Volker Heinemann, and William Plunkett

Department of Medical Oncology, The University of
Texas M. D. Anderson Cancer Center, Houston, Texas
77030, U.S.A.

INTRODUCTION

The clinical success of arabinosylcytosine evoked an interest in other deoxycytidine analogues. One such congener is 2',2'-difluorodeoxycytidine (dFdC, Gemcitabine), which has to pass through a repertoire of steps catalyzed by several cellular enzymes to exert cytotoxicity (1). Once transported into the cell, dFdC is phosphorylated to its active nucleotide metabolites. Several kinases are involved in the intracellular synthesis of dFdC nucleotides, but the first step, the phosphorylation of dFdC to its monophosphate by dCyd kinase, is rate-limiting (1,2). dCyd kinase is therefore the most important enzyme in the activation of dFdC. The activity of this enzyme is tightly regulated by endogenous deoxynucleotide pools, in particular deoxycytidine triphosphate (dCTP) (3). dCTP is of further interest and importance because dFdCTP competes with it for incorporation into the growing DNA strand, a mechanism of dFdC cytotoxicity (4). Cellular dCTP pools are maintained by the de novo pathway, the principal step of which is catalyzed by ribonucleotide reductase. Another enzyme influenced by the cellular dCTP pool is dCMP deaminase, which plays a major role in the elimination of phosphorylated dFdC metabolites (5). The aim of the present investigation was to analyze the role of dFdC metabolites on these key enzymes. A positive influence of dFdC on these enzymes would augment its own cytotoxicity, an action termed self-potentiation (6).

MATERIALS AND METHODS

Chemicals and Cell Lines. dFdC and [5-^{3}H]dFdC were supplied by Lilly Research Laboratories (Indianapolis, IN). The dFdC triphosphate (dFdCTP) was synthesized as described (7). All other chemicals were reagent grade. The CEM and K562 cell lines were obtained from the American Type Culture Collection (Rockville, MD). Maintenance of the cultures, mycoplasma analysis, and doubling time of these cell lines have been described previously (2,8).

Effect of dFdC on Cell Clonogenicity. The reproductive viability of the cells was determined by cloning. The cells were incubated alone or with 10 μM dFdC for 3 hr, washed, and diluted in drug-free medium. After mixing with Iscove's medium (GIBCO, Grand Island, NY) supplemented with 30% fetal calf serum and 5% glutamine, 1000 to 4000 cells were plated in 35 x 10 mm tissue culture plates with 0.3% agar. The plates were incubated for 10 days at 37°C in a humidified incubator with 5% CO_2, and colonies of more than 50 cells were counted under a microscope. The cytotoxicity of dFdC was expressed as a percentage of survival relative to the untreated control.

Nucleotide Extraction and Analysis. After incubation with dFdC or ara-C, the cells were washed twice with ice-cold phosphate-buffered saline (8.1 g NaCl, 0.22 g KCl, 1.1 g Na_2HPO_4, and 0.27 g KH_2PO_4 per liter H_2O, pH 7.4), and centrifuged. The cell pellet was then extracted with $HClO_4$ (1). The nucleoside triphosphates in the neutralized acid-soluble extract were analyzed by HPLC (Waters Associates, Milford, MA) as described previously (1). The identity of dFdCTP and ara-CTP were confirmed by their coelution with authentic triphosphates, their resistance to periodate oxidation, and their UV absorbance. The nucleotides in $HClO_4$ extracts were quantitated by electronic integration and reference to preprogrammed response factors (1). Metabolites (dFdC and dFdU) in the medium were analyzed by reverse-phase μBondapak C_{18} column (Waters Associates, Milford, MA).

Determinations of dNTP. Neutralized $HClO_4$ extracts from treated or untreated cells were evaporated to dryness in an Evapomix volume-reduction apparatus (Buchler Instruments, Fort Lee, NJ). The ribonucleotides in the extracts were degraded with $NaIO_4$, and the dNTPs were separated by anion-exchange column chromatography on a Waters HPLC system as described (9).

Purification of dCMP deaminase. The enzyme was partially purified from CEM cell extracts by HPLC through a molecular sieve column (protein pak glass 300 SW; Nihon Waters Ltd., Japan). dCMP deaminase activity was measured by the HPLC method reported by Fridland and Verhoef (10). The assay mixture contained 0.1 mM dCMP or dFdCMP as a substrate.

DNA Primer Extension Assay. The 17-base M13 sequencing primer 5'GTAAAACGACGGCCAGT3' was labeled with ^{32}P at its 5'-end and annealed to its complementary site on an M13mp19(+) DNA template. The ability of purified human DNA polymerase alpha to extend the labelled primer/template was evaluated in the presence or absence of dFdCTP. The DNA primer extension assay and sequence analysis were carried out as described (11). The radioactivity in each DNA band in the sequencing gel was quantitated with a Betascope 603 blot analyzer (Betagen Corporation, Waltham, MA).

RESULTS AND DISCUSSION

More than 90% of the CEM cells and 80% of the K562 cells were killed by a 2- or 3-hr incubation with 10 μM dFdC. The intracellular concentrations of dFdCTP after these incubations were 450 and 230 μM in CEM and K562 cells, respectively. Although the cytotoxic action of dFdC is thought to be exerted on S phase cells, a greater proportion of the population lost clonogenic capacity than would be in S phase during this short incubation. Thus, the active nucleotides of dFdC must be retained long enough to exert this effect or the effect of dFdC on normal metabolites or the processes is sustained until most of the cells progress in S phase. Experiments with CEM cells treated with dFdC demonstrated a linear relationship between cytotoxicity and the amount of dFdCMP incorporation into DNA ($r=0.954$, $p<0.0001$). Thus, it appears that the cytotoxicity of dFdC arises from its incorporation into DNA (4).

Because dCTP is a DNA precursor and competes with dFdCTP for incorporation into DNA, inhibition of ribonucleotide reductase should

Table 1. Effect of dFdC and ara-C on Deoxynucleotide Pools.

Cell lines	Drug	dNTP, μM* dCTP	dTTP	dATP	dGTP
CEM	untreated	28.0	64.0	76.0	51.0
CEM	dFdC	3.1	33.9	45.6	30.1
CEM	ara-C	56.0	106.9	133.8	45.4
K562	untreated	3.5	9.9	8.6	3.3
K562	dFdC	1.0	10.6	3.2	1.4
K562	ara-C	5.0	15.3	11.3	3.8

* The results are the average of three and four separate experiments for CEM and K562 cells, respectively. S.D. was less than 15%.

increase dFdCTP incorporation, resulting in greater cytotoxicity. The effect of dFdC on ribonucleotide reductase was determined by measuring the deoxynucleotide pool imbalances in CEM and K562 cells after a 2- or 3-hr treatment with 10 μM dFdC. After dFdC treatment, there was a 40% reduction in dGTP in both cell lines and reductions of 40% and 63% in dATP and 89% and 71% in dCTP in CEM and K562 cells, respectively (Table 1). The dTTP pool was not altered in K562 cells, but it was reduced 47% in CEM cells. Similar incubations with ara-C, nucleotides of which do not inhibit ribonucleotide reduction, tended to increase dNTPs in both cell lines.

These results indicate that in both cell lines, dFdC incubation affected dCTP the most by dFdC incubation. Since dCTP is a negative regulator of dCyd kinase, one would expect that a decrease in cellular dCTP from dFdC treatment would release dCK from feed-back inhibition, potentiating the enzyme activity.

To evaluate the hypothesis that dFdC incubation would activate dCyd kinase by decreasing dCTP pools, we studied the phosphorylation of dCyd and ara-C in untreated and dFdC-treated cells. The cells were incubated with 10 μM dFdC, washed, and incubated with 100 μM [^{3}H]dCyd. The results show that the cellular accumulation of [^{3}H]dCTP from exogenous [^{3}H]dCyd was greater in cells preincubated with dFdC (2). These results, which are consistent with the hypothesis that modulation of cellular dNTPs by dFdC affects dCyd kinase activity, suggest that the metabolism of other substrates of this enzyme would be also influenced after dFdC treatment. Because ara-C is also dependent on the activity of dCyd kinase for its initial phosphorylation, we examined the effect of dFdC preincubation on the metabolism of ara-C. As observed for dCTP, the accumulation of ara-CTP was significantly enhanced by dFdC pretreatment. The peak ara-CTP in control cells was 100 μM, whereas dFdC-treated cells accumulated nearly 300 μM ara-CTP.

The elimination of dFdCTP was dependent on the cellular concentration of dFdCTP, with biphasic kinetics at intracellular levels of 100 μM or greater (5). At low dFdCTP concentrations, dFdU was the predominant metabolite in the extracellular medium (Table 2). However, at high cellular dFdCTP concentrations, relatively low levels of dFdU were excreted; the major metabolite in the extracellular medium was dFdC (Table 2). Thus, it appears that at high dFdCTP concentrations, the dephosphorylation of dFdCTP to dFdC was the major catabolic pathway. Because CEM cells lack dCyd deaminase (10), these findings suggest that dCMP deaminase may play an important role in dFdCTP elimination.

Table 2. Extracellular Accumulation of Catabolites of dFdCTP Elimination.

Hours after dFdC washout	0.1 μM dFdC		10 μM dFdC	
	dFdC*	dFdU*	dFdC	dFdU
1	0.61	0.01	11.7	0.2
2	0.54	1.02	9.5	3.0
3	0.60	2.42	33.5	5.3
4	0.88	6.58	48.4	9.9
5	1.05	6.76	48.6	13.7
6	0.89	7.07	48.1	18.2

* μM dFdC or dFdU in the medium.

We further investigated this possibility and found that the activity of partially purified dCMP deaminase was inhibited 50% by 0.5 mM dFdCTP (Table 3). Moreover, the deamination of dFdCMP requires dCTP, indicating that the endogenous dCTP pool has an important effect on the activity of dCMP deaminase (Table 3). Because the activity of dCMP deaminase is regulated by the intracellular dCTP:dTTP ratio, the dFdC-mediated perturbation of dNTP pools should play a major role in the regulation of this enzyme. Since dFdC lowers dCTP pools more than dTTP pools, it decreases the ratio of dCTP to dTTP, which consequently inhibits the activity of dCMP deaminase. This in turn slows elimination of dFdCTP, potentiating dFdC cytotoxicity (1,5).

Table 3. Effect of dFdCTP and dCTP on the Activity of dCMP Deaminase.

substrate	dFdCTP, mM	dCTP, μM	dCMP deaminase nmole/mg/min
dCMP	0	5	315
dCMP	0.5	5	161
dCMP	1.0	5	135
dFdCMP	--	0	1.1
dFdCMP	--	4	45
dFdCMP	--	10	250

One of the most prominent effects of dFdC is its potent inhibitory activity on DNA synthesis in cultured cells. We used a DNA primer extension assay to evaluate the action of dFdCTP on DNA synthesis by purified human DNA polymerase α. The incorporation of dFdCTP into the growing DNA strand and the consequence of this incorporation were visualized by analyzing the reaction products on a DNA sequencing gel. As illustrated in Fig. 1, lane C^- is the reaction with 30 μM each of dATP, dGTP, and dTTP, but no dCTP. The intense band at the site before the first C indicates that most of the extending primer stopped before the first C site because of the absence of dCTP. The faint band at the first C site reflects a small amount of unmatched dNTP that was misincorporated. When 100 μM dFdC was added to the reaction without dCTP, the intense band before the first C site diminished and the primer was extended beyond it, indicating that the dFdCTP served as an alternate substrate for incorporation into the C site (lane dCTP 0/dFdCTP 100). Only a faint band was observed at the first C site, whereas the site after it had an intense

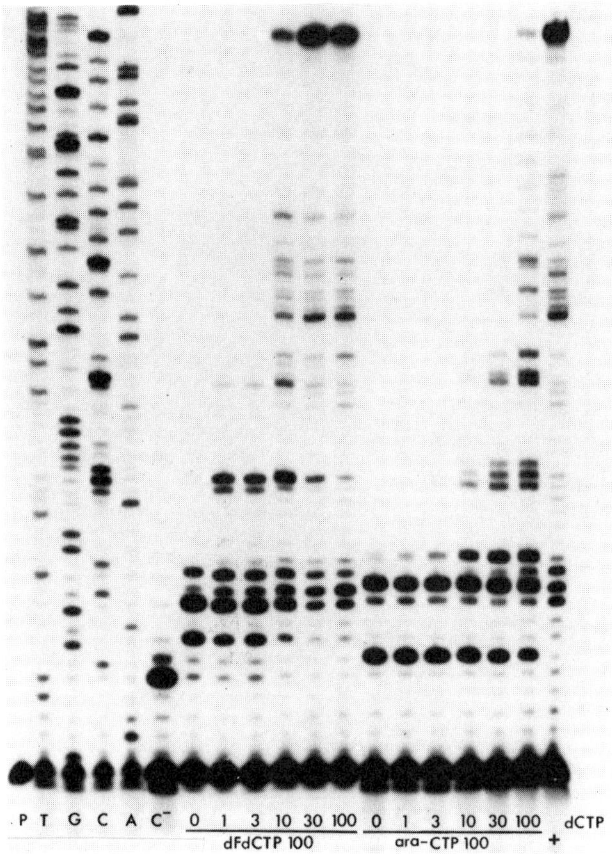

Figure 1. Effects of dFdCTP and ara-CTP on DNA primer extension by DNA polymerase α.

band. This indicated that the incorporated dFdCMP residue was further extended by one nucleotide and then caused DNA polymerase α to pause at the post-C site. This action was different from that of ara-CTP, which was incorporated into the C-site and caused the enzyme to pause at this site (lane dCTP 0/ara-CTP 100). Neither dFdCTP nor ara-CTP could substitute for dCTP to support the synthesis of full-length DNA. When increasing concentrations of dCTP were added to the reaction, dCTP competed with either dFdCTP or ara-CTP for incorporation into the C sites to support the synthesis of longer DNA strands by DNA polymerase alpha. In reactions that included dFdCTP, the pause bands caused by the incorporated dFdCMP at both the pre-C and post-C sites diminished with increasing dCTP concentrations, consistent with competition between dCTP and dFdCTP (lanes dCTP 1-100/dFdCTP 100). Ten μM dCTP effectively competed with 100 μM dFdCTP to support the synthesis of full-length DNA strands (lane dCTP 10/dFdCTP 100). One hundred μM dCTP was required to achieve a similar level of competition with 100 μM ara-CTP (lane dCTP 100/ara-CTP 100).

In conclusion, the cytotoxicity of dFdC arises from its incorporation into DNA and from its inhibition of ribonucleotide reductase. These actions are not independent. Accumulation of dFdC nucleotides is enhanced by activation of deoxycytidine kinase, which occurs when the dCTP levels are lowered by dFdC. Similarly, the deamination of dFdC monophosphate by dCMP deaminase is inhibited at low dCTP concentrations, slowing its elimination. Finally, the decrease in dCTP pools leads to a greater incorporation of dFdC 5'-triphosphate into DNA, an action correlated with cytotoxicity. Thus, the cytotoxicity of dFdC is potentiated by its own metabolites, which enhance dFdC nucleotide accumulation and incorporation into DNA.

ACKNOWLEDGMENTS

This work was supported in part by grants CA28596 and CA53311 from the National Cancer Institute, DHHS, and grant CH-130 from the American Cancer Society. The authors are grateful to Maureen Goode for critical editorial assistance.

REFERENCES

1. Heinemann, V., Hertel, L.W., Grindey, G.B., and Plunkett, W. Comparison of the cellular pharmacokinetics and toxicity of 2',2'-difluorodeoxycytidine and 1-β-D-arabinofuranosylcytosine. Cancer Res., 48:4024-4031, 1988.
2. Gandhi V. and Plunkett W. Modulatory activity of 2',2'-difluorodeoxycytidine on the phosphorylation and cytotoxicity of arabinosyl nucleosides. Cancer Res., 50: 3675-3680, 1990.
3. Liliemark, J.O. and Plunkett, W. Regulation of 1-β-D-arabinofuranosylcytosine 5'-triphosphate accumulation in human leukemia cells by deoxycytidine 5'-triphosphate. Cancer Res., 46:1079-1083, 1986.
4. Huang, P., Chubb, S., Hertel, L.W., and Plunkett, W. Mechanism of action of 2',2'-difluorodeoxycytidine triphosphate on DNA synthesis. Proc. Am. Assoc. Cancer Res., 31:426, 1990.
5. Heinemann, V., Hertel, L.W., Grindey, G.B., and Plunkett, W. Cellular elimination of 2',2'-difluorodeoxycytidine 5'-triphosphate. Proc. Am. Assoc. Cancer Res., 29: 504, 1988.
6. Chang, C-H., and Cheng, Y-C. Effects of deoxyadenosine triphosphate and 9-β-D-arabinofuranosyladenine 5'-triphosphate on human ribonucleotide reductase from Molt-6F cells and the concept of "self potentiation". Cancer Res., 40: 3555-3558, 1980.
7. Novotný, L. and Plunkett, W. Synthesis of 1-β-D-arabinofuranosyl-2-fluoroadenine 5'-triphosphate. In: L. B. Townsend and R. S. Tipson (eds.), Nucleic Acid Chemistry, Part 4, pp. 337-340, John Wiley & Sons, New York, 1990.
8. Heinemann, V., Xu, Y-Z., Chubb, S., Sen, A., Hertel, L.W., Grindey, G.B., and Plunkett, W. Inhibition of ribonucleotide reduction in CCRF-CEM cells by 2',2'-difluorodeoxycytidine. Mol. Pharmacol., 38: 567-572, 1990.
9. Gandhi, V. and Plunkett, W. Modulation of arabinosyl nucleoside metabolism by arabinosylnucleotides in human leukemia cells. Cancer Res., 48:329-334, 1988.
10. Fridland, A. and Verhoef, V. Mechanisms for ara-CTP catabolism in human leukemic cells and effect of deaminase inhibitors on this process. Sem. Oncol., 14 (Suppl 1): 262-268, 1987.
11. Huang, P., Chubb, S., and Plunkett, W. Termination of DNA synthesis by 9-β-D-arabinofuranosyl-2-fluoroadenine: a mechanism for cytotoxicity. J. Biol. Chem., 265:16617-16625, 1990.

EFFECT OF LEUCOVORIN ON 5-FLUOROURACIL INDUCED INHIBITION OF THYMIDYLATE SYNTHASE IN PATIENTS WITH COLON CANCER

Godefridus J. Peters*, Cees J. van Groeningen*, Clasina L. van der Wilt*, Kees Smid*, Sybren Meijer**, Herbert M. Pinedo*,#

Depts. Oncology* and Surgery**, Free University Hospital, PO BOX 7057, 1007 MB Amsterdam; #, Netherlands Cancer Institute Amsterdam, the Netherlands

INTRODUCTION

5-Fluorouracil (5FU) is often used in the treatment of patients with advanced colorectal cancer [1], but most tumors show intrinsic or acquired resistance; the objective response rates are below 20%. Currently leucovorin (LV; 5-formyl-tetrahydrofolate) is added to this therapy leading to improved response rates generally between 30-40% [2]. Resistance against 5FU in patients may be related to a reduced inhibition of thymidylate synthase (TS) by FdUMP, the active metabolite of 5FU [1]. TS catalyzes the conversion of dUMP to TMP for which 5,10-methylene-tetrahydrofolate (CH_2-THF) is the methyldonor. Inhibition of TS by FdUMP is mediated by the formation of a covalent ternary complex between FdUMP, TS and CH_2-THF. The extent and retention of inhibition of TS, depending on the concentrations of FdUMP and CH_2THF, are considered to be limiting factors. In addition, TS can have different characteristics in tumors of different patients, determining whether inhibition by FdUMP is effective or not. It has been demonstrated by others [3,4] and ourselves [5] that TS shows a very large heterogeneity in colorectal cancer; this includes a very large variation in enzyme activity (more than 20-fold) and variations in kinetic properties (affinity of FdUMP and of the natural substrate dUMP for TS). Also different affinities for the folate-cofactor have been demonstrated [6]. Spears et al [7] postulated that a number of these factors may determine resistance to inhibition by FdUMP. Several authors determined inhibition of TS in tumors of patients after treatment with 5FU [7-9]. However, generally the time of biopsies was limited to a few hr after 5FU administration, but never longer than 24 hr. Since a long retention may be a crucial factor in determining the antitumor effect we measured the inhibition of TS in tumors and normal tissues of patients 48 hr after treatment with 5FU. In addition we determined whether enhancement of the intra-tumoral folates pools by administration of LV would lead to a more pronounced inhibition of TS. We not only measured TS with the FdUMP ligand bindings assay but also with the catalytic assay, i.e. the conversion of dUMP to TMP.

Purine and Pyrimidine Metabolism in Man VII, Part A
Edited by R.A. Harkness *et al.*, Plenum Press, New York, 1991

MATERIALS AND METHODS

Patients Only patients with histologically proven colorectal cancer (both primary and metastasized), were entered into the study. Biopsy specimens from 19 patients (9 males/10 females; median age 57, range 37-72) were obtained after about 2 days (range 40-49 hr) after treatment. All patients were treated with an i.v. injection of 5FU at 500 mg/m^2, and 11 patients received 5FU in the middle of a 2-hr infusion of LV (500 mg/m^2). Tissues were frozen in liquid nitrogen as soon as possible after surgical removal, dissected and stored at -80°C until analysis. Before analysis tissues were pulverized using a micro-dismembrator [10].

Enzyme assays Since no pretreatment samples could be obtained, the amount of TS inhibition had to be evaluated through dissociation of FdUMP from the ternary complex; after this process the total number of FdUMP binding sites (TS-*tot*) and the total TS catalytic activity (TS-*total*) could be determined. The uninhibited activity in the non-dissociated samples was termed TS-*free* (remaining number of FdUMP binding sites) and TS-*res* (residual catalytic TS activity). After pulverization of the tissues the frozen powder was suspended in the "sonication" buffer (pH 7.4) and centrifuged twice (400 and 7000g). Of the 7,000 g supernatant one part (for TS-*free* and TS-*res*) was frozen at -80°C and to the other part "dissociation" buffer (pH 7.8 containing CMP, NaF and excess dUMP) was added and incubated at 30°C for 3 hr [11]. Nucleotides were removed by a neutral charcoal wash [11]. In both parts we measured FdUMP binding and TS catalytic activity. TS-*tot* and TS-*free*, the amount of free FdUMP binding sites which can be bound by [6-^{3}H]-FdUMP, were determined according to a modification [5,11] of published methods [7]. TS-*total* and TS-*res* were measured with the tritium release assay [5]. The catalytic TS activity was determined at 1 and 10 μM dUMP, being suboptimal (*Km*) and saturating dUMP concentrations. As discussed elsewhere evaluation of TS inhibition can sometimes be hampered by the experimental conditions [11]. Optimization and standardization minimized variation; dUMP was added to optimize dissociation and stabilization of the enzyme; NaF and CMP were added to inhibit phosphatase and 5'-nucleotidase mediated nucleotide degradation; the assay time for the bindings assay was limited to minimize exchange of labeled and non-labeled FdUMP. Tissues were considered as evaluable when the radioactivity was at least 1.5 *x* higher than in the blank.

RESULTS

In tissue biopsies (primary tumors, liver metastases, colon mucosa and liver) of patients we studied several characteristics of TS using the ligand binding assay for FdUMP and the catalytic assay for TS. The total activity of TS (both FdUMP binding and catalytic activity) in tumors and metastases obtained after 48 hr, showed a large variation, and was generally higher (Table 1) than in normal mucosa and liver from the same patient. Because of the higher sensitivity of the catalytic assay more samples could be evaluated with this assay than with the FdUMP binding assay, this difference was more pronounced in samples of the liver and normal mucosa. In addition the sensitivity of the catalytic assay at a substrate concentration of 1 μM was higher than at 10 μM. The ratio between the activity at 10 and 1 μM dUMP varied between 2 and 8. From 4 patients both primary tumors and liver metastases could be obtained. Since no significant difference was observed in values obtained from patients treated with either 5FU alone or with the combination of LV and 5FU, these values have been pooled (Table 1).

Table 1. FdUMP Binding (TS-*tot*) and Total Catalytic TS Activity (TS-*total*); effect of 5FU and LV/5FU treatment in tumors.

Parameter	TS-*tot* and TS-*total*			Ratio of TS-*free/tot* and TS-*res/total* in tumors	
	Mucosa	Liver	Tumor	5FU	LV-5FU
FdUMP binding	0 (24)	21 (21)	83 (24)	0.37 [6]	0.24 [12]
Catalytic activity					
1 μM dUMP	26 (4)	5 (14)	34 (24)	0.68 [8]	0.49 [12]
10 μM dUMP	105 (4)	25 (12)	129 (22)	0.59 [7]	0.31 [13]

Values (from tissues removed 2 days after treatment) for TS-*tot* (fmol/mg protein) and TS-*total* (pmol/hr/mg protein) represent medians of the number of tissues indicated within ( ). Values below detection limit (< 1.5 *x* blank) are denoted as 0. From several patients both primary tumors and metastases were obtained. Ratios were calculated from the separate tumors and represent means of the number of evaluable tumors indicated within [ ].

Inhibition of TS was evaluated by comparison of the TS-*free* and TS-*res* among the patients with TS-*tot* and TS-*total*, respectively. Ratios could be only calculated in those patients in which both assays were evaluable. In tumors from patients treated with 5FU alone the ratios were higher than in tumors from patients treated with LV-5FU. In normal liver no difference between 5FU and LV-5FU treatment was observed. Several patients were subsequently treated with intra-arterial 5FU administration (1000 mg/m^2/day for 5 days) and the total activity of TS and the inhibition of TS could be related to the outcome of therapy. In 4 patients from the 5FU pretreated group with a low TS-*free* and TS-*res* (below or at detection limit) disease stabilization (2 *x*) and a partial response (2 *x*) were observed; from the LV/5FU pretreated group 3 patients with such a low TS activity had a disease stabilization after subsequent 5FU therapy. In one patient with high TS values progression was observed. The other patients were not yet evaluable for response.

DISCUSSION

In this study we demonstrate that the TS activity in tumors is still inhibited up to 2 days after treatment with single agent 5FU. In patients treated with the combination of LV and 5FU this inhibition is more pronounced. In other studies evaluation of the inhibition of TS in tumors from patients has been limited to shorter time periods using the FdUMP bindings assay [7-9]. Swain et al [8] demonstrated that a partial TS inhibition in patients with 5FU-resistant breast cancer could be converted to a complete inhibition after LV-5FU treatment. In addition to the FdUMP bindings assay used by others we also evaluated TS inhibition using the catalytic assay. The assay appeared to be more sensitive than the bindings assay and gives an estimate of the capacity of tumors to convert dUMP to TMP. Since inhibition of TS can lead to an increase in dUMP levels [7], the use of low and high substrate concentrations gives some insight in the potential inhibition of the enzyme in the tumor. Relative inhibition of TS at 10 μM dUMP concentration was usually higher than at low dUMP.

An attempt was made to correlate the inhibition of TS with the response to subsequent 5FU treatment. The present data and evaluation of TS inhibition at shorter time periods [11] demonstrate that a complete inhibition of TS at 24 or 48 hr after 5FU treatment is likely to be related with effective 5FU treatment (partial response and disease stabilization). In addition high TS-*tot* or TS-*total* values have been associated with ineffective treatment. The activity of TS after dissociation is considered to represent pretreatment TS levels, so, evaluation of TS in patients before treatment might give indication whether subsequent treatment will be succesful [5]. In addition to enzyme measurements TS can also be evaluated by determining the amount of TS protein using specific antibodies [12] and by determining the expression of TS-mRNA using PCR [13]. Both approaches are currently being evaluated.

ACKNOWLEDGEMENTS

This work was supported by grants of the Dutch Cancer Society (grant IKA 88-20), and of Lederle, the Netherlands. Dr. G.J. Peters is the recipient of a senior research fellowship of the Royal Netherlands Academy of Sciences.

REFERENCES

1 Pinedo HM, Peters GJ (1988). 5-Fluorouracil: biochemistry and pharmacology. J Clin Oncol **6**: 1653-1664

2 Peters GJ, Van Groeningen CJ (1991). Clinical relevance of biochemical modulation of 5-fluorouracil. Ann Oncol; **2**: in press

3 Danenberg PV, Bapat AR (1989). Thymidylate synthases isolated from human colon tumors have varied affinity towards 5-fluoro-2'-deoxyuridylate. Reg Cancer Treatm **2**: 1-4

4 Bardot V, Luccioni C, Lefrancois D, Muleris M, Dutrillaux B (1991). Activity of thymidylate synthetase, thymidine kinase and galactokinase in primary and xenografted human colorectal cancers in relation to their chromosomal patterns. Int J Cancer **47**: 670-674

5 Peters GJ, Van Groeningen CJ, Laurensse EJ, Pinedo HM (1991). Thymidylate synthase from untreated human colorectal cancer and colonic mucosa; enzyme activity and inhibition by 5-fluoro-2'-deoxyuridine-5'-monophosphate. Eur J Cancer **27**: 263-267

6 Berger SH, Barbour KW, Berger FG (1988). A naturally occurring variation in thymidylate synthase structure is associated with a reduced response to 5-fluoro-2'-deoxyuridine in a human colon tumor cell line. Mol Pharmacol **34**: 480-484

7 Spears CP, Gustavsson BG, Berne M, Frosing R, Bernstein L, Hayes AA (1988). Mechanisms of innate resistance to thymidylate synthase inhibition after 5-fluorouracil. Cancer Res **48**: 5894-5900

8 Swain SM, Lippman ME, Egan EF, Drake JC, Steinberg SM, Allegra CJ (1989). Fluorouracil and High-Dose Leucovorin in Previously Treated Patients with metastatic Breast Cancer. J Clin Oncol **7**: 890-899

9 Trave F, Rustum YM, Petrelli NJ, Herrera L, Mittelman A, Frank C, Creaven JP (1988). Plasma an tumor tissue pharmacology of high-dose intravenous leucovorin calcium in combination with fluorouracil in patients with advanced colorectal carcinoma. J Clin Oncol **6**: 1184-1191

10 Peters GJ, Laurensse E, Leyva A, Pinedo HM (1986). Tissue homogenization using a microdismembrator for the measurement of enzyme activities. Clin Chim Acta **158**, 193-198.

11 Peters GJ, Van Groeningen CJ, Van der Wilt CL, Meijer S, Smid K, Laurensse EL, Pinedo HM (1991). Time course of inhibition of thymidylate synthase in patients treated with 5-fluorouracil and leucovorin. Sem Oncol **18** (suppl), in press

12 Freemantle SJ, Aherne GW, Hardcastle A, Lunec J, Calvert AH (1991). Increases in thymidylate synthase protein levels measured using newly developed antibodies. Proc Amer Assoc Cancer Res **32**: 360 (Abstr 2138)

13 Horikoshi T, Danenberg KD, Stadlbauer THW, Shea LC, Aigner K, Gustavsson B, Danenberg PV (1991). Quantitation of gene expression in human tumors by RNA-to-RNA amplification using the polymerase chain reaction (PCR). Proc Amer Assoc Cancer Res **32**: 362 (Abstr 2151)

A NOVEL XANTHINE DEHYDROGENASE INHIBITOR (BOF-4272)

Seiji Sato, Kunihiko Tatsumi* and Takeshi Nishino**

Otsuka Pharmaceutical Factory, Inc., Naruto, Tokushima
*Otsuka Pharmaceutical Co., Ltd, Otsu, Shiga
**Department of Biochemistry, Yokohama City University School of Medicine, Yokohama, Kanagawa, Japan

INTRODUCTION

An inhibitor of xanthine oxidase (XO)/xanthine dehydrogenase (XDH), an enzyme catalyzing the last step of purine catabolism, might be expected to be effective as remedy for hyperuricemia and possibly for ischemia-reperfusion injury. However, no clinically effective XO/XDH inhibitor except allopurinol have been used since it was introduced for clinical use in 1962 (1, 2).

The complicated mechanism of XO/XDH or the difference of uric acid metabolism in species especially between human and mammals except for chimpanzee, such as the tissue distribution of XO/XDH and the presence of uricase, might be a reason why no compound could follow allopurinol.

BOF-4272 (Sodium(±)-8-(3-methoxy-4-phenylsulfinylphenyl)pyrazolo [1, 5-a]-1, 3, 5-triazine-4-olate monohydrate), a new inhibitor for XO/XDH, was discussed with respect to the efficacy in vivo.

METHODS

XO/XDH inhibition in mice

Six-week-old male ICR mice were fed and given water ad lib. The animals were divided into groups of 5 or 8. BOF-4272 or allopurinol suspended in 0.5% CMC sodium solution was orally administered at dose of 1 or 5 mg/kg. A 0.5% CMC sodium solution was administered for control.

Blood, liver, small intestine and lung were collected by sacrifice before and 1, 4, 8, 12 and 24 hours after administration of test drugs.

Blood samples were centrifuged after stand on ice for about 30 minutes. Other organs were homogenized in ice-cooled phosphate buffer, then centrifuged. Uric acid (UA), hypoxanthine (HX) and xanthine (XA) in serum and supernatants of organs were determined by HPLC method for the index of XO/XDH inhibition.

RESULTS AND DISCUSSION

Kinetic analyses with purified enzyme from bovine milk, chicken liver, and rat liver revealed that BOF-4272 bound tightly to the

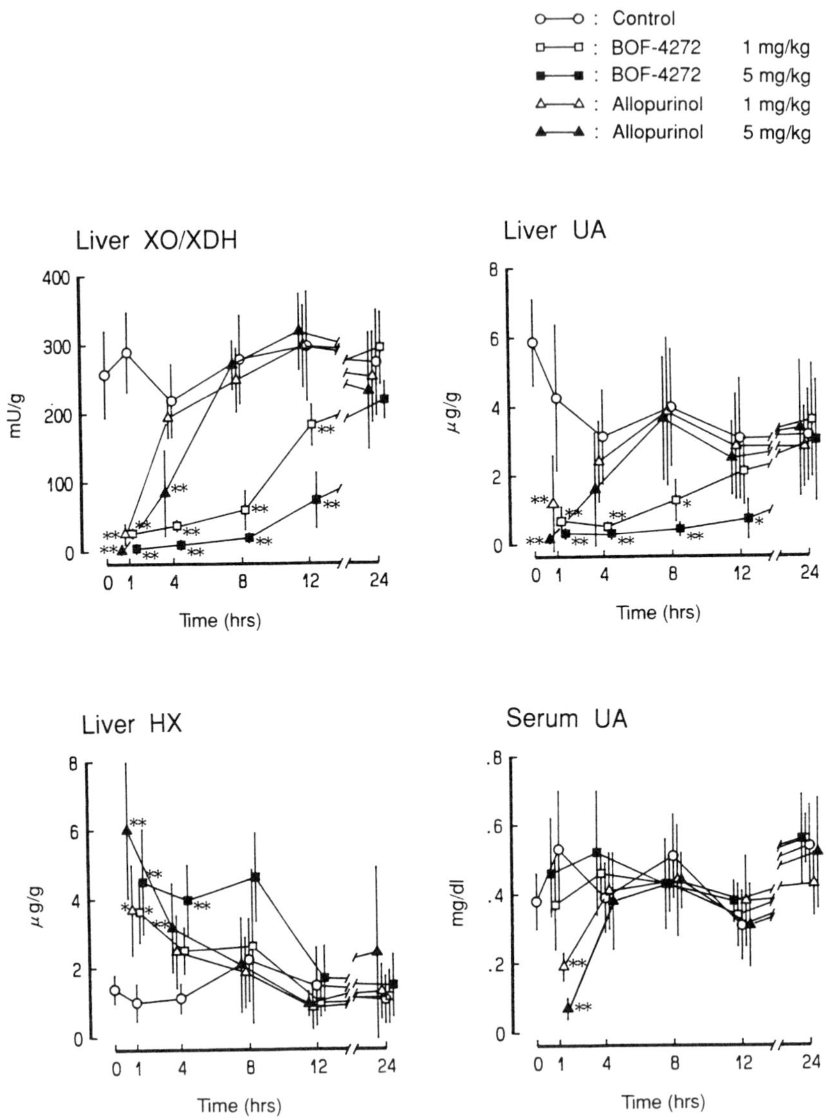

Fig.1. Effects of BOF-4272 and Allopurinol on Uric Acid Metabolism in Intact Mice.
mean±SD, n=5~8
*P<0.05, **P<0.01 vs Control

Molybdenum site of the enzyme both in oxidized and reduced forms (for oxidized form, $Ki=5\times10^{-9}M$). The inhibition mechanism of BOF-4272 was thought to differ from that of oxipurinol (3) (manuscript in preparation).

In vivo results are shown in Fig.1. BOF-4272 had organ specificity for XO/XDH inhibition. BOF-4272 strongly decreased UA and increased HX in liver and small intestine, the main organs of UA production in human. The effect of BOF-4272 lasted longer than that of allopurinol. However, in lung and serum the levels of UA and HX were not changed by BOF-4272. Allopurinol had not such organ specificity.

The organ specificity of XO/XDH inhibition of BOF-4272 was thought to be attributable to the difference of drug distribution in organs. BOF-4272 is distributed mainly in the upper part of digestive organs and liver, but less than one-tenth in serum and other organs except kidney.

Table 1. Difference in Uric Acid Metabolism among Animal Species.

Species	Serum UA (mg/dl)	XO/XDH				Uricase
		Blood	Liver	Small Intestine	Lung	Liver
Human	4～7	−	+	+	−	−
Chimpanzee	2～4		+	+		−
Monkey	0.1～0.2		+	+		+
Dog	0.2～0.3	+	+	+	+	+
Rat	0.3～0.6	+	+	+	+	+
Mouse	0.3～0.6	+	+	+	+	+

Metabolism of UA differs among animal species (Table 1). Mammals except for chimpanzee have uricase in liver, that degradates UA to allantoin, and have XO/XDH activity in serum and lung. But, human has no uricase activity in liver, and is also thought to have no or very weak XO/XDH activity in serum and lung. These facts suggest that the source of serum UA is different between human and mammals except for chimpanzee (4). So, it may be considered that in mammals except for chimpanzee serum UA is derived from serum, lung or other organs but not liver and small intestine. On the other hand, the serum UA in human is mainly derived from liver and small intestine. Thus, for the index of serum UA decreasing efficacy in human, serum UA level in experimental animals seems to be not always valuable, while the inhibition of XO/XDH in target organs of UA synthesis such as liver and small intestine seems to be the most important.

Although serum UA decreasing effect of BOF-4272 in intact animals is far weak than that of allopurinol, BOF-4272 has strong XO/XDH inhibition in liver and small intestine, i.e., the target organs of UA production in human. So, it is expected that such a drug may effectively decrease serum level of UA in human.

REFERENCES

(1) G. B. Elion, S. Callahan, H. Nathan, S. Bieber, R. W. Rundles and G. H. Hitchings, Potentiation by inhibition of drug degradation: 6-substituted purines and xanthine oxidase, Biochem. Pharmacol. 12, 85-93 (1963)

(2) R. Hille and V. Massey, Tight binding inhibitors of xanthine oxidase, Pharmac. Ther. 14, 249-263 (1981)

(3) V. Massey, H. Komai and G. Palmer, On the mechanism of inactivation of xanthine oxidase by allopurinol and other pyrazolo[3, 4-d] pyrimidines, J. Biol. Chem. 245 (11), 2837-2844 (1970)

(4) D. A. Parks and D. N. Granger, Xanthine oxidase : biochemistry, distribution and physiology, Acta Physiol. Scand. Suppl. 548, 87-99 (1986)

CHANGES CAUSED BY ETHANOL INTAKE ON METABOLISM OF HYPOURICEMIC AGENTS (COMBINATION OF ALLOPURINOL AND BENZBROMARONE)

Kaneko K [1], Fujimori S [2] and Akaoka I [2]

Central Laboratory of Analytical Biochemistry [1], Department of Internal Medicine [2], Teikyo University School of Medicine, Tokyo, Japan

INTRODUCTION

Several agents have been employed for treatment of hyperuricemia. Allopurinol, an inhibitor of xanthine oxidase (1) is often used as a hypouricemic agents. Benzbromarone, which is reported to have a uricosuric effect (2), has also been employed. It has been recently reported that combination of these two drugs has been favorable in gouty patients (3).

The metabolism of both allopurinol and benzbromarone has been influenced by ethanol intake in our previous work (4). In this study, the changes caused by ethanol intake on metabolism of these two agents in combined administration was investigated.

MATERIALS AND METHODS

Subjects

Three normal adult men aged 29, 46 and 47 years old volunteered. They had normal baseline hepatic and renal function (Ccr: 103.5 ± 4.6 ml / min).

Protocol

Drug (200 mg of allopurinol and 100 mg of benzbromarone) was administered at 0 hr. Oral loading of ethanol was achieved by intake of 200 ml (80 g of ethanol) of whisky (SUNTORY LTD., Osaka, Japan) over 30 minutes around 0 hr. Studies were done two weeks apart from in time with and without ethanol.

Blood samples were drawn intravenously 1 hr before and 2, 3, 6, and 22 hr after drug administration. During the experiment the subjects were maintained at a normal level of oral hydration after the first 6 hr. Urine samples were collected 1 hr before and 1, 2, 3, 4.5, 6, 9, 12, 18, 24 and 30 hr after drug administration.

Biochemical Determination

Uric acid, creatinine and plasma ethanol levels were measured by the same method in previous report (4). Lactic acid levels in blood were measured by a blood lactic acid test (Boehringer Mannheim GmbH, Mannheim, W.Germany).

Plasma samples were deproteinized and diluted to proper concentration for analysis of drugs. Concentration of allopurinol, oxypurinol and benzbromarone were determined by HPLC according to the method previously reported (4).

Statistical Procedure

Statistical analysis were carried out using the two-tailed Student's t-test and Welch's procedure when variance was not equal.

RESULTS

Effect of ethanol on metabolism of allopurinol

The plasma alcohol concentrations at 2, 3 and 6 hr after oral ethanol loading were 26.5 ± 2.7, 23.1 ± 2.9, 11.5 ± 1.9 mM (n=3, mean ± S.E.M.), respectively. After ethanol intake, lactic acid level was not elevated (before: 1.24 ± 0.2, after: 1.34 ± 0.1 mM).

The plasma levels of allopurinol and oxypurinol are shown in Fig.1(a) and Fig 1(b), respectively. After oral ethanol intake, the plasma allopurinol level was increased from 5.3 ± 0.7 to 10.8 ± 1.6 μM ($P<0.05$) at 3 hr after administration of allopurinol (Fig.1(a)). The plasma level of oxypurinol was reduced to about one half of the value observed with no alcohol ingestion (Fig.1(b)). Those of oxypurinol at 3 hr with and without ethanol intake were 26.4 ± 1.5 and 12.2 ± 4.1 μM ($P<0.05$), respectively.

The urinary excretion of allopurinol and oxypurinol is illustrated in Fig.2(a) and Fig.2(b). The excretion of allopurinol after ethanol intake was reduced for first 1 hr ($P<0.05$), and later became increased (Fig.2(a)). The amounts of excreted allopurinol without and with ethanol ingestion were 20.3 ± 3.5 mg and 31.9 ± 6.6 mg over 23 hours (from 1 hr to 24 hr after administration of allopurinol), respectively. On the other hand, the excretion of oxypurinol appeared to be reduced by intake of ethanol (Fig.2(b)).

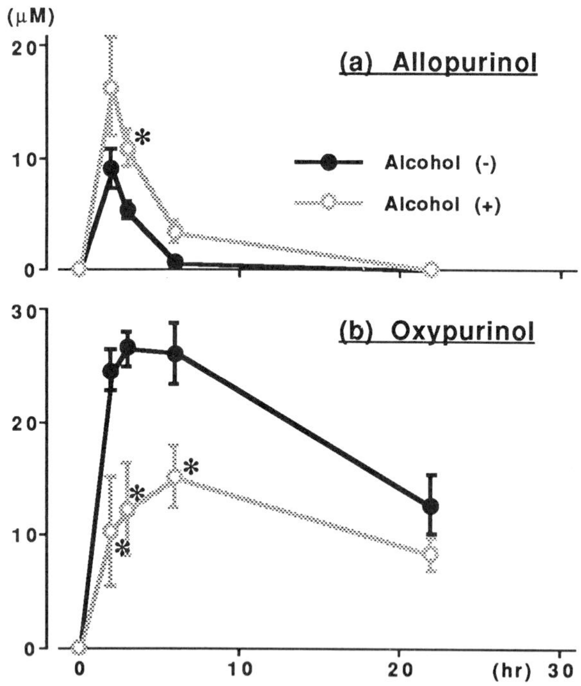

Fig. 1 Plasma levels of allopurinol (a) and oxypurinol (b) after ethanol intake. Plasma levels are shown as means ± S.E.M. (μM) (n=3). Closed circles indicate results obtained without ethanol and open circles indicate those with ethanol intake. ✻ Significant at $P<0.05$

Effect of ethanol on metabolism of benzbromarone

The plasma levels of benzbromarone are shown in Fig.3. When alcohol was ingested, the plasma level of benzbromarone was reduced in all three subjects, from 7.8 ± 1.0 to 2.3 ± 0.9 μM ($P<0.05$) at 2 hr after administration of both agents. At 6 hr, however, plasma benzbromarone levels with and without ethanol intake were 3.0 ± 0.3 and 5.9 ± 0.7 mM, respectively. Peak of plasma level was shifted at 6 hr after ethanol ingestion.

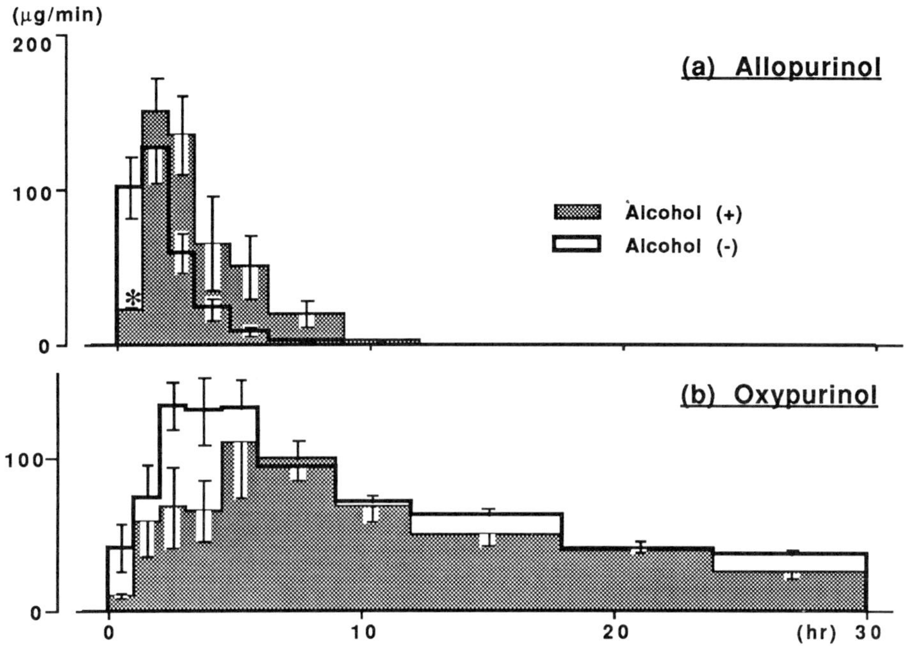

Fig. 2 Urinary excretion of allopurinol (a) and oxypurinol (b) after ethanol intake. Excretion is shown as means ± S.E.M. (μg/min) (n=3). Bold line indicates amount obtained without ethanol and shadow shows that with ethanol ingestion. ✻ Significant at P<0.05

Effect of ethanol on purine metabolism

The plasma level of uric acid was reduced slightly from 308.9 ± 37.5 to 150.5 ± 34.4 μM at 22 hr after administration of both compounds. The plasma level of uric acid with ethanol was also decreased and was not influenced by ethanol intake. However, the urinary excretion of uric acid was delayed after ethanol ingestion. The peak of excretion without ethanol intake was shown at 3 hr, whereas, that of with ethanol was shifted at 6 hr (Fig. 4). As benzbromarone is an uricosuric agent, this effect of ethanol is considered to correspond to the shift of plasma benzbromarone level shown in Fig. 3.

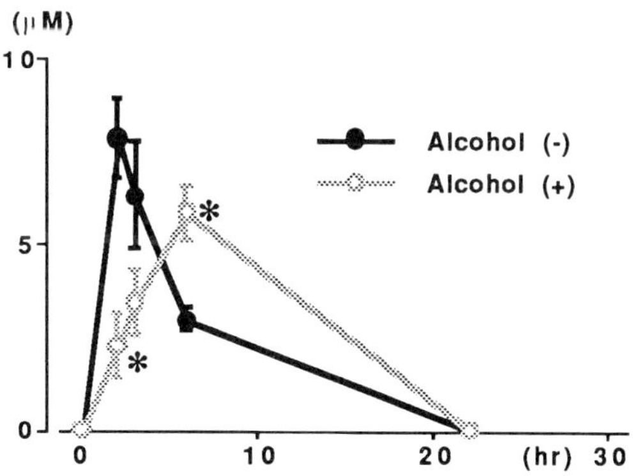

Fig. 3 Plasma level of benzbromarone after ethanol intake. Plasma level is shown as means ± S.E.M. (μM) (n=3). Closed circles indicate results obtained without ethanol and open circles indicate those with ethanol intake. ✻Significant at P<0.05

DISCUSSION

Changes caused by ethanol intake on metabolism of allopurinol and benzbromarone in combined administration was investigated in this study. The amount of ethanol ingested (80 g of ethanol) was considered to be the usual amount that a patient with a history of excessive alcohol consumption

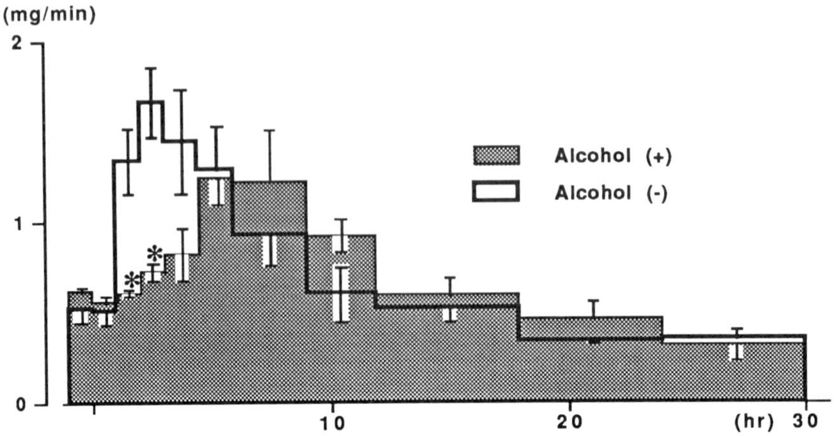

Fig. 4 Urinary excretion of uric acid after ethanol intake. Excretion is shown as means ± S.E.M. (mg/min) (n=3). Bold line indicates amount obtained without ethanol and shadow shows that with ethanol ingestion. * Significant at P<0.05

could drink at one time (less than 1 hour). The content of purine bases in this whisky was found to be very low (5). Therefore, the results demonstrated in this study were considered to be caused not by the purines in the whisky but by its content of ethanol.

The metabolism of both compounds was affected by ethanol ingestion. The plasma level of allopurinol was elevated, and that of oxypurinol was reduced. Excretion of unchanged allopurinol was delayed, and was increased compared with the amount excreted without ethanol. These results suggests that metabolism of allopurinol is inhibited and is excreted in greater amounts as an unchanged form without conversion to its active metabolite, oxypurinol. This may be the cause of the reduced effectiveness of allopurinol in gouty patients who consume excessive amounts of alcohol. The plasma level of benzbromarone was altered. This change with ethanol may be due to changes in the absorption rate of benzbromarone. Similar change with food intake in the plasma level of benzbromarone was reported (6). The change caused by ethanol intake was reflected in excretion of uric acid.

Understanding these effect of ethanol may contribute to more effective medication to gouty patients with a history of heavy alcohol consumption.

REFERENCES

1. Kelley WN, Palella TD: Gout and other disorders of purine metabolism. in Braunwald E, Isselbacher KJ, Petersdorf RG, et al (ed): Harrison's principles of internal medicine (eleventh ed). New York, McGraw-Hill, 1987, pp 1623-1632
2. Zollner N, Griebsch A, Fink JK: The effect of benzbromarone on serum uric acid level and uric acid excretion in gouty persons. Dtsch Med Wochenschr 95: 2405-2412, 1970
3. Mertz DPI: Reducing the risks in the treatment of gout and hyperuricemia. Dtsch Med Wochenschr 101: 1288-1292, 1976
4. Kaneko K, Fujimori S, Ishizuka I, et al: Effect of ethanol on metabolism of the hypouricemic agents allopurinol and benzbromarone. Clin.Chim.Acta 193: 181-186, 1990.
5. Fujimori S, Nakayama Y, Kaneko K, et al: Contents of purine bases in alcoholic beverages. Uric Acid Research 9: 128-133,1985
6. Walter-Sack I, Vries JX, Ittensohn A, et al: Benzbromarone disposition and uricosuric action; evidence for hydroxilation instead of debromination to benzarone. Klin Wochenschr 66: 160-166, 1988

ALLOPURINOL IN BREAST MILK

Irmingard Kamilli[1], Ursula Gresser[1], Christof Schaefer[2] and Nepomuk Zöllner[1]

[1] Medizinische Poliklinik der Universität München Pettenkoferstraße 8a, D-8000 München 2
[2] Teratology Information Service of Berlin, FRG

INTRODUCTION

This is the first published case report with measurement of allopurinol and oxipurinol in a breast milk sample and a plasma sample of a breast fed infant.

SUBJECTS AND METHODS

We studied the concentration of allopurinol and oxipurinol in plasma and breast milk of a hyperuricemic woman (A.H., born 1963) taking allopurinol (Wellcome) 300 mg/day and in plasma of her breast fed 5 weeks old male infant (D.H.). After a period of 4 weeks of regular intake of allopurinol, samples were taken from maternal blood, breast milk and infant's blood (table).

Allopurinol and oxipurinol levels were measured by HPLC. The apparatus consisted of Beckman diode array detector module 168, Beckman autosampler 507, Beckman programmable solvent module 126 and Hypersil Shandon ODS column. The mobile phase was 0.01 M phosphate buffer, pH 4.0 at a flow-rate of 1.0 ml/min. The detector wavelength was 254 nm.

RESULTS

The results are summarized in the table. Plasma allopurinol in the infant was below the limit of detection. The normal breast milk intake of infant D.H. was 150 ml/kg/day. The average daily dose ingested by the infant was 8 mg oxipurinol. This almost corresponds to the therapeutical dose for children of 10-20 mg/kg/day.

TABLE.1 Allopurinol and oxipurinol levels in maternal's and infant's plasma and in breast milk of the mother.

hours	samples	allopurinol (μg/ml)	oxipurinol (μg/ml)
7:00	300 mg allopurinol		
9:00	maternal plasma	1.0	13.8
9:00	breast milk	0.9	53.7
9:00	breast feeding		
11:00	maternal plasma	1.0	19.9
11:00	breast milk	1.4	48.0
11:00	infant plasma	not found	6.6

CONCLUSIONS

The data indicate that allopurinol and oxipurinol are excreted in breast milk. In particular, the oxipurinol level in breast milk (53.70 μg/ml) was surprisingly high. In steady state conditions after chronic allopurinol intake (300mg/day) plasma levels of oxipurinol are in the range of 13-19 μg/ml. The allopurinol levels in breast milk and maternal plasma were nearly the same.

Breast milk of mothers on allopurinol therapy contains therapeutical doses of allopurinol and oxipurinol. Therefore breast fed infants are suggested to have the same risk of side effects as infants on allopurinol therapy.

REFERENCE

Kamilli I, Schaefer C, Gresser U, Kölle P, Zöllner N (1991) Allopurinol in breast milk. 7th International/3rd European Joint Symposium on Purine and Pyrimidine Metabolism in Man. Bournemouth 30.6-5.7.1991. J Pur Pyr Res 2 (Suppl 1): abs 79

PHARMACOKINETICS OF BENZBROMARONE: EVIDENCE FOR A BILIARY-INTESTINAL-BILIARY RECIRCULATION

Ursula Gresser, Helmut Empl, Irmingard Kamilli and Nepomuk Zöllner

Medizinische Poliklinik der Universität München
Pettenkoferstraße 8a, D-8000 München 2

The uricosuric substance benzbromarone initially was reported to be debrominated to bromobenzarone and benzarone (Yü 1976). Recently hydroxilation was shown to be the primary metabolic pathway (Walter-Sack et al 1987). In addition, in a few individuals pharmacokinetic studies showed evidence for a deficient benzbromarone elimination from plasma, characterized by a longer persistance of benzbromarone in the plasma (Walter-Sack et al 1988; Walter-Sack et al 1990a), which is likely to be a familial disorder (Zöllner et al 1990).

PATIENTS AND METHODS

We did detailed pharmacokinetic and pharmacodynamic investigations (Gresser 1990) of 5 individuals with deficient benzbromarone elimination and of 6 individuals with regular benzbromarone elimination. An isocaloric formula diet with defined purine intake was given for 12 to 16 days in order to reach a steady state of uric acid metabolism. All volunteers received a single dose of 100 mg benzbromarone, blood and urine samples were collected in short intervals. Uric acid in plasma and urine was determined enzymatically according to the method of Praetorius and Poulson (1953) as modified by Zöllner (1963). Benzbromarone was determined by high performance liquid chromatography (HPLC) with isocratic elution according to Ittensohn and de Vries (1989).

RESULTS

The uricosuric effect of benzbromarone shows no differences between individuals with deficient benzbromarone elimination and individuals with regular benzbromarone elimination. The additional renal uric acid excretion after administration of 100 mg benzbromarone p.o. was between 307.8 to 1020.4 mg. There was no relationship to the rate of the elimination of benzbromarone.

Individuals with regular benzbromarone elimination show maximal plasma benzbromarone levels 1-4 h after drug ingestion followed by an exponential elimination of benzbromarone from the plasma. In 3 individuals the plasma benzbromarone levels reach a first maximum (2-4 h after drug ingestion), then decrease to a minimum (8 h) and reach a second maximum which is 70-80 % of the first maximum (9.5-11.5 h), then a slow decrease follows. Graphically a plot with two peaks results (table).

TABLE.1 Relationship of the 1st maximum (= 100 %),the 2nd maximum and the 1st minimum (lying in between of the both maxima) of the plasma levels of benzbromarone in 3 volunteers after administration of 100 mg benzbromarone p.o.

	time after benzbromarone was taken (h), relationsship to the 1st maximum (%)		
initials	1st maximum	1st minimum	2nd maximum
S.Al.	2 h, 100 %	8 h, 19.7 %	11.5 h, 76.2 %
A.O.	4 h, 100 %	8 h, 63.9 %	9.5 h, 70.0 %
B.S.	4 h, 100 %	8 h, 73.6 %	11.5 h, 81.7 %

CONCLUSIONS

Until now the relative uricosuric potency of benzbromarone and its metabolites is still unknown. Investigations with bromobenzarone and benzarone showed that both are uricosuric. The uricosuric potency of the really existing metabolites, one of them 1-hydroxy-benzbromarone (Walter-Sack et al 1990b), the other still unidentified, has never been tested, nor in vitro neither in vivo. Our observations show, that there is no difference in the uricosuric effect of 100 mg benzbromarone in individuals with different slow elimination of benzbromarone from the plasma. In addition the pharmacokinetic differences suggest, that there is a biliary-intestinal-biliary recirculation of benzbromarone.

REFERENCES

Gresser U (1990) Therapie von Hyperurikämie und Gicht mit Urikosurika. Mediscript-Verlag München, ISBN 3-88320-202-9

Ittensohn A, de Vries JX (1989) A simplified method for a quantitative determination of benzbromarone in human plasma by HPLC. Eur J Clin Pharmacol 36 (Suppl): A228

Praetorius E, Poulson H (1963) Enzymatic determination of uric acid with detailed directions. Scand J clin Lab Invest 5: 273-279

Walter-Sack I, de Vries JX, Ittensohn A, Weber E (1987) Benzbromarone disposition following single dose application. Evidence for lack of debromination to benzarone. Klin Wochenschr 65 (Suppl X): 5

Walter-Sack I, de Vries JX, Ittensohn A, Kohlmeier M, Weber E (1988) Benzbromarone disposition and uricosuric action; evidence for hydroxilation instead of debromination to benzarone. Klin Wochenschr 66: 160-166

Walter-Sack I, Gresser U, Adjan M, Kamilli I, Ittensohn A, de Vries JX, Weber E, Zöllner N (1990a) Variation of benzbromarone elimination in man - a population study. Eur J Clin Pharmacol 39: 173-176

Walter-Sack I, de Vries JX, Ittensohn A, Weber E (1990b) Rapid and slow benzbromarone elimination phenotypes in man: benzbromarone and metabolite profiles. Eur J Clin Pharmacol 39: 577-581

Yü T-F (1976) Pharmacokinetic and clinical studies of a new uricosuric agent - benzbromarone. J Rheumatol 3: 305-312

Zöllner N (1963) Eine einfache Modifikation der enzymatischen Harnsäurebestimmung. Normalwerte in der deutschen Bevölkerung. Z klin Chem 1: 178-183

Zöllner N, Gresser U, Walter-Sack I (1990) Deficient benzbromarone elimination: a familial disorder? Klin Wochenschr 68: 101

ANALYSIS OF URIC ACID TRANSPORT IN RENAL TUBULES USING BENZBROMARONE AND PYRAZINAMIDE

Yuji Moriwaki, Tetsuya Yamamoto, Sumio Takahashi, Michio Suda, Oluyemi E Agbedana, Toshikazu Hada, Kazuya Higashino

Third Department of Internal Medicine, Hyogo College of Medicine, Mukogawacho 1-1, Nishinomiya, Hyogo 663, Japan

INTRODUCTION

Four components are said to be involved in the uric acid transport system of the human kidney, but the definite mechanism of uric acid transport has not yet been clarified. We conducted both the benzbromarone (BZB) loading test and the combined pyrazinamide (PZA) and BZB test 1) to quantify the contribution of the four components involved in the renal handling of uric acid, 2) to determine whether BZB selectively inhibits postsecretory reabsorption of uric acid.

MATERIALS AND METHODS

Three healthy male volunteers, aged 28-45 years, were included in this study. Any medication which might have interfered with the renal handling of uric acid was withheld for at least two weeks prior to this study. The BZB loading test was performed as follows ; Urine and serum were obtained during a one-hour control period. At the end of the control period, 200mg BZB was given orally and from 90 min later, three 60 min clearances were performed at 60 min intervals. The combined PZA and BZB test was performed at least a week following the BZB loading test. Urine and serum were obtained during control period as in the BZB loading test. At the end of the control period, 3.0 grams PZA were administered and from 90 min later, two 60 min clearances performed. Two hundred and ten minutes after the administration of PZA, 200mg BZB was administered and thereafter, urine and sera were obtained as in the BZB loading test. Urine flow rate of 1 ml/min was maintained by drinking water during both tests. Fractional uric acid clearance (Cua/Ccr) was calculated for each of the clearance periods. Uric acid concentration was measured by the uricase method. Serum and urine creatinine concentrations were measured using an autoanalyser.

RESULTS

BZB-loading test and combined PZA and BZB test

The results of the BZB loading test and combined PZA and BZB test are summarized in Table 1. The uricosuric effect of BZB appeared in two hours and reached a maximum (4.1 to 6.8 fold compared to control values) at four hours in all three cases (Cua/Ccr; 0.083±0.018 vs. 0.415±0.035, $p<0.001$). Three grams of PZA almost completely abolished tubular secretion of uric acid after 180 minutes (Cua/Ccr; 0.050±0.023 vs. 0.0065±0.0023, $p<0.001$) and almost suppressed the uricosuric action of 200mg of BZB, as well (Cua/Ccr; 0.050±0.023 vs. 0.031±0.021, NS).

Table 1. Changes in fractional uric acid clearance following BZB loading test (above) and combined PZA and BZB test (below)

1) Results of BZB loading test

subject	1	2	3	mean ± SD	
Cua/Ccr					
(1)	0.076	0.065	0.108	0.083±0.018	
(2)	0.107	0.300	0.186		*
(3)	0.365	0.439	0.440	0.415±0.035	
(4)	0.250	0.414	0.415		

2) Results of combined PZA and BZB test

subject	1	2	3	mean ± SD	
Cua/Ccr					
(1)	0.0340	0.0340	0.0830	0.050±0.023	
(2)	0.0056	0.0035	0.0121		*
(3)	0.0061	0.0039	0.0096	0.0065±0.0023	
(4)	0.0580	0.0290	0.0074	0.031±0.021	
(5)	0.1130	0.0550	0.0087		

* $p<0.001$

Formulation and calculations of presecretory reabsorption, tubular secretion and postsecretory reabsorption

According to the four component model (Levinson and Sorenson, 1980), we can postulate unknown quantities as indicated in Fig. 1 and quantify the relative roles of these components in uric acid handling. The role of each component is hypothesized as follows: glomerular filtration is complete; uric acid is reabsorbed in presecretory reabsorption "in 'X' of the filtered uric acid"; tubular secretion "in 'Y' of the presecretory reabsorption" and postsecretory reabsorption "in 'Z' of the tubular secretion plus uric acid which escapes presecretory reabsorption";, respectively. X, Y, Z are factors which allow quantification of the amount of uric acid transported for each component of the system. In addition, the coefficient "a" is assumed in order to take into consideration the effect, if any, of BZB on presecretory reabsorption.

The following formulae are constructed theoretically based on the four component theory (Fig. 1).

$$(1-X+XY)(1-Z) = Cua/Ccr \;\ldots\ldots\; (\text{control}) \qquad (1)$$
$$(1-aX+aXY) = Cua/Ccr \;\ldots\ldots\ldots\; (\text{BZB}) \qquad (2)$$
$$(1-X)(1-Z) = Cua/Ccr \;\ldots\ldots\ldots\; (\text{PZA}) \qquad (3)$$
$$(1-aX) = Cua/Ccr \;\ldots\ldots\ldots\ldots\; (\text{PZA+BZB}) \qquad (4)$$

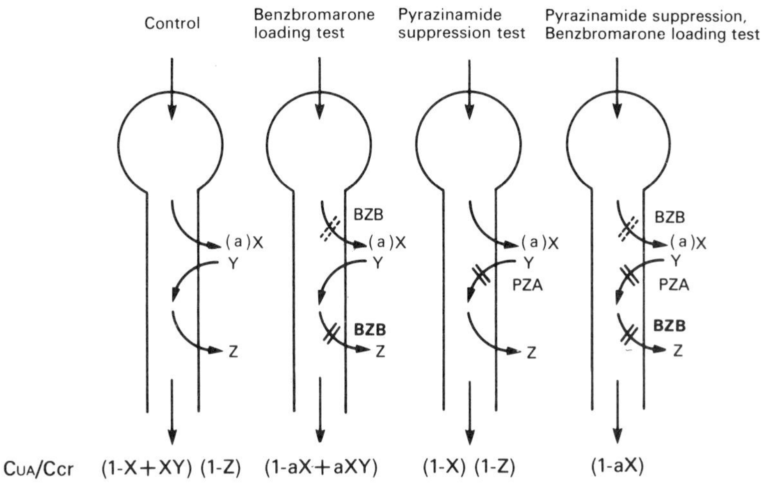

Fig. 1. Hypothetical uric acid transport system and schematic representation of fractional uric acid clearance under the state of control, benzbromarone loading, pyrazinamide suppression and pyrazinamide suppression combined with benzbromarone loading.
Presecretory reabsorption, tubular secretion, post-secretory reabsorption are indicated as "X", "Y" and "Z", respectively. The effect of benzbromarone on presecretory reabsorption was taken into consideration by assuming the coefficient "a".

(where Cua/Ccr ..... ( ): fractional uric acid clearance in control period (1), and in periods under the effect of BZB (2), PZA (3) and PZA and BZB (4))

Cua/Ccr was calculated from the two tests described above and the results were substituted into the formulae. By solving the set of the linear system, "X", "Y", "Z" and "a" were calculated and the results are shown in Table 2. The contribution of presecretory reabsorption, tubular secretion and postsecretory reabsorption in uric acid transport became 0.96±0.02 (0.94-0.98), 0.40±0.05 (0.33-0.44) and 0.84±0.05 (0.79-0.91), respectively. "a" becomes almost 1 in each of the three cases.

Table 2. Relative role of presecretory reabsorption, tubular secretion, postsecretory reabsorption and coefficient "a"

subject	1	2	3	mean	SD
X	0.94	0.98	0.95	0.96	0.02
Y	0.33	0.42	0.44	0.40	0.05
Z	0.91	0.83	0.79	0.84	0.05
a	1.00	0.99	1.04	1.01	0.02

DISCUSSION

The four component theory has been widely accepted as a model of renal handling of uric acid: that is 1) glomerular filtration, 2) presecretory reabsorption, 3) tubular secretion and 4) postsecretory reabsorption. However, renal transport of uric acid is not yet completely clarified despite having been investigated. Clinically, several drugs have been employed to explore the relative role of each component in the uric acid transport system.

High doses of PZA almost abolish the urinary excretion of uric acid through the suppression of tubular secretion (De Vries et al. 1979). Hence, high doses of PZA are often employed to investigate presecretory reabsorption of uric acid. However, when some uric acid escaping presecretory reabsorption undergo reabsorption at the postsecretory site, an overestimation of presecretory reabsorption may occur. Several uricosuric agents have been employed to determine the relative role of each component in the renal transport of uric acid. Two grams of probenecid (PB) markedly increase uric acid clearance by the inhibition of postsecretory reabsorption (Spirota et al., 1952), while a low dose of PB causes uric acid retention, as explained by the inhibition of tubular secretion (Gutman and Yu 1958). It is therefore suggested that PB inhibits not only reabsorption but also tubular secretion of uric acid and generally acts as a uricosuric agent. BZB, a potent uricosuric drug, has been considered to exert a uricosuric effect by inhibiting the postsecretory reabsorption of uric acid (Politta et al. 1973, Sinclair and Fox, 1975). Because BZB is not excreted by a renal tubular organic acid transport system, it is suggested that it does not interfere with the secretion of uric acid. However, whether the uricosuric action of BZB is exerted only by inhibition of postsecretory reabsorption or together with presecretory reabsorption is a matter which has not been completely settled. We, therefore, performed the BZB loading test both alone and after pretreatment with PZA 1) to allow a better evaluation of the relative role of each component in the renal transport of uric acid and, in addition, 2) to determine whether BZB selectively blocks postsecretory reabsorption of uric acid.

The uricosuric effect of BZB was evident within two hours with the maximal effect at four hours, as shown in Table 1 and was comparable with the results reported for BZB (Tofuku et al., 1982). The anti-uricosuric effect of PZA was apparent within two hours and continued for at least 24 hours (data now shown). We, therefore, could clearly define the period when both PZA and BZB exert their action on the renal tubules and thus, establish four equations as indicated in Fig. 1. As a result of the tests, the relative role of each of the components was quantified. In addition, BZB was demonstrated to selectively inhibit postsecretory reabsorption.

This quantification may represent only an approximation of the events, as regard to 1) quantitation (the inhibition of uric acid reabsorption by uricosuric agent is not complete) and 2) localization of uric acid transport (secretion and reabsorption are likely to occur simultaneously in some of the proximal tubules (Rieselbach et al. 1974). Further, specificity of the action site of the pharmacological agents used has been questioned (PZA might increase uric acid reabsorption (Guggino et al. 1983), as for PB, described above). However, as there has been no alternative explanation for the renal transport of uric acid other than the four component model, BZB loading test and combined PZA and BZB test seems, at present, a better means to assess the relative role of each component clinically.

REFERENCES

1) De Vries et al., Inborn hypouricemia due to isolated renal tubular defect, Biomedicine 30:75 (1979).
2) Guggino et al., Paradoxical effects of pyrazinamide (PZA) on urate transport in dog renal brush border membrane vesicles (BBMV). Kidny Int 23:256(A) (1983).
3) Gutman and Yu, Renal regulation of uric acid secretion in normal and gouty man: Modification by uricosuric agents, Bull NY Acad Med 34:387 (1958).
4) Levinson and Sorenson, Renal handling of uric acid in normal and gouty subjects: evidence for a 4-component system, Ann Rheum Dis 39:173 (1980).
5) Politta et al., Mecanisme de laction uricosrique de la benzbromarone. Schweiz Rundschau Med. (Praxis) 62:1345 (1973).
6) Rieselbach et al., Influence of the kidney upon urate homeostasis in health and disease, Am J Med 56:665 (1974).
7) Sinclair and Fox, The pharmacology of hypouricemic effect of Benzbromarone, J Rheumatol 2:437 (1975).
8) Spirota et al., Effect of Benemid on urate clearance and discrete renal functions in gouty subjects, J Clin Invest 31:696 (1952).
9) Tofuku et al., Hypouricemia due to renal urate wasting, Nephron 30:39 (1982).

DEFICIENT BENZBROMARONE ELIMINATION FROM PLASMA: EVIDENCE FOR A NEW GENETICALLY DETERMINED POLYMORPHISM WITH AN AUTOSOMAL RECESSIVE INHERITANCE

Ursula Gresser, Mohamed Adjan, Nepomuk Zöllner
Medizinische Poliklinik der Universität München
Pettenkoferstraße 8a, D-8000 München 2

INTRODUCTION

Benzbromarone is one of the three most important uricosuric drugs and widely used in Europe. Initially it was reported to be debrominated to bromobenzarone and benzarone, as shown by Yü (1976). Recently Ingeborg Walter-Sack from Heidelberg and her coworkers (Walter-Sack et al 1987; Walter-Sack et al 1990b) could show, that hydroxilation is the primary metabolic pathway. When the elimination of benzbromarone was determined by measuring plasma concentration in healthy subjects one unusual case was found. In this individual plasma concentration 12 hours after oral administration was ten times higher than in the others. 24 hours after drug application this person had still a measurable benzbromarone plasma concentration, which was higher than the 12 hours plasma concentration of the others (Walter-Sack et al 1988). The elevated 24 hours plasma concentration was thought to be the characteristic metabolic feature of deficient benzbromarone elimination.

In a population study we screened 153 subjects in search of further deficient eliminators (Walter-Sack et al 1990a). Using the plasma benzbromarone level 24 hours after ingestion of 100 mg benzbromarone as measure of drug elimination we found a wide distribution of the elimination phenotypes.

SUBJECTS AND METHODS

To determine whether deficient benzbromarone elimination is a familial disorder (Zöllner et al 1990) and to shed light on its inheritance, we investigated 46 family members of 8 individuals with different slow benzbromarone elimination rates. A single oral dose of 100 mg benzbromarone was given with 250 ml of tap water one to five hours after a continental breakfast. An additional 500 ml of water had to be ingested within 2 hours after drug administration. Further fluid intake and diet were ad libitum. Prior to and 24 hours after benzbromarone administration a 10 ml venous blood sample was drawn into a heparinized tube and centrifuged at 600 g. The plasma was stored at -23 degree celcius. Benzbromarone was determined by high performance liquid chromatography according to the method of Ittensohn and de Vries (1989).

RESULTS

Within the family of the deficient eliminator P.B. we found two brothers with deficient benzbromarone elimination. The rest of the family eliminated benzbromarone rapidly within 24 hours. The individual with deficient benzbromarone elimination P.B. had a benzbromarone plasma concentration of 1.36 µg/ml. In his father P.F. and his two sisters K.M. and P.R. either no benzbromarone was detectable after 24 hours or the concentration was less than 0.02 µg/ml. His mother P.A. and his nonidentical twin brother P.Fr. had a concentration of 0.03 µg/ml. In the younger brother P.W. the benzbromarone plasma concentration was 2.39 µg/ml. In the family of the mother of P.B. none had concentrations more than 0.03 µg/ml, in the family of the father the highest concentrations were 0.15 µg/ml in the father's brother P.O. and 0.13 µg/ml in his sister M.E..

In the other 7 families, partly investigated in three generations, we found plasma benzbromarone concentrations between 0.00 µg/ml to 0.78 µg/ml, but the results were not as clear as in P.B.'s family. For the complete list of data see Gresser (1990).

CONCLUSIONS

The population study in 153 individuals showed only one subject with clearly deficient benzbromarone elimination. Within a family of parents and five children we found two brothers with deficient benzbromarone elimination whereas the rest of the family eliminated benzbromarone rapidly within 24 hours. Therefore the results of the family examination suggest, that deficient benzbromarone elimination is a genetically controlled disorder.

An inspection of the combined results of the population study and the family studies offers evidence, that deficient benzbromarone elimination follows the pattern of autosomal recessive inheritance like other genetically determined polymorphisms, e.g. sparteine or debrisoquine. Using the plasma benzbromarone level 24 h after drug ingestion as distinguishing feature heterozygotes differ not or only little from healthy persons. The deficient benzbromarone elimination may be the marker of a new genetically determined polymorphism.

REFERENCES

Gresser U (1990) Therapie von Hyperurikämie und Gicht mit Urikosurika. Mediscript-Verlag München, ISBN 3-88320-202-9

Ittensohn A, de Vries JX (1989) A simplified method for a quantitative determination of benzbromarone in human plasma by HPLC. Eur J Clin Pharmacol 36 (Suppl): A228

Walter-Sack I, de Vries JX, Ittensohn A, Weber E (1987) Benzbromarone disposition following single dose application. Evidence for lack of debromination to benzarone. Klin Wochenschr 65 (Suppl X): 5

Walter-Sack I, de Vries JX, Ittensohn A, Kohlmeier M, Weber E (1988) Benzbromarone disposition and uricosuric action; evidence for hydroxilation instead of debromination to benzarone. Klin Wochenschr 66: 160-166

Walter-Sack I, Gresser U, Adjan M, Kamilli I, Ittensohn A, de Vries JX, Weber E, Zöllner N (1990a) Variation of benzbromarone elimination in man - a population study. Eur J Clin Pharmacol 39: 173-176

Walter-Sack I, de Vries JX, Ittensohn A, Weber E (1990b) Rapid and slow benzbromarone elimination phenotypes in man: benzbromarone and metabolite profiles. Eur J Clin Pharmacol 39: 577-581

Yü T-F (1976) Pharmacokinetic and clinical studies of a new uricosuric agent - benzbromarone. J Rheumatol 3: 305-312

Zöllner N, Gresser U, Walter-Sack I (1990) Deficient benzbromarone elimination: a familial disorder? Klin Wochenschr 68: 101

NUCLEOTIDE STATUS IN ERYTHROCYTES OF RATS INFECTED WITH PLASMODIUM BERGHEI

Gisela Jacobasch, Andreas Werner, Werner Siems and Christa Gerth

Institute of Biochemistry, Medical Faculty, Humboldt University, Hessische Str. 3-4, D(O)-1040 Berlin, F.R.G.

INTRODUCTION

Red blood cells (RBC) as well as plasmodia are not able to synthesize purines; therefore salvage pathways are essentiell for both. The source of purines is plasma adenosine. It enters the plasma membrane of erythrocytes via a specific transport protein which is located in band 4.5 together with the glucose transporter[1]. The parasites provide a high influx of adenosine into their infected host by altering the expression of the nucleoside transporter. In the erythrocyte adenosine is converted to AMP and to hypoxanthine. Last compound is preferentially translocated into parasites. They are characterized by a high capacity of purine nucleotide metabolizing enzymes including those of the salvage pathways. Parasites use hypoxanthine as substrate for the synthesis of all purine nucleotides. The enzymes of these reactions are associated with the plasmodial membrane. Therefore the question arises which interactions do exist in the purine nucleotide metabolism between the two compartments parasite and red blood cell. Is was our aim to clarify it by an analysis of the purine nucleotide status.

MATERIALS AND METHODS

Rat RBC infected with p.berghei strain LNK65 were obtained by cyclic passage of the parasites through mice, rats (150 g body weight). Purine nucleotide concentrations were measured in samples of infected whole blood and of washed RBC fractions. They was obtained by density centrifugation using a discountinous percoll sorbitol gradient. Three fractions were collected and washed with a medium containing RPMI 1640, 25 mM HEPES, 32 mM $NaHCO_3$ and 25 mM glucose, pH 7,2 and incubated one hour at 37°C. The nucleotide status was determined by HPLC using an ion-pair reversed-phase gradient elution system[2]. The degree of infection was determined microscopically after staining according to May-Gruenwald-Giemsa. The rate of glycolysis were measured photometrically according to Beutler.[3]

RESULTS AND DISCUSSION

Glucose which is the most important energy source of plasmodia is preferably transformed into lactate. The rate of glucose consumption in p.berghei infected RBC was found to be 170 ± 32 (n = 7) and that of lactate

Purine and Pyrimidine Metabolism in Man VII, Part A
Edited by R.A. Harkness *et al.*, Plenum Press, New York, 1991

Table 1. Nucleotide concentrations of p.berghei infected blood of rats (values are given as µmoles/ml cells)

Metabolite	Parasites in % of infected cells			
	65	58	47	36
ATP	1.620	1.072	1.490	1.040
ADP	0.170	0.070	0.144	0.171
AMP	0.210	0.213	0.224	0.122
GTP	0.360	0.319	0.264	0.149
GDP	0.100	0.051	0.109	0.043
UTP	-	0.189	0.114	-

formation 293 ± 31 (n = 10) µmoles/ml RBC . h at pH 7,2 and 37°C. These values are two orders of magnitude higher than in uninfected cells. A special feature of parasite energy metabolism is the fact that no Pasteur-effect does occur because no interaction does exist between glycolysis and respiration via adenine nucleotides.[4]

What enables the host to safeguard its energy status under conditions of such a high proton formation? In contrast to the parasite, PFK inhibition by ATP is also more than two orders of magnitude stronger in RBC and more sensitive against pH decrease[5]. Therefore it is to expect that RBC hexokinase is inhibited by an accumulation of Fru-6-P and G6P respectively resulting in a lower ATP generation by glycolysis in the host compartment. Nevertheless the ATP level remains approximately constant during the course of infection (Table 1). No significant differences were found between the ATP level in RBC and parasites (Table 2). In contrast to that the concentrations of GTP and UTP increase with the rate of infection. This result was examined at different stages of plasmodia development (Table 3).

Parasite infected RBC lose very easely adenine nucleotides during in vitro procedures such as density gradient centrifugation. The percentage of ATP loss grows with the degree of infection. Therefore the ATP level is lowered in these fractions. But the data in Table 3 confirm our assumption that the concentrations of GTP and UTP but not of ATP rise with the number of parasites. Furthermore, the concentrations of hypoxynthine, adenine and adenosine increase in highly infected RBC. This could be related to the importance of the salvage pathway in parasites.

We conclude from this study GTP and UTP are mainly located in the parasites but no differences do exist between the ATP concentrations in parasite and host. This seems to be an important regulatory phenomenon by which parasites can supply ATP from their own energy metabolism for the use in energy dependent reactions catalyzed in the host thus stabilizing favourable conditions for their own reproduction.

Because UTP and GTP cannot be translocated into the RBC compartment one can postulate that ATP is regulated by a specific adenine nucleotide translocase which must be located in the parasite membrane. Results of Kanaani and Ginzburg support this conclusion; they observed a decline in host cytosolic ATP and an increase in parasite ATP when inhibitors of translocase are applied to infected RBC.[6] In order to verify that the parasite makes ATP available to its host various inhibitors well known to act on steps of the respiratory chain have been applied. The ATP level of non-infected RBC does not decrease when uncubating the cells with oligomycine rotenone or carboxylcyanide-m-chlorophenylhydrazone (CCCP).

Table 2. ATP concentration in infected red blood cells and in separated host cell compartments (Values are given as μmoles/ml cells, mean ± S.D.).

Parasites (%)	Intact infected cells		Hemolysate (parasites removed)		n
0	1.737	0.136	1.660	0.135	5
26	1.264	0.130	1.364	0.145	3
43	1.227	0.038	1.200	0.235	3

Table 3. Nucleotide concentrations in red blood cells with different stages of p.berghei development (Values are given as μmoles/ml cells. Fraction 1: 84 % infected cells with schizonts; fraction 2: 30 % with trophozoites; fraction 3: 2 % with merozoites).

Metabolite	Fraction 1	2	3
ATP	0.607	0.673	0.891
ADP	0.052	0.064	0.044
Adenosine	0.037	0.004	0
Adenine	0.047	0.002	0.003
GTP	0.403	0.092	0.088
GDP	0.064	0.083	0.075
IMP	0.208	0.108	0.138
UTP	0.443	0.045	0
UDP	0.089	0	0
Hypoxanthine	0.016	u.d.	0.005
Inosine	0.008	u.d.	0.020

But a time-dependent decline of the ATP concentrations results in infected cells. As had to be expected this decrease never caused an activation of the glycolytic flux. From first compartment analysis experiments it could be concluded that a lowering of the ATP level resulting from the above mentioned inhibitors is probably not confined to the parasites but occurs also in the host compartment. This finding is in a good agreement with the postulation of the existence of an adenine nucleotide translocator regulatering the ATP level in both compartments parasite and RBC respectively.

REFERENCES

1. S. M. Jarvis, J. C. Ellory, J. D. Young, Radiation inactivation of the human erythrocyte nucleoside and glucose transporters, Biochim. Biophys. Acta, 855:312 (1986)
2. A. Werner, W. Schneider, W. Siems, T. Grune, C. Schreiter, Ion-pair reversed-phase HPLC determination of nucleotides, nucleosides and nucleobases - Application to nucleotide metabolism in hepatocytes, Chromatographia, 27:639 (1989)
3. E. Beutler, Red cell metabolism. A manual of biochemical methods. Grune & Stratton, New York (1975)
4. G. Jacobasch, D. Buckwitz, Ch. Gerth and R. Thamm, Regulation of the energy metabolism of plasmodium berghei, Biomed. Biochim. Acta, 49:289 (1990)

5. D. Buckwitz, G. Jacobasch, Ch. Gerth, Phosphofructokinase from p.berghei; a kinetic model of allosteric regulation, Molecular and Biochem. Parasitol., 40:225 (1990)
6. J. Kanaani, H. Ginsburg, Metabolic Interconnection between the Human Malarial Parasite Plasmodium falciparum and Its Host Erythrocyte, J. Biol. Chem., 264:3194 (1989)

HYPOXANTHINE DEPLETION INDUCED BY XANTHINE OXIDASE INHIBITS MALARIA PARASITE GROWTH IN VITRO

P.A. Berman and L. Human

Department of Chemical Pathology
University of Cape Town Medical School
Observatory, 7925, South Africa

INTRODUCTION

While *Plasmodium falciparum*, cause of the most virulent form of human malaria, synthesizes its pyrimidine requirements *de novo*, it lacks the ability to synthesize purines *de novo*, and depends on the erythrocyte host for its purine supply (1). Although ATP forms about 90% of the total erythrocyte purine pool, there is considerable evidence that the purine base, hypoxanthine, is the immediate purine precursor utilized by these parasites (2). Previous studies by ourselves (3) and others (4), have shown that the steady state concentration and nett rate of formation of hypoxanthine under physiological conditions is relatively low, and, apparently, inadequate to support parasite nucleic acid synthesis. We showed, however, that hypoxanthine formed in erythrocytes was recycled via a three component oxypurine cycle, comprising hypoxanthine, IMP and inosine, and that such recycling limited the rate of hypoxanthine release (5). Irreversible oxidation of hypoxanthine to uric acid by xanthine oxidase interrupted the cycle and enhanced the nett rate of ATP catabolism. During the course of these experiments, we noted that xanthine oxidase markedly decreased intracellular hypoxanthine levels. We therefore investigated whether xanthine oxidase could deplete intracellular hypoxanthine to the point where it could no longer sustain the growth and replication of malaria parasites *in vitro*.

METHODS

A chloroquine resistant strain of *P. falciparum*, FCR-3, was maintained in continuous culure in human O^+ erythrocytes, suspended in RPM1-1640 medium supplemented with 10% human serum, in an atmosphere of 4% O_2, 4% CO_2, and 92% N_2 (6). Parasite growth was assessed microscopically on Giemsa-stained blood films, and quantitated by incorporation of [^{14}C]glycine into parasite protein, or by incorporation of [^{14}C]purine, in cells in which the ATP pool had been prelabelled with [^{14}C]adenine (5), into parasite nucleic acid. Infected erythrocytes were lysed in 0.15% saponin, liberated parasites trapped on glass-fibre discs, washed, and the radioactive content measured by liquid scintillation photometry. Nucleic acid content of infected erythrocytes was determined by lysing the cells and digesting the parasite pellet with Proteinase K. Residual protein was 'salted out' with saturated sodium acetate, nucleic acid isolated by isopropanol precipitation, and quantitated by its absorbance at 260 nm.

The nucleic acid extract was electrophoresed on 2% agarose, and stained with ethidium bromide to assess its composition and provide visual semiquantitation. Xanthine oxidase, superoxide dismutase and catalase were used at concentrations of 0.1, 0.4 and 0.15 mg/ml respectively. Chloroquine sulphate was added at the concentrations indicated, and 5 mM hypoxanthine used in the 'rescue' experiments. L-lactate was measured by a spectrophotometric end-point assay, using NAD^+ and lactate dehydrogenase.

RESULTS

As can be seen in Fig. 1A and B, xanthine oxidase markedly impaired both incorporation of [^{14}C]glycine and [^{14}C]ATP derived purine into parasite protein and nucleic acid respectively. Heating the xanthine oxidase at 100°C for 10 minutes abolished inhibitory activity, while inclusion of superoxide dismutase and catalase had no effect on growth inhibition, as assessed by [^{14}C]glycine uptake.

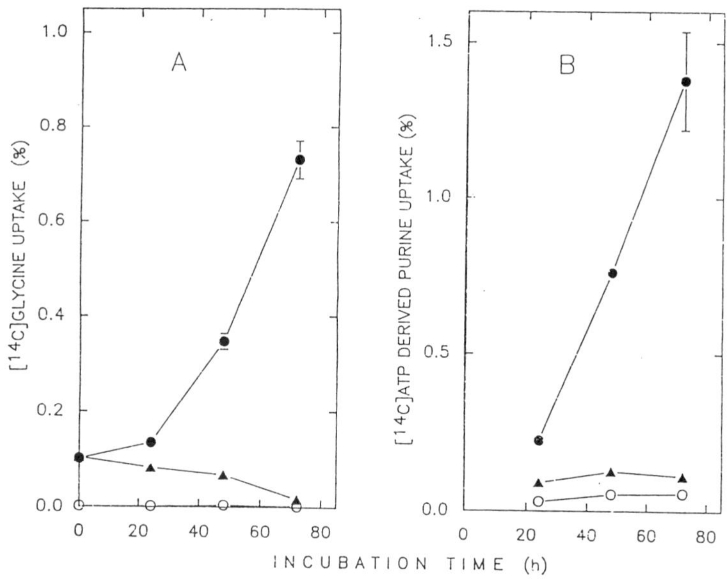

Fig. 1. Time course of [^{14}C]glycine (A) and [^{14}C]ATP derived purine (B) incorporation into parasite protein and nucleic acid, respectively (● infected erythrocytes, ▲ infected erythrocytes plus xanthine oxidase, o non-infected erythrocytes).

Fig. 2 shows the effect of chloroquine and xanthine oxidase on nucleic acid content of infected erythrocytes. Non-infected erythrocytes contain negligible nucleic acid, while infected cells show a progressive increase in their nucleic acid content. This accumulation is markedly inhibited by xanthine oxidase, but can be restored by addition of excess hypoxanthine. The effect of chloroquine is shown for comparison. 10^{-7}M chloroquine is insufficient to inhibit nucleic acid synthesis, while 5 x 10^{-7}M chloroquine is effective. This confirms a degree of chloroquine resistance. Agarose electrophoresis (Fig. 3) provides direct visual confirmation of these findings, and indicates that the predominant nucleic acid isolated is high molecular weight DNA. Microscopic examination of parasites showed that xanthine oxidase caused arrest of parasite maturation in the late trophozoite stage, that is, at the commencement of replication. Hypoxanthine restored the normal distribution of parasite stages, with ring forms predominating.

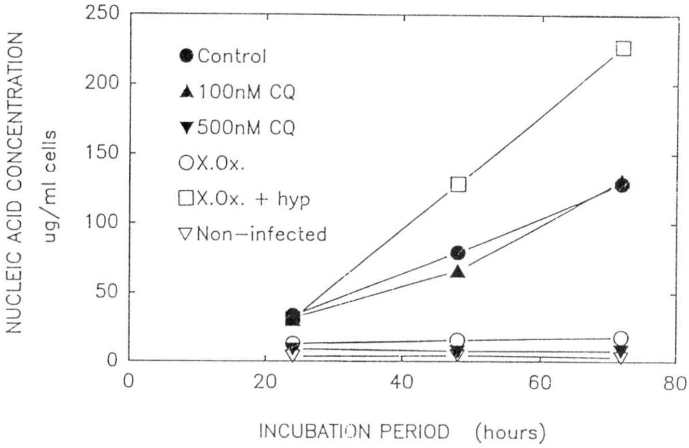

Fig. 2. Time course of nucleic acid accumulation in infected erythrocytes incubated for 72 hours under the conditions indicated. CQ chloroquine, X.Ox. xanthine oxidase, hyp hypoxanthine.

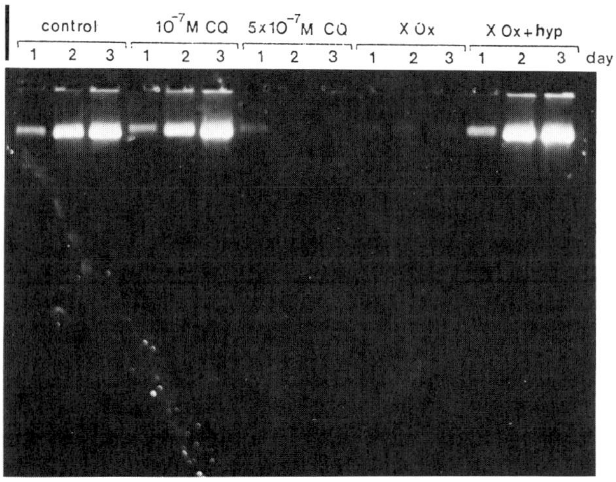

Fig. 3. Agarose gel electrophoresis of the nucleic acid extracts corresponding to each time point in Fig. 2.

Fig. 4 shows that parasite metabolic activity, as reflected by L-lactate release, shows the same pattern as nucleic acid accumulation in its response to chloroquine and to xanthine oxidase, with or without hypoxanthine rescue.

DISCUSSION

We have shown, by a variety of independent biochemical parameters, that xanthine oxidase markedly impairs parasite growth, and that this inhibition can be reversed by excess hypoxanthine but not by active oxygen radical scavengers including superoxide dismutase and catalase. This argues strongly that xanthine oxidase toxicity results from hypoxanthine depletion, rather than from the generation of active oxygen metabolites. The study confirms that hypoxanthine is indeed an obligatory intermediate in purine utilization by the malaria parasite, and raises the possibility of treating malaria infection *in vivo* with xanthine oxidase. This is of interest in view of the recent emergence of strains of Plasmodium showing multiple drug resistance to conventional chemotherapy. Modification of the enzyme with polyethylene-glycol (7) or by encapsulation in red cell ghosts (8), may shield the enzyme from immune detection and prolong its circulating life, while leaving its catalytic action intact.

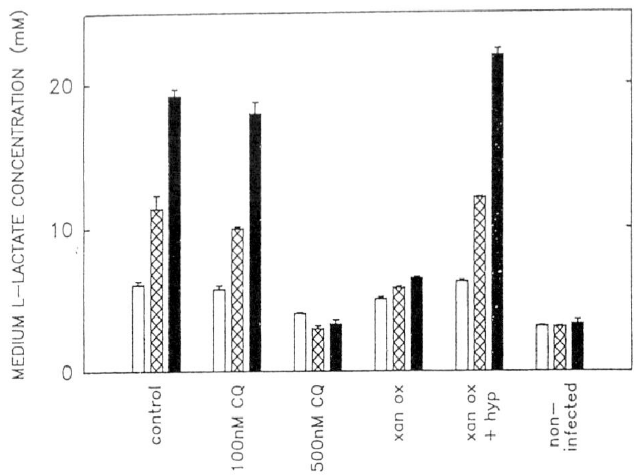

Fig. 4. Accumulation of lactate in the medium of cells incubated under the conditions described in Fig. 2 (□ day 1, ▧ day 2, ■ day 3).

ACKNOWLEDGEMENTS

I would like to thank the Medical Reserach Coucil for financial support and L. Odes for typing the manuscript.

REFERENCES

1. WHO Scientific Group, Nucleic acid metabolism. *In* The biology of malaria parasites. WHO Tech. Rep. Ser. Geneva. 743: 125 (1987).
2. H.K. Webster and J.M. Whaun, Purine metabolism during continuous erythrocyte culture of human malaria parasites (*P. falciparum*). Prog. Clin. Biol. Res. 55: 557 (1981).
3. P.A. Berman, D.A. Black, L. Human and E.H. Harley, An oxypurine cycle in human erythrocytes regulated by pH, inorganic phosphate and oxygen. J. Clin. Invest. 82: 980 (1988).
4. F. Bontemps, G. Van den Berghe and H.G. Hers, Pathways of adenine nucleotide catabolism in erythrocytes. J. Clin. Invest. 77: 824 (1986).
5. P.A. Berman and L. Human, Regulation of PRPP and of hypoxanthine uptake and release in human erythrocytes by oxypurine cycling. J. Biol. Chem. 265: 6562 (1990).
6. W. Trager and J.B. Jensen, Human malaria parasites in continuous culture. Science 193: 673 (1976).
7. A. Abuchowski, J.R. McCoy, N.C. Palczuk, T. van Es and F.F. Davis, Effect of covalent attachment of polyethylene glycol on immunogenicity and circulating life of bovine liver catalase. J. Biol. Chem. 252: 3582 (1977).
8. K. Adriaenssens, D. Karcher, A. Lowenthal and H.G. Terheggen, Use of enzyme-loaded erythrocytes in *in vitro* correction of arginase deficient erythrocytes in familial hyperargininaemia. Clin. Chem. 22: 323 (1976).

NEW NUCLEOSIDE TRANSPORT PATHWAYS INDUCED IN THE HOST ERYTHROCYTE MEMBRANE OF MALARIA AND BABESIA INFECTED CELLS

Annette M. Gero and Andrew M.Wood

School of Biochemistry, University of New South Wales P.O.Box 1, Kensington, NSW Australia, 2033

INTRODUCTION

Within hours after the human malarial parasite *Plasmodium falciparum* invades the host erythrocyte, new permeability pathways appear in the host cell membrane[1]. We have demonstrated that the intraerythrocytic malarial parasite appears to induce an additional component of nucleoside transport in the host erythrocyte membrane and this comprises about 40% of the total transport in the infected cells. It is stage specific; it occurs mainly at the trophozoite stage of the malarial parasite development. This induced component is insensitive to the classical inhibitors of mammalian nucleoside transport, nitrobenzylthioinosine (NBMPR), nitrobenzylthioguanosine (NBTGR), dilazep and dipyridamole, indicating that the properties of the induced transporter are significently different to that of the host erythrocyte[2,3]. Similarly, nucleoside transport, which is inoperative in normal bovine erythrocytes, is induced upon infection of the bovine erythrocyte with *Babesia bovis*. In a manner similar to transport in malaria infected cells, transport inhibitors, such as NBMPR, showed significantly less inhibition[4].

A logical approach to the control of diseases such as malaria and babesiosis is to develop agents which will specifically inhibit the growth and reproduction of parasites in their host cells, the erythrocyte. The ideal treatment would specifically target the parasite-infected cell and leave the normal host tissues unaffected, thus resulting in few side effects. One way to achieve this is to exploit the presence of transport systems in infected cells which are different from those present in normal cells. The induced nucleoside transporters in parasite infected erythrocytes present the possibility of the utilization of cytotoxic nucleosides against *P. falciparum* or *B.bovis* infection in conjunction with a nucleoside transport inhibitor to protect the host tissue. We present here the toxicity data for a number of synthetic nucleosides which have been screened for possible antimalarial and antibabesial activity.

EXPERIMENTAL

Babesia bovis (Samford isolate) and *Plasmodium falciparum.* (FCQ-27 isolate) were maintained in culture as previously described[5,6]. Growth *in vitro* was monitored by the incorporation of [G-^{3}H]hypoxanthine into parasitic nucleic acids [3]. The results were confirmed by microscopic analysis of Giemsa-stained blood smears. IC_{50} value for each drug equals the concentration of drug which reduced [^{3}H] hypoxanthine incorporation to 50% of the control. *In vivo* tests were based on the work of Gati et al[7].

Deoxyadenosine and adenosine analogues were supplied by Dr R.L. Blakley, St Judes Childrens Hospital, Memphis, TN. USA, Dr. P. Parsons, Queensland Institute of Medical Research, Brisbane, Australia and the National Cancer Institute, Bethesda, MD., U.S.A. NBMPR-P (nitrobenzylthioinosine 5' monophosphate) was a gift from Dr A.R.P. Paterson, University of Alberta, Canada.

RESULTS AND DISCUSSION

Of the adenosine analogues tested, the chemical structures fall into two main categories: (a) the modification of the purine ring to include substituents on positions 1-8 and 7 deaza purine analogues (i.e.tubercidin); and (b) analogues of group (a) containing deoxy or dideoxy sugar moieties. Dideoxy compounds were tested because of their antiviral and anticancer activities.

All compounds were tested in *in vitro* cultures of *B. bovis* and *P. falciparum* over a range of concentrations and IC50 values obtained (Table 1). Tubercidin was shown to be the most toxic compound towards *B. bovis* with an IC_{50} value of 0.07 µM, whilst sangivamycin was most toxic against malaria (0.2 µM). For the bromoadenosine series, 2-bromo-adenosine was approximately eight-fold more toxic than 8-bromo-adenosine against *B. bovis in vitro.* Substitution of the ribose with deoxyribose (2-bromo-deoxyadenosine) decreased the toxicity. Conversely, the 8-bromo-deoxyadenosine analogue showed slightly greater potency when compared with the corresponding 8-bromo-adenosine. Linking the sugar via the N^3 of the purine ring (i.e. 8-bromo-3-ribosyl adenine), increased the toxicity almost five-fold. However, in the case of the deoxyadenosine analogue, attachment of the sugar via the N^3 position (i.e. 8-bromo-3-(2'-deoxyribosyl)adenine) rendered the compound non-toxic towards Babesia.

However, in most instances, Babesia and malaria had different susceptibilities to the compounds, for instance, sangivamycin 2-amino-triazolo-deoxyadenosine and 6- methylamino-deoxy-adenosine were toxic against malaria, but were not toxic against Babesia. Clearly there are extensive differences in drug specificities between the two organisms, although in both cases the drugs enter by the via the nucleoside transporters. The fact that some compounds have little toxicity against *B. bovis* but toxicity against malaria or vice versa, suggest that there are significant differences in either the transport of these compounds into each of the infected cells and/or that the metabolism of such compounds in each parasite is different. These differences are the subject of further investigation.

TABLE 1. IC50 Values for Adenosine Analogues Against *B. bovis* and *P. falciparum*

COMPOUND	*B. bovis* $IC_{50}(\mu M)$	*P. falciparum* $IC_{50}(\mu M)$
Tubercidin	0.07	0.42
2-bromo-adenosine	5.5	35
8-bromo-3-(2'-ribosyl) adenine	6.0	≤ 75
6-phenylamino-deoxyadenosine	8	>10
2-CH_3S-deoxyadenosine	16	>10
8-bromo-deoxyadenosine	23	>100
Sangivamycin	40	0.2
8-bromo-adenosine	40	≤ 15
2-bromo-deoxyadenosine	40	≤ 30
2-CF_3-deoxyadenosine	55	>50
8-bromo-3-(2'-deoxyribosyl) adenine	58	>100
2-amino-deoxytriazolo-adenosine	> 100	10
6-methylamino-deoxyadenosine	> 100	8

In order to determine antiparasitic activity *in vivo*, mouse trials were conducted using NBMPR-P and dilazep as the nucleoside transport inhibitors and tubercidin and sangivamycin as the toxic compounds in combination therapy in mice infected with malaria. Mice were injected with 10^6 *P. berghei* parasites on day 1 and treatments (i.p. injections)were on days 4,5,6 and 7. Weight loss and parasitaemias were monitored daily. The results are shown in Table 2.

Treatment with 25mg/kg NBMPR-P or dilazep + 6 or 7mg/kg tubercidin on days 4,5,6,and 7 resulted in lower parasitemia and longer survival (13 days to the first death) compared to the untreated mice (7 days) or for those treated with only NBMPR-P (7 days) or dilazep alone (6 days). The uninfected control groups of mice treated with the above drug combinations showed no ill effects (not shown). In further experiments NBMPR-P + sangivamycin at 2mg/kg also slowed parasitaemia, and mice had a survival time of 13 days to the first death. The results show that all 3 combinations greatly slowed the progression of parasitemia, with all survival times increased to a similar extent.

These results confirm the potential use of cytotoxic nucleosides in parasite infected erythrocytes,in which the nucleoside enters the infected cell through the induced NBMPR insensitive transporters. These compounds have potential as antiparasitic agents, when used in combination with a transport blocker which allows protection to the host cell from the toxic effect of the nucleoside.

TABLE 2. Effect of nucleosides in combination with NBMPR-P or dilazep against *P. berghei* infected mice.

Analogue (mg/kg)	NMBPR-P (25 mg/kg)	Dilazep (25 mg/kg)	No. of mice alive Day 7	Day 11	Parasitemia (%) Day 7	Day 11	Day of 1st Death
None	-	-	4/6	1/6	52±20	45	7
-	-	+	2/5	0/5	70±16	NA	6
-	+	-	5/5	0/5	61±12	NA	7
Tubercidin							
7	-	+	5/5	5/5	4±1	15±3	13
6	+	-	4/4	4/4	4±1.5	18±3	13
4	+	-	5/5	3/5	12±10	17±9	11
2	+	-	5/5	3/5	21±12	18±9	10
2	-	+	5/5	4/5	12±4	31±5	11
Sangivamycin							
2	+	-	6/6	6/6	21±14	24±20	13
1.4	+	-	5/5	3/5	15±3	34±22	11

Further characterisation of the altered properties of nucleoside transport and the further identification of the biochemical differences that exist between the infected cell and its host with respect to metabolism and the altered permeability site will contribute to the more complete understanding in the design of new antiparasitic drugs which can be selectively directed into the malarial or babesial infected cell.

ACKNOWLEDGEMENT

This work was supported by UNDP/World Bank/WHO Special Programme for the Research and Training in Tropical Diseases.

REFERENCES

1. I.W Sherman. Mechanisms of molecular trafficking in Malaria. Parasitol. 96:s57 (1988).
2. A M. Gero, E. M. A. Bugledich, A. R. P. Paterson and G. Jamieson. Stage specific alterations of nucleoside permeability and nitrobenzylthioinosine insensitivity in Plasmodium falciparum infected erythrocytes, Mol.Biochem. Parasitol, 27:159 (1988).
3. A.M. Gero, H. V. Scott, W. J. O'Sullivan and R. I. Christopherson, Antimalarial action of nitrobenzylthioinosine in combination with purine nucleoside antimetabolites. Mol. Biochem Parasitol 34:87 (1989).
4. A M. Gero. Alteration in nucleoside permeability in Babesia bovis infected erythrocytes. Mol Biochem Parasitol. 35:269 (1989).
5. S.E.Nott, W.J. O'Sullivan, A.M.Gero and A.S Bagnara. Routine screening for potential babesicides using cultures of Babesia bovis. Int J. Parasitol 20:797 (1990).
6. W.Trager, and J.B. Jensen. Human malaria parasites in continuous culture. Science 193: 673 (1976)
7. W. P. Gati, A.F. Stoyke, A. M Gero and A. R. P. Paterson. NBMPR insensitive nucleoside permeation in mouse erythrocytes infected with Plasmodium yoelii, Biochim. Biophys. Res. Comm. 145:1134 (1987).

EFFECT OF POLYETHYLENE GLYCOL-MODIFIED ADENOSINE DEAMINASE (PEG-ADA) THERAPY IN TWO ADA-DEFICIENT CHILDREN: MEASUREMENT OF ERYTHROCYTE DEOXYADENOSINE TRIPHOSPHATE AS A USEFUL TOOL

C. Bory[1], R. Boulieu[2], G. Souillet[3], C. Chantin[1], P. Guibaud[4], M.S. Hershfield[5]

Laboratoire de Biochimie[1], Unité Hors-Germe[3], Service de Pédiatrie[4], Hôpital Debrousse, Lyon, France. Laboratoire de Pharmacie Clinique[2], Faculté de Pharmacie, Lyon, France
Medecine Department[5], Duke University, Durham, USA

INTRODUCTION

Adenosine deaminase (ADA) deficiency is usually associated with severe combined immunodeficiency (SCID) and results in clinical, immunological and biochemical abnormalities. Biochemical abnormalities are characterized by accumulation of purine metabolites ; adenosine, deoxyadenosine, dADP and dATP in biological fluids and cells[1,2]. Untreated this disease is usually fatal during the first few months of life. The therapy of choice is bone-marrow transplantation from histocompatible donors[3]. Unfortunately in most instances, HLA-identical donors are not available . Thus others therapeutic approaches including gene therapy have to be considered. Enzyme replacement therapy in the form of repeated red cell transfusions has been of clinical benefit in some children but the response is often unsustained[4]. Recently, a new enzyme replacement therapy, polyethylene glycol-modified bovine ADA has been tested in some children with ADA deficiency and SCID[5] .
In this paper, the effect of PEG-ADA therapy on biochemical abnormalities, clinical status and immunological function in two ADA-deficient children is studied. Using a specific HPLC method for the determination of dATP concentrations, a correlation between dATP levels and the evolution of immunologic functions is shown.

METHODS

HPLC Analysis of purine metabolites

Blood sample collection and treatment are described elsewhere in detail[6]. Adenine, adenosine and deoxyadenosine were measured by a sensitive and specific reversed-phase HPLC method[7]. The method also allows the determination of S-adenosyl homocysteine (SAH). Ribonucleotides and deoxyribonucleotides of adenine were analysed using an anion-exchange HPLC method[7]

ADA assay

ADA activity in erythrocytes, lymphocytes and plasma was determined by a spectrophotometric method.

Immunologic methods

Lymphocyte sets were analyzed by a flow-cytometric method using monoclonal antibodies. Lymphocyte blastogenic response to non specific mitogens, concanavalin A, phytohemaglutinin and pokeweed mitogen, was performed by ^{3}H-thymidine incorporation.

CASE REPORTS

Patient 1 is a 4 years old girl hospitalized at birth for cyanosis and respiratory distress. On admission to the hospital, she weighted 2.9 kg and her height was 47 cm. At 2 months of age, she was diagnosed as an ADA-deficient patient with SCID. ADA activity in erythrocytes was not detectable (< 2 nmol/h/mg Hb). A severe lymphopenia was found (65 to 100 cells/mm^3). At the age of 3 months, the child presented severe pulmonary candidiasis requiring intensive care with repeated episodes of apnea and a cardiac arrest. An hypotonia and a postanoxic renal failure persisted. The child was put into germ-free environment until the age of 28 months when she could be discharged from the hospital. After diagnosis, the child received six foetal thymus/ liver transplants between the ages of 4 and 10 months, without clinical and biochemical improvement. Transfusions with irradiated red cells were started at 12 months and performed monthly for 8 months . Despite of clinical improvement associated with a decrease of erythrocyte dATP concentrations, transfusion therapy was not sufficiently effective to reduce and to maintain erythrocyte dATP levels at values compatible with restoration of immune function[8].
A clinical trial with PEG-ADA was started at 21 months according to Hershfield protocole . On the basis of the biochemical and immunological responses of the patient, she received a single intramuscular injection of PEG-ADA every week with an increase of PEG-ADA dose during the first 4 weeks of treatment from 10 to 30 U/kg. The child received 30 U/kg twice a week from april 1990 to november 1990 and then a single dose of 30 U/kg per week. Plasma ADA activity ranged from 30-40 µmol/h/ml.

Patient 2 is a 1.5 years old adopted boy diagnosed at birth as phenylcetonuric. He was hospitalized for recurrent infections (respiratory tract and digestive infections) and slow growth rate (-3DS). A lymphopenia and a reduced number of T lymphocytes were found. He was diagnosed as ADA-deficient with SCID at 11 months. The child was transferred to a germ-free unit. PEG-ADA treatment was started on july 1990 according to Hershfield protocole. He received a single weekly injection of PEG-ADA at a dose of 20 U/kg and then at a dose of 30 U/kg since august 1990. Plasma ADA activity was between 30-40 µmol/h/ml.

RESULTS

At diagnosis , four abnormal purine metabolites, adenosine, deoxyadenosine , dADP and dATP were found in biological fluids and cells from patient 1. In erythrocytes, dATP level was 479 nmol/ml packed cells and dADP was 100 nmol/ml packed cells whereas these compounds were not detectable from healthy subjects[1]. The determination of purine nucleotides in lymphocytes was not possible as the peripheral lymphocytes count was too low. In plasma samples, adenosine was present at a concentration of 1.9 µmol/l (normal values 0.8 ± 0.4 µmol/l, mean ± SD)[8], whereas deoxyadenosine and SAH were not detectable. In urine, high levels of deoxyadenosine were found.

In patient 2, before PEG-ADA treatment, high levels of dATP and dADP (951 and 126 nmol/ml packed cells respectively) were found in erythrocytes. In plasma, adenosine and deoxyadenosine levels were 2.5 and 3.1 µmol/l respectively whereas SAH was undetectable.

Following PEG-ADA therapy, in both children, the concentrations of erythrocyte dATP decreased rapidly as shown in Fig. 1. The levels of dADP were undetectable after one month of treatment. Similarly, in plasma, adenosine in both patients and deoxyadenosine in patient 2 became undetectable.
Regarding to the tests of cellular immune function we noted that before treatment, both patients present severe lymphopenia with reduced number of T lymphocytes. During the twelve first months of treatment , circulating T lymphocytes increased regularly in patient 1 (Fig. 1a) associated with a significant increase of lymphocyte blastogenic response to mitogens (Fig. 2a). Then, during the second year, a decline of T lymphocyte counts was observed with a low blastogenic response to non specific mitogens.
In patient 2, T lymphocyte subsets and lymphocyte response to mitogens increased to normal range within 3 months of treatment and remained near normal values over 7 months of PEG-ADA treatment (Fig. 1b and 2b).

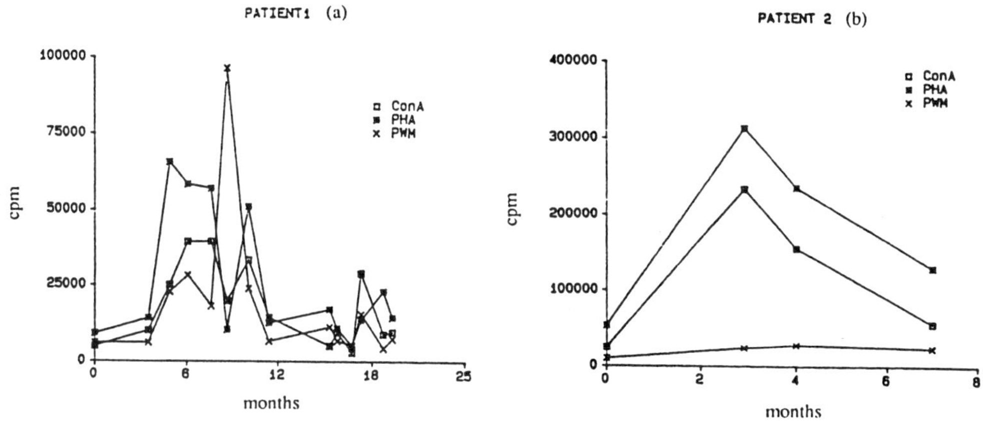

Fig.1.Effect of PEG-ADA therapy on erythrocyte dATP concentrations and circulating counts of T lymphocyte subsets.

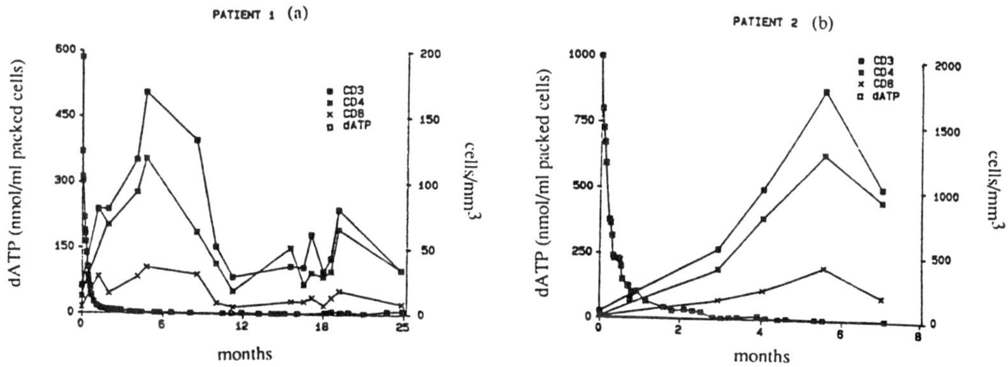

Fig.2. Lymphocyte response to non specific mitogens during PEG-ADA treatment.

On the clinical level, there have been no serious infections in either patient following PEG-ADA treatment. Patient 1 regained appetite with a significant weight gain. However, the height-weight development remained slow. On the neurologic and psychomotor level, a marked improvement was observed (the child looked alert, there was an improvement in muscular tonus) but progress was slow and a psychomotor retardation persisted. Patient 2 has gained 4 kg within 7 months and is doing very well clinically. Both children live at home.

DISCUSSION

Following PEG-ADA therapy, the concentrations of abnormal purine metabolites decreased to near undetectable levels associated with an increase of T lymphocyte counts and lymphocyte blastogenic response to mitogens. Concomitantly, a significant improvement of clinical status, indicated by resistance to infection and weight gain, was observed. In the two patients, the results obtained show that the start up of immune functions, albeit partial in patient 1, is correlated with dATP erythrocyte concentrations. Immune functions were restored only when dATP levels were below 15 nmol/ml packed cells. In a previous study[9] from an ADA-deficient bone marrow transplanted patient we found that normal immune functions are compatible with low erythrocyte dATP levels (14 nmol/ml packed cells).

In patient 1, a decline of T lymphocyte count has been observed during the second year of PEG-ADA treatment. However, dATP levels remained low (< 3 nmol/ml packed cells) and no clinical deterioration was observed. So, it appears that even if the lymphocyte count is low, the cells provide adequate immune protection, so long as PEG-ADA is given to prevent dATP accumulation.

Compared to red cell transfusion, PEG-ADA therapy is a suitable treatment to reduce and maintain dATP levels to near undetectable levels allowing the immune function to be restored.In conclusion, our results show that the measurement of dATP levels is very useful in monitoring and evaluating the degree of efficiency of ADA-deficiency therapy.

REFERENCES

1. G. Morgan, G. Levinsky, K. Hugh-Jones, L.D. Fairbanks, G.S. Morris, H.A. Simmonds. Heterogeneity of biochemical, clinical and immunological parameters in severe combined immunodeficiency due to adenosine deaminase deficiency. Clin. Exp. Immunol. 70 : 491 (1987).
2. N.A. Kredrich, M.S. Hershfield. Immunodeficiency diseases caused by adenosine deaminase deficiency and purine nucleoside phosphorylase deficiency. In : "The Metabolic basis of Inherited Disease", (Edition McGraw-Hill, New-York (1989).
3. R.M. Blaese, D.B. Kohn, R.C. Moen. The treatment of adenosine deaminase deficiency. Immunol. Today 8 : 296 (1987).
4. R. Hirchhorn. Inherited enzyme deficiencies and immunodeficiency : adenosine deaminase (ADA) and purine nucleoside phosphorylase (PNP) deficiencies. Clin. Immunol. Immunopathol. 40 : 157 (1986).
5. M.S. Hershfield, R.H. Buckley, M.L. Greenberg, A.L. Melton, R. Schiff, C. Hatem, J. Kurtzberg, M.L. Markert, R.H. Kobayashi, A.L. Kobayashi, A. Abuchowski. Treatment of adenosine deaminase deficiency with polyethylene glycol-modified adenosine deaminase. N. Engl. J. Med. 316 : 589 (1987).
6. R. Boulieu, C. Bory, C. Gonnet. High performance liquid chromatographic method for the analysis of purine and pyrimidine bases, ribonucleosides and deoxyribonucleosides in biological fluids. J. Chromatogr. 339 : 380 (1985).
7. R. Boulieu, C. Bory, C. Gonnet. Liquid chromatographic measurement of purine nucleotides in blood cells. Clin. Chem. 31 : 727 (1985).
8. C. Bory, R. Boulieu, G. Souillet, C. Chantin, M.O. Rolland, M. Mathieu, M.S. Hershfield. Comparison of red cell transfusion and polyethylene glycol-modified adenosine deaminase therapy in an adenosine deaminase-deficient child : measurement of erythrocyte deoxyadenosine triphosphate as a useful tool. Pediatr. Res. 28 : 127 (1990).
9. R. Boulieu, C. Bory, G. Souillet. Purine metabolism in a bone marrow transplanted adenosine deaminase deficient patient. Clin. Chim. Acta 178 : 349 (1988).

URATE TRANSPORT IN THE PROXIMAL TUBULE OF HUMAN KIDNEY

Dominique Werner, Barbara Guisan
and Françoise Roch-Ramel

Institute of Pharmacology and Toxicology
University of Lausanne, CH-1005 Lausanne
Switzerland

INTRODUCTION

Urate is transported bidirectionally by the renal proximal tubule of mammals. The direction of the net flux is, however, highly species dependent, so that some mammals are urate "reabsorbers" while others are urate "secreters" (Roch-Ramel and Weiner, 1980).

Urate reabsorption as well as secretion involves transport across two barriers: the brush border membrane (BBM) and the basolateral membrane. Previous studies, reviewed by Kahn and Weinman (1985), have shown that urate is transported across the BBM of rats and dogs, two urate "reabsorbers", by a urate/anion exchanger which has affinity for a wide variety of inorganic and organic anions, such as OH^-, Cl^-, HCO_3^-, pyrazinoate, probenecid, L-lactate, p-aminohippurate. On the other hand, such urate/anion exchanger was found to be absent from the BBM of pigs (Werner et al., 1990) and rabbits (Martinez et al., 1990), which are considered as urate "secreters". Thus, as postulated by Guggino et al. (1983), the luminal urate/anion exchanger might play a role in urate reabsorption. If this is the case, such exchanger should be present in the BBM of man, since he is a urate "reabsorber". We have tested this hypothesis by using BBM vesicles isolated from human renal cortex.

METHODS

Membrane Preparation. Renal material was obtained from patients undergoing nephrectomy for malignancy and only the tumor-free renal cortex was used for the isolation of BBM vesicles. Kidney cortex slices (0.8 g) were homogenized in 16 ml of isolation buffer (300 mM mannitol, 5 mM EGTA, 12 mM Tris-HCl, pH 7.4) with a Polytron and the purification of the membranes was achieved by the EGTA/Mg^{2+}-precipitation method described previously by Biber et al. (1981). The final BBM vesicles suspension was frozen and stored in liquid nitrogen until use.

Protein Determination. Protein content was measured according to Bradford (1976) using the Bio-Rad protein assay kit with bovine plasma gamma globulin as a standard.

Purine and Pyrimidine Metabolism in Man VII, Part A
Edited by R.A. Harkness *et al.*, Plenum Press, New York, 1991

Enzyme Assays. The purification of the BBM vesicle preparations was evaluated by measuring the specific activities of leucine-aminopeptidase (a typical brush border enzyme) and of Na^+/K^+-ATPase (a specific basolateral marker) according to Haase et al. (1978) and Berner and Kinne (1976), respectively. Compared to the initial homogenate, the enrichment factors in the membrane vesicles were 16.8 ± 1.5 (mean ± SE; n=3) for leucine-aminopeptidase and 0.97 ± 0.14 (n=3) for Na^+/K^+-ATPase. This means that the brush border membranes were enriched at least 17-fold with respect to the basolateral membranes.

Uptake Studies. The BBM vesicles, suspended in 40 mM KCl, 60 mM K^+ gluconate, 100 mM mannitol, 20 mM HEPES-Tris, pH 7.4, were diluted with their intravesicular buffer in order to give a protein content of 15 to 22.5 mg/ml. Ethanolic solutions of valinomycin (20 mg/ml) and of FCCP (3 mg/ml) were added to the membrane suspension at a concentration of 0.5% (v/v). The vesicles were rehomogenized by passing them 20 times through a 25-gauge needle with a syringe and kept on ice until the beginning of the experiment. The uptake of [^{14}C]-urate (54 mCi/mmol; Amersham International) was studied at 25°C by the rapid filtration technique as previously described (Werner et al., 1991). The precise experimental conditions (composition of uptake medium, volumes of membrane suspension and of uptake medium) will be specified in the legends to the Figure and to the Table.

RESULTS AND DISCUSSION

In order to investigate if a urate/anion exchanger is present in the BBM of the human proximal tubule, we studied, under both pH and voltage clamp conditions, the effect of an outwardly directed Cl^- gradient (Cl^-in = 40 mM; Cl^-out = 6.7 mM) on urate uptake in BBM vesicles (Fig. 1). Compared to the control condition (Cl^- equilibrium: Cl^-in = Cl^-out = 40 mM), urate uptake was markedly enhanced in presence of the Cl^- gradient and a transient accumulation above the equilibrium value could be observed. This stimulation was not secondary to an alkalinization of the intravesicular space due to a Cl^-/OH^- exchange, since the pH was clamped. The creation of an electropositive intravesicular space by the diffusion of Cl^- ions along their concentration gradient was not responsible either for the stimulation of urate uptake, since the voltage was clamped. These results are thus consistent with the presence of a transport system allowing direct Cl^-/urate exchange in human BBMV.

The affinity of a few organic anions for this Cl^-/urate exchanger was investigated as described in the legend of Table 1. All the inhibitors tested here have been shown to have affinity for the urate/anion exchanger of rats and dogs. The data of Table 1 show that cold urate when treated as an inhibitor produced a dose-dependent inhibition of [^{14}C]-urate uptake indicating a saturation process. The transport system was completely saturated with 1 mM cold urate. Pyrazinoate, probenecid and L-lactate inhibited the Cl^-/[^{14}C]-urate exchange in a dose dependent fashion with the following relative power sequence: pyrazinoate › probenecid › L-lactate. In strike contrast to what has been observed in rats and dogs, p-aminohippurate seemed to have only a very low affinity for the human exchanger. Indeed, even at a concentration (5 mM) which was more than 100 times higher than that of [^{14}C]-urate (40 µM), p-aminohippurate induced an inhibition of less than 20%.

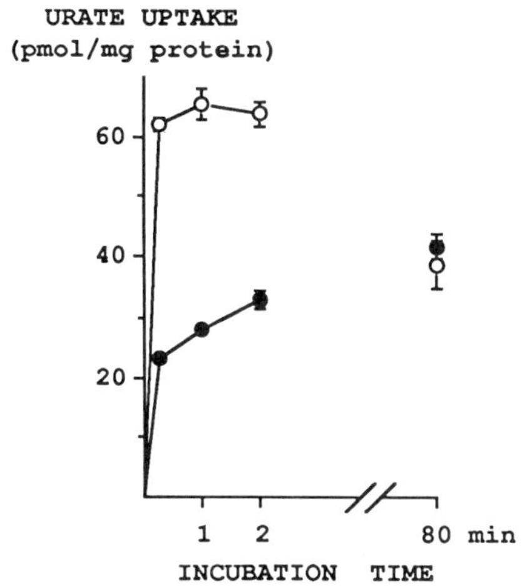

Fig. 1. Effect of an outwardly directed Cl^- gradient on urate uptake in human BBM vesicles.
17 µl BBM vesicles were mixed with 85 µl uptake medium containing 100 mM K^+ gluconate (Cl^- gradient: ○) or 40 mM KCl, 60 mM K^+ gluconate (Cl^- equilibrium: ●), 100 mM mannitol, 20 mM HEPES-Tris, pH 7.4 and 40 µM [^{14}C]-urate. Data are the means ± SD of one experiment performed in triplicate.

Table 1. Cis-inhibition of Cl^-/[^{14}C]-urate exchange by various organic anions

Anion	mM	% Inhibition
Urate	0.1	44.7 ± 3.3
	1.0	95.8 ± 1.9
Pyrazinoate	0.1	61.6 ± 3.5
	1.0	100.7 ± 4.0
Probenecid	1.0	80.5 ± 0.8
	2.0	95.5 ± 2.2
PAH	5.0	18.4 ± 4.5
L-Lactate	1.0	20.4 ± 4.7
	5.0	66.3 ± 2.8

The uptake of 40 µM [^{14}C]-urate was measured after 15 sec incubation. BBM vesicles (3.33 µl) were mixed with 16.67 µl uptake medium containing either 100 mM K^+ gluconate (Cl^- gradient) or 40 mM KCl, 60 mM K^+ gluconate (Cl^- equilibrium), 100 mM mannitol, 20 mM HEPES-Tris, pH 7.4, in absence or in presence of various organic anions at different concentrations, as indicated. The component of uptake due to Cl^-/[^{14}C]-urate exchange was obtained from the difference between the Cl^- gradient condition and the Cl^- equilibrium condition. In absence of inhibitor (control), it averaged 28.49 ± 0.14 pmol/mg protein. Data represent the mean ± SE of three separate experiments.

In summary, we have demonstrated that urate transport in human BBM vesicles isolated from human cortex can occur in exchange of Cl^-. This transport mechanism is saturable and can be cis-inhibited by pyrazinoate, probenecid and L-lactate. These results are consistent with the presence of a urate/anion exchanger in the BBM of the human proximal tubule. Although sharing many similarities with the urate/anion exchangers of rats and dogs, the human transport system differs because it has only a very low affinity for p-aminohippurate.

ACKNOWLEDGMENTS

We are grateful for the expert secretarial assistance of Mrs Nicole Skarda-Coderey. This study was supported by the Swiss National Foundation for Scientific Research (grant # 31.25647.88).

REFERENCES

Berner, W., and Kinne, R., 1976, Transport of p-aminohippuric acid by plasma membrane vesicles isolated from rat kidney cortex, Pflügers Arch., 361:269.

Biber, J., Stieger, B., Haase, W., and Murer, H., 1981, A high yield preparation for rat kidney brush border membranes. Different behaviour of lysosomal markers, Biochim. Biophys. Acta, 647:169.

Bradford, M.M., 1976, A rapid and sensitive method for the quantitation of microgram quantities of protein utilizing the principle of protein-dye binding, Anal.Biochem., 72:248.

Guggino, S.E., Martin, G.J., and Aronson, P.S., 1983, Specificity and modes of the anion exchanger in dog renal microvillus membranes, Am. J. Physiol., 244:F612.

Haase, W., Schäfer, A., Murer, H., and Kinne, R., 1978, Studies on the orientation of brush-border membrane vesicles, Biochem. J., 172:57.

Kahn, A.M., and Weinman, E.J., 1985, Urate transport in the proximal tubule: in vivo and vesicle studies, Am. J. Physiol., 249:F789.

Martinez, F., Manganel, M., Montrose-Rafizadeh, C., Werner, D., and Roch-Ramel, F., 1990, Transport of urate and p-aminohippurate in rabbit renal brush-border membranes, Am. J. Physiol., 258:F1145.

Roch-Ramel, F., and Weiner, I.M., 1980, Renal excretion of urate: factors determining the actions of drugs, Kidney Int., 18:665.

Werner, D., and Roch-Ramel, F., 1991, Indirect Na^+ dependency of urate and p-aminohippurate transport in pig basolateral membrane vesicles, Am. J. Physiol., 261: in press.

Werner, D., Martinez, F., and Roch-Ramel, F., 1990, Urate and p-aminohippurate transport in the brush border membrane of the pig kidney, J. Pharmacol. Exp. Ther., 252:792.

THE GENETICS OF RENAL EXCRETION OF URATE IN MAN

BT Emmerson*, SL Nagel*, J O'Connor*, DL Duffy** and NG Martin**

*Department of Medicine, The University of Queensland, Princess Alexandra Hospital, Ipswich Road, Woolloongabba, Brisbane, Queensland, 4102, Australia;
**Queensland Institute of Medical Research, Bramston Terrace, Herston, Brisbane, Queensland, 4006, Australia

INTRODUCTION

Several studies have shown that hyperuricaemia tends to occur in families and this has been attributed to both inherited and acquired factors (1). However, except in over-producers such as patients with deficiency of hypoxanthine-guanine phosphoribosyltransferase, the precise mechanism whereby the genetic factor determining hyperuricaemia is manifest has not been defined. It is becoming increasingly clear, however, that defective excretion of urate by the kidney is one of the most important and frequent contributors to gout (2). This could be demonstrated either by a reduced urate clearance or by a reduced fractional excretion of urate (FE urate), so that excreted urate would comprise a smaller proportion of filtered urate than normal. Some family studies (3) have shown significant correlations between these components in patients with gout and their relatives. However, this type of study cannot exclude environmental influences on the serum urate concentration, particularly those introduced by a common diet.

The hypothesis to be tested, therefore, was that the genetic predisposition to gout is mediated by the renal clearance of urate, which is one of the most important determinants of the degree of hyperuricaemia.

In this regard, the urate clearance was calculated as the urate excretion in unit time divided by the plasma urate concentration, whereas the fractional excretion of urate was calculated as the ratio of the urate clearance to the creatinine clearance. It was proposed therefore to measure the urate and creatinine clearances and their ratios in healthy twins. If a significant genetic influence were to exist, the correlation between these measures should be significantly higher in monozygotic twins than in dizygotic twins. Further analysis of the results is able to measure the variance due to genetic and environmental influences, whether shared as in a family or unshared.

METHODS

Thirty-seven pairs of healthy twins from the Australian Twin Registry who lived in Brisbane consented to be studied. Twenty-five were monozygotic and 12 were dizygotic. All were healthy and none had ever suffered from gout. Their zygosity was confirmed both by questionnaire and by blood grouping.

The renal clearances of urate and creatinine were calculated from 3 consecutive 30 minute urine specimens collected sequentially commencing at 8.00 am. The patient was fasting but had taken an initial water load of 500 ml with 150 ml of fluid being taken half-hourly subsequently (4). Blood was sampled at the mid-points of the first and third collection periods. The patients were taking their normal diet and had taken no drugs at all. All clearances were corrected to the standard surface area of $1.73m^2$ (5). Urate was measured by an automated modification of the differential spectrophotometric uricase method and creatinine by the standard auto-analyser technique.

The urate clearance, the creatinine clearance and the fractional excretion of urate in any one of the twins was correlated with the same parameter in the other twin and the correlation coefficients in the groups of monozygotic and dizygotic twins were compared. If a characteristic is under significant genetic control, one would expect the correlation to be significantly higher in the monozygotic twin pairs than in the dizygotic pairs (6). Using the methods of genetic path analysis (LISREL 7.16) (7,8) the broad heritability of the characteristic was measured. It was appreciated that for such studies, the numbers available for study were relatively small and that the results should not be over-interpreted.

RESULTS

Thirty-seven males and 37 females were studied, the mean serum urate in the males being 0.36 mmol/L and that for the females being 0.29 mmol/L. The mean urate clearance in all subjects was 10 ml/min (confidence interval 4.4 - 22.2) with a correlation between any 2 of the 3 clearances in each individual being highly significant (0.81). For the creatinine clearance, the mean was 124 ml/min with a correlation between any 2 of the clearances in any one individual being 0.71 ($P<0.001$). This study did not demonstrate any significant differences between the two sexes for either urate or creatinine clearance so that the two sexes were pooled for the genetic analysis.

Intertwin correlations for the creatinine clearance for the monozygous twins were 0.80 (CI 0.33 to 0.92) whereas that for the dizygotic twins was 0.55 (CI -0.32 to 0.86). For the urate clearance, the intertwin correlations for monozygotic twins was 0.85 (CI 0.65 to 0.94), whereas for the dizygotic twins the correlation coefficient was 0.54 (CI 0.15 to 0.86) (9). For the fractional excretion of urate, the correlation coefficient was 0.85 (CI 0.66 to 0.95) for the monozygotic twins, whereas for the dizygotic twins it was 0.66 (CI 0.02 to 0.87) (9). The differences between these clearances were significant for the urate clearance and the fractional excretion of urate but not for the creatinine clearance.

The LISREL models of urate clearance (8) also confirmed a major genetic contribution to twin resemblance with the heritability of the urate clearance being estimated at 69%. However, the standard confidence limits of this heritability were quite wide at 40%-100% but did not include zero. For the fractional excretion of urate, the heritability was 87% with confidence limits of 45%-100%. The data did not support a significant role for shared environmental effects.

DISCUSSION

We therefore concluded that the study gave some support to the concept that the common inherited component of the plasma urate concentration was likely to be determined by the renal handling of urate and that this could be reflected in the urate clearance or by the fractional excretion of urate.

It should be noted that the analysis involved a comparison of one twin with a second twin so that they were always of the same age and had received an identical water load. The major limitation in this study was brought about by the relatively small sample size. Accordingly, we would like to see a larger study undertaken to attempt to replicate these results since larger numbers should narrow the confidence intervals of the various correlation coefficients which would tend to increase the significance of the differences. Although the measurement of the renal clearances of urate and creatinine is a relatively cumbersome and tedious procedure, it appears to give results of real value in interpreting the renal handling of urate. These results can be analysed to determine the contribution by genetic and environmental factors to the control of the renal handling of urate and its secondary effect upon the development of hyperuricaemia.

REFERENCES

1. Boyle JA, Greig WR, Jasani MK, Duncan A, Diver M, Buchanan WW. Relative roles of genetic and environmental factors in the control of serum uric acid levels in normouricaemic subjects. Ann Rheum Dis 26:234-237 (1967).

2. Rieselbach RE, Sorensen LB, Shelp WD, Steele TH. Diminished renal urate secretion per nephron as a basis for primary gout. Ann Intern Med 73:359-366 (1970).

3. Scott JT, Pollard AC. Uric acid clearance in patients with gout and normal subjects. Ann Rheum Dis 30:285-288 (1971).

4. Richardson JA, Philbin, PE. The one-hour creatinine clearance rate in healthy men. JAMA 216:987-990 (1971).

5. McIntosh JF, Moller R, Van Slyke DD. Study of urea excretions: the influence of body size on urea output. J Clin Invest 6:467 (1928).

6. Joreskog KG, Sorbom D. "PRELIS Users' Guide," 1st ed. Scientific Software, Moorsville (1986).

7. Heath AC, Neal MC, Hewitt JK, Eaves LJ, Fulker DW. Testing structural equation models for twin data using LISREL. Behav Genet 19:9-35 (1989).

8. Joreskog KG, Sorbom D. "LISREL 7. A Guide to the Program and Applications," 2nd ed. SPSS Inc, Chicago (1989).

9. Efron B, Tibshirani R. Bootstrap methods for standard errors, confidence intervals, and other measures of statistical accuracy. Statist Sci 1:54-77 (1986).

WHAT IS THE PATHOGENESIS OF FAMILIAL GOUTY NEPHROPATHY?

J. Stewart Cameron * Fernando Moro *+ and H. Anne Simmonds +

+ Purine Research Laboratory and * Renal Unit
Guy's Campus UMDS, London, UK

Familial juvenile gouty nephropathy (FJGN: McCusick 16200) was first noted by Duncan and Dixon in 1960 (1). We first observed the condition in 1968-72 (2), and by 1990 18 papers describing 38 affected kindreds containing 130 affected individuals had been described (3), plus three new kindreds described more recently (4-7). We are also aware of a number of unpublished kindreds, from some of whom material has been available to us for study. The purpose of this communication is to summarise what information we have accumulated: to speculate from these data, published and unpublished, as to what the pathogenesis of FJGN may be; and to suggest further studies which may illuminate the problem in the future.

Clinical features of FJGN which require explanation

The characteristics of FJGN are now well-established: hyperuricaemia and consequent gout arise in young individuals irrespective of sex because of a low fractional excretion of urate, FEur. Renal failure has appeared in 78 of 130 affected individuals in the available kindreds (leading to end stage disease in 32), but the finding that the low FEur and consequent hyperuricemia can occur before the appearance either of renal dysfunction or of gout (5-7) suggests that this is a prior event. Purine enzymes are normal and renal urate excretion (UVur) within normal limits. Secondary hypertension may be found (in 13 of 38 kindreds), but is usually a late event, appearing in those with renal dysfunction. Onset is usually in childhood, adolescence, or early adult life, and renal failure usually appears between 20 and 40 years of age. Untreated hypertension of course may accelerate the entrance to renal failure. Until and if gout appears in one or more members, the condition may be missed and a diagnosis of a "familial nephritis" made (5).

The disease is familial, inheritance following a dominant pattern, (consecutive generations affected, 78 males and 32 females). One kindred was reported as being consistent with an X-linked inheritance (8), but it is notoriously difficult to prove or exclude this in small studies. Male to male transmission was present in several kindreds studied by us and other workers (e.g. ref 4), which together with abundant affected females makes a simple dominant inheritance certain in the majority of families.

What is the relationship between the hyperuricaemia and the renal failure? Clearly we can relate the hypertension to the renal failure, and the clinical gout to the hyperuricaemia. But how do the renal failure and the

Purine and Pyrimidine Metabolism in Man VII, Part A
Edited by R.A. Harkness *et al.*, Plenum Press, New York, 1991

low FEur/hyperuricemia relate? One obvious possibility is that the renal disease might be a form of urate nephropathy. However, a number of observations exclude his possibility, even though allopurinol may slow or prevent progression (see Moro et al. 6, this volume).

First, prolonged hyperuricaemia of itself, although an independent risk factor for vascular disease, does not lead to renal dysfunction in the presence of a normal FEur in (for example) idiopathic symptomless hyperuricaemia (9), hereditary fructose intolerance or PRPP synthetase overactivity observed over 10 years or more (Simmonds, unpublished); in HGPRT deficiency of course, the massive renal excretion of urate/uric acid may lead to massive crystal nephropathy. However, in FJGN the UVur has been observed to be within normal limits in all cases, and in the kidneys studied in the 44 renal biopsies and 6 autopsies known to us, no or only occasional urate crystals have been noted. Thus it seems impossible that the non- specific interstitial nephropathy seen in FJGN with renal failure is the result of urate deposition nephropathy

Could the low FE ur damage the tubules in some way through an increased transtubular flux of urate? We do not know what, in functional detail, underlies the decreased fractional excretion, and it would be worth doing pharmacologic studies using probenecid/benzbromarone and pyrazinamide; but the detailed interpretation of such data in terms of secretion and reabsorption is now heavily criticised. However, given present models of urate handling in the mammalian tubule (see Weiner, this volume) it seems likely that some defect in a proximal tubular secretory mechanism is at fault, suggesting that FJGN may be the human analogue of the inherited gouty chicken, in which a defect in a basolateral urate transporter has been described (10).

What is the nature of the defect in proximal tubular urate secretion? The most likely explanation is that there is a defect in the gene coding for one of the anion transporters in the proximal tubular epithelium, either in the basolateral or the brush border membranes. Since none of these have yet been cloned the search for candidate genes (in which are presently engaged) is difficult, and non-specific search techniques must be used. Whether a single mutation, or many different mutations, might be present in the gene can only be speculated at the moment; however it is relevant that many mutations at the same site have been discovered in genetic disorders of other transporters (e.g. in cystic fibrosis) which can lead to phenotypically identical disease. Also, even at the level of pharmacologic studies, the inherited defect leading to isolated hyperuricosuria shows great diversity between the rather small numbers of kindreds studied (11,12). Therefore, we might predict both genetic and phenotypic diversity to underlie the low FEur of FJGN. Of interest to subsequent discussion, hypouricaemia with reduced tubular absorption can be acquired, following the appearance of some product from malignant tissue (13).

The renal failure might arise as a direct consequence of this genetic alteration, as outlined above, even if we do not understand the mechanism; but it might also result from a concurrently inherited (but structurally separate) gene which in some other way determines the appearance of renal failure from interstitial nephritis. Inherited interstitial nephritis may occur in the kd.kd mouse (14) and in man, in many of whom cystic changes occur, as they do in the mice (15). The finding of medullary cysts in 7 patients with FJGN (4 at autopsy and 3 by biopsy)- as occurs also in the mice, should not lead to a diagnosis of nephronophthisis (uraemic medullary cystic disease) (16). In addition this condition usually shows an autosomal recessive inheritance.

Could a toxic metabolite lead to the low FEur?

One analogous disease worth considering here, in which a low FEur is found together with hyperuricemia, gout, normal UVur and progressive renal damage, is of course lead poisoning. Here, the urate handling changes seem to be secondary to the action of the toxic compound: lead. Could there be an inherited analogue of lead in FJGN, its production induced by the genetic mutation?

A parallel possibility is that the abnormal gene might code, not for an anion transporter, but for an enzyme whose mutation leads to the production of an abnormal pyrazinamide- like metabolite, and thus to a low FEur. Although there is no lack of compounds which inhibit net urate secretion, this is usually on a short term or acute basis; we do not know of any studies of long term renal function in any condition where a low FEur and hyperuricaemia results from co-transport of a competing compound or metabolite. However, FJGN could be the first disease of this type. There are, however, the difficulties in implicating urate directly in the renal damage, as discussed above.

Lessons from renal transplantation

In distinguishing between these possibilities and that of a stuctural defect in an anion transporter, renal transplantation holds hope of definitive information: if a structural transport defect is present, then renal transplantation should effect a permanent cure of the defect. On the contrary, if a toxic metabolite of whatever type is the source of the disease, then the transplanted kidney itself would be affected eventually. One interesting observation is that these patients are extraordinarily sensitive to the marrow depressant effects of azathioprine, and two of our patients died of pulmonary infection- now a very rare ocurrence with modern multi-drug immunsuppressive regimes, and in two others severe pneumonia led to graft loss from reduced immunosuppression (4). The metabolism of azathioprine in FJGN needs to be studied in detail.

Unfortunately, in practice it has been difficult to extract any information from the few patients (7 known to us) transplanted for the renal failure of FJGN. First, there has been a high failure rate because of infection, as just mentioned. Second, ischaemic damage to the graft during transplantation often leads, in those with excellent function, to a proximal renal tubular defect including a high FEur which may persist six months or more. Third, in contrast drugs given today to transplant recipients- particularly cyclosporine and diuretics- lead to a high incidence of hyperuricaemia and gout even in young women (17). Finally the process of chronic graft rejection, present in almost all cadaveric unrelated grafts to some extent, manifests itself in allograft biopsies as a non-specific interstitial nephritis which cannot be distinguished from he changes of FJGN in the patient's own kidneys. The evolution in the native kidneys is over many years and is likely to be similarly slow in the allograft. Nevertheless, more observations on histology and urate handling in transplanted FJGN patients would be welcome.

Lessons from allopurinol treatment

Obviously if the renal lesion of FJGN were the result of urate crystal deposition disease, then one would expect improvement on allopurinol therapy. Despite the convincing evidence that the renal damage is not the result of intrarenal urate deposition discussed above, allopurinol seems to us to ameliorate the disease at least in some subjects (see Moro, this volume (6)). How could it do this? One explanation is that, if there were an abnormal metabolite, then its production might be the result of a reaction inhibitable by allopurinol.

Moreover, if an allopurinol-inhibitable pyrazinamide-like metabolite were responsible for the low FEur, then the low FEur should rise towards normal on treatment, independently of any reduction in UVur, whereas if a structural transport defect is present the low FEur would persist. We have only preliminary data on this important point as yet.

Conclusions

FJGN is certainly more common than at present realized, especially in renal units, because not all families or individuals show clinical gout. Better understanding of its pathogenesis may not only help a number of unfortunate patients, but is almost certain to illuminate further the mechanisms of urate transport in renal epithelium.

Acknowledgments

This work was supported by the Arthritis and Rheumatism Council of the UK (JSC).

REFERENCES

1. H.Duncan, A.StJ.Dixon. Gout, familial hyperuricaemia, and renal disease.Q J Med 29:127 (1960).
2. H.A.Simmonds, D.J.Warren, J.S.Cameron, C.F.Potter, D.A.Farebrother. Familial gout and renal failure in young women.Clin Nephrol 14:176(1980).
3. G.Calabrese, H.A.Simmonds, J.S.Cameron, P.M.Davies. Precocious familial gout with reduced fractional urate excretion and normal purine enzymes.Q J. Med 75:441 (1990).
4. J.W.Foreman, M.Yudkoff. Familial hyperuricemia and renal insufficency.Child Nephrol Urol 10: 115 (1990).
5. F.Moro, C.S.Ogg, H.A.Simmonds, J.S.Cameron, C.Chantler, M.B.McBride, J.A.Duley, P.M.Davies. Familial juvenile gouty nephropathy with renal urate hypoexcretion precding renal disease.Clin Nephrol 35:263 (1991).
6. F.Moro, H.A.Simmonds, M.B.McBride, J.S.Cameron, D.G.Williams, C.S.Ogg. Does allopurinol ameliorate progression in familial juvenile gouty nephropathy (FJGN)? (this volume).
7. M.B.McBride, V.Raman, C.S.Ogg, C.Chantler, J.S.Cameron, J.Duley, H.A. Simmonds. A new kindred with familial juvenile gouty nephropathy (this volume).
8. N.G.Westberg, E.Rosen, J.Waldenstrom. Recessive X-linked hyperuricemia with gout and renal damage, normal activity of hypoxanthine - guanine ribosyl transferase and resistance to azaguanine.Acta Med Scand 205:163 (1979).
9. W.J.Fessel. Renal outcomes of gout and hyperuricemia.Am J Med 67:74 (1979).
10. M.J.Zmuda, A.J.Quebbemann. Localisation of the renal tubular uric acid defect in gouty chickens.Am J Physiol 229:820 (1975).
11. I.Akaoka, T.E.Nishizawa, N.Yano, Y.Kamatani, S.Nishida. S.Sasaki. Renal urate secretion in five cases of hypouricemia with an isolated renal defect of urate transport.J Rheumatol, 4:86 (1977).
12. M.Shichiri, H.Iwamoto, T.Shiigai. Hypouricemia due to increased tubular urate secretion.Nephron 45:31 (1987).
13. A.Tykarski. Mechanism of hypouricemia in Hodgkin's disease. Isolated defect in post secretory reabsorption of uric acid.Nephron 50:217(1988).
14. E.G.Neilson, E.McCafferty, A.Feldman, M.D.Clayman, B.Zakheim, R.Korngold. Spontaneous interstitial nephritis in kd.kd mice I. A new model of autoimmune renal disease.J Immunol 119:1552 (1984).
15. C.J.Kelly, E.G.Neilson. Medullary cystic disease: an inherited form of autoimmune interstitial nephritis?Am J Kidney Dis 10:389 (1987).

16. J.Burke, J.A.Inglis, P.Craswell et al. Juvenile nephronophthisis and medullary cystic disease.Clin Nephrol 18:1 (1982).
17. H-Y.Lin, L.L.Rocher, M.A.McQillan, S.Schmaltz, T.D.Palella, I.H.Fox. Cyclosporine- induced hyperuricemia and gout.N Engl J Med 321: 287(1989).

RENAL URATE HYPOEXCRETION PRECEDING RENAL DISEASE IN A NEW KINDRED WITH FAMILIAL JUVENILE GOUTY NEPHROPATHY (FJGN)

Mary B. McBride, Venkat Raman, Chisholm S. Ogg,
Cyril Chantler, J.S. Cameron, John A. Duley,
and H. Anne Simmonds

Purine Research Laboratory, Renal Units at St Mary's
Hospital Portsmouth and Guy's Hospital, London

INTRODUCTION

We have reported the syndrome of FJGN in numerous kindreds as well as isolated young subjects of either sex (reviewed in 1). The latter have generally had a strong family history of gout and sometimes renal disease. Subsequent investigations have invariably revealed FJGN in other seemingly healthy family members. The hallmark of the disorder has been a grossly reduced fractional uric acid clearance [FE_{ur}] for the young age, sex and renal function, as discussed elswhere in this Symposium (Cameron).

This paper summarises investigations in a large hitherto unreported kindred of a male patient originally considered to have a familial renal disease, who subsequently presented with a single attack of gout. The findings are compared with results in two other kindreds reported earlier by us, in which at least one member was also found with renal urate hypoexcretion preceeding renal disease (1-3)

PATIENTS AND METHODS

Kindred R: The propositus II 3, whose father had died of renal disease at 43, was considered to have a familial renal disease until an isolated attack of gout focussed our attention on other kindred members. A family history revealed that 3 of the 4 siblings of the propositus were on treatment for varying degrees of renal impairment (Figure 1), one of whom had received a successful renal allograft. This prompted investigations in the 14 offspring of these 5 subjects (Fig 1) ranging in age from 4-23 years, eight of whom (6f:2m) have been studied so far.

Kindred W has been reported previously (1). When originally studied in 1983 it comprised 6 subjects from two generations (I 1, II 1-5; Figure 1). Since that time 5 subjects belonging to a third generation have been born, but studies have been possible in only one of these (III 2).

Kindred B has been previously reported (2).

RESULTS

Kindred R: Renal function was apparently normal in all 8 young subjects available for study (plasma creatinine 45-89umol/l). However, one subject (III 4: f, aged 16, creatinine 89umol/l) was found to be hyperuricaemic for age and sex (345umol/l), associated with an FE_{ur} of 6.2%. Table 1.

Kindred W: Follow-up investigations in this kindred revealed that II 1 had a

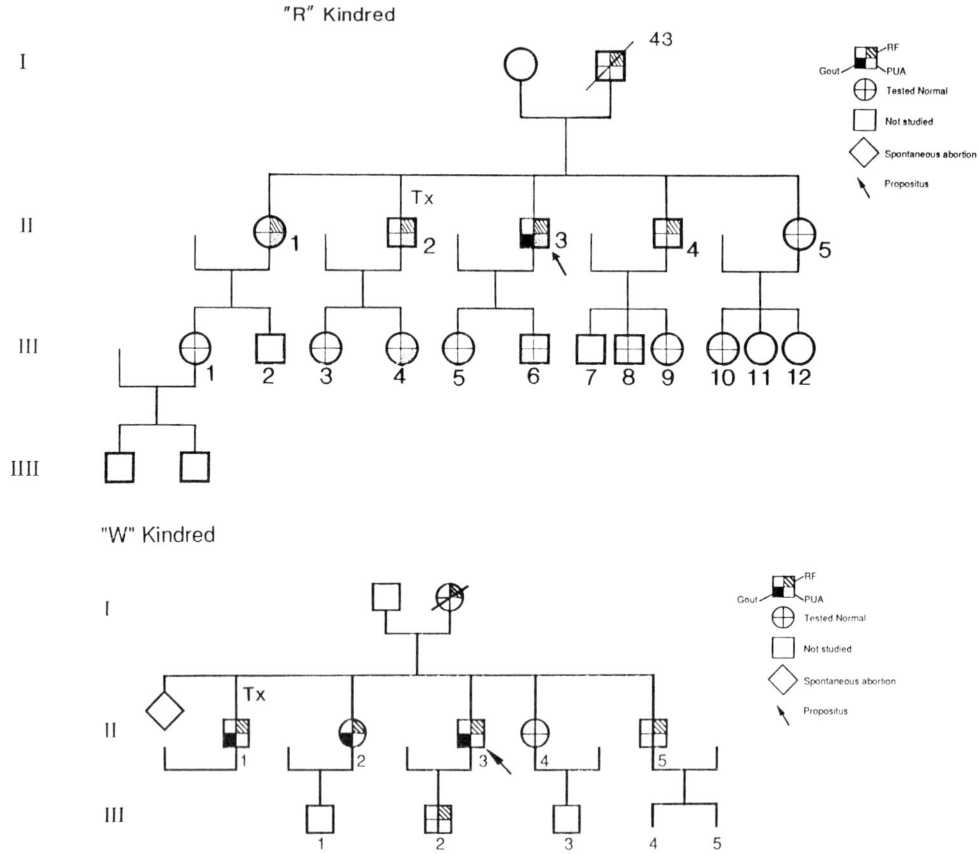

Figure 1 Family tree of R kindred (above) and W kindred (below)

successful renal allograft in 1986, that renal function in II 2 had remained stable on allopurinol, but that GFR in II 3, normal when first studied at age 26 and not compliant to allopurinol therapy, had fallen to 52ml/min. Further studies have not been possible in II 5 or his children. Studies in III 2, (5y) revealed an FE_{ur} of 6.9 and GFR of 74 ml/min.

Both R III 4 and W III 2 have been commenced on allopurinol.

DISCUSSION

Studies of healthy young members of the large R kindred have revealed one young girl with hyperuricaemia associated with a reduced FE ur, but no evidence of renal dysfunction. These results are similar to those reported elswhere (2) in the B kindred where two subjects were also found with hyperuricaemia associated with a reduced FE ur, but a normal GFR. Interestingly, this kindred, likewise, was considered originally to have familial renal disease, until an isolated attack of gout prompted family studies which revealed other members with FJGN (2).

Prior to the above studies, the W kindred (1) was the only one of the many studied in which we had identified hyperuricaemia associated with a reduced FE ur, but normal renal function. This was found in a single subject, a young man of 26. Unfortunately, there is now clear evidence of a 50% reduction in GFR in this patient, who was not compliant with allopurinol therapy. A son, born subsequent to the first study, also has FJGN. The results in the "W" kindred thus provided the first clear evidence that a reduced FE_{ur} does preceed the onset of renal disease.

Table 1. Summary of data from subjects with renal urate hypoexcretion compared with unaffected kindred members and healthy controls of same age

Kindred	PUA umol/l	PCr umol/l	FE_{ur} %	age (yr)	sex
R III.4	345	89	6.2	16	f
R III (n=4)	213 (210-220)	72 (59-80)	12.6 (8.9-17.7)	(12-24)	f
Control	(140-280)	(70-130)	(12.2+1.9)	adult	f
R III 5	210	49	12.5	7	f
R III 6	170	45	15.1	5	m
W III.2	340	77	6.9	5	m
Control	(130-250)	(36-65)	18.5+3.2	child	m,f
R III 8	230	89	6.7	17	m
Control	(180-340)	(70-130)	8.2+1.2	adult	m

All other young subjects so far studied in kindred R, apart from III 4, have normal fractional urate clearances, although some results are at the bottom of the normal range. However, we have been able to study only 50% of this generation so far. Continuing studies in this and other similar kindreds are obviously vital and should assist our understanding of the evolution of this potentially lethal disorder. The need for a genetic marker is evident.

These results confirm that an isolated attack of gout can be extremely important in focusing attention on a familial disorder, which can only be identified through detailed family histories, and purine studies in all members. Low fractional urate clearance for age and sex should indicate the need for frequent re-investigation and the instigation of allopurinol therapy, which in our experience appears to be beneficial in FJGN (Moro et al this Symposium).

Acknowledgements: These studies were supported by the Arthritis and Rheumatism Research Council.

REFERENCES

1. G. Calabrese, H.A.Simmonds, and J.S. Cameron. Precocious familial gout with reduced fractional urate clearance and normal purine enzymes. Quart J Med 75:441-450 (1990).
2. F. Moro, C.S. Ogg, J.S. Cameron, H.A. Simmonds, J.A. Duley, and M.B. McBride, P.M. Davies. Familial juvenile gouty nephropathy with renal urate hypoexcretion preceding renal disease. Clin Nephrol 35:263 (1991).
3. J.S. Cameron, C.S.Ogg, F. Moro, C. Chantler, and H.A. Simmonds. Precocious familial gout (letter). Lancet ii;745 (1990).

FAMILIAL NEPHROPATHY AND GOUT: WHICH COMES FIRST?

Juan G. Puig, Mª.E. Miranda Carús, Felícitas A. Mateos, Mª. Luz Picazo, M. López Jiménez and J. Ortíz Vázquez

Divisions of Internal Medicine, Clinical Biochemistry and Pathology
"La Paz" Hospital, Universidad Autónoma, Madrid, Spain

INTRODUCTION

The kidney has been traditionally considered as one of the two primary target organs in gout (1), but long-term follow-up studies have failed to show a deleterious effect of gout on renal function (2). However, it has been reported that some patients with uncomplicated gout have developed rapidly progressive renal failure (2). Moreover, several families have been described with hyperuricemia or gout, usually appearing at an early age and often in both sexes, associated with progressive renal failure (3-15). The pathogenesis of this syndrome has not been clearly elucidated. We describe eight patients in two families who had hyperuricemia or gout, renal insufficiency, and bilaterally small kidney size.

PATIENTS AND METHODS

The propositus of Family M (subject III-1, age 34-years) had gout, arterial hypertension, renal insufficiency and small kidney size (Figure 1). His mother, aged 59, had hyperuricemia, renal insufficiency and small kidney size, but normal blood pressure. The two brothers of the propositus were examined: one (subject III-2, aged 33) had gout, mild hypertension, renal insufficiency, and small kidney size. In contrast, the youngest sibling (subject III-3, aged 28) only had gout, with blood pressure, glomerular filtration rate (GFR) and kidney size within normal limits. The patient´s 4-year-old son (subject IV-1) was healthy and renal echography showed normal kidney size.

The propositus of Family T (subject II-2, age 46-years) had gout, hypertension, renal failure and reduced kidney size (Figure 1). The patient´s sister, aged 47, had hyperuricemia, arterial hypertension, renal insufficiency and small kidney size. His younger brother, aged 43, was normotensive, had normal kidney size, serum creatinine and urate levels. The patient´s two asymptomatic daughters, aged 15 and 12, and were both found to have the characteristics of the syndrome. Both siblings were normotensive but showed hyperuricemia, increased serum creatinine and small renal indexes for their age. Kidney damage was less severe in the younger than in the elder sibling.

Subjects II-1, III-1, III-2 and III-3, from Family M, and subjects II-2, III-1 and III-2, from Family T (Figure 1) were admitted to the Clinical Research Unit. Medications were discontinued four weeks before the studies, except for colchicine (0.5 mg per day), which was given as prophylaxis against acute arthritis. All subjects were placed on a weight-maintenance, isocaloric, purine-restricted diet with a protein content of 10% to 15%. On the fourth to the ninth hospital day purine metabolism was examined by

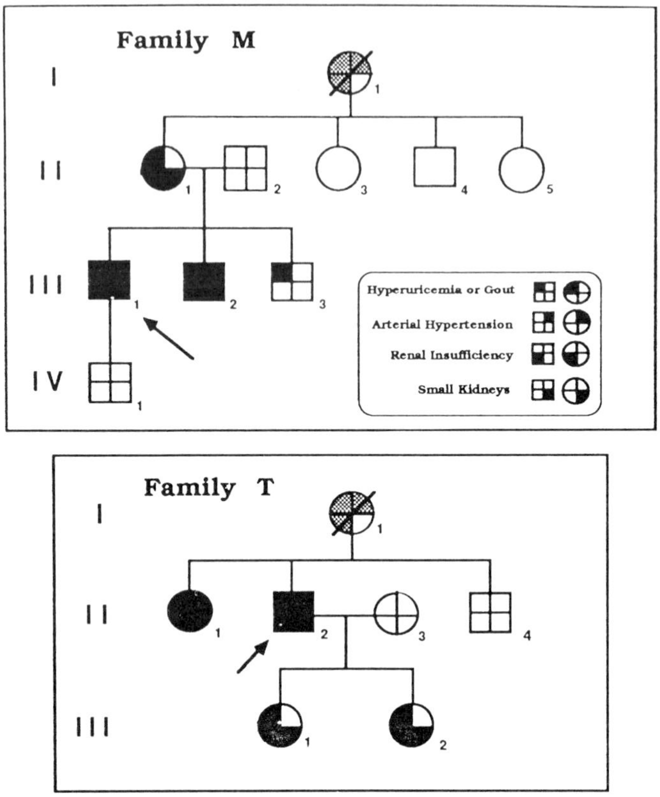

Figure 1. Pedigree of Two Families with Nephropathy Associated to Hyperuricemia. Circles denote female subjects, squares male subjects, and symbols with diagonals dead subjects. Solid areas within symbols denote pathological characteristics, empty areas normal features, and hatched areas suspected pathological features. Symbols without quadrants denote subjects not available for the study.

determining plasma and 24-hour urinary hypoxanthine, xanthine and uric acid concentrations (16). Twenty-four-hour urine samples were collected daily under toluene (3 mL) over the ensuing 5 days. Venous blood was obtained at the end of 24-hour urine collections following overnight fast and rest (16). The means of three determinations for plasma purines and five for urinary purines were used. The results of purine metabolism were compared to those obtained in 25 patients with primary gout (age, 67±13-years; range 39 to 83-years) and moderate renal insufficiency due to hypertension or atherosclerosis, and in 25 normal subjects (age, 62±14-years, range 41 to 84-years), all of whom were submitted to an identical protocol to assess purine metabolism.

RESULTS

Purine Metabolism

Patients with familial nephropathy showed a markedly decreased 24-hour urinary uric acid, hypoxanthine and xanthine excretion rates compared to normal subjects, despite significantly increased plasma purine concentrations (Table 1). Remarkable similarities in purine metabolism were observed between the patients with familial nephropathy and subjects with gout and renal failure. However, the urate/creatinine ratio, and daily hypoxanthine and xanthine excretion rates were significantly lower in patients with familial nephropathy than in gout patients with renal failure.

Table 1. Purine Metabolism in the Studied Subjects.

	Familial Nephropathy (n=7)	Gout and Renal Failure (n=25)	Normal Subjects (n=25)
PLASMA			
Creatinine (μmol/l)	175±54[a]	212±88[a]	80±18
Urate (μmol/l)	429±42[a]	476±95[a]	292±12
Hypoxanthine (μmol/l)	4.4±2.1[a]	4.4±3.2[a]	2.3±1.1
Xanthine (μmol/l)	1.2±0.5[b]	0.9±0.5	0.7±0.2
URINE			
Uric acid (mg/day/1.73 m^2)	294±64[a]	295±89[a]	442±86
Urate/creatinine (mg/mg)	0.20±0.04[a,c]	0.26±0.08[a]	0.34±0.11
Hypoxanthine (μmol/g cr.)	9±4[a,c]	18±12[a]	33±9
Xanthine (μmol/g cr.)	6±3[a,c]	10±7[a]	22±9
CLEARANCES (ml/min/1.73 m^2)			
Creatinine	58±25[a]	38±15[a]	101±21
Uric acid	2.8±0.8[a]	2.6±0.8[a]	6.4±1.3
Hypoxanthine	4.5±1.8[a]	5.0±3.8[a]	13.3±5.5
Xanthine	6.5±3.9[a]	10.6±7.0[a]	29.7±13.1

[a] P<0.01 versus normal subjects.
[b] P<0.05 versus normal subjects.
[c] P<0.05 versus gout and renal failure.

Renal Studies

PRA and aldosterone concentrations were normal in all patients. With the exception of subject III-3 of Family M, all patients showed a diminished GFR. Renal pathology was similar in the three patients of Family T, being more severe in the father than in his two daughters. The thickness of the cortex was diminished and the lesions were preferentially localized in the subcapsular and medium cortex as compared with the juxtamedullar cortex. The patchy areas of glomerular obsolescence and tubular atrophy were similar in all patients. Some glomeruli showed wrinkling of the capillary basement membrane, suggesting an ischemic mechanism for the renal damage. Biopsy sections from the three patients fixed in absolute alcohol, prior to staining, did not show birefringent crystals.

Follow-Up Studies

Low sodium-diet and enalapril normalized blood pressure in hypertensive patients. Allopurinol maintained serum urate levels below 298 μmol/l in all 7 hyperurcemic subjects. Serum creatinine increased in all patients. Patient III-3 of Family M showed initially normal serum creatinine levels (80 to 106 μmol/l) and GFR (104 ml/min/1.73 m^2). Four-years later, serum creatinine increased to 133 to 150 μmol/l and GFR decreased to 72 ml/min/1.73 m^2. Simultaneously, kidney size declined over time in 6 patients, but remained within normal limits in subject III-3 of Family M. During the 4-year period serum urate, creatinine, blood pressure and kidney size remained within normal limits in subject IV-1.

DISCUSSION

Chronic hyperuricemia has long been considered deleterious for the kidney, and according to some authors this familial disease represents a severe form of gouty nephropathy (8-10). However, three lines of evidence argue against this hypothesis. First, the concept of gouty nephropathy as a complicating renal disease secondary to increased serum urate levels has vanished (17,18). Second, urate crystal deposits have been thought to be a distinctive feature of gouty nephropathy (1) but despite intensive search no uric acid crystals could be identified in the three kidney biopsies of our patients. With the exception of one case (10) in no other previously reported patients have uric acid crystals been encountered (4,7,11,12). Third, the results of our follow-up studies have indicated that despite adequate control of hyperuricemia renal deterioration could not be prevented nor halted. We conclude that this syndrome is a distinctive interstitial nephropathy that is (a) presumably inherited as an autosomal dominant trait, (b) has a non-specific pathology, (c) plasma purine concentrations are increased due to a severe purine underexcretion, hyperuricemia allowing early identification of affected subjects, and (d) the disease progresses to renal failure and is not halted nor prevented by an effective control of hyperuricemia.

ACKNOWLEDGEMENTS

We are indebted to Dr. T. Ramos, Javier Díaz, and Mrs. Mª Paz Canencia for valuable technical assistance. This work was supported by grants from Caja de Madrid and Fondo de Investigaciones Sanitarias de la Seguridad Social (FISS, 90/0575), Spain.

REFERENCES

1. Talbott JH, Terplan KL. The kidney in gout. Medicine (Balt) 1960; 39:405-7.
2. Berger L, Yü T-F. Renal function in gout. IV. An analysis of 524 gouty subjects including long-term follow-up studies. Am J Med 1975; 59:605-13.
3. Duncan H, Dixon AS. Gout, familial hyperuricemia, and renal disease. Q J Med 1960; 29:127-35.
4. Rosenbloom FM, Kelley WN, Carr AA, Seegmiller JE. Familial nephropathy and gout in a kindred (abstract). Clin Res 1967; 15:270.
5. Treadwell BLJ. Juvenile gout. Ann Rheum Dis 1971; 30:279-84.
6. Van Goor W, Kooiker CJ, Mees EJD. An unusual form of renal disease associated with gout and hypertension. J Clin Path 1971; 24:354-9.
7. Thompson GR, Weiss JJ, Goldman RT, Rigg GA. Familial occurrence of hyperuricemia, gout, and medullary cystic disease. Arch Intern Med 1978; 138:1614-1617.
8. Massari PU, Hsu CH, Barnes RV, Fox IH, Gikas PW, Weller JM. Familial hyperuricemia and renal disease. Arch Intern Med 1980; 140:680-4.
9. Simmonds HA, Warren DJ, Cameron JS, Potter CF, Farebrother DA. Familial gout and renal failure in young women. Clin Nephrol 1980; 14:176-82.
10. Farebrother DA, Pincott JR, Simmonds HA, Warren DJ, Dillon MJ, Cameron JS. Uric acid crystal-induced nephropathy: Evidence for a specific renal lesion in a gouty family. J Pathol 1981; 135:159-68.
11. Richmond JM, Kincaid-Smith P, Whitworth JA, Becker GJ. Familial urate nephropathy. Clin Nephrol 1981; 16:163-8.
12. Leumann EP, Wegmann W. Familial nephropathy with hyperuricemia and gout. Nephron 1983; 34:51-7.
13. Fleckenstein JL, Simkin PA. Corrected clearance identifies underexcretion of uric acid in a gouty kindred. Adv Exp Med Biol 1986; 195A:295-7.
14. Calabrese G, Simmonds HA,Cameron JS, Davies PM. Precocious familial gout with reduced fractional urate clearance and normal purine enzymes. Q J Med 1990; 75:441-50.
15. Yokota N, Yamanaka H, Yamamoto Y, Fujimoto S, Eto T, Tanaka K. Autosomal dominant transmission of gouty arthritis with renal disease in a large Japanese family. Ann Rheum Dis 1991; 50:108-11.
16. Puig JG, Mateos FA, Jiménez ML, Ramos TH: Renal excretion of hypoxanthine and xanthine in primary gout. Am J Med 1988; 85:533-537.
17. Reif MC, Constantiner A, Levitt MF. Chronic gouty nephropathy: A vanishing syndrome? N Eng J Med 1981; 304:535-6.
18. Beck LH. Requiem for gouty nephropathy. Kidney Int 1986; 30:280-7.

DOES ALLOPURINOL AFFECT THE PROGRESSION OF FAMILIAL JUVENILE GOUTY NEPHROPATHY?

Fernando Moro, H. Anne Simmonds, J. Stewart Cameron, Chisholm S. Ogg, Gwyn D. Williams, Mary B. McBride, and Phillip M. Davis

Purine Research Laboratory and Renal Unit
UMDS Guy's Hospital, London, UK

INTRODUCTION

Familial juvenile gouty nephropathy (FJGN, McKusick 16200) first described by Duncan and Dixon (1), is an autosomal dominant condition with high penetrance and characteristic features. It affects a sub-group of young people who develop early progressive renal dysfunction. A common hallmark is hyperuricaemia disproportionate to age, sex and degree of renal failure. The nature of the biochemical defect remains unknown and there is debate as to whether the abnormal tubular handling of urate causes the damage (2,3), or if the treatment for the hyperuricaemia affects the evolution of the disease (2,4). Histologically, a non specific interstitial nephritis (NSIN) is usually seen without evident deposits of sodium urate (1,2,3).

This study details our experience in 12 patients from a total of 47 subjects belonging to 25 different kindreds referred to Guy's for investigation of hyperuricaemia or renal disease, of whom 41 had a family history of gout and renal failure (2,5,6,7). Known hereditary or secondary causes of hyperuricaemia have been excluded. As discussed in companion papers presented at this Symposium, a characteristic finding was a low clearance of uric acid relative to creatinine FE_{ur}.

PATIENTS AND METHODS

Long-term follow-up studies, or detailed investigations of other family members have been possible in only seven of the above kindreds, from which the 12 subjects in the present study are drawn (Table I). The length of follow-up has varied from 5 to 23 years. 10 of the patients studied have been treated with allopurinol in a dose adjusted to their renal function (RF), but one of these was not compliant and is considered as not treated. Of the other 2 patients who were not given allopurinol, one (patient 10), was being treated only for "familial renal disease" until an attack of gout drew attention to the true nature of the disorder (7), the other (patient 12), whose mother had died at 34 with FJGN (5), had been studied regularly since 1974 and only recently shown clear evidence of the familial disorder.

RESULTS

The length of follow-up and treatment are listed in Table 1 and the

Purine and Pyrimidine Metabolism in Man VII, Part A
Edited by R.A. Harkness *et al.*, Plenum Press, New York, 1991

TABLE I. **FJGN patients.** Follow-up and evolution

Patient	age 1st study	Follow up yrs	GFR ml/min* Rx began	Pcr µmol/l Rx began	Current GFR	Current Pcr	Evolution
1 RS	(f 17)	7	?	178	?	195	Stable
2 MN	(m 29)	7	50	186	?	154	Stable
3 AW	(m 31)	5	44	152	48	148	Stable
4 PaW	(f 26)	6	56	124	?	110	Stable
5 YD	(f 16)	22	20	310	-	-	Death. OIH
6 LK	(f 16)	23	60	148	25	240	Stbl last 9yrs OIH
7 TY	(f 39)	23	35	270	33	190	Renal TX. OIH
8 DeY	(f 14)	13	68	170	36	195	Stbl last 9yrs
9 DaY	(f 12)	13	78	104	74	91	Stable last 12 yrs
10 LJ	(f 20)	9.5	35	145	ESRD	-	Renal TX.
11 PtW	(m 27)	5.5	102	115	52		Non compliant
12 PC	(m 25)	21	78	108	78	108	Rx just began

*GFR by ^{51}Cr-EDTA or Ccr. Rx:treatment with allopurinol. OIH:oestrogen induced hypertension TX: transplant. ESRD end stage renal disease. PCr (normal 50-120 µmol/l).

findings summarised in Figures 1 and 2. Five patients (1,2,3,4 and 9, Table I) all have had stable RF from 5 to 12 years following the instigation of therapy. Patient 9 suffered the loss of one kidney after a renal biopsy in another hospital. Her creatinine clearance (C_{cr}) fell initially to 78 ml/min (Fig 2) and has remained the same for 12 years. The biopsy showed a NSIN.

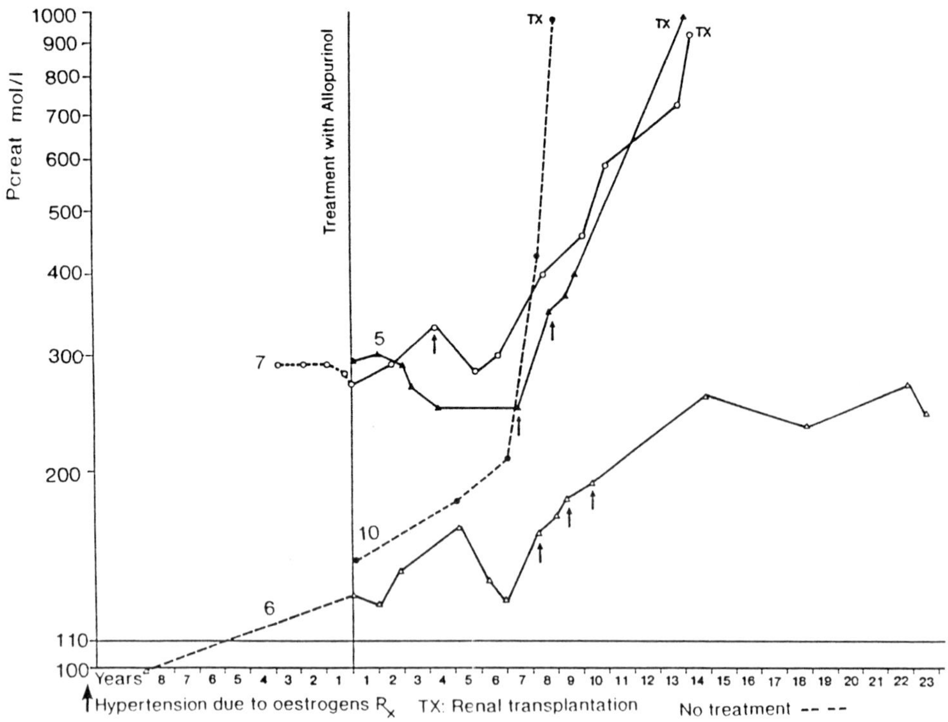

Figure 1. shows the importance of allopurinol in patient 6 where a sharp fall in RF over 6 years prior to any therapy was halted following the start of therapy. Identical results have been reported in patient 8 (2).

Two of the 9 patients taking allopurinol (Fig 1) had plasma creatinines (P_{cr})in excess of 300umol/l when treatment began and progressed to end stage renal disease (ESRD). Although P_{cr} remained constant initially in patient 5, a rapid deterioration occurred following the introduction of contraceptive therapy and the onset of hypertension (2). Death occurred after nine years on haemodialysis and 2 failed renal transplants. RF in patient 7, who was also hypertensive initially and who progressed to ESRD, is stable following a renal allograph. Patient 6 (twin sister of patient 5) began allopurinol when her C_{cr} was 58 ml/min (Table I). Her RF remained stable for 6 years and then deteriorated sharply when she developed hypertension following the institution of oestrogens (Fig 1). Once oestrogens were stopped and the hypertension controlled, her P_{cr} levelled off at a higher value and has remained stable for 9 years. After an initial decline in the C_{cr}, patient 8 (DeY) has maintained her RF stable for 9 years. Altogether she has lost 46% of her C_{cr} in 13 years (Fig 2).

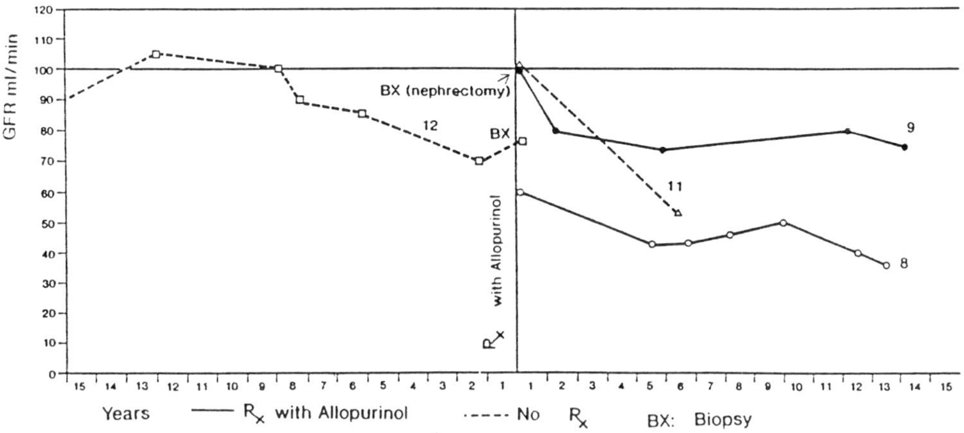

Figure 2. shows GFR plotted against time

Patient 11, the non-complier who had normal renal function when first studied (6), has suffered a 50 % reduction in GFR over 5 years (Fig 2). Results in patient 10, treated only for renal disease who progressed to ESRD in 8 years (7) and is now on dialysis having failed a transplant, give perhaps the most clear-cut demonstration of the importance of allopurinol, particularly when compared with patient 6 of similar age and sex. GFR in patient 12, who has been investigated frequently since 4 years of age (5), has fallen 20% in the past seven years and he has now been put on allopurinol after a biopsy showing a NSIN.

DISCUSSION

There is still no definite idea of the basic lesion in this dominant disorder with high penetrance. However, in the few cases which have been studied early in the evolution of the disease, a prominent feature is a low FE_{ur} preceding the fall in GFR (2,6,7). In the past there has been considerable debate as to whether urate per se plays a role in the damage to the renal parenchyma or whether this represents primary renal disease with the uric acid handling problems secondary, or supplementary, to this (2,3,6). Whatever the pathogenic mechanism, these subjects clearly differ from the middle aged-male with primary gout whose RF today is normal for age. However, it is noteworthy that renal disease was common in such subjects 30 years ago, prior to awareness of the role of associated diseases, hypertension, diet or the introduction of allopurinol (8).

There has been equal debate as to whether allopurinol is useful in FJGN (2,3,4,6). Our results, covering periods from 5-23 years support the concept that allopurinol may have an ameliorating effect on the progression of the renal lesion in FJGN. All 3 patients not on allopurinol have shown a decline in GFR of from 20% to ESRD over 8 years. This contrasts with 7 subjects, with evidence of a reduction in GFR of up to 50% at diagnosis, where RF has remained relatively stable following the introduction of the drug.

From the results in the two subjects on allopurinol who progressed to ESRD, the important elements in the equation appear to be the degree of RF at the time therapy was instigated and the presence of hypertension triggered by the introduction of oestrogens. We suggest that prognosis for those with a GFR of 40 ml/min or less is not good, although compared with the rapid decline in patient 11, here too allopurinol may have been beneficial. Hypertension has been an uncommon feature in the earlier phases of the disease in patients studied by us, but it is evident that when it develops it has a deleterious effect on the evolution of the disorder.

CONCLUSION

The present results support the concept that allopurinol ameliorates the progression of the renal damage in FJGN, although a consistent pattern is not evident. Important factors determining the efficacy of therapy are a) the degree of renal function at the time of commencing therapy, b) the presence of hypertension or its adequate control. These findings underline the importance of detailed familial studies in order to identify other affected members sufficiently early in the evolution of their disease.

AKNOWLEDGEMENTS

This study was supported by the Arthritis and Rheumatism Council for Research and the Special Trustees of Guy's Hospital.

REFERENCES

1. H. Duncan and A.St.J. Dixon. Gout, familial hyperuricaemia, and renal disease, Quart.J.Med. 29:127 (1960).
2. H.A. Simmonds, D.J. Warren, J.S. Cameron and D.A. Farebrother, Familial gout and renal failure in young women, Clin Nephrol. 14:176 (1980).
3. J.M. Richmond, P Kinkaid Smith, J.A. Withworth and G.J. Becker. Familial Urate Nephropathy, Clin Nephrol. 16:163 (1981).
4. T. Mukarami, H. Kawakami, K. Nakatsuda, K. Jojima, H. Nohno and H. Matsutzaki. Underexcretory-type hyperuricemia, disproportionate to the reduced glomerular filtration rate in two boys with mild proteinuria, Nephron. 56:439 (1990).
5. H.A. Simmonds, J.S. Cameron, C.F. Potter, D. Warren, T. Gibson and D. Farebrother, Renal Failure in young subjects with familial gout. Adv. Exp. Med. Biol. 122A:15 (1980).
6. G. Calabrese, H.A. Simmonds, J.S. Cameron, P.M.Davies. Precocious familial gout with reduced fractional urate clearance and normal purine enzymes, Quart.J.Med. 75:441 (1990).
7. F. Moro, C.S. Ogg, H.A. Simmonds, J.S. Cameron, C. Chantler. Familial juvenile gouty nephropathy with renal urate hypoexcretion preceding renal disease, Clin Nephrol. 35:263 (1991).
8. L. Berger and T-F Yu, Renal function in gout IV. An analysis of 524 gouty subjects including long-term studies, Am. J. Med. 59:605 (1975).

CONGENITAL CHLORIDE LOSING ENTEROPATHY ASSOCIATED WITH TOPHACEOUS GOUTY ARTHRITIS

G.Nuki, ML.Watson, BC.Williams, HA.Simmonds and RC.Wallace

Rheumatic Diseases Unit and Department of Medicine, University of Edinburgh, Scotland, UK

INTRODUCTION

Congenital chloride losing enteropathy (familial chloride diarrhoea, congenital chloridorrhea)(1) is a recessively inherited disorder which results from a defect in chloride/bicarbonate exchange in the ileum. It is characterised by hydramnios in utero followed by lifelong osmotic diarrhoea. Inadequate fluid and electrolyte replacement can result in chronic hypokalaemic alkalosis, progressive renal insufficiency, drug resistant hyperuricaemia and gout.

CASE REPORT

Miss AT was born at 35 weeks gestation on 3 1 67 (birth weight 3.2kg) following a pregnancy complicated by gross hydramnios. She was the 4th child of non-consanguinous Scottish parents.

Her mother's first children were 32 week non-identical twins, one of whom died at 48 hours with hilar membrane disease and the other at the age of two and a half years following persistent profuse diarrhoea. AT developed abdominal distension and watery diarrhoea at birth and underwent a laparotomy in the neonatal period to exclude mechanical obstruction before it was ascertained that she had chloridorrhea with low urine chloride excretion. After IV rehydration she was treated with oral supplements of 50-200mmol chloride/metre2/day as a combination of KCl and NaCl but she continued to pass 6-7 profuse watery motions daily and to have a persistent hypokalaemic, hypochloraemic alkalosis despite maintenance therapy. Between the years 1967-1983 she required numerous admissions to hospital for rehydration with IV fluids following episodes of vomiting, increased diarrhoea, dehydration and muscular weakness. Plasma electrolytes were in the range: Na+ 122-130 mmol/l, K+ 1.4-1.8 mmol/l, Cl- 60-70 mmol/l, CO2 40-48 mmol/l and typical stool and urine electrolytes are shown in Table 1.

Table 1. Stool and urine electrolytes

	pH	Volume ml/24hours	Na+	K+	Cl-
			mmol/24hours		
Stool	5.5	1500-2000	136	93	250
Urine	7.5	350-600	<10	30	<15

Investigation of renal function in 1972 showed a creatinine clearance of 36ml/min. Renal concentrating potential, intravenous pyelogram and micturating cystourethrogram were all normal but renal biopsy showed evidence of glomerular hyalinisation and shrinkage with mesangial hypertrophy and juxtaglomerular hyperplasia.

In 1989, at the age of 22, AT had a typical attack of podagra. Investigations showed a plasma urate of 0.86 mmol/l with a creatinine of 0.25 mmol/l, Na+ 122 mmol/l, K+ 1.6 mmol/l, CO_2 61 mmol/l. From 1989-1991 she continued to have recurrent episodes of acute gouty arthritis affecting the knees and elbows as well as the metatarsophalangeal joints of the great toes despite continuing treatment with allopurinol 300mgs/day and Colchicine 0.5mg bd. There were early tophi at the left elbow and x-rays of the feet showed erosions in the first MTP joints. The results of biochemical investigations are shown in Table 2.

Table 2. Investigations 1990

Plasma	Plasma (resting, supine)
Na+ 130mmol/l	Aldosterone 360pmol/l
K+ 1.7mmol/l	Renin activity 10.8ng/ml/hour AI
Cl- 69mmol/l	Angiotensin I 1.45ng/ml
CO_2 47mmol/l	Angiotensin II 58.8pg/ml
Creatinine 0.14mmol/l	
Urate 0.8mmol/l	Total Body K+
Oxypurinol 3umol/l	60.6g (predicted 83.1g)
Urine	
Uric Acid 0.7mmol/24hours	Urate clearance
Hypoxanthine +	1ml/min
Xanthine 0.008mmol/24hrs	Creatinine clearance 33 ml/min

The persistent hyperuricaemia with low plasma concentrations of oxypurinol and low urinary concentrations of the oxypurines hypoxanthine and xanthine during treatment with allopurinol suggested impaired absorption of allopurinol and there was no significant response to careful increase in the dose of allopurinol or treatment with probenecid (Figure 1).

The current approaches to therapy are to:-

1. Improve chronic hypovolaemia and electrolyte depletion by supplemtary oral replacement of KCl and NaCl with overnight infusions via nasogastric tube.

2. Add benzbromarone to continuing treatment with allopurinol 400mgs/day and Colchicine 0.5mg bd.

DISCUSSION

Congenital chloride losing enteropathy (CCLE) was first described in 1945 (2, 3) as "congenital alkalosis with diarrhoea". It is a rare disorder inherited as an autosomal recessive (4) that results from a defect in chloride/bicarbonate exchange in the ileal brush border. Larger series of patients from families with high levels of consanguinity have been reported from Finland (5) and Kuwait (6) where the estimated frequency of CCLE is 7.6/100000 live births. The disorder should be suspected in premature infants presenting with abdominal distension, profuse watery diarrhoea and a hyponatraemic, hypokalaemic, hypochloraemic alkalosis following hydramnios in

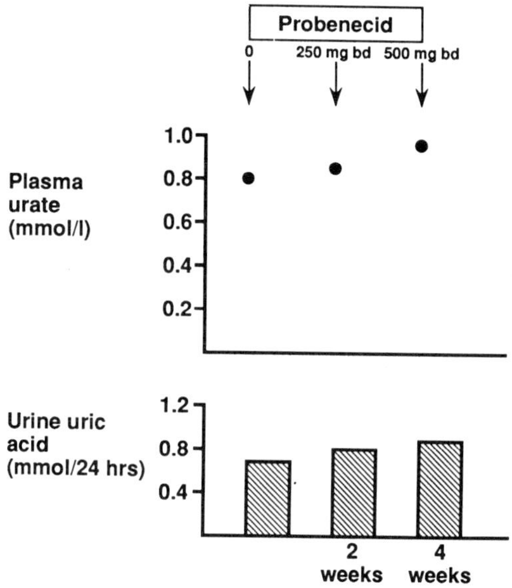

Figure 1. Treatment with probenecid. No significant effect on SUA or urinary uric acid excretion. Drug discontinued because of nausea.

pregnancy and the diagnosis can be confirmed by stool electrolyte analysis when faecal chloride content exceeds the sum of faecal sodium and potassium. In the majority of patients complications can be avoided by prompt and adequate oral replacement of electrolytes (5, 6) but delay in diagnosis or inadequate electrolyte replacement leads to juxtaglomerular hyperplasia with increased renin, angiotensin II and aldosterone secretion. Excessive angiotensin II can lead to secondary glomerular damage and chronic hyperaldosteronism results in renal potassium wasting, renal tubular damage and nephrocalcinosis (7). It is clear from this case report that severe drug resistant hyperuricaemia and gout may also be a clinical problem. Hyperuricaemia was reported in a 21 year old male with CCLE, hyperaldosteronism and renal impairment (8) as well as in Darrow's (3) original patient who was reinvestigated at the age of 31 (9), but the basis for the hyperuricaemia was not firmly established.

The results of investigations in AT suggested that her hyperuricaemia and marked reduction in fractional urate clearance are a direct consequence of chronic salt, water and volume depletion leading to activation of the renin-angiotensin system (Fig 2).

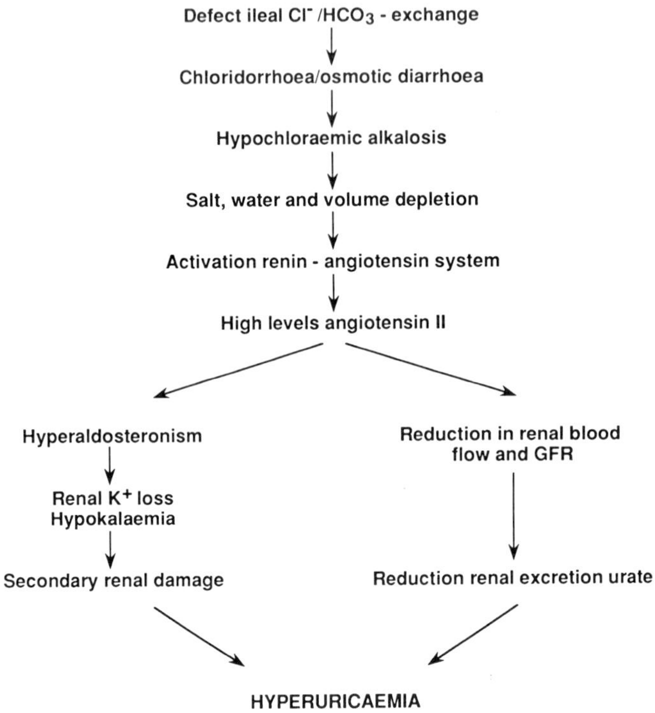

Figure 2 Proposed mechanisms for hyperuricaemia.

Infusion of angiotensin II is associated with a decrease in the fractional clearance of urate (10) and many of the features seen in AT (juxtaglomerular hyperplasia, hypereninaemia, hyperaldosteronism, urinary potassium loss and hypokalaemia) resembled Bartter's syndrome which is also associated with hyperuricaemia (11). Although treatment of CCLE with a prostaglandin synthesis inhibitor has been shown to be of benefit in some patients (12) as it is in Bartter's syndrome, an attempt to treat acute gouty arthritis in AT with Indomethacin (Fig. 3) resulted in a dramatic fall in urine production suggesting that her renal blood flow was critically prostacyclin dependent.

Some support for the hypothesis that the decrease in fractional clearance of urate may result in part from a direct effect of angiotensin II on renal blood flow came from a careful trial of treatment with the angiotensin converting enzyme inhibitor Captopril following IV electrolyte and volume repletion (Fig. 4). Unfortunately however treatment with the ACE inhibitor could not be continued as maintenance of blood pressure in AT appeared to be critically dependent on the pressor effect of angiotensin II.

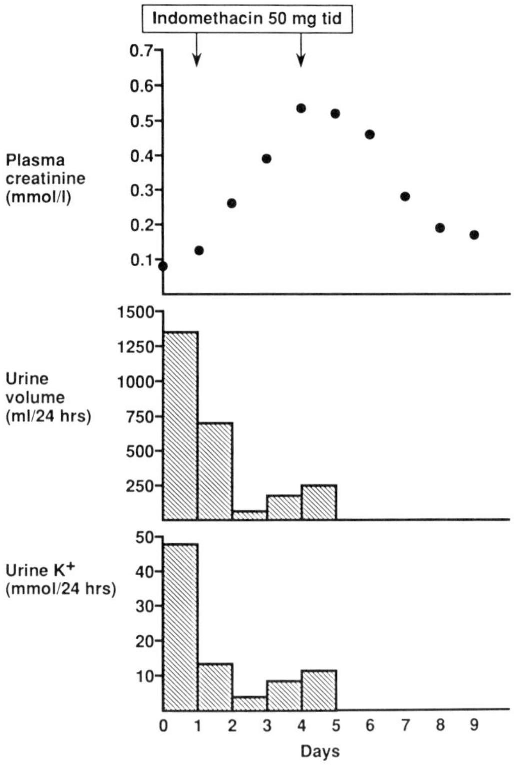

Figure 3. Effect of treatment with NSAID.

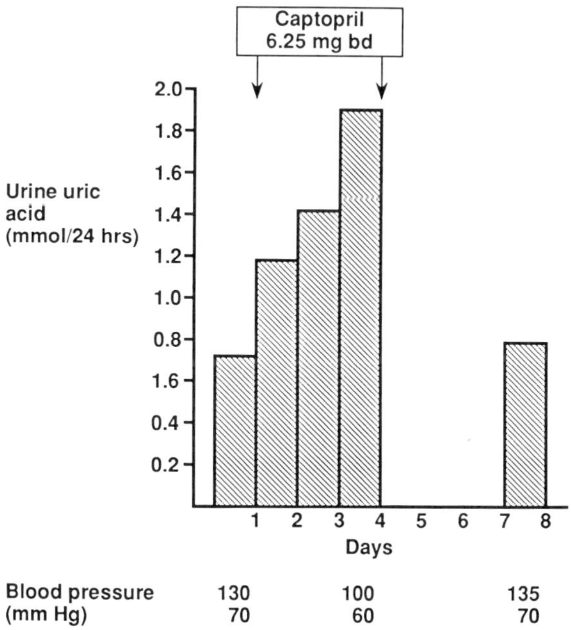

Figure 4 Effect of treatment with the ACE inhibitor Captopril on urine uric acid excretion.

REFERENCES

1. McKusick VA, 214700 Familial chloride diarrhoea in Mendelian inheritance in man. 8th Ed. John Hopkins Univ Press Baltimore 1988.
2. Gamble JL, Fahey KR, Appleton J and MacLachlan EA: Congenital alkolosis with diarrhoea. J.Paediat. 26; 509 (1945)
3. Darrow DC: Congenital alkolosis with diarrhoea. J.Paediat. 26; 519 (1945)
4. Norio R, Perheentupa J, Launiala K et al: Congenital chloride diarrhoea, and autosomal recessive disease. Genetic Study of 14 Finish and 12 other families. Clin.Genet. 2; 182 (1971)
5. Holmberg C, Perheentupa J, Launiala K et al: Congenital chloride diarrhoea; clinical analysis 21 Finish patients. Arch. Dis. Child 52; 255 (1977)
6. Lubani MM, Doudin KI, Sharda DC et al: Congenital chloride diarrhoea in Kuwaiti children. Euro J. Paediat 148; 333 (1989)
7. Holmberg C, Perheentupa J, Pasternack A: The renal lesion in congenital chloride diarrhoea. J.Paediat 91; 738 (1977)
8. Pearson AJG, Sladen GE, Edmonds CJ et al: The pathophysiology of congenital chloridorrhoea. Quart J Med 42; 453 (1973)
9. Gorden P, Levitin H: Congenital alkolosis with diarrhoea. A sequel to Darrow's original description. Ann Intern Med 78; 876 (1973)
10 Ferris TF and Gorden P. Effects of angiotensin and norepinephrine upon urate clearance in man. Amer J Med 44; 359 (1968)
11 Meyer WJ III, Gill JR jr, Barter FC: Gout as a complication of Bartter's syndrome. A possible role for alkolosis in the decreased clearance of uric acid. Ann Intern Med 83; 56 (1975)
12 Minford AMB and Barr DGD: Prostaglandin synthetase inhibitor in an infant with congenital chloride diarrhoea. Arch Dis Child 55; 70 (1980)

THE ROLE OF LEAD IN GOUT NEPHROPATHY REVIEWED: PATHOGENIC OR ASSOCIATED FACTOR?

Mª.Eugenia Miranda, Juan G. Puig, Felícitas A. Mateos, Teresa H. Ramos, Elisa Herrero, Antonio González, Antonio Gil and Julio O. Vázquez

Division of Internal Medicine and Clinical Biochemistry, "La Paz" Hospital, Universidad Autónoma, Madrid, Spain

INTRODUCTION

The relationship between lead and gout has long been recognized (1). Lead has been responsible for several epidemics of gout (2-4) and may cause an interstitial nephropathy (5,6) characterized by severe uric acid underexcretion (7). In the past decade, and despite no direct apparent exposure to lead, several reports have documented increased lead body stores in patients with gout and renal insufficiency (8-11). Renal failure by itself could not account for the increased body lead stores since non-gout patients with chronic renal disease had normal urinary lead excretion rates (8-11). In contrast, other authors were unable to demonstrate that lead was important in the pathogenesis of renal failure in patients with gout (12). The present study was undertaken to determine: (a) body lead stores in patients with gout at our institution, (b) whether an increased body lead content could be related to the derangement in purine metabolism characteristic of gout, and (c) if renal function and purine metabolism are modified following chelation therapy and normalization of increased mobilizable lead.

PATIENTS AND METHODS

We examined 70 patients with gout (64 men and 6 postmenopausal females) who ranged in age from 32 to 86-years (mean±SD, 63±13-years). Gout was diagnosed in all patients according to the ARA criteria (13). The only selection criterion was willingness to come into the hospital for 10 days. Patients were divided into two groups: (a) patients with renal insufficiency (n=25), defined as a serum creatinine concentration above 133 μmol/L or a simultaneous 24-hour creatinine clearance below 40 mL/min/1.73 m^2, and (b) patients (n=45) with normal glomerular filtration rate (GFR). These patients were matched for age and body mass index to patients with renal insufficiency. One patient in each group had had symptoms attributable to lead intoxication and 68 subjects denied apparent exposure to lead. No patient had basophilic stippling. Renal insufficiency developed in all patients after their first episode of acute gout. Eighteen patients with renal insufficiency (72%) had some manifestation of atherosclerosis. Renal failure was attributed to nephrosclerosis in the majority of the patients.

All subjects were admitted to the Clinical Research Unit of "La Paz" University Hospital and placed on a weight-maintenance, isocaloric, purine-restricted diet with a protein content of 10% to 15%. Four days after hospital admission, patients underwent an

EDTA lead mobilization test. Urinary lead was measured over 5 consecutive days after administration of 2 g of EDTA (with procaine) on the first day (8). Purine metabolism was examined during the same 5-day period by determination of plasma and 24-hour urinary creatinine, hypoxanthine, xanthine and uric acid concentrations (14). The mean of at least three determinations for plasma and five for urinary creatinine and purines was used. Results of the EDTA test and purine metabolism were compared with those obtained in 25 normal subjects (20 males, 5 postmenopausal females; age 62±14-years, range 41 to 84-years), all of whom were submitted to an identical protocol.

Patients with a cumulative 5-day urinary lead excretion above 650 µg were treated with 1 g of EDTA, twice per week. Patients with a post-treatment cumulative lead excretion below 600 µg/120 hours were admitted to the hospital and a second study was performed. Lead in blood and urine was measured by atomic absorption spectrophotometry (Perkin Elmer 460, spectrophotometer). Uric acid and creatinine in plasma and urine were measured in a multichannel autoanalyzer (Hitachi 737, Boehringer Mannheim). Plasma and urinary hypoxanthine and xanthine were determined by high-pressure liquid chromatography (14).

RESULTS

Blood lead levels were significantly higher in patients with renal insufficiency than in patients with gout and normal GFR (Table 1). Gout patients with renal failure showed an

Table 1. Mobilizable Lead and Purine Metabolism in the Studied Subjects.

	Gout and Renal Failure (n=25)	Gout without Renal Failure (n=45)	Normal Subjects (n=25)
Blood lead (µg/dL)	19±8 [a,c]	14±8	12±5
Mobilizable lead (µg/120-h)	859±804 [a,b]	330±383	215±150
Plasma (µmol/L)			
Creatinine	212±88 [a,b]	97±18	80±18
Urate	476±95 [a]	446±83 [a]	292±60
Hypoxanthine	4.4±3.2 [a]	4.4±2.1 [a]	2.3±1.1
Xanthine	0.9±0.5	1.2±0.6 [a]	0.7±0.2
24-h Urine			
Creatinine (mmol/24-h/1.73 m²)	10.5±3.4 [c]	12.4±2.9	12.1±3.1
Uric acid (mmol/24-h/1.73 m²)	1.75±0.53 [a,b]	2.52±0.92	2.63±0.51
Uric acid/Creatinine (mmol/mmol)	0.17±0.05 [a,b]	0.21±0.07	0.23±0.08
Hypoxanthine (µmol/24-h/1.73 m²)	18±12 [a]	21±10 [a]	33±9
Xanthine (µmol/24-h/1.73 m²)	10±7 [a]	15±9 [a]	22±9
Clearance (mL/min/1.73 m²)			
Creatinine	38±15 [a,b]	97±26	101±21
Uric acid	2.6±0.8 [a,b]	3.9±1.4 [a]	6.4±1.3
Hypoxanthine	5.0±3.8 [a]	5.2±3.6 [a]	13.3±5.5
Xanthine	10.6±7.0 [a]	11.6±5.8 [a]	29.7±13.1
Fractional excretion (%)			
Uric acid	7.6±3.8 [b]	4.3±1.5 [a]	6.4±1.5
Hypoxanthine	11.0±7.1 [b]	5.7±3.8 [a]	13.2±5.6
Xanthine	23.9±15.8 [b]	12.1±6.7 [a]	30.2±12.7

[a]P<0.01 versus normal subjects.
[b]P<0.01 versus gout without renal failure.
[c]P<0.03 versus gout without renal failure.

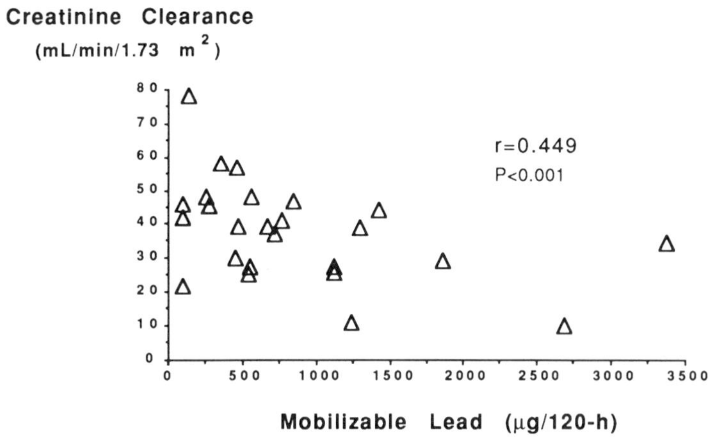

Figure 1. Relation between creatinine clearance and mobilizable lead in 25 patients with gout and moderate renal insufficiency.

increase in mobilizable lead in response to the EDTA test as compared to patients with normal GFR and normal controls. Among the 25 patients with impaired renal function, twelve (48%) showed a lead excretion rate above 650 µg/120 h (mean±SD, 1568±985 µg/120 h). In contrast, only 3 patients with normal GFR (7%) showed a mobilizable lead excretion above 650 µg/120 h (1329, 1044, and 2193 µg/120 h) ($P<0.001$). When patients with and without renal insufficiency were analyzed together cumulative lead excretion, as measured by the EDTA chelation test, was significantly related to blood lead concentrations ($r=0.516$; $P<0.0001$). The severity of renal disease, as determined by the creatinine clearance, correlated directly with the mobilizable lead ($r=0.449$; $P<0.001$) in gout patients with renal failure (Figure 1).

Patients with gout exhibited a severe renal underexcretion of purines (Table 1). Plasma hypoxanthine, xanthine and urate concentrations were significantly higher and 24-h urinary hypoxanthine xanthine and uric acid excretion rates were significantly lower in patients with gout as compared with normal subjects. Thus, clearance of hypoxanthine, xanthine and uric acid were markedly diminished in gout patients. Purine metabolism was remarkably similar in the two groups of patients except for a significantly lower uric acid excretion in subjects with impaired GFR.

Renal function and purine metabolism were reexamined in 6 patients with renal insufficiency following chelation therapy with EDTA. These patients received a mean of 57±34 g of EDTA (range, 29 to 120 g) over a period of 27±18-weeks (range, 15 to 60-weeks). Blood lead levels decreased from 23±11 to 15±4 µg/dL and cumulative lead excretion decreased from a baseline value of 1568±985 to 425±272 µg/120-h ($P<0.02$). GFR and plasma and urinary concentrations of hypoxanthine, xanthine and uric acid were not significantly modified by EDTA therapy.

DISCUSSION

This study shows that: (a) increased mobilizable lead was significantly more frequent in gout patients who also had impaired GFR (48%) than in subjects with normal GFR (7%) ($P<0.0001$), (b) purine underexcretion in patients with gout could not be related to increased lead body stores, and (c) GFR and plasma and urinary hypoxanthine, xanthine and uric acid concentrations were essentially unmodified by chelation therapy despite normalization of mobilizable lead. In the past 15 years our concepts about the pathogenesis of renal failure in patients with gout -thought to be the result of urate deposits in the renal interstitium ("gout nephropathy")- have substantially changed. The fact that urate deposits in the renal tissue can no longer be viewed as specific of "gout nephropathy" (15,16) and the findings of long-term follow-up studies demonstrating that gout and chronic hyperuricemia are not deleterious for the kidney (17-20), have led some authors to discard the concept of chronic gouty nephropathy (21,22). On the other hand, reports of increased mobilizable lead in patients with gout and renal failure (8-11) prompted the hypothesis that lead may be responsible for the observed decline in renal function associated with gout. Our results suggest that an increased lead mobilization test in gout patients with renal failure is a non-pathogenic factor but an associated condition.

ACKNOWLEDGEMENTS

We are indebted to Dr. T. Ramos, Javier Díaz, and Mrs. Mª Paz Canencia for valuable technical assistance. This work was supported by grants from Caja de Madrid and Fondo de Investigaciones Sanitarias de la Seguridad Social (FISS, 90/0575), Spain.

REFERENCES

1. Yü T-F. Lead nephropathy and gout. Am J Kidney Dis 1983; 2:555-8.
2. Ball GV, Morgan JM. Chronic lead ingestion and gout. South Med J 1968; 61:21-4.
3. Ball GV. Two epidemics of gout. Bull Hist Med 1971; 45:401-8.
4. Emmerson BT. Chronic lead nephropathy. Kidney Int 1973; 4:1-5.
5. Inglis JA, Henderson DA, Emmerson BT. The pathology and pathogenesis of chronic lead nephropathy occurring in Queensland. J Pathol 1978; 124:65-76.
6. Bennett WM. Lead nephropathy. Kidney Int 1985; 28:212-20.
7. Emmerson BT. The clinical differentiation of lead gout from primary gout. Arthritis Rheum 1968; 11:623-34.
8. Batuman V, Maesaka JK, Haddad B, Tepper E, Landy E, Wedeen RP. The role of lead in gout nephropathy. N Eng J Med 1981; 304:520-3.
9. Craswell PW, Price J, Boyle PD, et al. Chronic renal failure with gout: A marker of chronic lead poisoning. Kidney Int 1984; 26:319-23.
10. Behringer D, Craswell P, Mohl C, Stoeppler M, Ritz E. Urinary lead excretion in uremic patients. Nephron 1986; 42:323-9.
11. Colleoni N, D´Amico G. Chronic lead accumulation as a possible cause of renal failure in gouty patients. Nephron 1986; 44:32-5.
12. Wright LF, Saylor RP, Cecere FA. Occult lead intoxication in patients with gout and kidney disease. J Rheumatol 1984; 11:517-20.
13. Wallace SL, Robinson H, Masi AT, Decker JL, McCarty DJ, Yü T-F. Preliminary criteria for the classification of the acute arthritis of primary gout. Arthritis Rheum 1977; 20:895-900.
14. Puig JG, Mateos FA, Jiménez ML, Ramos TH. Renal excretion of hypoxanthine and xanthine in primary gout. Am J Med 1988; 85:533-537.
15. Verger D, Leroux-Robert C, Ganter P, Richet G. Les tophus goutteux de la medullaire renale des uremiques chroniques. Nephron 1967; 4:356-70.
16. Linnane JW, Burry AF, Emmerson BT. Urate deposits in the renal medulla. Nephron 1981; 29:216-22.
17. Berger L, Yü T-F. Renal function in gout. IV. An analysis of 524 gouty subjects including long-term follow-up studies. Am J Med 1975; 59:605-13.
18. Fessel WJ. Renal outcomes of gout and hyperuricemia. Am J Med 1979; 67:74-82.
19. Yü T-F, Berger L, Dorph DJ, Smith H. Renal function in gout. V. Factors influencing the renal hemodynamics. Am J Med 1979; 67:766-71.
20. Yü T-F, Berger L. Impaired renal function in gout. Am J Med 1982; 72:95-100.
21. Reif MC, Constantiner A, Levitt MF. Chronic gouty nephropathy: a vanishing syndrome. N Eng J Med 1981; 304:535-6.
22. Beck LH. Requiem for gouty nephropathy. Kidney Int 1986; 30:280-7.

MYOGENIC HYPERURICEMIA IN HYPOPARATHYROIDISM

Takamichi Nishimura , Ikuo Mineo , Takao Shimizu , Masanori Kawachi , Akira Ono , Hiromu Nakajima , Masamichi Kuwajima , Norio Kono and Seiichiro Tarui

The Second Department of Internal Medicine,
Osaka University Medical School, Osaka, Japan

INTRODUCTION

We previously reported that in genetic metabolic myopathies including glycogenosis types III, V and VII the defective energy generation resulted in excessive purine degradation in working muscle, to develop hyperuricemia after exercise ("myogenic hyperuricemia") (Kono et al., 1986; Kono et al., 1987; Mineo et al., 1985; Mineo et al., 1987). Myopathic symptoms similar to those of muscle glycogenoses sometimes develop in patients with hypoparathyroidism. The present study was aimed to disclose the development of myogenic hyperuricemia in this acquired myopathy secondary to hypoparathyroidism.

PATIENTS AND METHODS

Patients

Six patients with hypoparathyroidism participated in this study. Hypocalcemia (ranging from 2.3 to 3.7 mEq/L) and hyperphosphatemia (ranging from 5.1 to 6.7 mg/dL) were revealed in all the patients. Two patients complained of myopathic symptoms such as stiffness and cramp associated with high serum creatine kinase levels (4255 and 414 U/L). Calcification of the basal ganglia was observed on brain CT scanning. Other four patients did not exhibit such signs and symptoms. As a representative disease of myogenic hyperuricemia, 4 patients with previously diagnosed glycogenosis type V, muscle phosphorylase deficiency, also participated in this study.

Exercise Tests

Semi-ischemic forearm exercise test was performed as reported previously (Mineo et al., 1984). The patients also exercised on a bicycle ergometer as described previously,

Purine and Pyrimidine Metabolism in Man VII, Part A
Edited by R.A. Harkness *et al.*, Plenum Press, New York, 1991

according to various protocols depending on their individual ability to exercise. The workload and the net duration of exercise were 15 to 35 W and 60 min in the patient with glycogenosis type V, and 50 to 100 W and 40 min in the patient with hypoparathyroidism. Blood samples were obtained from cuvital vein at the time indicated. The patients were tested again with glucose infusion under the equivalent exercise protocol. Solution containing 10% glucose and 22.5 mU/mL insulin was infused intravenously at a rate of 6.7 mL/min throughout from -15 min to the end of the exercise.

<u>Assays</u>

Blood lactate (Gutmann et al., 1974), plasma ammonia (van Anken et al., 1974) and serum uric acid (Lorentz et al., 1967) were assayed enzymatically. Hypoxanthine was assayed using high performance liquid chromatography as described previously (Mineo et al., 1985).

RESULTS AND DISCUSSION

In the patients with glycogenosis type V, semi-ischemic forearm exercise caused no appreciable increases of venous lactate but excessive increases of venous ammonia and hypoxanthine. The results indicated that energy crisis due to the defective glycogen breakdown led to excessive purine degradation in forearm muscles (data not shown).

The 4 hypoparathyroid patients without myopathy showed essentially normal responses of venous lactate, ammonia and hypoxanthine. However, the 2 hypoparathyroid patients with myopathic symptoms exhibited great responses of ammonia and hypoxanthine despite normal lactate responses (Fig.1), indicating that excessive purine degradation occurred in the working muscle. Both the patients with myopathy were tested again after therapy with vitamin D_3. The abnormal responses of ammonia and hypoxanthine were completely normalized, suggesting that disturbed energy metabolism leading to excessive purine degradation is reversible in myopathy associated with hypoparathyroidism.

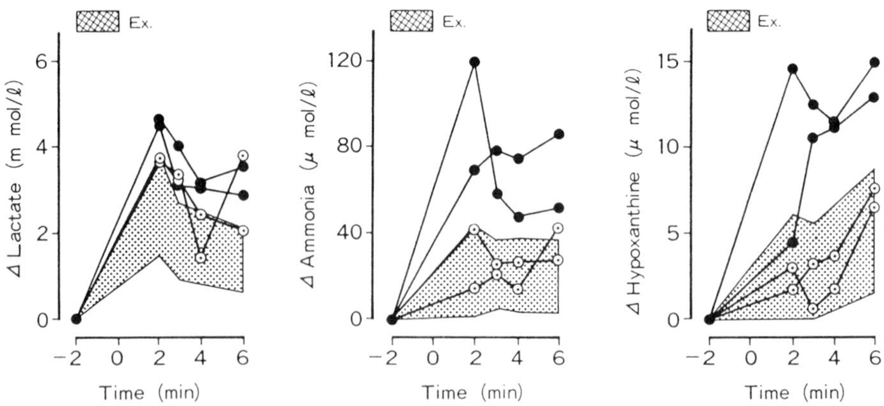

Fig. 1. Semi-ischemic forearm exercise tests on the patients with hypoparathyroid myopathy.
● : before vitamin D3 treatment
⊙ : after vitamin D3 treatment
▒ : normal controls (mean ± SD , n=25)

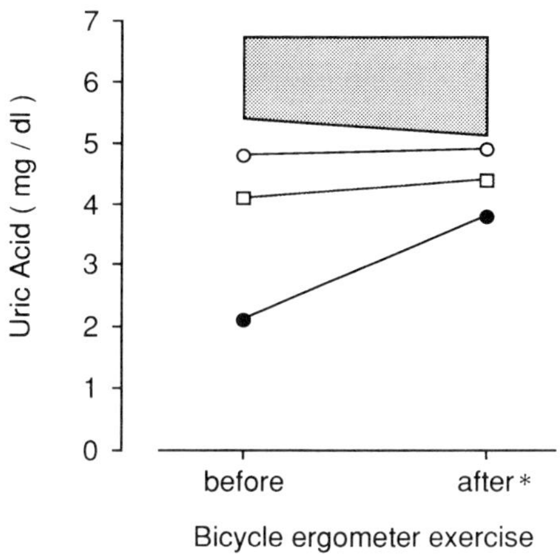

Fig. 2 Changes in serum uric acid levels after exercise in the patient with hypoparathyroid myopathy.
The patient exercised on a bicycle ergometer under three conditions :
no treatment (●), with infusion of glucose plus insulin (□), and after vitamin D3 treatment (○). The shaded area indicates means ± SD of normal subjects (n=4).
*,Blood samples obtained 120 min after the end of exercise.

In the patient with glycogenosis type V, the plasma hypoxanthine levels increased greatly after bicycle ergometer exercise. Glucose infusion during exercise abolished this increase in hypoxanthine. Increased utilization of blood glucose may bypass the metabolic block at glycogenolysis and thus restore the energy expenditure of the muscle.

In the patient with hypoparathyroid myopathy, similar to glycogenosis type V, bicycle ergometer exercise caused great increases of plasma hypoxanthine followed by a significant rise of serum uric acid (Fig. 2). Theses abnormal increases in hypoxanthine and uric acid were normalized with glucose infusion. Therapy with vitamin D3 was also effective to prevent myogenic hyperuricemia.

CONCLUSION

In the patients with hypoparathyroid myopathy, excessive purine degradation occurs in working muscle, causing hyperuricemia after exercise ("myogenic hyperuricemia"). This altered purine metabolism is reversible and is probably due to an impaired glycogenolysis. These findings would add further knowledge on the metabolic basis of myogenic hyperuricemia.

REFERENCES

Gutmann, I., Wahlefeld, A. W., 1974, L-lactate determination with lactate dehydrogenase and NAD, in: "Methods of enzymatic analysis," H. U. Bergmeyer, ed., Academic Press, New York.

Kono, N., Mineo, I., Shimizu. T., Hara, N., Yamada, Y., Nonaka, K., and Tarui, S., 1986, Increased plasma uric acid after exercise in muscle phophofructokinase deficiency, Neurology, 36:106.

Kono,N., Mineo, I., and Tarui, S., 1987, Hyperuricemia in type V glycogenosis, Neurology, 37:278.
Lorentz, K., Berndt, W., 1967, Enzymatic determination of uric acid by a colorimetric method, Anal. Biochem., 18:58.
Mineo, I., Kono, N., Shimizu, T., Sumi, S., Nonaka, K., and Tarui, S., 1984, A comparative study on glucagon effect between McArdle disease and Tarui disease, Muscle Nerve, 7:552.
Mineo, I., Kono, N., Shimizu, T., Hara, N., Yamada, Y., Sumi, S., Nonaka, K., and Tarui, S., 1985, Excess purine degradation in exercising muscles of patients with glycogen storage disease types V and VII, J. Clin. Invest., 76:556.
Mineo, I., Kono, N., Hara, N., Shimizu, T., Yamada, Y., Kawachi, M., Kiyokawa, H., Wang, Y.L. and Tarui, S., 1987, Myogenic hyperuricemia. A common pathophysiologic feature of glycogenosis types III,V and VII, N. Engl. J. Med., 317:75.
van Anken, H. C., Schiphorst, M. E., 1974, A kinetic determination of ammonia in plasma, Clin. Chim. Acta., 56:151.

LIPOPROTEIN METABOLISM IN PRIMARY GOUT - INFLUENCE OF ALCOHOL INTAKE AND BODY WEIGHT

Sumio Takahashi, Tetsuya Yamamoto, Yuji Moriwaki, Michio Suda, Oluyemi E Agbedana, Toshikazu Hada, Kazuya Higashino

Third Department of Internal Medicine, Hyogo College of Medicine, 1-1, Mukogawa-cho, Nishinomiya, Hyogo 663, Japan

INTRODUCTION

It has been reported that a raised serum uric acid is related to the development of atherosclerosis, but the direct association between hyperuricemia and atherosclerosis remains uncertain[1]. Therefore, in gout it seems important to investigate serum lipids which are among the known risk factors for atherosclerosis. In previous studies, some investigators suggested that hyperlipidemia in primary gout was secondary to excessive alcohol intake[2] and/or obesity[3], while others suggested that hyperlipidemia was a consequence of essential factors in primary gout[4].

Therefore, we attempted to clarify whether or not the impairment of lipid metabolism was secondary to alcohol intake and/or body weight in 158 males with primary gout.

SUBJECTS AND METHODS

The subjects were 158 male patients with primary gout (Gout) aged 49.2±1.0 (mean±SEM) and 164 male control subjects (Control) aged 50.3±0.9 (mean±SEM), selected randomly.

Any medication affecting serum lipid was withheld for at least 8 months before the study, except for anti-hyperuricemic agents. The oral glucose tolerance test showed a normal pattern in these patients. Blood samples were drawn after an overnight fast and serum was separated for the determination of lipids and apolipoproteins. Fractionation and subfractionation of HDL were performed by the method of Kostner et al[5]. The cholesterol content of the supernatants and apolipoproteins were measured by commercially available kits. Information concerning ethanol consumption was obtained by questionnaire on frequency, quantity and type.

STATISTICAL ANALYSIS

Data were expressed as the mean±SEM and Student's t test was used to evaluate the observed differences ($P<0.05$).

RESULTS

There was no difference between Gout and Control with regard to age, body mass index (BMI) and alcohol intake. The serum triglyceride (S-TG) level was significantly higher in Gout than in Control. However, the serum cholesterol (S-TC,S-FC) and phospholipid (S-PL) levels did not differ significantly between Gout and Control. The total cholesterol in HDL fraction (HDL-C) and in its subfraction, HDL2 (HDL2-C) were significantly reduced in Gout.

The serum concentrations of apolipoprotein AI (apo AI), apolipoprotein AII (apo AII) and apolipoprotein E (apo E) in Gout and Control were similar, while the concentrations of apolipoprotein B (apo B), apolipoprotein CII (apo CII) and apolipoprotein CIII (apo CIII) were significantly higher in Gout than in Control (Table 1).

Table 1. Clinical features and serum lipid, apolipoprotein profiles in Gout and Controls.

	Gout (n=158)	Controls (n=164)
Age (years)	49.2±1.0	50.3±0.9
Body Mass Index (kg/m^2)	23.3±0.2	23.4±0.2
Alcohol Intake (g/day)	17.7±1.4	16.8±1.5
S-TC	185.6±2.7	188.1±2.5
S-PL	219.7±3.7	213.1±3.4
S-TG	176.1±11.0*	132.3±8.1
S-FC	48.3±0.8	49.2±0.7
HDL-C	43.1±1.0 *	46.7±1.0
HDL2-C	11.0±0.4 **	13.7±0.5
HDL3-C	32.0±0.7	33.1±0.7
apo AI	130.5±2.0	133.8±1.6
apo AII	34.8±0.6	34.6±0.5
apo B	100.8±2.0 **	90.3±1.8
apo CII	4.8±0.2 **	3.8±0.1
apo CIII	11.7±0.4 **	9.6±0.3
apo E	4.2±0.1	4.0±0.2

Values are expressed as mean±SEM (mg/dl)
*$P<0.05$, **$P<0.01$

Table 2 presents the clinical features and serum lipid concentrations in all the subjects subdivided according to alcohol intake. In non-drinkers with Gout the serum triglyceride was increased, while both HDL-C and HDL2-C were reduced when compared with corresponding values in Control non-drinkers. On the other hand, HDL-C, HDL2-C and HDL3-C were higher in drinkers with Gout than those in non-drinkers with Gout.

Table 3 shows the clinical features and serum lipid concentration in all the subjects subdivided according to BMI. The serum concentrations of HDL-C, HDL2-C and HDL3-C were higher in the below the mean value BMI subjects with Gout than in the above the mean value BMI subjects with Gout. In addition, the serum concentrations of HDL-C and HDL2-C were lower in the below the mean value BMI subjects with Gout than in the below the mean BMI subjects with Control.

Table 2. Subjects, clinical features, serum lipid concentrations, total cholesterol concentrations, in HDL fraction and its subfraction in all the subjects subdivided according to alcohol intake

	Gout		Controls	
	Non-Drinkers (n = 63)	Drinkers (n = 95)	Non-Drinkers (n = 74)	Drinkers (n = 90)
Age (years)	50.7±1.7	48.1±1.2	50.4±1.5	50.2±0.9
BMI (kg/m^2)	23.5±0.3	23.2±0.3	23.4±0.2	23.2±0.3
Alcohol intake (g/day)	0	29.5±1.43	0	30.7±1.7
S-TC	179.4±4.6	189.7±3.4	188.7±3.6	187.7±3.4
S-PL	205.9±5.2	228.9±4.8^{a}	207.4±3.7	217.7±5.3
S-TG	179.6±17.1^{f}	173.8±14.4^{g}	129.6±7.2	134.6±13.6
S-FC	47.0±1.4	49.1±1.0	50.0±1.1	48.6±1.0
HDL-C	37.8±1.3^{f}	46.6±1.3^{b}	43.1±1.2	49.6±1.4^{f}
HDL2-C	9.0±0.5^{f}	12.4±0.6^{bh}	11.7±0.6	15.3±0.8^{f}
HDL3-C	28.8±1.1	34.2±0.9^{b}	31.4±0.9	34.6±1.1^{e}

Values are expressed as mean±SEM (mg/dl)
a P<0.05 and b P<0.01 when compared to the non-drinkers with Gout.
c P<0.05 and d P<0.01 when compared to the drinkers with Gout.
e P<0.05 and f P<0.01 when compared to the non-drinkers with Control.
g P<0.05 and h P<0.01 when compared to the drinkers with Control.

Table 3. Subjects, clinical features, serum lipid concentrations, total cholesterol concentrations, in HDL fraction and its subfraction in all the subjects subdivided according to BMI

	Gout		Controls	
mean BMI (kg/m^2)	23.3±0.2		23.4±0.2	
	< mean±BMI (n = 84)	> mean±BMI (n = 74)	< mean±BMI (n = 81)	> mean±BMI (n = 83)
BMI (kg/m^2)	21.3±0.1	25.6±0.2	21.5±0.1	25.1±0.2
Age (years)	49.7±1.5	48.5±1.4	50.3±1.2	50.3±1.2
Alcohol Intake (g/day)	17.0±1.8	18.6±2.2	18.9±2.3	15.5±2.0
S-TC	185.7±3.8	185.5±4.0	181.6±3.8	194.5±3.1^{f}
S-PL	218.2±5.1	221.4±5.4	211.6±4.0	214.5±5.4
S-TG	163.3±15.6^{f}	190.6±15.3^{h}	115.9±9.2	148.3±13.1^{e}
S-FC	48.0±1.1	48.6±1.3	47.1±1.1	51.3±1.0^{f}
HDL-C	45.8±1.4^{e}	40.0±1.3^{b}	51.2±1.5	42.3±1.1^{f}
HDL2-C	12.0±0.7^{f}	9.9±0.5^{ga}	15.6±0.8	11.8±0.6^{f}
HDL3-C	33.8±1.0	30.1±1.0^{a}	35.6±1.1	30.8±0.9

Values are expressed as mean±SEM (mg/dl)
a P<0.05 and b P<0.01 when compared to the below the mean value BMI subjects with Gout.
c P<0.05 and d P<0.01 when compared to the above the mean value BMI subjects with Gout.
e P<0.05 and f P<0.01 when compared to the below the mean value BMI subjects with Control.
g P<0.05 and h P<0.01 when compared to the above the mean value BMI subjects with Control.

DISCUSSION

The most commonly reported lipid abnormality in Gout is hypertriglyceridemia[2),4)], which has been suggested to be correlated with alcohol intake and/or obesity[2)-4)]. However, the present study suggested that hypertriglyceridemia was associated with an essential factor(s) in Gout because hypertriglyceridemia was demonstrated in Gout whenever alcohol and influence of the BMI were considered.

Few studies have previously evaluated HDL changes in gout. Some demonstrated that HDL2-C level was normal although HDL-C and HDL3-C levels were decreased in gout[6)], while others found reduced HDL-C and HDL2-C levels and a normal HDL3-C level in gout[7)]. However, these studies were performed in a small group of gout and the influence of alcohol intake was not considered. We have therefore investigated a relatively large group of Gout, where the influence of alcohol intake and BMI were considered.

Our results showed significant reduction in both HDL-C and HDL2-C. In addition, apo B is increased in Gout and this together with a decrease in HDL-C and in HDL2-C may contribute to more prevalence of atherosclerosis in Gout.

A positive relation between serum triglyceride, HDL-C level and alcohol intake is well known[8)-9)]. Alcohol intake also raised the serum triglyceride in Gout in the present study. Our results showing increased HDL-C, HDL2-C and HDL3-C with alcohol intake, seem consistent with the results in alcoholics as reported by Dai et al[9)], and in normal subjects as reported by Castelli et al[10)]. Therefore, in Gout the mechanism of serum lipid changes induced by alcohol may be the same as in normal subjects.

Obesity is a heterogeneous group in terms of etiology, time of development, adipose tissue characteristics and metabolic abnormalities, and so on. In the present study, a decrease in HDL-C, HDL2-C and HDL3-C levels in the above the mean value BMI subjects with Gout was consistent with the result of Ford et al[11)], indicating that the BMI was related to a decrease in HDL-C in Gout.

In conclusion, the deceased concentration of HDL2-C in Gout may be associated with an essential factor(s), although the factor(s) remains to be elucidated. Alcohol intake increased HDL-C level and BMI decreased HDL-C level in Gout, indicating that the same mechanism works in Gout as in Control.

REFERENCES

1. F. N. Brand, D. L. McGee, W. B. Kannel, J. Stokes III and W. P. Castelli, Hyperuricemia as a risk factor of coronary disease: The Framingham study, Am J Epidemiol 121:11-8 (1985).
2. T. J. Gibson and R. Grahame, Gout and hyperlipidaemia, Ann Rheum Dis 33:298-303 (1974).
3. W. J. Fessel, High uric acid as an indicator of cardiovascular disease: independence from obesity, Am J Med 68:401-4 (1980).
4. S. Jiao, K. Kameda, Y. Matsuzawa and S. Tarui, Hyperlipoproteinaemia in primary gout: Hyperlipoproteinaemic phenotype and influence of alcohol intake and obesity in Japan, Ann Rheum Dis 45:308-13 (1986).
5. G. M. Kostner, E. Molinari and P. Pichler, Evaluation of a new HDL2/HDL3 quantitation method based on precipitation with polyethylene glycol, Clin Chim Acta 148:139-47 (1985).

6. A. Ulreich, G. M. Kostner, K. P. Pfeiffer, P. Sedlmayr and F. Rainer, Serum lipids and lipoproteins in patients with primary gout. Rheumatol Int 5:73-7 (1985).
7. G. A. A. Fern, J. Lanham, J. Stocks, C. Ritchie, J. Katz and D. J. Galton, The measurement of high density lipoprotein sub-fractions in patients with primary gout using a simple precipitation method, Ann Clin Biochem 22:526-32 (1985).
8. P. Belfrage, B. Berg, I. Hagerstrand, E. P. Nilsson and H. Tornqvist, Alteration of lipid metabolism in healthy volunteers during long-term ethanol intake, Eur J Clin Invest 7:127-31 (1977).
9. W. S. Dai, R. E. Laporte and D. L. Hom et al, Alcohol consumption and high density lipoprotein cholesterol concentration among alcoholics, Am J Epidemiol 22:620-7 (1985).
10. W. P. Castelli, J. T. Doyle and T. Gordon et al, Alcohol and blood lipids; The cooperative lipoprotein phenotyping study, Lancet 11:153-5 (1977).
11. E. Ford, R. Copper, B. Simmons, S. Katz and R. Patel, Sex differences in high density lipoprotein cholesterol in urban blacks, Am J Epidemiol 127:753-61 (1988).

SERUM URATE AND URIC ACID EXCRETION

N. Di Sciasco*, C. Crognale*, M. Golato*, C.P. Quaratino°, M.B. Di Sciascio*, M. Lucarelli+, and A. Giacomello°

°Institute of Medical Physiopathology, University of Cieti
*Renzetti Hospital of Lanciano, Chieti
+Department of Human Biopathology
University of Rome, Italy

In the course of an epidemiological survey, the following laboratory tests were performed on morning urine and blood specimens obtained from fasting patients seen during a five-month period in the Renzetti Hospital of Lanciano (Chieti) : serum uric acid (S.U.A.) (1), serum creatinine (S.CR.)(2), serum total cholesterol (S.CH.) (3), serum triglyceride (S.T.)(1), serum glucose (S.G.) (4), urine uric acid (U.U.A.) (1) and urine creatinine (U.CR.) (2) concentrations. All assays were performed using an automated analyzer (Monarch - Instrumentation Laboratory -). Age, sex and department of admission were also recorded. To quantify renal excretion of urate, uric acid excretion per deciliter of glomerular filtrate (ER) (5) and fractional excretion of uric acid (FE) (urate clearance x 100 / creatinine clearance) were employed in order to avoid problems inherent in a 24-hours urine collection and to correct the excretion rate for differences in body size.

Although more than 5000 patients have been studied, in the present paper the preliminary results obtained by a computer assisted analysis of the data from the first 1780 subjects are reported. Serum urate concentrations have been measured systematically in many different ethnic populations, occupations, diseases, and other epidemiologic groupings (6-8).

Although such studies have revealed intriguing associations, the excretion of uric acid has rarely been examined. In the present paper the correlation between [S.U.A.], ER, FE and some of the other examined variables has been studied. Since renal insufficiency affects uric acid excretion (9), data from patients with a serum creatinine concentration higher than 2 mg/dl were not included.

RESULTS AND DISCUSSION

Means and standard deviations of some variables were considered for each group of subjets classified on the basis of sex, admission to the Hospital (medical and surgical departments) or ambulatory visits of the population.

Mean values were compared by using Student's t test. In good agreement with the current literature (6-8), serum urate concentration was higher in men than in women ($p<0.01$) in all groups. Although in patients admitted to surgical divisions no significant sex differece in ER was observed, in the other groups the mean value of this variable was higher in women ($p<0.01$). The high coefficient of variation obtained for ER and FE could be, at least in part, attributed to differences in dietary intake of purines (chiefly in the ambulatory population) and in the associated diseases. Possible effects of diurnal variations in the excretion rate of uric acid should also be considered. Correlation analysis between serum uric acid concentration and some of the other variables has been carried out (Table 2).

TABLE 1. MEAN VALUE AND STANDARD DEVIATION OF THE VARIABLES EXAMINED FOR EACH GROUP OF SUBJECTS

SUBJECTS	SEX	NR	AGE (years)		[S.U.A.] (mg/dl)		[S.CR.] (mg/dl)		[S.CH.] (mg/dl)		[S.T.] (mgldl)		[S.G.] (mg/dl)		ER (mg/dl)		FE	
			M	DS	M	DS	M	DS	M	DS	M	DS	M	DS	M	DS	M	DS
TOTAL	M,F	1552	49	20	4,88	1,71	1,08	0,23	169,0	44,1	142,6	132,6	102,5	37,2	0,47	0,29	10,66	7,93
OUTPATIENTS	F	261	50	18	4,19	1,16	1,02	0,17	194,1	47,7	174,2	166,1	101,9	32,8	0,49	0,33	12,69	8,19
	M	220	50	16	5,28	1,45	1,12	0,22	184,8	40,5	199,8	166,3	107,5	43,6	0,42	0,17	8,39	4,08
INPATIENTS	F	263	48	20	4,26	2,00	1,02	0,21	166,5	42,1	125,6	123,4	102,3	36,1	0,44	0,31	11,62	8,26
(surg.div)	M	316	52	29	5,42	1,59	1,17	0,21	160,7	39,9	126,0	82,2	101,1	32,6	0,48	0,28	9,67	6,28
INPATIENTS	F	227	47	23	4,55	1,50	1,03	0,23	159,1	38,5	116,2	90,2	104,2	38,5	0,45	0,23	11,23	10,65
(med.div)	M	265	45	24	5,47	1,86	1,24	0,28	152,0	40,3	122,9	133,0	99,3	40,1	0,51	0,36	10,27	8,19

In all groups, serum uric acid was positively correlated with serum creatinine and serum triglyceride concentration. A positive correlation with serum glucose concentration, especially in women, and with serum cholesterol concentration in men admitted to Hospital was also observed. Serum uric acid concentration was positively correlated with age, but for ambulatory men, in all groups. In this group of subjects, those ranging in age between 39 and 49 years, a negative correlation of serum uric acid concentration with age was observed. Correlation analisys between ER, FE, [S.U.A.], [S.CR.], [U.U.A.], [U.CR.] and age is shown in Tables 3 and 4.

TABLE 2. SAS CORRELATION ANALYSIS BETWEEN [S.U.A.] AND THE VARIABLES LISTED BELOW

SUBJECTS	Nr	SEX		[S.G.] (mg/dl)	[S.CR.] (mg/dl)	[S.TR.] (mg/dl)	[S.CH.] (mg/dl)	AGE (years)
TOTAL	1552	F,M	r	0,122	0,395	0,224	0,073	0,237
			p	0,0001	0,0001	0,0001	0,0037	0,0001
	261	F	r	0,13	0,219	0,165	0,014	0,282
			p	0,036	0,0004	0,0074	0,8194	0,0001
OUTPATIENTS								
	220	M	r	-0,019	0,280	0,224	0,123	0,092
			p	0,7338	0,0001	0,0008	0,0674	0,1724
	263	F	r	0,219	0,406	0,355	0,122	0,337
			p	0,0003	0,0001	0,0001	0,0475	0,0001
INPATIENTS (surg.div.)								
	316	M	r	0,038	0,310	0,196	0,144	0,155
			p	0,4978	0,0001	0,0004	0,0099	0,0057
	227	F	r	0,217	0,328	0,206	0,070	0,298
			p	0,0001	0,0001	0,0018	0,2920	0,0001
INPATIENTS (med.div.)								
	265	M	r	0,190	0,426	0,351	0,320	0,279
			p	0,0019	0,0001	0,0001	0,0001	0,0001
			r (PEARSON CORRELATION COEFFICIENTS)					
			p (PROB. >R UNDER Ho: Rho = 0)					

ER is correlated positively with [S.CR.], [U.U.A.], FR, and negatively with [U.CR] in all groups. A positive correlation of the variable with age is seen especially in women. ER is correlated with [S.U.A.] only in women admitted to medical departments and in ambulatory men. In these groups of patients the correlation is positive . A positive correlation of FE with [U.U.A.], ER, and a negative correlation with [S.U.A.] and [U.CR.] is evident in all groups. FE does not appear to be correlated with age. It has been suggested (5) that uric acid excretion per deciliter of glomerular filtrate (ER) could be a useful screening test to detect increased urinary excretion. As has been shown in the present paper ER is not indipendent of age and sex. Several standards are therefore required to employ this method. Furthermore, ER does not evaluate the efficiency of the kidney in handling the plasma urate load. On the other hand FE is indipendent of age and is always negatively correlated to [S.U.A.].

TABLE 3. SAS CORRELATION ANALYSIS BETWEEN ER AND THE VARIABLES LISTED BELOW

SUBJECTS	Nr	SEX		[S.U.A.] (mg/dl)	[S.CR.] (mg/dl)	[U.U.A.] (mg/dl)	[U.CR.] (mg/dl)	FE	AGE (years)
TOTAL	1552	F,M	r	0,084	0,190	0,407	-0,226	0,745	0,096
			p	0,0009	0,0001	0,0001	0,0001	0,0001	0,0001
	261	F	r	0,084	0,215	0,304	-0,289	0,817	0,005
			p	0,1756	0,0005	0,0001	0,0001	0,0001	0,9340
OUTPATIENTS									
	220	M	r	0,175	0,336	0,387	-0,27	0,796	0,097
			p	0,0090	0,0001	0,0001	0,0001	0,0001	0,1510
	263	F	r	0,052	0,246	0,654	-0,215	0,860	0,153
			p	0,4007	0,0001	0,0001	0,0004	0,0001	0,0131
INPATIENTS (surg.div.)									
	316	M	r	0,026	0,136	0,402	-0,214	0,846	0,116
			p	0,6398	0,0157	0,0001	0,0001	0,0001	0,0383
	227	F	r	0,244	0,238	0,478	-0,275	0,431	0,153
			p	0,0002	0,0003	0,0001	0,0001	0,0001	0,0212
INPATIENTS (med.div)									
	265	M	r	0,068	0,146	0,268	-0,193	0,865	0,091
			p	0,2675	0,0167	0,0001	0,0016	0,0001	0,1382

r (PEARSON CORRELATION COEFFICIENTS)
p (PROB. > R UNDER Ho : Rho= 0)

TABLE 4. SAS CORRELATION ANALYSIS BETWEEN FE AND THE VARIABLES LISTED BELOW

SUBJECTS	Nr	SEX		[S.U.A.] (mg/dl)	[S.CR.] (mg/dl)	[U.U.A.] (mg/dl)	[U.CR.] (mg/dl)	ER	AGE (years)
TOTAL	1552	F,M	r	-0,363	0,002	0,357	-0,202	0,745	-0,031
			p	0,0001	0,9169	0,0001	0,0001	0,0001	0,2175
	261	F	r	-0,36	0,122	0,370	-0,204	0,817	-0,091
			p	0,0001	0,0476	0,0001	0,0009	0,0001	0,1419
OUTPATIENTS									
	220	M	r	-0,394	0,145	0,323	-0,238	0,796	0,057
			p	0,0001	0,0308	0,0001	0,0004	0,0001	0,3984
	263	F	r	-0,288	-0,005	0,664	-0,183	0,860	-0,071
			p	0,0001	0,9361	0,0001	0,0028	0,0001	0,2484
INPATIENTS (surg.div.)									
	316	M	r	-0,419	-0,03	0,365	-0,208	0,846	0,034
			p	0,0001	0,5918	0,0001	0,0002	0,0001	0,5443
	227	F	r	-0,366	0,157	0,233	-0,122	0,431	0,012
			p	0,0001	0,0180	0,0004	0,0652	0,0001	0,8579
INPATIENTS (med.div.)									
	265	M	r	-0,339	-0,051	0,280	-0,21	0,865	-0,06
			p	0,0001	0,4031	0,0001	0,0006	0,0001	0,3376

r (PEARSON CORRELATION COEFFICIENTS)
p (PROB. > R UNDER Ho : Rho= 0)

REFERENCES

1) K.Tamaoku, Y.Murao, K.Akiura. Anal.Chim.Acta, 32:485 (1971)
2) J.C.H.Cook, Clin.Chim.Acta, 32:485 (1971)
3) R.Friedman et al. Clin.Chem. 26: 4 (1980)
4) P.Sharp. Clin.Chim.Acta, 40:115 (1972)
5) P.A.Simkin, P.L.Hoover, C.S.Paxon and W.F.Wilson Ann.Int.Med. 91:44 (1979)
6) L.A.Healy, A.P.Hall. Bull.Reum.Dis. 20:600 (1970)
7) S.L.Wallace. Arthritis Rheum. 18:721 (1975)
8) J. Zalokar, J. Lellouch, J.R. Claude and D. Kuntz J. Chron. Dis. 25:305 (1972)
9) R.E. Rieselback, T.H. Steele. Am. J. Med. 56:665 (1974)

GOUT AND MAIN RHEUMATIC DISEASES IN MAN

M.L. Sorghi, C.P. Quaratino, F. Savini, A. Zoppini, and A. Giacomello

Institute of Rheumatology, University of Rome
Institute of Medical Physiopathology, University of Chieti
Italy

During the last 20 years 4.755 patients (2045 males) were admitted to the Institute of Rheumatology of the University of Rome. The prevalence of gout among these patients was 0.04 % in women. Table 1 shows the values of some variables considered in the most prevalent diseases affecting man.

TABLE 1. GROUPS OF HYPOURICEMIC SUBJECTS CLASSIFIED ACCORDING TO ER VALUE

						VARIABLES		
DISEASE	PREVALENCE (%)	AGE (years)		DURATION OF DISEASE (years)		SERUM URATE CONCENTRATION (mg/dl)		FRACTION OF PATIENTS RF pos (%)
		M.	SD.	M.	SD.	M.	SD.	
RHEUMATOID ARTHRITIS (RA)	22,1	49	17	6	7	5.80	1.41	61,1
GOUT (G)	11,2	53	12	8	8	9.15	2.56	8,5
OSTEOARTHRITIS (OA)	10,3	57	12	9	11	6.30	1.49	4,1
ANKYLOSING SPONDYLITIS (AS)	8,4	38	12	12	9	5.26	1.38	3,4
PSORIATIC ARTHRITIS (PA)	5,4	44	10	5	5	5.80	1.73	8,7

As expected serum uric acid concentration was higher in gout and percentage distribution of positive tests for Rheumatoid Factor was higher in Rheumatoid Arthritis. A computer assisted analysis of patients with gout has been already described (1). The purpose of the present paper is to evaluate the diagnostic significance of serum urate levels.

The distribution of 134 patients with nontophaceous gout, rheumatoid arthritis, osteoarthritis, ankylosing spondylitis and psoriatic arthritis in 14 test classes of serum urate concentration (class width: 1 mg/dl) was evaluated. When gout was compared with the other diseases (Figs 1 to 4), the fraction of gouty subjects in each test class (class prevalence of disease , CPD) (2) gradually increased with increasing serum urate levels in the range of 4 to 10 mg/dl. In the same Figures sensitivity, specificity and predictive values (2) are plotted as a function of discriminator position (serum uric acid levels). The inverse relationship between sensitivity and specificity and between positive predictive value (PVpos) and negative predictive value (PVneg) is evident. Moving the discriminator from higher toward lower test values increases the sensitivity and the PVneg, but also decreases specificity and PVpos.

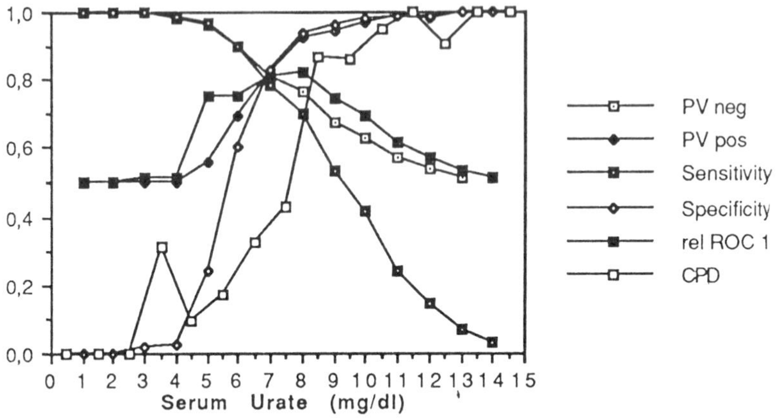

Fig.1 Comparison between serum urate levels in gout and in rheumatoid arthritis. The dependent variables sensitivity, specificity, predictive values, as well as relative ROC1 have been plotted as a function of the discriminator (serum urate concentration) position. The CPD curve is also reported.

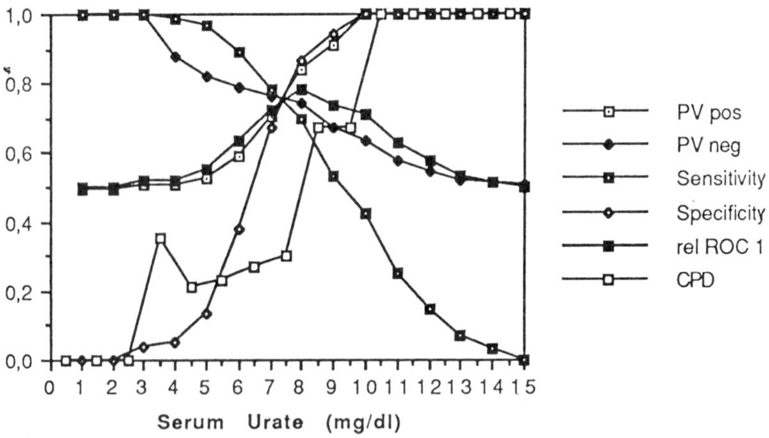

Fig.2 Comparison between serum urate levels in gout and in osteoarthritis. The dependent variables sensitivity, specificity, predictive values, as well as relative ROC1 have been plotted as a function of the discriminator (serum urate concentration) position. The CPD curve is also reported.

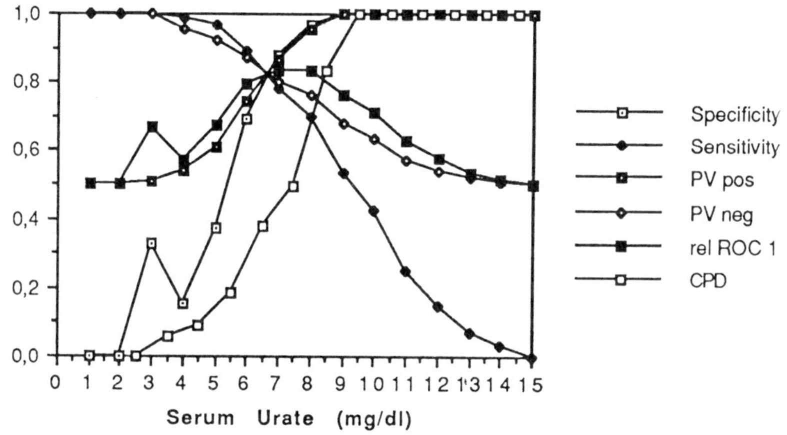

Fig.3 Comparison between serum urate levels in gout and in ankylosing spondylitis. The dependent variables sensitivity, specificity, predictive values, as well as relative ROC1 have been plotted as a function of the discriminator (serum urate concentration) position. The CPD curve is also reported.

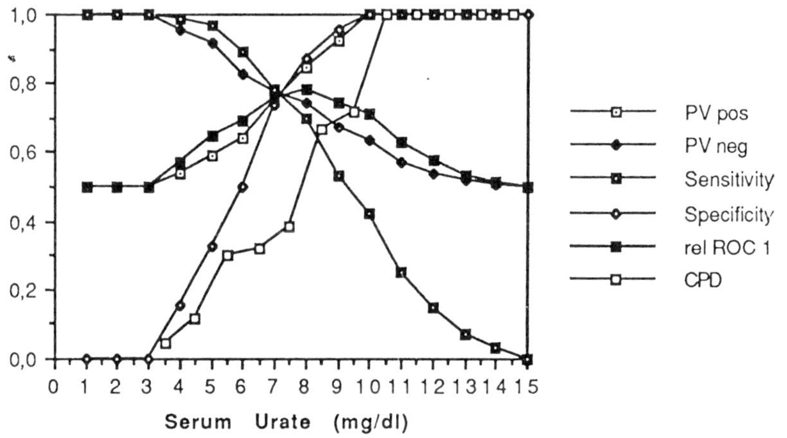

Fig.4 Comparison between serum urate levels in gout and in psoriatic arthritis. The dependent variables sensitivity, specificity, predictive values, as well as relative ROC1 have been plotted as a function of the discriminator (serum urate concentration) position. The CPD curve is also reported.

Moving the discriminator from lower toward higher values has the opposite effect. The discriminator position should be selected with respect to the clinical purpose. In the same Figures it is also shown that the "optimal" discriminator position (maximum of the relative ROC 1 curve) (2) is attained at a serum urate level of about 8 mg/dl. At this discriminator position the fraction of patients correctly classified is maximal.

REFERENCES

1) A. Spaccarelli, A. Giacomello, M. L. Sorgi, and A. Zoppini, 1984, Clinical survey of 200 patients with gout, in "Purine Metabolism in Man-IV, Part A", C. H. M. M. De Bruyn, H. A. Simmonds, and M. M.Muller, eds., Plenum Publishing Corporation, New York.
2) W. Gerhardt and H. Keller, Evaluation of test data from clinical studies, Scand. J. Lab. Invest. 46 (Supplement 11) : 1-74 (1986).

IMPORTANCE OF THE CONFOUNDING FACTORS AGE AND SEX IN THE CORRELATION OF SERUM URIC ACID, CHOLESTEROL AND TRIGLYCERIDE LEVELS

Birgit S Gathof, Martin A Schreiber,
Ursula Gresser, Irmingard Kamilli, Nepomuk Zöllner

Medizinische Poliklinik der Universität München
Pettenkoferstraβe 8a, D-8000 München 2

INTRODUCTION

A correlation of serum uric acid (UA) to serum cholesterol (CHOL) and/or triglyceride (TRI) levels was discussed in several previous studies from other groups. A strong influence of age and sex on these parameters has also been reported. In this study we aimed to determine whether there is a real correlation between cholesterol, triglycerides and uric acid or a simulated one, created by the influence of sex and age.

MATERIAL AND METHODS

Serum of 3200 healthy nonfasting Bavarian blood donors was examined in autumn 1989. Uric acid, cholesterol and triglycerides were determined enzymatically (Boehringer, Mannheim, FRG). Methods, distribution of uric acid levels (Gresser et al. 1990) and of serum total cholesterol, HDL-cholesterol and triglycerides (Gathof et al. in press) have been described in detail. Statistical analysis was performed using BMDP programm (Statistical Software, Los Angeles, USA). Partial correlation coefficents were calculated to eliminate confounding factors.

RESULTS

The blood donor population examined (3200 persons) consisted of 2097 men (65.5%) and 1103 women (34.5%) between 18 and 65 years of age. Cholesterol, triglyceride and uric acid levels differed significantly in men and women (table 1). Triglycerides were not normally distributed. In men no age related increase of uric acid levels was observed (Gresser et al. 1990), cholesterol and triglyceride levels varied in the different age groups (Gathof et al. in press). Only in women cholesterol and triglyceride levels rose constantly with age. Uric acid levels were lower in younger women and increased steadily in older age groups ($p < 0.05$), not so in men.

Table 1. Serum levels of uric acid, cholesterol and triglycerides (the difference between men and women is significant in each parameter, $p < 0.001$)

	uric acid		cholesterol		triglycerides	
	$\bar{x}\pm s$ (mg/dl)	$\bar{x}$ (µmol/l)	$\bar{x}\pm s$ (mg/dl)	$\bar{x}$ (µmol/l)	$\bar{x}\pm s$ (mg/dl)	$\bar{x}$ (µmol/l)
men	5.90±1.16	350.7	223±46	5.8	236±163	2.7
women	4.16±0.96	247.3	211±40	5.5	144± 88	1.6

The single correlations calculated (table 2) were all significant ($p < 0.001$). Elimination of confounding influence of age (by partial correlation) lowered the degree of correlation (correlation coefficent "r"), but they still remained significant ($p < 0.001$).

Table 2. Correlation coefficents of uric acid and cholesterol, and of uric acid and triglycerides (* without confounding factor age)

	uric acid - cholesterol		uric acid - trigly.	
	simple	partial*	simple	partial*
men	r=0.208	r=0.215	r=0.272	r=0.271
women	r=0.237	r=0.199	r=0.346	r=0.323

DISCUSSION

The correlation coefficients corrected (after elimination of confounding factor) were lower than the original ones. Especially in women the influence of age to the strength of correlation between cholesterol and uric acid is illustrated by the difference in the correlation coefficent "r" after correction by age. The correlation of uric acid and triglycerides is not influenced to the same extend by age. Also other risk factors (e.g. body weight, which we could not examine) should be discussed. In literature (table 3) partial correlations were not often calculated (only Reed et al. 1972).

Table 3. Correlation of uric acid - cholesterol and uric acid - triglyceride in other populations (* These values were calculated by partial correlation)

Reference	Population	Correlation Coefficents UA-Chol	UA-Tri
Reed et al. 1972*	1214 Chamorros	♂ r=0.04 ♀ r=0.00	r=0.22 r=0.13
Bräuer et al. 1976	421 patients in general practice	r=0.127	r=0.42
Yano et al. 1977	7921 men of Japanese ancestry in Hawaii	not done	r=0.33
Helgeland et al.1978	785 middle aged hypertensive men	r=0.15	r=0.26

To our opinion high correlation coefficents may be due to an artefact ("mixed correlation") if the influences of sex and age were not eliminated. Only thereafter differences in height of correlation factor "r" may be attributable to population based differences.

CONCLUSIONS

We observed a rather weak but significant correlation of serum uric acid and cholesterol levels and a stronger correlation of serum uric acid and triglyceride levels in men and in women, which is in good agreement with the literature. After elimination of the factor age (by partial correlation) the correlation coefficents dropped, especially in women, but still remained significant.

REFERENCES

Bräuer R, Clémencon G, Otto C (1976) Serumharnsäurewerte im Patientengut der Allgemeinpraxis. Schweiz Rundschau Med 65: 1094-1098

Gathof B, Gresser U, Kamilli I, Zöllner N (1991) Serum cholesterol and triglyceride levels in a blood donor population from southern Germany. Ann Nutr Metab (in press)

Gresser U, Gathof B, Zöllner N (1990) Serum uric acid levels in Bavarian blood donors. Klin Wochenschr 68: 122-1228

Helgeland A, Hjermann I, Holme I, Leren P (1978) Serum triglycerides and serum uric acid in untreated and Thiazide-treated patients with mild hypertension. The Oslo Study. Am J Med 64: 34-38

Reed D, Labarthe D, Stallones R (1972) Epidemiologic studies of serum uric acid levels among Micronesians. Arthr Rheum 15: 381-391

Yano K, Rhoads G, Kagan A (1977) Epidemiology of serum uric acid among 8000 Japanese-American men in Hawaii. J Chron Dis 30: 171-184

CHANGES IN URATE METABOLISM AFTER CASTRATION OF PATIENTS SUFFERING FROM CARCINOMA OF THE PROSTATE

L. G. Darlington, J. G. Ainsworth, A. Blight, M. D. Khong, T. A. N. Mann, R. D. Waldon, and E. A. Warburton

Rheumatology Unit, Epsom General Hospital, Surrey, U.K.

Summary

13 patients with carcinoma of the prostate were investigated on a low purine diet before and after surgical castration, performed to slow progression of their disease. A control group of herniorrhaphy patients was similarly studied. The 2 groups were very similar pre-operatively except for urea which was higher pre-operatively in patients than controls (p0.004). Testosterone levels fell significantly in patients soon after surgery (W0.002). There was a significant fall in serum uric acid in the early post-operative phase in both patients (W0.004) and controls (W0.01) but the fall only remained significant one month after surgery in castrated patients (W0.007). Urinary uric acid levels in patients fell significantly soon after surgery (W0.04).

This is a small study but results suggest that endogenous male hormones are at least partially responsible for serum uric acid levels since castration, with reduction in circulating testosterone, resulted in a fall in these levels.

Aim

To investigate prospectively whether purine metabolism is affected by surgical castration.

Method

13 patients with adenocarcinoma of the prostate, 6 of whom had metastases on bone scanning, underwent bilateral, total, subcapsular orchidectomy to slow progression of their disease. They were given a low purine diet for 9 days before surgery and, for the last 3 pre-operative days, serum uric acid, urea, creatinine, testosterone and urinary uric acid/24 hr were measured. On the day of operation and for every post-operative day until discharge these measurements continued and patients remained on a low purine diet. One month after discharge, patients returned to a low purine diet for 9 days and measurements were repeated. A control group of patients undergoing herniorrhaphy was similarly studied. Serum and urinary uric acid were measured by a uricase method (1), coefficients of variation 2-3% and 3-4%; testosterone by a radioimmunoassay (Personal communication D. Fry), coefficient of variation 10%, urea and creatinine by methods described in (2) and (3) respectively. Haematuria was sought in all patients and controls by routine ward testing.

Purine and Pyrimidine Metabolism in Man VII, Part A
Edited by R.A. Harkness *et al*., Plenum Press, New York, 1991

Results

By examination of mean differences in pre-operative average and peak readings for patients and controls, the two groups were shown to be comparable pre-operatively, except for urea, for which patients had higher pre-operative readings than controls (p0.004).

Longitudinal analysis was performed to compare early and late post-operative data with pre-operative values for patients and controls (Table 1).

Table 1

LONGITUDINAL COMPARISON OF EARLY AND LATE POST-OPERATIVE DATA WITH PRE-OPERATIVE VALUES FOR PATIENTS AND CONTROLS (WILCOXON SIGNED RANK TEST)

Early Post-Operative Data

Average pre-op. compared with minimum post-op. data	Patients			Controls		
	n	$\bar{d}$(SE)	W	n	$\bar{d}$(SE)	W
Serum uric acid mmol/L	13	0.06(0.02)	0.04 *	11	0.04(0.01)	0.01 **
Urinary uric acid mmol/24 hr	10	1.32(0.61)	0.04 *	10	0.13(0.16)	0.60
Urea mmol/L	7	1.83(0.43)	0.01 **	11	0.26(0.36)	0.55
Creatinine mmol/L	12	7.13(4.89)	0.22	11	3.86(2.54)	0.21
Testosterone nmol/L	10	8.62(1.88)	0.002 **	11	1.38(1.19)	0.09

$p \leqslant 0.05$ = * $p \leqslant 0.01$ = **

Late Post-Operative Data

Average pre-op. compared wtih minimum post-op. data	Patients			Controls		
	n	$\bar{d}$(SE)	W	n	$\bar{d}$(SE)	W
Serum uric acid mmol/L	8	0.12(0.03)	0.007 **	11	-0.01(0.01)	0.38
Urinary uric acid mmol/24 hr	9	0.48(0.57)	0.50	10	-0.19(0.26)	0.71
Urea mmol/L	5	2.00(0.62)	0.04 *	11	-0.50(0.46)	0.26
Creatinine mmol/L	7	12.58(3.43)	0.01 **	11	-7.86(6.25)	0.18
Testosterone nmol/L	5	2.77(1.64)	0.27	11	0.11(0.93)	0.73

$p \leqslant 0.05$ = * $p \leqslant 0.01$ = **

Testosterone levels fell significantly in patients soon after surgery (W0.002). There was a significant fall in serum uric acid in the early post-operative phase in both patients (W0.04) and controls (W0.01) but the fall only remained significant one month after surgery in castrated patients (W0.007). Urinary uric acid levels in patients fell soon after surgery but statistical significance was marginal (W0.04). Urea fell significantly in patients soon after castration (W0.01) and remained lower one month after surgery (W0.04), while creatinine fell significantly in patients only one month after surgery (W0.01).

2-sample t-tests which compared changes pre- and post-operatively in patients and controls, for each variable were significantly greater in patients than controls for testosterone (p0.004) and urea (p0.01) soon after surgery, and for serum uric acid (p0.001), urea (p0.007) and creatinine (p0.03) one month after surgery (Table 2).

Table 2

2-SAMPLE t-TESTS COMPARING CHANGES IN PATIENTS AND CONTROLS

Average pre-op. compared with minimum post-op. data	Early Post-Operative Data						
	Patients		Controls				
	n	$\bar{x}$(SEM)	n	$\bar{x}$(SEM)	t	p	
Serum uric acid mmol/L	13	0.06(0.02)	11	0.04(0.01)	0.51	0.62	
Urinary uric acid mmol/24 hr	10	1.32(0.61)	10	0.13(0.16)	1.90	0.07	
Urea mmol/L	7	1.83(0.43)	11	0.26(0.36)	2.75	0.01	**
Creatinine mmol/L	12	7.13(4.89)	11	3.86(2.54)	0.58	0.57	
Testosterone nmol/L	10	8.62(1.88)	11	1.38(1.19)	3.32	0.004	**

$p \leqslant 0.05$ = * $p \leqslant 0.01$ = **

	Late Post-Operative Data						
	n	$\bar{x}$(SEM)	n	$\bar{x}$(SEM)	t	p	
Serum uric acid mmol/L	8	0.12(0.03)	11	-0.01(0.01)	3.91	0.001	***
Urinary uric acid mmol/24 hr	9	0.48(0.57)	10	-0.19(0.26)	1.11	0.28	
Urea mmol/L	5	2.00(0.62)	11	-0.50(0.46)	3.13	0.007	**
Creatinine mmol/L	7	12.58(3.43)	11	-7.86(6.25)	2.44	0.03	*
Testosterone nmol/L	5	2.77(1.64)	11	0.11(0.93)	1.51	0.15	

$p \leqslant 0.05$ = * $p \leqslant 0.01$ = ** $p \leqslant 0.001$ = ***

Discussion

Timed urine collections are known to be difficult to achieve accurately but, in this study, patients and controls were supervised carefully and it is believed that errors in collection were minimised.

Significant reduction in urea in patients, soon after surgery and persisting, may reflect rehydration and the lack of similar early changes in creatinine makes a significant renal factor unlikely to influence results.

Intravenous fluids given at surgery to some patients could theoretically interfere, as a result of dilution of sera or by provoking a diuresis, but such effects are not likely to persist for one month even if they affected early post-operative results. Similarly, any effects from anaesthetic agents would be unlikely to persist beyond the perioperative phase. 3 patients and 3 controls were on diuretics, 1 patient and 2 controls took anti-hypertensives and 1 patient was on low-dose aspirin. No patients received oestrogens. Since numbers on drugs were similar in the two groups and since medication was unchanged throughout the study, such medication is unlikely significantly to have affected results.

Haematuria will increase urinary purine. Orchidectomy does not itself cause haematuria but prostatic carcinoma with cystoscopy, biopsy and/or prostatectomy may do so. Fortunately, in this study, microscopic haematuria was only recorded in 3 patients and not in any controls so any effect will, hopefully have been small.

Testosterone may be affected by age (4,5), by circannual and circadian rhythms. Age was controlled, since patients and controls were not significantly different in age ($p>0.05$). Since patients needed urgent orchidectomy it was not possible to control the time of year at which surgery took place. They acted, however, as their own controls for testosterone levels pre- and post-surgery and, since only a few weeks were involved for the study, little circannual effect would have been observed. Circadian effects were also minimised since blood and urine collections took place at approximately the same time of day throughout the study.

All these effects on testosterone, however are likely to be small in contrast with the major impact of castration itself.

In spite of all the factors mentioned above, the very significant fall in serum uric acid levels following castration suggests that endogenous male hormones are at least partially responsible for serum uric acid levels.

Acknowledgements

The authors gratefully acknowledge the help of Mr. M. Bailey, Mr. R. McFarland and Mr. S. Miller in allowing their patients to be studied. They wish to thank Miss P. Street for dietetic advice, Mrs S. Huxtable for typing this paper and Miss A. Petrie for performing the statistical analysis.

References

1. D. Barham, P. Trinder. An improved colour reagent for the determination of blood glucose by the oxidase system. Analyst. 97 (151): 142 (1972).

2. G. Paulson, R. Ray and J. Sternberg. A Rate-sensing Approach to Urea Measurement, Clin. Chem. Abstract 63. 17: 644 (1971).

3. D. L. Fabiny and G. Ertingshausen. Automated reaction-rate method for determination of serum creatinine with the Centrifi-Chem, Clin. Chem. 17 (8): 696 (1971).

4. R. Rubens, M. Dhont, and A. Vermeulen. Further studies on Leydig cell function in old age, J. Clin. Endocr. Metab. 39: 40 (1974).

5. S. M. Harman and P. D. Tsitouras. Reproductive hormones in ageing men.1. Measurement of sex steroids, basal luteinizing hormone, and Leydig cell response to human chorionic gonadotropin, J. Clin. Endocr. Metab. 51: 35 (1980).

RENAL HYPOURICEMIA ASSOCIATED WITH HYPEROXYPURINEMIA DUE TO DECREASED RENAL EXCRETION OF OXYPURINES: A NEW DEFECT IN RENAL PURINE TRANSPORT

Masanori Kawachi, Norio Kono, Ikuo Mineo, Hiroaki Kiyokawa, Hiromu Nakajima, Takao Shimizu, Akira Ono, Takamichi Nishimura, Shiro Yorifuji, Masamichi Kuwajima and Seiichiro Tarui

The Second Department of Internal Medicine
Osaka University Medical School, Osaka, Japan

INTRODUCTION

A decrease in uric acid level associated with an increase in oxypurine (xanthine and hypoxanthine) level in both plasma and urine is known as a principal feature of hereditary xanthinuria due to a defect of xanthine oxidase (EC. 1.2.3.2) (Holmes and Wyngaarden, 1989). In this paper, we report a case of renal hypouricemia associated with increased plasma levels of oxypurines caused by renal tubular dysfunction. We examined renal handling of urate and oxypurines by loading probenecid and pyrazinamide. The results suggest that the renal handling of oxypurines as well as urate is affected in this patient.

PATIENT AND METHODS

Patient: The patient, a 50-year-old man, had symptoms of Parkinson's disease such as tremor at rest and muscular rigidity. Routine laboratory examinations including renal function tests were all normal except for low serum urate levels. There was a consanguineous marriage of his parents. His two elder brothers and his nephew had also been suffering from Parkinson's disease.

Effect of dietary purine intake: The patient took low (70 mg/day) or high (240 mg/day) purine diet for 3 days, and blood samples were collected at rest in the early morning. The total caloric intake was set at about 2400 Kcal/day for both diets.

Probenecid and pyrazinamide loading tests: After the patient fasted overnight, 1.0 g (3.5 mmol) of probenecid or 3.0 g (24.4 mmol) of pyrazinamide was administered orally at 0 min. Blood was drawn at 0, 60 and 120 min, and hourly urine was collected from -60 min to120 min.

Assay: Uric acid was determined enzymatically (Lorentz et al., 1967). Xanthine and hypoxanthine were separated and quantitated by high performance liquid chromatography (HPLC) as described previously (Mineo et al., 1985; Kawachi et al., 1990). Creatinine was

Purine and Pyrimidine Metabolism in Man VII, Part A
Edited by R.A. Harkness *et al*., Plenum Press, New York, 1991

assayed by an automated modification of the Jaffe reaction. Xanthine oxidase activity in the liver specimen was determined by the UV method using hypoxanthine as a substrate (Kawachi et al., 1990).

RESULTS

Serum urate level was very low in the patient, 42 μmol/l (0.7 mg/dl), as compared to healthy men, 302 ± 36 μmol/l (mean ± SD, n = 5). The ratio of uric acid clearance to creatinine clearance (C_{UA}/C_{Cr}) was four times higher in the patient than in the controls (32 vs. 8 ± 2 %). From these findings we concluded that the hypouricemia is of renal origin (renal hypouricemia). In contrast to the hypouricemia, his plasma levels of xanthine and hypoxanthine were three times higher than those of the controls (Table 1). Both the ratio of xanthine clearance to creatinine clearance (C_X/C_{Cr}) and the ratio of hypoxanthine clearance to creatinine clearance (C_{Hx}/C_{Cr}) were decreased to about half of the control values (Table 1). The xanthine oxidase activity and xanthine and hypoxanthine contents in his liver were normal. Blood concentrations of uric acid, xanthine and hypoxanthine were not changed significantly by low and high purine diets.

A probenecid loading test revealed no response of C_{UA}/C_{Cr} but normal responses of C_X/C_{Cr} and C_{Hx}/C_{Cr} (Fig. 1). However, C_{UA}/C_{Cr}, C_X/C_{Cr} and C_{Hx}/C_{Cr} did not respond at all in a pyrazinamide loading test (Fig. 2).

DISCUSSION

Two mechanisms are known for the development of hypouricemia: decreased uric acid synthesis or increased uric acid excretion (Kelly, 1975). A representative disease of the former is hereditary xanthinuria, and that of the latter renal hypouricemia. In hereditary xanthinuria uric acid production is blocked by a defect of xanthine oxidase, resulting in increases in xanthine and hypoxanthine, precursors of uric acid. On the other hand, xanthine and hypoxanthine metabolism is not primarily affected in renal hypouricemia. Our patient with renal hypouricemia, however, showed high plasma levels of xanthine and hypoxanthine (Table 1). His renal clearance of xanthine and hypoxanthine were decreased to about half of the control levels (Table 1). The enzymatic analyses showed normal xanthine oxidase activity in the liver specimen from this patient, excluding an association of hereditary xanthinuria. From these results we concluded that the increased plasma concentrations of xanthine and hypoxanthine are due to decreased renal excretion of xanthine and hypoxanthine.

Table 1. Plasma and urinary oxypurine levels and renal oxypurine clearance

	Patient	Controls
plasma levels (μmol/l)		
xanthine	1.3	0.4 ± 0.2
hypoxanthine	5.4	1.7 ± 0.3
urinary excretion (μmol/day)		
xanthine	59.6	57.2 ± 21.2
hypoxanthine	50.4	51.6 ± 12.4
renal clearance (%)		
C_X/C_{Cr}	23	44 ± 6
C_{Hx}/C_{Cr}	11	22 ± 3

(mean ± SD, n = 5)

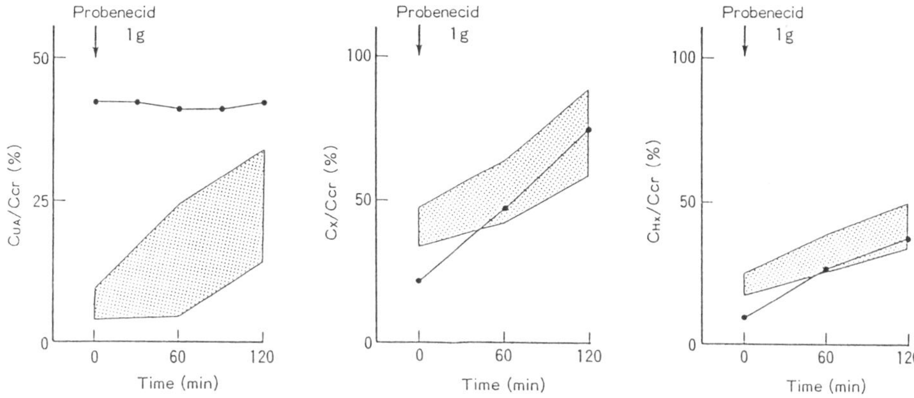

Fig. 1. Effects of probenecid on fractional clearance of uric acid, xanthine and hypoxanthine. UA denotes uric acid, X xanthine, Hx hypoxanthine and Cr creatinine. The shaded area indicates the mean ± SD in five healthy controls.

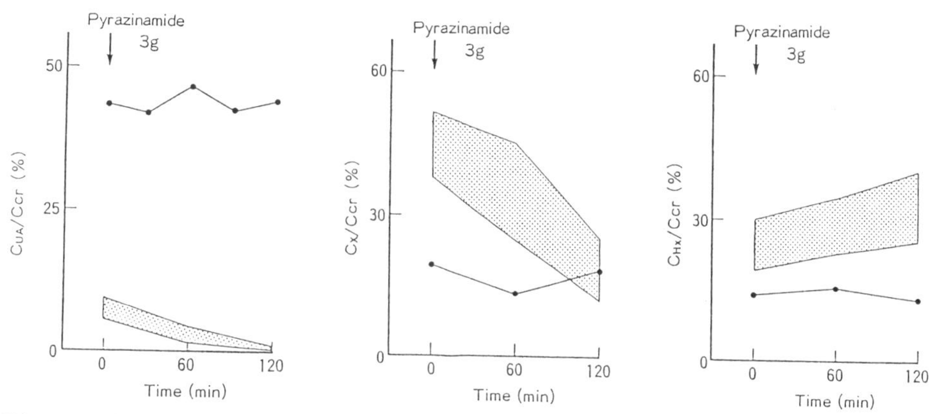

Fig. 2. Effects of pyrazinamide on fractional clearance of uric acid, xanthine and hypoxanthine. UA denotes uric acid, X xanthine, Hx hypoxanthine and Cr creatinine. The shaded area indicates the mean ± SD in five healthy controls.

Based on studies with probenecid, which suppresses tubular urate reabsorption, and pyrazinamide, which blocks tubular urate secretion (Diamond and Meisel, 1977; De Vries and Sperling, 1979), renal hypouricemia has been classified into the following types: 1) presecretory reabsorption defect (Benjamin et al., 1977), 2) postsecretory reabsorption defect (Sorensen and Levinson, 1980), 3) combined defect (Simkin et al., 1973), 4) hypersecretory type (Nakajima et al., 1987), 5) subtotal defect (Shichiri et al., 1987). Since renal urate excretion did not respond to either probenecid or pyrazinamide in the present patient (Figs. 1 and 2), the defect leading to hypouricemia could be similar if not identical to the subtotal defect type in which both tubular secretion and reabsorption of urate are disturbed.

Functional linkage between renal urate and oxypurine transport is recently suggested by the findings that probenecid increases urinary oxypurine excretion and pyrazinamide decreases it (Yamamoto et al., 1988). The probenecid loading test in the present patient revealed that oxypurine excretion increased normally despite the lack of urate response (Fig. 1), suggesting that tubular oxypurine reabsorption may not necessarily be coupled to tubular urate reabsorption. On the other hand, pyrazinamide did not suppress either urate or oxypurine excretion in the patient (Fig. 2), suggesting that tubular secretion of both urate and oxypurines was affected.

Whatever a primary defect of this patient is, the hyperoxypurinemia and hypouricemia are of renal origin, providing a new type of defect in renal purine transport.

ACKNOWLEDGMENTS

This study was supported, in part, by Grants-in-Aid from the Ministry of Education, Science and Culture of Japan, a grant from the Gout Research Foundation of Japan and a Basic Research Grant from the Muscular Dystrophy Association of the United States.

REFERENCES

Benjamin, D., Sperling, O., Weinberget, A., Pinkhas, J., and de Vries, A., 1977, Familial hypouricemia due to isolated renal tubular defect, Nephron, 18: 220.

De Vries, A., and Sperling, O., 1979, Inborn hypouricemia due to isolated renal tubular defect, Biomedicine, 30: 75.

Diamond, H. S., and Meisel, A. D., 1977, Classification of uricosuric states based upon response to pharmacological inhibitors of urate transport, Adv. Exp. Med.Biol., 76B: 63.

Holmes, E. W., and Wyngaarden, J. B., 1989, Hereditary xanthinuria, in: The metabolic basis of inherited disease, J. B. Stanbury, ed., McGraw-Hill Inc, New York: 1085.

Kawachi, M., Kono, N., Mineo, I., Yamada, Y., and Tarui, S., 1990, Decreased xanthine oxidase activities and increased urinary oxypurines in heterozygotes for hereditary xanthinuria, Clin. Chim. Acta., 188: 137.

Kelly, W. N., 1975, Hypouricemia, Arthritis. Rheum., 18: 731.

Lorentz, K., and Berndt, W., 1967, Enzymic determination of uric acid by a colorimetric method, Anal. Biochem., 18: 58.

Mineo, I., Kono, N., Shimizu, T., Hara, N., Yamada, Y., Sumi, S., Nonaka, K., and Tarui, S., 1985, Excess purine degradation in exercising muscles of patients with glycogen storage disease type V and VII, J. Clin. Invest., 76: 556.

Nakajima, H., Gomi, M., Iida, S., Kono, N., Moriwaki, K., and Tarui, S., 1987, Familial renal hypouricemia with intact reabsorption of uric acid, Nephron, 45: 40.

Shichiri, M., Iwamoto, H., Maeda, M., Kanayama, M., and Shiigai, T., 1987, Hypouricemia due to subtotal defect in the urate transport, Clin. Nephrol., 28: 300.

Simkin, P. A., Skeith, M. D., and Healey, L. A., 1973, Suppression of uric acid secretion in a patient with renal hypouricemia, Isr. J. Med. Sci., 9: 1113.

Sorensen, L. B., and Levinson, D. J., 1980, Isolated defect in post-secretory reabsorption of uric acid, Ann. Rheum. Dis., 39: 180.

Yamamoto, T., Moriwaki, Y., Takahashi, S., Hada, T., and Higashino, K., 1988, Renal excretion of purine bases---Effects of probenecid, benzbromarone and pyrazinamide, Nephron, 48: 116.

SIGNIFICANCE OF HYPOURICEMIA

*G.Riario Sforza, *C.Di Cosmo, *A.L. Di Mele, *E.Morelli, *C.Romano, °C.P. Quaratino and °A.Giacomello

*Hospital SS Annunziata - Chieti - and °Institute of Medical Physiopathology - University of Chieti -, ITALY

A few studies, reviewed for the most part in reference 1, have focused on the relevance of hypouricemia due to either a decreased production of urate or to an enhanced elimination of this compound. The only recognized causes of decreased uric acid production are those related to reduced activity of xanthine oxidase (xantinuria, severe liver disease, allopurinol administration). Increased renal excretion of uric acid is the other important cause of hypouricemia. This can be due to an isolated or to a generalized defect in renal tubular transport. The most common cause of hypouricemia in a hospital settings appears to be the administration of uricosuric drugs (2).

The purpose of the present investigation was to determine the clinical significance of hypouricemia and to identify hypouricemic subjects who are hyperexcretors of uric acid in hospital and ambulatory environment.

A total of 10,127 consecutive serum urate (S.U.A.) determinations were monitored at a large hospital in Chieti (Abruzzo, Italy). Hypouricemia was defined as the serum urate concentration lower than or equal to 3.5 mg/dl in men and lower than or equal to 2.5 mg/dl in women. Urine samples from 135 hypouricemic subjects were available. In these patients serum creatinine (S.CR.), urine uric acid (U.U.A.) and urine creatinine (U.CR.) concentrations were also measured. To quantify renal excretion of urate, uric acid excretion per deciliter of glomerular filtrate (ER) (3) and fractional excretion of uric acid (FE = uric acid clearance x 100 / creatinine clearance) were employed.

RESULTS AND DISCUSSION

Hypouricemia was present in 6.4% of women and in 4.7% of men. Although in women pregnancy seemed to be the main cause of hypouricemia, hypouricemic men had different associated diseases. The data obtained in hypouricemic patients with a known urine excretion of uric acid are summarized in the table. Simkin et al.(3) have proposed a screening test for hyperuricosuria that relates uric acid excretion to glomerular filtration rate. The excretion rate (ER) is determined from the product of (urine uric acid) and the (serum creatinine) divided by the (urine creatinine) with all values expressed in mg/dl. This screening test for hyperuricosuria has proved useful in the study of uric acid excretion in adult subjects and in children (3,4). A mean excretion of 0.4 $\pm$ 0.1 mg/dl (SD) has

TABLE 1. VALUES OF SOME VARIABLES CONSIDERED IN THE MOST PREVALENT RHEUMATIC DESEASES AFFECTING MEN ADMITTED TO THE INSTITUTE OF RHEUMATOLOGY OF THE UNIVERSITY OF ROME.

GROUP 1 :	ER > 0,6 mg/dl (NR. SUBJECTS = 4) (F/M = 1/1)	NAME	S E X	A G E	[S.U.A.]	[S.CR.]	[U.U.A.]	[U.CR.]	E R	F E	DIAGNOSIS
				(years)	(mg/dl)	(mg/dl)	(mg/dl)	(mg/dl)	(mg/dl)		
		D.I.	F	50,0	2,2	0,9	90,3	103,0	0,79	36	Cardiopaty
		D.F.	F	41,0	2,4	0,8	54,6	37,8	0,72	26	Diabetes
		D.A.	M	62,0	3,0	1,2	65,1	50,4	0,83	27	Diabetes
		R.A.	M	57,0	3,3	1,0	77,7	48,3	0,62	19	Cancer
GROUP 2 : 0,40 mg/dl <	ER < 0,6 mg/dl (NR. SUBJECTS = 12) (F/M = 1/2)	NR	SEX	AGE	(years)	E R	(mg/dl)	F E			MAIN ASSOCIATED CLINICAL CONDITIONS
				mean	range	mean	range	mean	range		
		4	F	55,2	13 74	0,44	0,43 0,46	23,0	18 36		Liver, Cardiovascular diseases
		8	M	59,2	42 80	0,46	0,42 0,53	18,0	15 20		Cardiovascular diseases, Carcinoma
GROUP 3 :	ER < 0,4 mg/dl (NR. SUBJECTS = 120) (F/M = 3,8/1)	NR	SEX	AGE	(years)	E R	(mg/dl)	F E			
				mean	range	mean	range	mean	range		
SUBGROUP 3 A :	FE > 15	16	F	35	21 65	0,36	0,24 0,43	17,7	15,0 21,6		Pregnancy (9 cases), Others
SUBGROUP 3 B :	FE < 15 (NR . SUBJECTS = 103) (F/M = 3,12/1)	78	F	36	6 83	0,196	0,07 0,34	9,1	3,3 14,8		Pregnancy (32 cases), Diabetes, Renal diseases, Others
		25	M	54,68	8 89	0,24	0,04 0,35	8,1	1,5 13,2		Diabetes, Carcinoma, Renal diseases, Others

been found in normal subjects. If the uric acid excretion is less than 0.5 mg/dl of glomerular filtrate, Simkin et al. regard the excretion as normal. Any person consistently excreting more than 0.7 mg of uric acid per deciliter of glomerular filtrate is regarded as an hyperexcretor (3). As shown in the table only 4 subjects (group 1) have an uric acid excretion of more than two standard deviations above normal. In the 12 subjects of group 2 and in 16 women (subgroup 3a), although the ER value appears to be normal, an increased fractional excretion of uric acid has been observed. In the other 103 patients (subgroup 3b) uric acid hyperexcretion does not appear to be the cause of hypouricemia. As shown in the same table the main clinical conditions associated with hypouricemia are usually those previously reported in the literature (1). The only recognized causes of decreased uric acid production are those related to reduced activity of the enzyme xanthine oxidase (xantinuria, liver disease and treatment with allopurinol). Although serum and urine oxypurines have not been determined in our patients, in xantinuria a serum urate concentrations lower than 1 mg/dl is usually observed but may range from 0.05 to 1.6 mg/dl (1). Such low serum urate concentrations have been observed in only 9 of our 135 patients. Liver diseases, which are sometimes associated with uric acid hyperexcretion (1), were present in 9 of our 135 patients. Although we had no information about the drugs administered to the patients, serum urate concentration usually does not drop into the hypouricemic range during allopurinol therapy.

In summary, it is clear that hypouricemia is a relatively common findings, as would be expected on statistical grounds. In less than 1/3 of the patients hypouricemia can be attributed to uric acid urine hyperexcretion. Other factors, such as low dietary purine intake, hemodilution (especially in pregnancy), may have a role in the development of hypouricemia in some of these patients. The possibility that, at least in some patients, increased secretion of uric acid through the gastrointestinal tract could have a role in the development of hypouricemia cannot be excluded.

REFERENCES

1) W.N.Kelley. Arthritis and Rheum. 18:721 (1975)
2) C.Ramsdell and W.N.Kelley. Ann.Int.Med. 78:239 (1973)
3) P.A.Simkin, P.L.Hoover, C.S.Paxon and W.F.Wilson. Ann.Int.Med. 91:44 (1979)
4) F.B.Stapleton and D.A.Nash. The J. of Ped. 102:88 (1983)

ORGANIC DISTRIBUTION AND INTERRELATIONSHIPS OF LESIONS OCCURRING IN LAYING HENS SUFFERING FROM GOUT AND UROLITHIASIS*

Gürsel Sönmez

Department of Pathology
Faculty of Veterinary Medicine
Uludağ University, Bursa, TURKEY

SUMMARY

Gout as a multifactorial syndrome can cause the death of chicken. In this study, the lesions formed were studied macro-, and microscopically in 134 gouty chicken. The gross and microscopic changes were found in all kidneys and renal capsules, though the other internal organs were also involved in many cases. Their serosal surfaces were more severely affected than the parenchymal parts. These organs were liver, lungs, heart, spleen and synovial sacs. Uroliths were found in all cases and were formed in ureters. This finding indicated that the gouty lesions were the results of urolithiasis. Physical characteristics and the chemical composition of the stones were studied, and the results confirmed that they were made up by urates of calcium, ammonium cations.

INTRODUCTION

Avian visceral gout is a multifactorial syndrome assaciated with the mismanagement, malnutrition, mycotoxicosis, infectious agents, genetic susceptibility and urolithiasis.[1] Uroliths formed in avian urolithiasis vary in size according to diameter of ureters, and they are composed of Calcium-Sodium or Calcium-Ammonium urates.[2] Increasing plasma uric acid level is reflected by necropsy findings including urate precipitates on organs and serous membranes. They resemble white-yellow chalky sediments on the surface and deeper parts of organs. Histologically, they appear as focal or disseminated granulomas composed of heterophil leucocytes and mononuclear cells infiltrated around the uric acid crystals.[3,4]

*This manuscript describes the summary of a project that was accepted as a fulfilment of a Ph.D. thesis.

In this study, organic distribution and interrelationships of lesions occurring in laying hens suffering from both gout and urolithiasis were comparatively investigated.

MATERIALS AND METHODS

Ten surviving and 124 dead chicken (total 134) collected from 9 different locations around Bursa, Turkey were used as study material.

At necropsy, organic distribution of uric acid crystals was recorded on previously prepared lists. For histological examination, samples of kidneys, heart, liver, lungs and spleen were fixed in buffered formalin solution before dehydrating in Autotechnicon (Duo,2A), then embbeded in paraffin blocks. And these were cut 5-6 micron thick with a rotary microtome (American optical, 820), and then stained with routine Hematoxylin and Eosin. In order to demonstrate the uric acid crystals in typical cases, they were stained with De Galantha's method, and to demonstrate the calcium salts Kossa's method was used.[5]

The hydroxylamine-phosphotungustate method was used to determine the plasma uric acid level in affected chicken. A sum of 25 blood samples collected from 4 different locations, and another 10 healty chicken samples collected from a disease-free location were used.

At necropsy, uroliths were present in all cases. But those stones collected from 42 chicken were used for picturing and chemical analysis. These samples were dried at room temperature before analyses. The physical characteristics as weight, shape, consistency and color were recorded. The chemical analysis included the tests for calcium, ammonium, cystine, xanthine, uric acid, phosphate, carbonate and oxalate.[6]

RESULTS

Uric acid precipitations were seen in kidneys and serosal surfaces of all other internal organs in all cases (100 %). In the cut surfaces of heart of 79.8 percent of cases, and in liver, lungs, spleen and joints 73.1, 71.6, 63.4 and 58.2 percent of cases respectively. Grey-white colored and 1-3 mm in diameter multiple urate granulomas (tofus) were seen in the lobules of hypertrofic kidneys. Microscopically, the urate precipitates were readily identifiable by their crystals structures precipitated in the necrotized area. These urate precipitates were surrounded by heterophil leucocytes and mononuclear cells, fibrocytes and fibroblasts which were increasing in the stromal tissues in atrophied kidneys. Grey-white colored, 0.5-1 mm in diameter, focal or disseminated urate granulomas were present both on serosal and on the cut surfaces of the heart, liver, lungs and spleen. These granulomas were centrally necrotized and consisted of uric acid crystals, heterophil leucocytes and mononuclear cells as seen in the kidneys and liver. The foreign-body giant cells were present in only those granulomas occurring in the kidneys. Uric acid crystals were seen in roset-like shape or needle-like masses by De Galantha's staining. Calcium was also present on the serosal surface of internal organs, and in uroliths stained with Kossa's method. Synovial capsules of affected chickens were slightly painted with colored exudate composed of uric acid and synovial liquid.

The plasma uric acid levels in the control chicken were 7.46-10.34 mg/100 ml, and were increased by a factor of five in affected chicken (23.25-75.89 mg/100 ml). At necropsy, uroliths were present in all cases in different segments of ureters. Their shape and numbers varried from case to case. Uroliths occurred in the left ureter in 34 (25.37 %), right ureter in 23 (17.16 %) and both ureters simultaneously in 77 (57.46 %) cases. Their weight was in the range of 10-1500 mg with an average of 525 mg in 42 cases. The lenght was in the range of 2-42 mm (avg. 22 mm). Those uroliths were ivory in color and rough, sylindiric or antlers' horn shape, spongy, and brittle in consistency. Chemical analysis of uroliths revealed that ammonia, uric acid, calcium were present in all samples (100 %), phosphate in 18 (42.8 %), carbonate in 13 (30.9 %), xanthine in 23 (59.5 %) and cystine in 32 (76.1 %) samples.

DISCUSSION AND CONCLUSION

A strong relationship between gout and urolithiasis was established. Uric acid crystals precipitate as the result of dehydration, and they devoid of water. Uroliths cause the death of animals by blocking the urine flow.[7,8] In this project, we determined that uroliths were present in all cases in the kidneys and ureters. Therefore, the gout in these cases was the result of urolithiasis. Our findings were in accored with the observation of typical and patognomonic uric acid crystals for the diagnosis of gout.[9,10] Gout and uroliths were occurring in those organs that have comparably slower blood flow to let enough time for the precipitation of uric acid crystals. Histological features were similar to those published in elsewhere.[11,12] Foreign-body giant cells were reported in granulomas (tofi) formed in the lungs and liver.[4] But we have not seen these in any organs but kidneys. Raportedly increasing plasma uric acid levels in gouty chicken were also strikingly high in our cases.[9,13]

Morphological features of the uroliths were similar to previous descriptions.[7,14] Chemical analysis of uroliths were reported to contain the several salts of Calcium[14] Calcium-Ammonium[2] and/or Calcium-Sodium.[15] In our study, uric acid, calcium and ammonium was present in all uroliths suggesting their composition as the urates of Calcium and Ammonium salts. Additionally, carbonate,xanthine and cystine were also present in some of our cases of which was contrary to the report of Mallinson et al.[7] These results showed that uroliths are the precipitates of salts of uric acid with calcium and ammonia excreted in the urine of dehydrated animals. The chemical composition of stones and the presence of organic debris help to explain the severity of renal lesions in gouty cases.

REFERENCES

1. A.I. Baba and S. Ghergariu, The gout syndrome in poultry, Institutul Agronomic, 501-508 (1987).
2. R.F. Wideman and R.P. Glahn, Nutrition, kidney dysfunction in poultry examined, Feedstuffs, December 28, 14-31 (1987).
3. O. Cotofan and A. Apetrei, Pathology of gout in the fowl, Institutul Agronomic, 99-100 (1975).
4. J.O. Okoye, Concurrent outbreak of visceral gout and lymphoid leucosis in domestic fowl in Nigeria, Rev.Elev.Med. Vet.Pays trop. 39 (3-4): 275-277 (1986).
5. L.G. Luna, "Manuel of Histologic Staining Methods of The Armed Forces Institute of Pathology", Mc Graw-Hill Book Company, New York (1968).
6. E. Ersoy and N. Bayşu, "Pratik Biyokimya", Ank.Üniv.Basımevi, Ankara (1981).
7. E.T. Mallinson, H. Rothenbacher, R.F. Wideman, Epizootiology, pathology and microbiology of an outbreak of urolithiasis in chickens, Avian Dis. 28 (1): 25-43 (1984).
8. R. Julian, Water deprivation as a cause of renal disease in chickens, Avian Path. 11: 615-617 (1982).
9. A. Minbay, H. Çamaş and H. Erdinç, Yumurta tavuklarında görülen gut hastalığının patolojisi, biyokimyası ve sağaltımı üzerinde bir araştırma, A.Ü.Vet.Fak.Derg. 32 (3): 493-506 (1985).
10. S.H. Fitz-Coy, S.A. Edgar and F.J. Hoerr, An outbreak of urolithiasis in Single Comb White Leghorn (SCWL) pullets, Avian Dis. 32: 563-566 (1988).
11. J.D. Wallach and C. Hoessle, Visceral gout in a captive reptiles, J.A.V.M.A. 151 (7): 897-899 (1967).
12. S.M. Shane and R.J. Young, Renal and parathyroid changes produced by high calcium intake in growing pullets, Avian Dis. 13: 558-567 (1969).
13. R.F.Wideman, E.T. Mallinson and H. Rothenbacher, Kidney function of pullets and laying hens during outbreak of urolithiasis, Poultry Sci. 62: 1954-1970 (1983).
14. T.P. Brown, J.R. Glisson, G. Rosales, Studies of avian urolithiasis associated with an Infectious Bronchitis virus, Avian Dis. 31: 629-636 (1987).
15. N.O. Oldroyd, Characterization and composition of uroliths from domestic fowl, Poultry Sci. 65: 1090-1094 (1986).

A BIOCHEMICAL INVESTIGATION ON CHICKEN GOUT OBSERVED IN THE MARMARA REGION IN TURKEY

Nihat Mert

Department of Biochemistry
Faculty of Veterinary Medicine
Uludağ University, Bursa, 16173, Turkey

SUMMARY

Poultry breeding has economical importance for the people living in Bursa and surrounding villiages. Similar to some sporadic disease, gout can also be reason for production loses and death in laying and broiler chickens. Gout cases have been observed in some pens of a big poultry company. In this study 20 healthy control and 40 sick Studler Iso-Brown chickens were used as research materials. Blood were taken from heart by syringe to plastic tubes with EDTA. Plasma were collected and analyzed for total protein, uric acid, bicarbonate, vitamin A, calcium and orotic acid. The blood levels of healthy and sick groups were; total protein % 4.60-5.98 gr, uric acid % 10.19-38.09 mg, bicarbonate 27.37-35.73 mEq/l, vitamin A % 42.20-35.55 mcg, calcium % 8.38-8.25 mg and orotic acid % 1.42-1.97 mg, respectively. Statistical analysis was done by t test. There were statistical importance for total protein, bicarbonate ($p < 0.01$) and for uric acid ($p < 0.001$) differences. Feeding program analysis was done and a feeding disturbance was determined. Chicken in gout observed pens have fed with chicken developping feed for two weeks earlier. For this case diet was the reason of gout.

INTRODUCTION

Uric acid (urate) is the end product of purine metabolism in man, higher apes and birds.[1,2] Chicken contains no tissue uricase therefore uric acid represent an end product of nitrogen metabolism, resembles the human. Except only a small fraction of plasma urate is eliminated by the biliary route and oxidized to allantoin by the uricase of gut bacteria.[3]

Gout in bird occurs in two forms which appear to have different origins.[4] Visceral gout is the more common form in which

Purine and Pyrimidine Metabolism in Man VII, Part A
Edited by R.A. Harkness *et al.*, Plenum Press, New York, 1991

plasma uric acid levels are increased and urates are deposited in the visceral organs.[4] In this form of gout results from pathological changes in the kidneys which obstruct kidney function in general. It has been produced experimentally in chicken by the feeding of rations which are deficient in vitamin A or, contain excess dietary calcium and nephrotoxic agents.[4,5] Articular gout is characterized by the accumulation of urates in the synovial capsules and tendon sheaths of the joints. It doesn't occur simultaneously with visceral gout. Articular gout has been produced experimentally by the feeding of rations containing excessive amount of protein.[4,6]

Articular gout occured in some individuals of sex-linked dwarf white leghorn population which had been fed a diet containing normal level of protein. The trait was found to respond genetic selection. After genetic stadies two lines were isolated as hyperuricemia (HUA) and normal plasma uric acid (LUA) lines.[7]

The present study was designated to determine the reason of gout in a big poultry farm has more than 200.000 laying Studler Iso-Brown chickens. The additional goal of this study was to determine the biochemical values of blood in gouty chickens.

MATERIAL AND METHODS

Materials have obtained from a poultry farm in Bursa. 40 gouty and 20 nongouty chickens, total 60, of same breed (Studler Iso-Brown) were used as research materials. All chickens were clinically examined. Blood samples were obtained by heart puncture using EDTA as an anticoagulant. The chickens were then killed and gross pathological observation recorded.

Total plasma protein was estimated by the biuret method,[8] plasma uric acid was determined by Hydroxylamine-phosphotungstate.[8] The levels of plasma bicarbonate were measured titrimetrically,[9] plasma vitamin A levels were estimated spectrophotometrically at 620 nm using Carr-Price Reagent.[8] Plasma calcium levels were measured by Glyoxal bis (2-4 Hydroxyanyl) method.[8] Finally the levels of plasma orotic acid were calculated spectrophotometrically at 458 nm.[10] All data analyzed according to Student t test.[11]

RESULTS

Data collected in presented research were summarized in Table 1. Differences between plasma total protein, bicarbonate ($p< 0.01$) and uric acid ($p< 0.001$) were statistically important. Plasma uric acid lȩvel of gouty chicken was nearly great 4 times than nongouty chicken. As the plasma uric acid level of gouty chicken increased the plasma protein levels were also increased.

Table 1. Blood plasma analysis of nongouty and gouty chickens.

	NONGOUTY			GOUTY			
	n	$\bar{x}$	$S\bar{x}$	n	$\bar{x}$	$S\bar{x}$	
Total protein % gr	18	4.60	0.38	39	5.98	0.39	**
Uric acid % mg	17	10.19	1.02	39	38.09	2.52	***
Bicarbonate mEq/l	19	27.37	2.49	20	35.75	1.64	**
Vitamin A % mcg	11	42.20	6.94	38	33.55	4.54	*
Calcium % mg	19	8.83	0.40	39	8.25	0.37	*
Orotic acid % mg	17	1.42	0.41	36	1.97	0.33	*

* Not significant, ** $p < 0.01$, *** $p < 0.001$

The vitamin A levels of gouty chicken were lower than nongouty chicken. But there was no statistical importance between their differences. Plasma bicarbonate levels were significantly different. This could be a result of kidney function of gouty chickens because of stone formation into kidney and urinary system. Orotic acid differences of both groups were not significantly important.

DISCUSSION

In this study, chickens of a big poultry farm have been examined after having gout in two pens. A detailed dietary analysis was done. Ration analysis couldn't be performed because of the ration changes. Blood sample analysis had indicated to high uric acid levels in gouty group.

The chicken is able to excrete excess dietary nitrogen when it is present at 2-3 times the dietary requirment of the chicken. Eighty percent or more of the urinary nitrogen excreted as uric acid.[4] Plasma uric acid concentration bears a direct relationship to dietary protein consumption in chicken.[12] Chicken with high uric acid excrete uric acid more slowly than those of low uric acid containing chicken. As a result, the increased plasma uric acid levels predispose chicken to gout.[4] The present data concerning plasma uric acid levels also indicated that high plasma uric acid resulted with gout in chicken.

Vitamin A deficiency produces metaplsia of urethral epithelium[13] and it is generally accompanied by visceral gout in chicken.[14] Hypovitaminosis A has been determined but the differences between two groups were not significantly important (Table 1). Some authors had indicated that the increased bicarbonate levels or toxications could be a reason for gout. In gouty chicken the plasma bicarbonate levels were significantly high.[15]

As conclusion, chickens in this case have visceral gout due to dietary reasons. High plasma uric acid bicarbonate and total protein levels were determined.

REFERENCES

1. R.E.Austic and R.K.Cole, Hereditary variation in uric acid transport by avian kidney slices, Am.J.Physiol. 231 (4), 1147-1151 (1956).
2. W.G.Siller, Avian nephritic and visceral gout, Lab. Invest. 8, 1319-1346 (1959).
3. M.J.Zmuda and A.J.Quebbeman, Localization of renal tubular uric acid transport defect in gouty chickens, Am.J.Physiol. 229 (3),820-825 (1975).
4. R.E.Austic and R.K.Cole, Impaired renal clearance of uric acid in chickens having hyperuricemia and articular gout, Am.J.Physiol. 233 (3), 525-530 (1972).
5. S.M.Shane, the effect of high calcium diets on the metabolism of growing pullets, Ithaca N.Y.Cornell University (1969).
6. G.H.Synoeyenbos, I.M.Reynolds and T.Tzianabos, Articular gout in turkeys. A case report. Avian Diseases (6), 32-36 (1962).
7. R.K.Cole and R.E.Austic, Hereditary uricemia and arcular gout in chickens. Poultry Sci. 59, 951-960 (1980).
8. N.Mert, Pratik Biyokimya, U.Ü.Vet.Fak.Yayn.Bursa (1991)
9. K.Aras ve G.Ersen, Klinik Biyokimya,Taş Yayn.Ank.(1975)
10. A.Stagner, J.Suva and F.Musial, The determination of orotic acid in the blood serum by means of the spectrophotometric method, Experienta 24 (2), 116-117 (1986).
11. G.V.Snedecar, Statistical methods, Ames Iowa State Coll. Press, (1958).
12. W.R.Featherson, Nitrogenous metabolites in the plasma of chicks adapted to high protein diets, Poultry Sci. 48, 646-652 (1969).
13. A.Kostantinov, Morphological changes in the ureters as a cause of gout in chicken.A avitaminosis. Cr.Acad. Bulg.Sci. 23, 461-464 (1970).
14. J.Bokori, Gout in poultry. IV. Experimental feeding to induce Vitamin A deficiency and gout, Magy.Allotory Lap. 21, 211-216 (1966).
15. M.Mazurkiewicz, A.Seniow and Z.Wachnik, Effects of animal protein and $NaHCO_3$ on the development of gout and on the level of serum proteins in fowl, Arch. Exp. Vet.Med. 24, 303-308 (1970).

PRODUCTION OF URIC ACID AND ALLANTOIN IN PERFUSED RAT LIVER

Roberto Pagani, Roberto Leoncini, Maria Pizzichini, Giuliano Cinci, Enrico Marinello and Rosella Fulceri*

Institute of Biochemistry and Enzimology, *Institute of General Pathology, University of Siena, Italy

INTRODUCTION

In a previous study on purine nucleotide catabolism in the rat, we reported that, after loading ^{14}C-formate, the specific radioactivity of rat liver allantoin was always higher than that of uric acid. This result was quite unexpected, and could not be ascribed to a different pool of the two compounds, since uric acid and allantoin content in the liver are quite similar: nor could the result be explained on the basis of the uricase catalysed trasformation:

uric acid ---> allantoin

because such a reaction, if the pools of the two compounds were similar, would lead to a comparable specific radioactivity.

Therefore, different factors have been considered: the turnover of both compounds, their clearance or uptake from the liver; the fact that the uric acid and allantoin found in the liver can originate in other organs, and thus, represent a mixed situation; some unknown side reaction in the sequence:

nucleotides ---> uric acid ---> allantoin

nor could the fact that a slightly modified form of allantoin might occur under such conditions be excluded.

In order to resolve these problems, we decided to verify, first of all, if the higher labelling of allantoin occurs only in the liver or if it is not the result of a mixed situation, the pool of allantoin deriving from the liver and from other organs. Thus, we followed the production of uric acid and allantoin in an isolated rat liver preparation perfused with ^{14}C-formate.

MATERIALS AND METHODS

We carried out two types of experiments:
a) on the rat itself which was killed after ^{14}C-formate administration;
b) on perfused rat liver.

Male Albino Wistar rats, 300 g b.w., were used for our experiments; they were kept at 25°C and allowed water and normal diet ad libitum.

<u>In vivo experiments</u>

The rat received one single dose (10 µCi/100 g b.w.) of ^{14}C-formate (54.8 mCi/mmol) intraperitoneally and were decapitated at different

Purine and Pyrimidine Metabolism in Man VII, Part A
Edited by R.A. Harkness *et al.*, Plenum Press, New York, 1991

times after injection. The liver was immediately excised, homogenized in 5% TCA (1 g + 5 ml) and centrifuged. Uric acid and allantoin were treated with mercuric acetate (1), according to the previously reported procedure (2) and were precipitated from the acid-soluble fraction, dissolved in water, adsorbed on a cationic resin (Dowex AG 50 W-X8) eluted with water, further separated on an anion exchange column (Dowex 1-X8), checked for purity and analysed for content and specific radioactivity (2).

Perfused rat liver

The liver was isolated from a normal rat as previously described (3) and perfused with Krebs-Henseleit buffer, without recirculation of the perfusate (open system) at 37°C. During the following 15 minutes, 10 µCi ^{14}C-formate was gradually added to the buffer. Perfusate flow was about 4 ml/min per g wet wt. and lactate dehydrogenase activity in the effluent perfusate was measured (4). Liver was excised and catabolic compounds were obtained as above reported.

All reagents were of the highest purity commercially available and were obtained either from Merck or Fluka. Standard compounds were from Sigma. Resins were from Bio Rad and ^{14}C-formate (54.8 mCi/mmol) was an Amersham product.

RESULTS AND DISCUSSION

Results are reported in Table 1, which shows the comparative specific radioactivity of uric acid and allantoin under different experimental conditions. When ^{14}C-formate was injected intraperitoneally and rats were killed at different times after labelled treatment, specific radioactivity of allantoin was fivefolds higher than that of uric acid at 15´, 30´ and 60´, but the difference tended to the decrease when the studies were carried out at 154´; when these parameters were observed in perfused rat liver 15´ after addition of ^{14}C-formate, behaviour was similar with the specific radioactivity of allantoin higher than that of uric acid.

We conclude that other organs probably do not essentially affect the specific radioactivity of purine catabolic products in rat liver. These findings suggest the need for further investigation of hepatic tissues.

Table 1. Content (mg/g) and specific radioactivity (dpm/µmol) of uric acid and allantoin in normal rat liver, at different times after ^{14}C-formate administration, and in perfused rat liver. We report here a typical experiment.

	URIC ACID		ALLANTOIN	
time	mg/g	S.A.	mg/g	S.A.
15´	0.014	4478	0.024	25356
30´	0.016	6606	0.022	35296
60´	0.016	9300	0.023	19617
154´	0.019	7683	0.022	10356
perfused rat liver				
15´	0.013	2216	0.023	13984

REFERENCES

1) Di Stefano A., Pizzichini M., Resconi G., Marinello E.: Determination of allantoin in mammalian tissues. Adv. Exp. Med. Biol. 253B: 511 (1989).
2) Pizzichini M., Di Stefano A., Marinello E.: The regulation of purine ribonucleotide biosynthesis by glucocorticoid hormones. It. J. Biochem. 34(5): 305 (1985).
3) Sies H.: The use of perfusion of liver and other organs for the study of microsomal electron-transport and cytochrome P-450 system. Methods Enzymol. 52: 48 (1978).
4) Bergmeyer H.U. (ed.): Methods of enzymatic Analysis, Verlag Chemie, Weinheim 1983.

DYNAMICS IN THE PURINE NUCLEOTIDES OF LIVER DURING VARIOUS PERIODS OF HYPOXIA/ISCHAEMIA AND REOXYGENATION

Gerhard Gerber, Werner Siems, Andreas Werner, Wolfgang Dubiel, Tilman Grune, Wolfgang Henke, Wolfram Schneider, and Mathias Ziegler

Institut für Biochemie, Medizinische Fakultät (Charité), Humboldt-Universität zu Berlin, O-1040 Berlin, Germany

INTRODUCTION

The constituents forming the purine pool play a pivotal role in virtually all cellular functions.[1,2] That implicates thermodynamic, kinetic as well as regulatory aspects. ATP, for instance, is best described as a kinetically versatile agent integrated in cell functions as an energy transducer, not as a source or store of energy.[3] There exists a mutual dependence:

- ATP and the related nucleotides need other constituents of the purine pool as precursors
- the synthesis de novo or by salvage reactions of constituents of the purine pool requires energy transduced by ATP.

The metabolic interconversions of purine compounds are taking place in various compartments of the cell. Therefore, cautiousness is strictly recommended in the interpretation of data on enzyme activities and nucleotide concentrations from whole cells or tissues, respectively.

This paper deals with the experimental determination of deteriorations in the purine pool of rat liver and hepatocytes during insufficient oxygen supply and reoxygenation; particular attention will be devoted to the purine nucleotide metabolism of mitochondria.

METHODS

In the perfusion experiments the rat liver was prepared and perfused in a recirculating mode as described.[4] Details of the experiments with parenchymal cells prepared from rat liver were presented in a previous paper.[5] The isolation of mitochondria from rat liver and the experimental procedure to study the catabolism of their endogenous adenine nucleotides were given elsewhere.[6,7] The analysis of nucleotides, nucleosides and nucleobases accomplished in this laboratory by ion-pair reversed-phase HPLC was reviewed.[8]

RESULTS AND DISCUSSION

The purine pool of rat liver and hepatocytes during hypoxia/anoxia and reoxygenation

Hepatocytes. The cells suspended glucose-free in Krebs-Henseleit bicarbonate buffer with albumin were initially incubated in presence of 95 % O_2 / 5 % CO_2; the exchange of the gas phase by 95 % N_2 / 5 % CO_2 causes dramatic perturbations in the pool of purine compounds. There is a precipitous fall of ATP within 10 min. A transient accumulation may be seen for ADP, AMP, adenosine and IMP followed by a sequential decline in the order given. Since allopurinol was present to inhibit xanthine oxidoreductase hypoxanthine accumulated (Fig. 1). This figure demonstrates the powerful capacity of the pathway of purine nucleotide degradation coming into play if the rephosphorylation of ADP is interrupted.

An overview on the degradation of purine nucleotides in liver and hepatocytes under various experimental conditions is presented in Table 1. O_2 deficiency as well as the presence of substrates which can be rapidly phosphorylated at the expense of ATP provoke a collaps of ATP, GTP and the sum of adenine nucleotides. The highest rates of degradation were observed if oxydative phosphorylation and glycolysis were inhibited or if the cells were prepared from fasted animals, resp.. During ischaemia or anoxic perfusion the degradation is retarded. Since the tension of O_2 at which ATP concentration and ATP:ADP ratio decline to about half lies in the range of 1 mm Hg,[21] it is conceivable that even traces of oxygen may contribute in maintaining to some extent the nucleotide pool.

Perfused Liver. The catabolism of purine bases and the formation of superoxide anion and hydrogen peroxide are related by the actions of xanthine oxidase and urate oxidase. The following estimations were performed in order to reckon the rate of cytosolic H_2O_2 generation via xanthine oxidase: (i) Determination of the activities of xanthine dehydrogenase and xanthine oxidase; (II) Calculation of the cytosolic NADH: NAD ratio using the lactate: pyruvate ratio measured; (iii) Calculation of the extent of inhibition of xanthine dehydrogenase by the hypoxia-mediated increase of NADH:NAD ratio

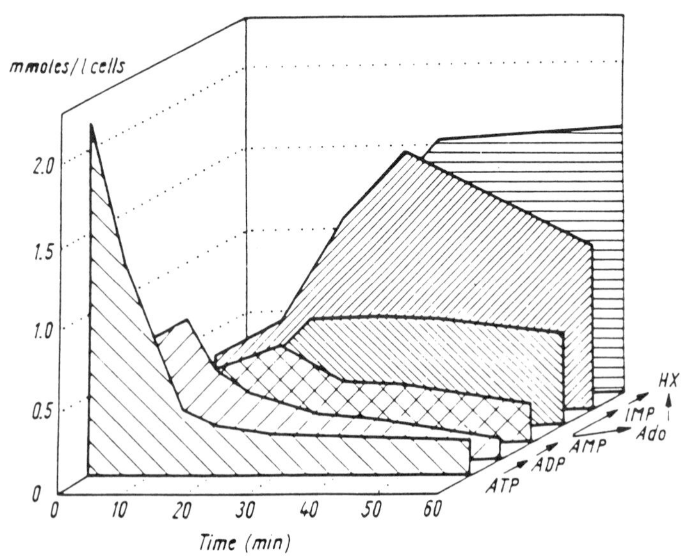

Figure 1. Adenine nucleotides and their products of degradation during anoxia of hepatocytes

Table 1. Degradation of Purine Nucleotides in Liver (L) and Hepatocytes (H) of Rat

Experimentals	τ min	References	
ATP			
(H) Starved animals N_2	2.3	Vincent	(1982)[9]
(H) KCN + 1AC pH 7.4	2.5	Gores	(1988)[10]
pH 6.4	2.5	Gores	(1988)[10]
Mathem. Modelling	2.7	Bartel	(1990)[11]
(H) Xylulose load	3	Vincent	(1989)[12,13]
(L) Fasted animals, anoxic perfusion, 30°C	4	Nishida	(1987)[14]
(H) Fasted chicken, N_2	4	Spychala	(1987)[15]
(H) Fructose load	5	Van den Berghe	(1980)[16]
(H) Tagatose load	5	Vincent	(1989)[12]
(L) Fasted animals ischaemia in-situ	5.5	Marubayashi	(1980)[17]
(L) Ischaemia, mouse, 20°C	6	McLaughlin	(1979)[18]
(H) Xylitol load	6	Vincent	(1989)[12,13]
(H) N_2	7.5	Schneider	(1989)[5]
(L) Ischaemia	9	Jaeschke	(1988)[19]
(H) Fed animals	20	Vincent	(1982)[9]
(L) Anoxic Perfusion	50	Jaeschke	(1988)[20]
GTP			
(H) N_2	3.5	Schneider	(1989)[5]
SUM OF ADENINE NUCLEODIES			
(H) N_2	15	Schneider	(1989)[5]
(L) Fasted animals ischaemia in-situ	25	Marubayashi	(1980)[17]
Mathem. Modelling	26	Bartel	(1990)[11]
(H) Fasted chicken N_2	29	Spychala	(1987)[15]
(L) Fasted animals anoxic perfusion 30°C	75	Nishida	(1987)[14]
(H) Starved animals N_2	>>40	Vincent	(1982)[9]
(H) Fed animals N_2	>>40	Vincent	(1982)[9]

assuming a competitive type of inhibition. The actual rate of the XOD reaction is then the difference between the total rate of (urate + allantoin) production minus the computed rate of the xanthine dehydrogenase reaction. The rat liver was isolated and perfused haemoglobin free at 25°C in a recirculating mode. The oxygen pressure in the perfusate was 80 kPa (normoxia) or 5.2 kPa (severe hypoxia). Details of the calculations and experimental determinations were communicated elsewhere.[22]

The results are shown in Fig. 2. Within two hours of hypoxia the contribution of XOD to the total activity of xanthine oxidoreductase increased only from 20 t 26 %. The flux of purine through both pathways catalyzed by XDH and XOD in the hypoxic organ is 4.5 times higher than under normoxia. The XOD pathway is about 6 times more active in comparision with normoxia. 53 % of the total amount of oxygen taken up in hypoxia is demanded by XOD and urate oxidase. Summing up, the prime mechanism of elevated H_2O_2 formation - and subsequent free radical species generation - is the excessive supply of hypoxanthine and not the dehydrogenase to oxidase transition of xanthine oxidoreductase.

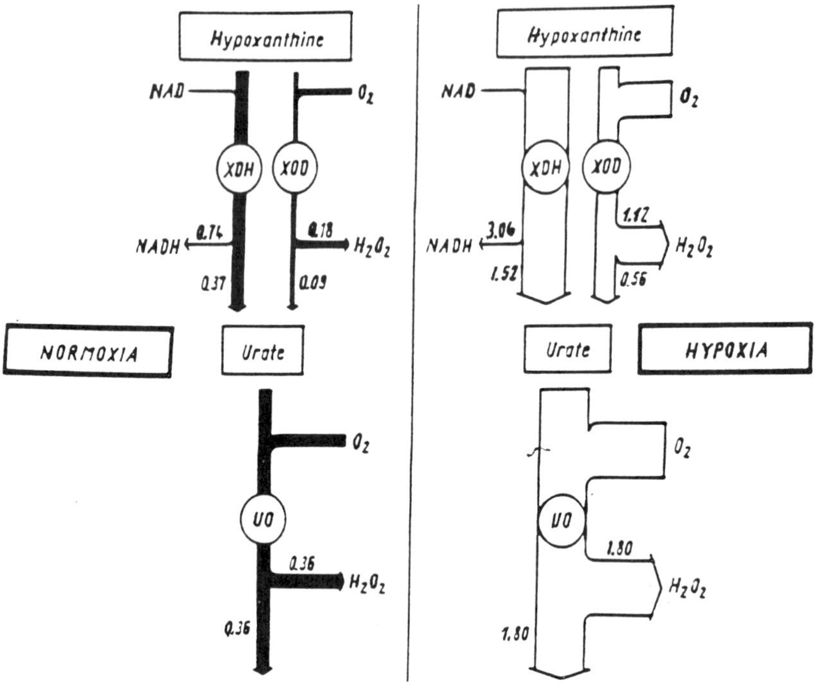

Figure 2. Rates of purine degradation and hydrogen superoxide formation by the rat liver in normoxia and severe hypoxia. XDH = NAD-dependent type of xanthine oxidoreductase: XOD = O_2-depedent type. The numbers indicate the flux rates of purine and oxygen equivalents as µmol/g wet weight x h.

The previous hypothesis of a extremely rapid Ca^{2+} mediated proteolytic conversion of the dehydrogenase type into the oxidase type[23] has, therefore, to be modified.[24] The relationship between purine catabolism and oxygen free generation in hypoxic liver and hepatocytes is discussed elsewhere.[25]

Some years ago a computer simulation of purine metabolism was made based on enzymatic rate laws taken from various tissues and species.[26] Thus, the application of that model to estimate concentrations of intermediates for a particular tissue or cell type may be of questionable significance. In this laboratory a model was developed which is based on rate expressions and kinetic data consequently from the rat hepatocyte.[11]

One of the goals of this modelling is the description of the stationary state and time-dependent changes of adenine nucleotides and their degradation products during anoxia and reoxygenation. The concentrations of NAD, NADH, R1P, Mg^{2+}, and PRibPP were "fixed" at an average value, since these metabolites are affected by other pathways which have not been taken into consideration. The model contains a system of 19 ordinary first-order differential equations; fast (equilibrium) and slow (irreversible) reactions were distinguished. Some very slow reactions were neglected(APRT, inosine kinase, GMP reductase). This model predicts the in vivo concentrations of selected nucleotides, nucleosides, and bases and their changes in anoxia. The variations initiated by reoxygenation following a 30 min anoxia are predicted, too. Further work is necessary to complete the model, to introduce also connections with other metabolic pathways and to check its usefulness for predictions of the changes in the purine pool in various experimental and clinical situations.

Purine metabolism in mitochondria

The mitochondria are highly important for the nucleotide metabolism since about 95 % of ATP are generated inside these organells. It is in discussion, however, whether mitochondria possess an autonomous metabolism of purine intermediates. There are observations favouring such a possibility: (i) determination of enzymes, nucleosides, and bases belonging to the purine metabolism, (ii) more than 30 % of the cellular adenine nucleotide pool is localized within the mitochondria, (iii) mitochondria possess a turnover of nucleic acids.

Experiments performed in our laboratory gave the following results:[6,7]

- Mitochondria from rat liver and kidney degrade endogenous adenine nucleotides producing adenosine and inosine.
- Adenosine is mainly accumulated in mitochondria; inosine is accumulated in the incubation medium. The concentration gradient between internal and external adenosine may approach 200.
- Externally added adenosine cannot be deaminated by rat liver mitochondria; therefore, it is postulated that adenosine from endogenous sources is deaminated intramitochondrially and inosine is then released into the surrounding medium.
- Liver mitochondria take up externally added inosine and guanosine. The existence of a nucleoside carrier in the mitochondrial membrane is suggested exchanging inosine and guanosine, but not adenosine.
- On the outer surface of the inner mitochondrial membrane of rat liver there exists an AMP-dephosphorylating activity. It shows an optimum at pH 7,5, the K_m value of AMP is 0.085 mM. Magnesium stimulates the activity ($K_{act.}$ = 0.05 mM). This enzyme may be implicated in cellular adenosine formation.

These data strongly favour the existence of a distinctive metabolism of purine compounds in mitochondria. It may supply precursors for the synthesis of nucleic acids. Besides the netto transport system for adenine nucleotides investigated by Aprille[27] it may stabilize the total adenine nucleotide content in mitochondria. The ATP/ADP translocator is strongly dependent on the size of the intramitochondrial adenine nucleotide pool. It is also exhausted during anoxia and the replenishment of this compartment may be of significance for the recovery of mitochondrial synthesis of ATP and its transport to extramitochondrial consumers.

REFERENCES

1. J.F. Henderson, and A.R.P. Paterson, "Nucleotide Metabolism. An Introduction", Academic Press, New York (1973).
2. T. W. Stones, and H. A. Simmonds, "Purines: Basic and Clinical Aspects", Kluwer Academic Publishers, Dordrecht (1991).
3. R. W. Hanson, The role of ATP in metabolism, Biochem. Educ. 17:86 (1989).
4. W. Siems, H. Schmidt, M. Müller, W. Henke, and G. Gerber, H_2O_2 formation during nucleotide degradation in the hypoxic rat liver: A quantitative approach, Free Rad. Res. Comms. 1:289 (1986).
5. W. Schneider, T. Grune, W. Siems, A. Werner, and G. Gerber, Nucleotide concentrations in hepatocytes during anoxia and reoxygenation in presence of allopurinol and oxypurinol, Biomed. Biochim. Acta 48:540 (1989).
6. W. Dubiel, W. Henke, Y. Miura, H.-G. Holzhütter, and G. Gerber, Identification and characteristics of a novel mitochondrial ATPase in rat liver, Biochem. Internat. 15:45 (1987).

7. M. Ziegler, W. Dubiel, A. M. Pimenov, Yu. V. Tikhonov, R. T. Toguzov, W. Henke, and G. Gerber, The catabolism of endogenous adenine nucleotides in rat liver mitochondria, Mol. Cell. Biochem. 93:7 (1990).
8. A. Werner, Analysis of nucleotides, nucleosides, nucleobases in cells by ion-pair reversed-phase HPLC, Chromatographia 31:401 (1991).
9. M.-F. Vincent, G. Van den Berghe, and H. G. Hers, The pathway of adenine nucleotide catabolism and its control in isolated rat hepatocytes subjected to anoxia, Biochem. J. 202:117 (1982).
10. G. J. Gores, A.-L. Nieminen, K. E. Fleishman, T. L. Dawson, B. Hermann, and J. J. Lemasters, Extracellular acidosis delays onset of cell death in ATP-depleted hepatocytes, Am. J. Physiol. 255:C315 (1988).
11. T. Bartel, and H.-G. Holzhütter, Mathematical modelling of the purine metabolism of the rat liver, Biochim. Biophys. Acta 1035:331 (1990)
12. M. F. Vincent, G. Van den Berghe, and H.-G. Hers, Increase in phosphoribosyl pyrophosphate induced by ATP and P_i depletion in hepatocytes, FASEB J. 3:1862 (1989).
13. M. F. Vincent, G. Van den Berghe, and H.-G. Hers, D-Xylulose-induced depletion of ATP and P_i in isolated rat hepatocytes, FASEB J. 3:1855 (1989).
14. T. Nishida, M. Koseki, W. Kamiike, M. Nakahara, K. Nakao, Y. Kawashima, T. Hashimoto, and K. Tagawa, Levels of purine compounds in a perfusate as a biochemical marker of ischemic injury of cold-preserved liver, Transplantation 44:16 (1987).
15. J. Spychala, and G. Van den Berghe, Adenine nucleotide metabolism in isolated chicken hepatocytes, Biochem. J. 242:551 (1989).
16. G. Van den Berghe, F. Bontemps, and H.-G. Hers, Purine catabolism in isolated rat hepatocytes. Influence of coformycin. Biochem. J. 188:913 (1980)
17. S. Marubayashi, M. Takenaga, K. Dohi, H. Ezaki, and T. Kawasaki, Adenine nucleotide metabolism during hepatic ischemia and subsequent blood reflow periods and its relation to organ viability, Transplantation 30:294 (1980)
18. A. C. McLaughlin, H. Takeda, and B. Chance, Rapid ATP assays in perfused mouse liver by ^{31}P NMR, Proc. Natl. Acad. Sci. USA 76:5445 (1979).
19. H. Jaeschke, C. V. Smith, And J. R. Mitchell, Reactive oxygen species during ischemia-reflow injury in isolated perfused rat liver, J. Clin. Invest. 81:1240 (1988)
20. H. Jaeschke, C. V. Smith, and J. R. Mitchell, Hypoxic damage creates reactive oxygen species in isolated perfused rat liver, Biochem. Biophys. Res. Commun. 150:568 (1988)
21. H. Hummerich, H. de Groot, T. Noll, and S. Soboll, Dependence of mitochondrial and cytosolic adenine nucleotide on oxygen partial pressure in isolated hepatocytes, Biochem. J. 250:641 (1988)
22. W. Siems, H. Schmidt, M. Müller, W. Henke, and G. Gerber, H_2O_2 formation during nucleotide degradation in the hypoxic rat liver: a quantitative approach, Free Rad. Res. Comms. 1:289 (1986)
23. J. M. McCord, and R. S. Roy, The pathophysiology of superoxide: roles in inflammation and ishemia, Can. J. Physiol. Pharmacol. 60:1346 (1982).
24. D. A. Parks, T. K. Williams, and J. S. Beckman, Conversion of xanthine dehydrogenase to oxidase in ischemic rat intestine: a reevaluation. Am. J. Physiol. 254:G768 (1988)
25. G. Gerber, W. Siems, and A. Werner, Purine nucleotide degradation and free radical generation in the hypoxic liver, in: "Membrane Lipid Oxidation", Vol. III, C. Vigo-Pelfrey ed., p. 116, CRC Press, Boca Raton (1991)
26. R. Franco, and E. I. Canela, Computer simulation of purine metabolism, Eur. J. Biochem. 144:305 (1984).
27. J. R. Aprille, Regulation of the mitochondrial adenine nucleotide pool size in liver: mechanism and metabolic role, FASEB J. 2:2547 (1988).

ADENOSINE: ORIGIN AND CLINICAL ROLES

Andrew C Newby

Department of Cardiology
University of Wales College of Medicine
Heath Park, Cardiff, CF4 4XN

In 1981 work from our laboratory[1] and that of Schrader[2] led us independently to the conclusion that two separate pathways, at least, were necessary to account for adenosine formation.

The extracellular pathway

A 5'-nucleotidase with a slight preference for AMP over IMP and a Km value around 20 µM is present with widely varying activity associated with the plasma membrane of most animal cells. This enzyme was rigourously shown to have its catalytic site exposed to the extracellular surface (i.e. it is an ecto-enzyme). Subsequent experiments using inhibitory antibodies or the selective inhibitor [α,β methylene]ADP have shown that the ecto-enzyme cannot hydrolyse cytoplasmic nucleotides (Table 1). Indeed, the recently determined primary structure of the enzyme shows it to be anchored by a phosphoinositol linkage and hence to have no cytoplasmic or membrane spanning domain[3]. The enzyme is present on the inner face of endocytic vesicles[4] and a solubilized fragment can be produced during tissue fractionation[5,6], but it is unlikely that either of these forms of the enzyme has access to cytoplasmic AMP or IMP. Ecto-5'-nucleotidase probably functions as the last member of a cascade together with separate ecto-ATPase and ecto-ADPase enzymes to achieve the hydrolysis to nucleoside and inorganic phosphate of any adenine nucleotides released from the cells (see Table 2). Adenine nucleotides are stored in granules of platelets and in some neurones together with neurotransmitters and can be released on excitation. Adenosine, which generally has a depressant effect either on release or action of the ATP or cotransmitted effector appears therefore to function as a feed-back modulator[7]. The potent inhibition of ecto-5'-nucleotidase by ATP and especially ADP[8] may delay the production of adenosine from released nucleotides and thereby provide a temporally separate cycle of cell activation and inhibition[9]. Another potential source of extracellular ATP is cell lysis for example during the action of phagocytes. It is therefore particularly interesting that adenosine also functions as an inhibitor of neutrophil activation[10].

The intracellular pathway

Net cellular ATP catabolism has been shown to result in intracellular adenosine formation a wide variety of tissues (Table 1). This conclusion has been reached either by simply separating cells and medium or by trapping cytoplasmic adenosine in the form of S-adenosylhomocysteine after infusion of L-homocysteine. In both types of experiments,

Table 1. Cell types in which adenosine release is totally or partly independent of ecto-5'-nucleotidase

Cell type	Stimulus	Intracellular formation	Blocked by transport inhibitor	Ref
Rat perfused heart	Hypoxia	Not tested	Yes	11
Guinea-pig perfused heart	Hypoxia	Yes	Yes	2
Pigeon perfused heart	Ischaemia	Not tested	Not tested	6
Rat neonatal myocytes	Poisoning	Yes	Yes	12
Rat adult myocytes	Poisoning	Not tested	Yes	
Chick embryo myocytes	Poisoning hypoxia	Yes	Yes	13
Rat liver cells	Poisoning	Not tested	Yes	14
Rat neutrophils	Poisoning	Yes	Not tested	1
Rat lymphocytes	Poisoning	Not tested	Not tested	
Guinea pig cortical synaptosomes	Electrical	Not tested	Yes	15,16
Chick embryo CNS and peripheral glia, ciliary ganglia	Poisoning	Yes	Yes	17

Cytosolic adenosine has also been detected by trapping with homocysteine in the rat brain[18] and dog heart[19], although the role of 5'-nucleotidase was not addressed in these studies.

Table 2. Release of ATP or ADP

Cell type	Stimulus	Control for breakage
Platelets	Aggregation	Yes
Cholinergic and adrenergic nerves	Depolarisation	Yes/no
NANC	Depolarisation	Yes/no
Endothelial cells	Trypsin, thrombin	No
Rabbit ear	Nerve stimulation	No
Rat perfused heart	Hypoxia	No
Rat adult myocytes	Hypoxia, Ca^{2+}	No
Human, frog skeletal muscle	Exercise, electrical	No

For citations see reference 20.

inhibition of the symmetric facilitative diffusion carrier for adenosine leads to greater intracellular accumulation (Table 1). This demonstrates directly that the transporter mediates adenosine efflux from cells. Released adenosine brings about vasodilatation, inhibition of cellular activity and inhibition of endogenous excitatory nerve firing. Since all these actions tend to restore the balance of cellular ATP formation and utilization, adenosine appears to function as a signal of cellular ATP depletion. We have proposed the name "retaliatory metabolite" for adenosine[21,22] because its actions provide a means for cells to retaliate against an excessive external drive. Respiratory stimulation by adenosine[23] and a local pain-generating (algogenic) effect[24] suggest furthermore that this retaliatory role may extend to the whole organism!

The pathway leading to cytoplasmic adenosine formation first involves disproportionation of 2 ADP molecules to one of ATP and one of AMP by the ubiquitously active adenylate kinase enzyme. The concentrations of ATP (5-10 mM), ADP (ca. 50 μM) and ATP (ca. 1 μM) in the cytoplasm of normal cells ensures that an undetectably small percentage fall in ATP concentration greatly elevates both ADP and particularly AMP concentrations. A cytoplasmic 5'-nucleotidase then catalyses adenosine formation. Two such enzymes have been well characterised (Table 3). The first shows a 20 fold preference for IMP over AMP, is activated by ATP and inhibited by inorganic phosphate. This suggests that it is most active when the ATP pool is replete and probably functions together with AMP deaminase to set the upper concentration limit of the adenine nucleotide pool. Nevertheless, if ATP depletion is rapid and severe as it may be pathologically in an infarcted organ, some adenosine may be generated by this enzyme[25,26]. More recently a cytosolic 5'-nucleotidase with a preference for AMP over IMP has been demonstrated in the hearts of pigeons, rats and rabbits (Table 3). The enzyme is activated by ADP but not ATP and is only weakly inhibited by inorganic phosphate. As a result, the enzyme should be activated during ATP catabolism and the rate of adenosine formation should be more than linearly related to the rise in cytosolic AMP concentration, consistent with experimental data[27]. The abundance of the AMP-specific 5'-nucleotidase in pigeon heart leaves little doubt of its role in this tissue. Direct demonstration of flux through the enzyme in other tissues must await the development of selective inhibitors or genetic manipulation.

Table 3. Properties of cytosolic 5'-nucleotidases

	IMP-specific	AMP-specific
Substrates	IMP/AMP = 20	AMP/IMP = 15
K_M for AMP	2-5 mM	1-4.5 mM
Activators	Mg^{2+}, ATP>ADP ionic strength	Mg^{2+}, ADP not ATP
Inhibitors	P_i, not [α,β methylene]ADP	Not P_i or [α,β methylene]ADP
Structure	Homotetramer (50-70 kD)	150 kD native 40 kD subunits
Location	Cytoplasm	Cytoplasm, with contractile proteins

Substrate preference is based on relative values of V_{max}/K_M. Data are taken from references 28,29,30,31,32,33,34,35 and 36.

Clinical uses

Adenosine receptors are present, and diverse physiological effects can be elicited on most cells. Adenosine is therefore an unpromisingly non-specific clinical tool! Fortunately adenosine itself, unlike most adenosine analogues, is metabolised within seconds in blood, and hence its effects are both very short-lived and potentially localised by infusion close to the site of action. Surprisingly, in view of this, adenosine apparently crosses the blood-brain barrier[46].

The most established clinical use of adenosine is in terminating supraventricular tachycardia[37,38,39], in adults and in children[40]. Adenosine is well known to prolong A-V conduction time and this action appears to be "retaliatory" because it helps restore the supply to demand ratio for oxygen to the ventricles[41]. Bolus injections of adenosine into patients apparently interrupt reentrant pathways of activation which involve the A-V node, and thereby rapidly terminate most supraventricular tachycardias. Adenosine does not terminate atrial fibrilation or flutter or prevent latent pre-exitation, but it can aid their diagnosis in the electocardiogram[42,43]. Adenosine administration is associated with several side-effects[38,40,43,44,45], but these are short-lived. The most alarming side-effects of bolus injections are up to several seconds of sinus arrest and partial or complete A-V conduction block. Chest pain, hypotension, facial flushing, respiratory stimulation, epigastric pain and psychological symptoms have also been reported[46]. These side effects limit the dose and duration of infusions into conscious patients. Brief infusions at up to 140 μg/ min/ kg have been used, despite side effects, to measure coronary vasodilator reserve[47], or in conjunction with thallium scintigraphy to reveal perfusion defects in patients with suspected coronary disease[44,48]. Local infusions of adenosine at lower doses into selected coronary arteries produce similar degrees of coronary vasodilatation with less side effects[47], but are more invasive. Prolonged low dose (70 μg/ min /kg) venous infusions may be sufficient to exert a useful pulmonary dilator effect[45]. In anaesthetised patients, high dose adenosine infusion is feasible and has been used during neurosurgery to produce controlled hypotension and as an 'antiadrenergic' agent in operations to remove pheochromocytomas[46]. Adenosine infusion may also be useful to preserve platelets during extracorporeal perfusion (i.e. cardiopulmonary bypass).

Future potential clinical uses include use of adenosine in cardioplegic solution, because it arrests rat hearts more rapidly than potassium alone[49,50]. Adenosine infusion shortly after reperfusion of ischaemic myocardium apparently improves perfusion and reduces infarct size[51,52] in dogs, although the effect may be lost after long periods of ischaemia[53]. These observations may be relevant to heart surgery and thrombolysis respectively, but have yet to be investigated in man.

REFERENCES

1. A. C. Newby, C. A. Holmquist, Adenosine production inside rat polymorphonuclear leucocytes, Biochem. J. 200:399-403 (1981).
2. W. Schutz, J. Schrader, E. Gerlach, Different sites of adenosine formation in the heart, Am. J. Physiol. 240:H963-H970 (1981).
3. Y. Misumi, S. Ogata, S. Hirose, Y. Ikehara, Primary structure of rat liver 5'-nucleotidase deduced from the cDNA - Presence of the COOH-terminal hydrophobic domain for possible post-translational modification by glycophospholipid, J. Biol. Chem. 265:2178-2183 (1990).
4. J. P. Luzio, E. M. Bailyes, M. Baron, K. Siddle, B. M. Bullock, H. J. Geuze, K. K. Stanley. The properties, structure, function, intracellular localization and movement of hepatic 5'-nucleotidase, in "Cellular Biology of Ectoenzymes," G. W. Kreutzberg, M. Reddington, H. Zimmerman, ed., Springer-Verlag, Berlin, Heidelberg, New York, Tokyo, pp 89-116 (1986).
5. P. Fritzson, T. B. Haugen, H. Tjernshaugen, The presence and activity in normal and regenerating rat liver postmicrosomal supernatant fraction of an enzyme with properties similar to those of membrane-bound 5'-nucleotidase, Biochem. J. 239:185-190 (1986).
6. P. Meghji, K. M. Middleton, A. C. Newby, Absolute rates of adenosine formation during ischaemia in rat and pigeon hearts, Biochem. J. 249:695-703 (1988).
7. P. J. Richardson, S. J. Brown, E. M. Bailyes, J. P. Luzio, Ectoenzymes control adenosine modulation of immunoisolated cholinergic synapses, Nature. 327:232-234 (1987).
8. R. M. Burger, J. M. Lowenstein, 5'-Nucleotidase from smooth muscle of small intestine and from brain. Inhibition by nucleotides., Biochemistry. 14:2362-2366 (1975).
9. E. L. Gordon, J. D. Pearson, L. L. Slakey, The hydrolysis of extracellular nucleotides by cultured cells from pig aorta: feed forward inhibition of adenosine production at the cell surface, J Biol Chem. 261:(1986).
10. B. N. Cronstein, S. B. Kramer, G. Weissman, R. Hirschhorn, Adenosine: a physiologic modulator of superoxide anion generation by human neutrophils, J Exp Med. 158:1160-1177 (1983).
11. G. P. Frick, J. M. Lowenstein, Studies of 5'-nucleotidase in the perfused rat heart, J. Biol. Chem. 251:6372-6378 (1976).
12. P. Meghji, C. A. Holmquist, A. C. Newby, Adenosine formation and release from neonatal-rat heart cells in culture, Biochem. J. 229:799-805 (1985).
13. P. Meghji, R. Rubio, R. M. Berne, Intracellular adenosine formation and its carrier-mediated release in cultured embryonic chick heart cells, Life Sci. 43:1851-1859 (1988).

14. F. L. Belloni, P. L. Elkin, B. Giannotto, The mechanism of adenosine release from hypoxic rat liver cells, Br J Pharmac. 85:441-446, (1985).
15. F. Pons, R. F. Bruns, J. W. Daly, Depolarization-evoked accumulation of cyclic AMP in brain slices: the requisite intermediate adenosine is not derived from hydrolysis of released ATP, J Neurochem. 34:1319-1323 (1980).
16. J. Daval, C. Barberis, Release of radiolabelled adenosine derivatives from superfused synaptosome beds, Biochem Pharmac. 30:2559-2567 (1981).
17. P. Meghji, J. B. Tuttle, R. Rubio, Adenosine formation and release by embryonic chick neurons and glia in cell culture, J Neurochem. 53:1852-1860695-703 (1989).
18. H. McIlwain, J. D. Poll, Adenosine in cerebral homeostatic role: appraisal through actions of homocysteine, colchicine and dipyridamole, J Neurobiol. 17:39-49 (1986).
19. A. Deussen, M. Borst, K. Kroll, J. Schrader, Formation of S-adenosylhomocysteine in the heart: A sensitive index for regional myocardial underperfusion, Circ Res. 63:250-261 (1988).
20. T. W. Stone, A. C. Newby, H. G. E. Lloyd. Adenosine release, in "Adenosine and adenosine receptors," M. Williams, ed., The Humana Press, Clifton, New Jersey, pp 173-223 (1990).
21. A. C. Newby, Adenosine and the concept of 'retaliatory metabolites', TIBS. 9:42-44 (1984).
22. A. C. Newby, Y. Worku, P. Meghji, M. Nakazawa, A. C. Skladanowski, Adenosine: a retaliatory metabolite or not., News in Physiol Sci. 5:67-70 (1990).
23. A. H. Watt, P. A. Routledge, Adenosine stimulates respiration in man, Br J Clin Pharmacol. 20:503-506 (1985).
24. B. Lagerqvist, C. Sylven, B. Beermann, G. Helmius, A. Waldenstrom, Intracoronary adenosine causes angina pectoris like pain - an inquiry into the nature of visceral pain, Cardiovasc. Res. 24:609-613 (1990).
25. A. C. Skladanowski, G. B. Sala, A. C. Newby, Inhibition of IMP-specific cytosolic 5'-nucleotidase and adenosine formation in rat polymorphonuclear leucocytes by 5'-deoxy-5'-isobutylthio derivatives of adenosine and inosine, Biochem. J. 262:203-208 (1989).
26. Y. Worku, A. C. Newby, The mechanism of adenosine production in rat polymorphonyclear leucocytes, Biochem. J. 214:325-330 (1983).
27. R. Bünger, S. Soboll, Cytosolic adenylates and adenosine release in perfused working heart. Comparision of whole tissue with cytosolic non-aqueous fractionation analyses., Eur. J. Biochem. 159:203-213 (1986).
28. R. Itoh, Purification and some properties of cytosol 5'-nucleotidase from rat liver, Biochim. Biophys. Acta. 657:402-410 (1981).
29. R. Itoh, Regulation of cytosol 5' nucleotidase by adenylate energy charge, Biochim. Biophys. Acta. 659:31-37 (1981).
30. R. Itoh, Studies on some molecular properties of cytosol 5'-nucleotidase from rat liver, Biochim. Biophys. Acta. 716:110-113 (1982).
31. R. Itoh, Regulation of rat heart cytosol 5'-nucleotidase by adenylate energy charge, Biochem. J. 235:847-851 (1986).
32. A. C. Newby, The pigeon heart 5'-nucleotidase responsible for ischaemia-induced adenosine formation, Biochem. J. 253:123-130 (1988).
33. A. C. Skladanowski, A. C. Newby, Partial purification and properties of an AMP-specific soluble 5'-nucleotidase from pigeon heart, Biochem. J. 268:117-122 (1990).
34. V. L. Truong, A. R. Collinson, J. M. Lowenstein, 5'-nucleotidases in rat heart. Evidence for the occurence of two soluble enzymes with different substrate specificities, Biochem.J. 253:117-121 (1988).
35. Y. Yamazaki, V. Truong, L. JM, 5'-nucleotidase I from Rabbit Heart, Biochemistry. 30:1503-1509 (1991).

36. J. Oka, H. Ozasa, R. Itoh, A reappraisal of the subunit molecular mass of chicken liver cytosol 5'-nucleotidase, Biochim. Biophys. Acta. 953:114-118 (1988).
37. L. Belardinelli, B. B. Lerman, Electrophysiological basis for the use of adenosine in the diagnosis and treatment of cardiac arrhythmias, Br. Heart J. 63:3-4 (1990).
38. J. P. DiMarco, W. Miles, M. Akhtar, S. Milstein, A. D. Sharma, E. Platia, B. McGovern, M. M. Scheinman, W. C. Govier, T. A. f. P. S. Group, Adenosine for paroxysmal supraventricular tachycardia: dose ranging and comparison with verapamil. Assessment in placebo-controlled, multicenter trials, Ann. Int. Med. 113:104-110. (1990).
39. G. A. Pantely, J. D. Bristow, Adenosine. Renewed interest in an old drug, Circulation. 82:1854-1856 (1990).
40. J. Till, E. A. Shinebourne, M. L. Rigby, B. Clarke, D. E. Ward, E. Rowland, Efficacy and safety of adenosine in the treatment of supraventricular tachycardia in infants and children, Br. Heart J. 62:204-211 (1989).
41. J. R. Jenkins, L. Balardinelli, Atrioventricular nodal accommodation in isolated guinea-pig hearts: physiological significance and role of adenosine, Circ Res. 63:97-116 (1988).
42. C. J. Garratt, A. Antoniou, M. J. Griffith, D. E. Ward, A. J. Camm, Use of intravenous adenosine in sinus rhythm as a diagnostic test for latent preexcitation, Am. J. Cardiol. 65:868-873 (1990).
43. A. C. Rankin, K. Goldroyd, E. Chong, A. P. Rae, S. M. Cobbe, Value and limitations of adenosine in the diagnosis and treatment of narrow and broad complex tachycardias., Br Heart J. 62:195-203 (1989).
44. M. S. Verani, J. J. Mahmarian, J. B. Hixson, T. M. Boyce, R. A. Staudacher, Diagnosis of coronary artery disease by controlled coronary vasodilation with adenosine and thallium-201 scintigraphy in patients unable to exercise, Circulation. 82:80-87 (1990).
45. P. Reid, A. G. Fraser, A. H. Watt, A. H. Henderson, P. A. Routledge, Acute haemodynamic effects of adenosine in concious man, Eur H J. 11:1018-1028 (1990).
46. A. Sollevi, Cardiovascular effects of adenosine in man; possible clinical implications, Progr. Neurobiol. 27:319-349 (1986).
47. R. F. Wilson, K. Wyche, B. V. Christensen, S. Zimmer, D. D. Laxson, Effects of adenosine on human coronary arterial circulation, Circulation. 82:1595-1606 (1990).
48. T. Nguyen, J. Heo, D. Ogilby, A. S. Iskandrian, Single photon emission computed tomography with thallium-201 during adenosine-induced coronary hyperemia: correlation with coronary arteriography, exercise thallium and two-dimensional echocardiography, J. Am. Coll. Cardiol. 16:1375-1383 (1990).
49. T. Schubert, H. Vetter, P. Owen, B. Reichart, L. H. Opie, Adenosine cardioplegia. Adenosine versus potassium cardioplegia: effects on cardiac arrest and postischemic recovery in the isolated rat heart, J. Thorac. Cardiovasc. Surg. 98:1057-1065 (1989).
50. J. W. de Jong, P. van der Meer, H. van Loon, P. Owen, L. H. Opie, Adenosine as adjunct to potassium cardioplegia: effect on function, energy metabolism, and electrophysiology, J. Thorac. Cardiovasc. Surg. 100:445-454 (1990).
51. D. G. Babbitt, R. Virmani, M. B. Forman, Intracoronary adenosine administered after reperfusion limits vascular injury after prolonged ischemia in the canine model, Circulation. 80:1388-1399 (1989).
52. C. J. Pitarys, R. Virmani, H. D. Vildibill, E. K. Jackson, M. B. Forman, Reduction of myocardial reperfusion injury by intravenous adenosine administered during the early reperfusion period, Circulation. 83:237-247 (1991).
53. D. G. Babbitt, R. Virmani, H. D. Vildibill, E. D. Norton, M. B. Forman, Intracoronary adenosine administration during reperfusion following 3 hours of ischemia: effects on infarct size, ventricular function, and regional myocardial blood flow, Am. Heart J. 120:808-813 (1990).

OXYPURINES IN EXTRACELLULAR FLUIDS FROM PIGLETS DURING HYPOXEMIA AND REOXYGENATION

Jan Peter Poulsen (1) and Ola Didrik Saugstad (2)

(1) Institute for Surgical Research
(2) Departments of Pediatrics and Pediatric Research
Rikshospitalet; Oslo , NORWAY

INTRODUCTION

Tissue hypoxia can represent a serious insult both in the neonatal and adult period of life.

Tissue hypoxia can be evaluated by a series of biochemical measurements. When diagnosing hypoxia with biochemical indicators, one is really interested in measuring the intra-cellular energy charge, but unfortunately few metabolites are suitable for this purpose. Further, to find a clear correlation between a biochemically measured value and the final outcome of hypoxia is difficult. A high negative correlation between the intracellular energy charge and the purine metabolite hypoxanthine has been found (1). Measurements of hypoxanthine can therefore be used as an indicator of hypoxia. Hypoxanthine is easily measured in extracellular fluids, and in the last 15- 20 years an extensive literature on this metabolite in hypoxia has been accumulated (For review see ref 2).

We have been interested in studying how reoxygenation or resuscitation after an hypoxemic episode will alter the concentrations of extracellular oxypurine metabolites such as hypoxanthine and xanthine. We wanted further to examine to what extent two concentrations of oxygen during the first hour after hypoxemia effect these metabolites. Is there any difference if one reoxygenates with 100 % oxygen compared to resuscitation with room air (21 % oxygen) ? The answer to this question could influence daily clinical practice, especially the resuscitation procedures in the newborn period.

METHODS

Ten piglets with a mean age of 10 ± 2 days were used. Their mean weight was 2528 ± 880 g. They were anesthetized with sodium pentobarbital, and given a continuous infusion of NaCl and Glucose (10 mL/kg/h). Artificial ventilation was given by a ventilator, keeping the $PaCO_2$ between 3.5 - 5.3 Kpa. Catheters were placed in the abdominal aorta and the inferior vena cava. Blood pressure was measured continuously by the intraarterial line. Hypoxemia was induced in all piglets by artificial ventilation with a mixture of 8% oxygen in nitrogen. During

Purine and Pyrimidine Metabolism in Man VII, Part A
Edited by R.A. Harkness *et al.*, Plenum Press, New York, 1991

hypoxemia PaO_2 remained between 3.5 - 6.5 Kpa.

After 2 hours hypoxemia, the piglets were randomly divided into two groups, group 1 (n = 5) was reoxygenated with 100 % oxygen for one hour, while group 2 (n = 5) was reoxygenated with room air (21 % oxygen) for one hour.

Arterial blood samples were withdrawn before hypoxemia and then every 30 min for measuring blood gases and acid-base parameters and for hypoxanthine and xanthine analyses.

Vitreous humor (VH) was sampled with a vacutainer before hypoxemia, after hypoxemia and at the end of reoxygenation.

Cerebrospinal fluid (CSF) was sampled with a lumbar puncture needle in Cisterna magna at the same times as the vitreous humor.

The hypoxanthine and xanthine levels were analyzed using an HPLC method (3).

Results are given as mean ± SD. $p < 0.05$ is significant (Wilcoxon's two-tailed rank test)

RESULTS

Table 1 shows the arterial PaO_2, O_2 saturation, pH, arterial base excess (BE), and mean arterial blood pressure (MAPB) in the two groups of pigs (Gr).

Table 1

	Gr	Basal level	120 min hypoxia	5 min resusc.	15 min resusc.	60 min resusc.
PaO_2 Kpa	1 2	9.7± 0.8 10.2± 1.8	5.0± 0.5 6.0± 3.0	51.1± 1.0 10.4± 1.1	53.7± 7.0 10.8± 1.1	58.2± 8.2 11.1± 0.9
$SatO_2$ %	1 2	94.1± 3.1 95.3± 2.3	29.5± 6.8 40.1± 18.2	99.6± 0.2 72.3± 16.8	99.6± 0.3 75.8±18.1	99.8± 0.1 83.8± 15.1
pH	1 2	7.40± 0.07 7.46± 0.05	6.87± 0.15 6.92± 0.30	6.81± 0.12 6.89± 0.36	6.87± 0.11 6.94± 0.34	7.13± 0.12 7.10± 0.32
BE mmol/L	1 2	-3.7± 3.3 -1.0± 2.7	-27.6± 5.2 -24.7± 13.0	-31.2± 4.4 -26.4±14.5	-29.4± 4.3 -25.1±13.8	-17.5± 5.4 -18.0±14.6
MABP mmHg	1 2	88 ± 13 95 ± 13	37 ± 26 37 ± 34	73 ± 21 61 ± 8	63 ± 17 60 ± 7	53 ± 18 55 ± 18

There were no differences between the two groups in any of the values before hypoxemia nor after the hypoxemic period. During reoxygenation the PaO_2 was significantly higher in group 1 than in group 2, but the PaO_2 returned to normal values within 5 min also in group 2. No other significant difference was found between the two groups.

Table 2 shows the hypoxanthine (Hx)(μmol/L) and the xanthine (X)(μmol/L) in the two groups of pigs (Gr). In both groups, hypoxanthine in the vitreous humor (VH) increased about two-fold during the hypoxemic period. Hypoxanthine continued to increase during reoxygenation as well, reaching a final concentration about 4 times the basal value. No difference was found between the two groups at any time.

The same pattern was found in the CSF concentrations of hypoxanthine in the 2 groups, but here the increase during reoxygenation was less marked than in the vitreous humor. No significant difference was found between the 2 groups at any time.

Both hypoxanthine and xanthine increased significantly in plasma during hypoxemia. In both groups during the first 5 min of reoxygenation, the two metabolites continued to increase. However, neither of these increases were significant when compared to the levels found at the end of the hypoxemic period. Both metabolites then decreased during the next 55 min of reoxygenation. No significant difference was found between the 2 groups at any time.

Table 2

	Gr	Basal level	120 min hypoxia	5 min resusc.	15 min resusc.	60 min resusc.
VH Hx	1	13.6± 3.2	19.8± 3.4*			38.0± 8.1#
	2	12.0± 4.0	22.0± 9.0*			36.4±16.4#
VH X	1	<1.0± 0.0	1.9± 0.8*			5.6± 4.2#
	2	<1.0± 0.0	1.1± 0.4*			3.4± 2.0#
CSF Hx	1	25.0± 12.9	60.0± 9.2*			72.8±19.3
	2	20.2± 8.4	55.3±26.1*			92.7±39.2#
CSF X	1	6.0± 2.5	16.4± 3.8*			21.8±5.8#
	2	6.7± 3.4	16.5± 4.6*			24.2±8.5#
Plasma Hx	1	26.6± 4.5	92.2±18.9*	102.3±38.0*	86.6± 27.4*	76.0±68.2#
	2	29.8±11.2	123.8±71.7*	134.4±59.0*	122.0±56.2*	77.7±49.8#
Plasma X	1	1.2± 0.4	33.5±12.5*	38.0±23.3*	34.7±20.9*	26.3±24.5*
	2	<1.0± 0.0	43.7±23.0*	51.6±26.9*	43.0±28.8*	36.7±34.2*

* $p < 0.05$ compared to basal values,
# $p < 0.05$ compared to values after hypoxemia

DISCUSSION

Elevated levels of hypoxanthine has been found in the CSF in children after hypoxia (4), and in the vitreous humor in victims of sudden infant death syndrome (5). We have earlier reported that both the degree and the duration of an hypoxemic episode influence hypoxanthine concentrations in these fluids (6,7). Our present findings in CSF and vitreous, of an increase in hypoxanthine and xanthine, both during hypoxemia and during reoxygenation, makes it even more complicated to assess hypoxemia by measuring purine metabolites in these extracellular fluids.

In plasma, hypoxanthine concentrations increased both during hypoxemia and during the first 5 min of reoxygenation. We believe that this reflects the clinical situation, an accumulation during hypoxemia with an additional wash out during the first minutes of resuscitation.

When oxygen is administered after a significant period of lack of oxygen, an explosive tissue injury is found. This phenomenon has been described both in clinical and experimental medicine, and has been designated the oxygen paradox since it seems paradoxical that administration of oxygen to tissues lacking oxygen can be deleterious. Obviously, something must happen to the tissues in the reoxygenation period causing this "post-hypoxic reoygenation injury". What is the damaging factor in this injury ? Is it accumulation of a substance in the tissue during hypoxia, that in the presence of oxygen will form harmful substances ? Is it the way we administer oxygen during the reoxygenation period that is harmful ? Is a combination of a potential harmful substance with high concentration of oxygen the crucial step ?

An explanation for the oxygen paradox was given by the hypothesis that oxygen radicals are formed by the hypoxanthine/ xanthine oxidase system in the post-hypoxic period (8,9). This post-hypoxic tissue injury could, at least in theory, be lessened if either the concentration of hypoxanthine, oxygen or xanthine oxidase in the reoxygenation period was lowered.

The present experiments indicate that room air is as efficient as 100% oxygen in the resuscitation procedure. If in addition less tissue damage is caused by room air, room air should be used instead of 100 % oxygen in a resuscitation situation.

Acknowledgements

The Norwegian Cancer Society, the Norwegian Research Council for Science and Humanities, the Norwegian Women's Public Health Asc.

REFERENCES

(1) Simmonds RJ, Coade SB, Harkness RA, Drury L, Hytten FE: Nucleotide, nucleoside and purine base concentrations in human placenta. **Placenta** 1982;3:29-38
(2) Saugstad OD: Hypoxanthine as an indicator of hypoxia: Its role in health and disease through free radical production. **Pediatr Res** 1988;23:143-150
(3) Wung WE, Howell SB: Simultaneous liquid chromatography of 5-Fluoro-uracil, Uridine, Hypoxanthine, Xanthine, Uric acid, Allopurinol and Oxypurinol in plasma. **Clin Chem** 1980;26:1704-1708
(4) Harkness RA, Lund RJ: Cerebrospinal fluid concentrations of hypoxanthine, xanthine, uridine, and inosine: high concentrations of the ATP metabolite; hypoxanthine, after hypoxia **J.Clin Pathol** 1983;36:1-8
(5) Rognum TO, Saugstad OD, Øyasæter S, Olaisen B: Elevated levels of hypoxanthine in the vitreous humor indicate prolonged cerebral hypoxia in victims of sudden infant death syndrome. **Pediatrics** 1988;82:615-618
(6) Poulsen JP, Øyasæter S, Sanderud J, Rognum TO, Saugstad OD: Hypoxanthine, Xanthine and Uric Acid Concentrations in the Cerebrospinal Fluid, Plasma and Urine of Hypoxemic Pigs **Pediatr Res** 1990;28:477-481
(7) Poulsen JP, Rognum TO, Øyasæter S, Saugstad OD: Changes in Oxypurine Concentrations in Vitreous Humor of Pigs during Hypoxemia and Post-Mortem. **Pediatr Res** 1990;28:482-484
(8) Saugstad OD, Aasen AO: Plasma hypoxanthine concentrations in pigs. A prognostic aid in hypoxia. **Eur Surg Res** 1980;12:123-129
(9) Mc Cord JM: Oxygen derived free radicals in postischemic injury. **N Engl J Med** 1985;312:159-163

REGULATION OF ADENOSINE CONCENTRATIONS BY ACADESINE (AICA-RIBOSIDE) IN HUMAN B-LYMPHOBLASTS

J. Barankiewicz, R. Jimenez, J. Uyesaka and H.E. Gruber

Gensia Pharmaceuticals, Inc., 11025 Roselle Street
San Diego, CA 92121, USA

A large number of physiological functions of adenosine including regulation of coronary vasodilation, (1) neurotransmission (2) and immune response (3) have been described and discussed (4-8). However, the mechanism(s) involved in adenosine formation remain controversial. Adenosine found in the extracellular environment can be formed by: a) the intracellular process of AMP dephosphorylation and SAH hydrolysis resulting in adenosine release across the cell membrane; b) extracellular degradation of adenine nucleotides released from the tissues; c) both intracellular and extracellular mechanisms occurring simultaneously.

Recently, it has been reported that acadesine (AICA-riboside) increases adenosine concentrations in the blood from the ischemic myocardium (9) and since then, it has attracted increased attention as a compound which can possibly regulate adenosine concentrations. We studied the possible mechanism(s) involved in the action of acadesine on raising extracellular adenosine concentrations surrounding B-lymphoblasts (10). B-lymphoblasts are known to be able to produce adenosine by both intracellular and extracellular mechanisms (11, 12).

Acadesine

Acadesine (5-amino-4-imidazole-carboxamide riboside; AICA-riboside), in a variety of cells is initially metabolized by adenosine kinase (13,14) to form acadesine monophosphate (ZMP) (13-16), the penultimate compound in purine biosynthesis de novo. In most cells, ZMP can be converted further to IMP which serves as a precursor of other purine compounds including adenine and guanine nucleotides (13, 14). Therefore, acadesine was first considered as a possible beneficial compound by increasing purine nucleotide synthesis, especially in postischemic myocardium (17). However, some authors (18-23) did not demonstrate enhancement of ATP levels by acadesine.

Besides the observations of acadesine increasing adenosine concentrations in blood draining ischemic myocardium (9), more recently acadesine has been found to also enhance adenosine mediated inhibition of platelets aggregation in human whole blood (24), to ameliorate reperfusion injury and oxidant damage in heart (25), to inhibit mast cell degradation (26) and to prevent pacing induced myocardial ischemia (27).

Effect of acadesine on adenosine formation from intracellular ATP

Two alternative pathways (via AMP deamination or via AMP dephosphorylation) are involved in intracellular ATP degradation, but only the latter results in adenosine formation. However, there is good evidence that intracellular ATP degradation proceeds predominantly via AMP deamination followed by IMP dephosphorylation (28-30). As a result, only a small part of AMP undergoes dephosphorylation leading to adenosine formation which if it is not deaminated to inosine by adenosine deaminase (ADA) and/or salvaged back to AMP by adenosine kinase (AK) can be released from the cells.

Purine and Pyrimidine Metabolism in Man VII, Part A
Edited by R.A. Harkness *et al.*, Plenum Press, New York, 1991

To determine whether acadesine increases the release of adenosine from intracellular ATP, we studied the effect of acadesine on adenosine formation from intracellular ATP prelabelled with radioactive adenine (10). It was found that during 2-deoxyglucose induced intracellular degradation of ATP, adenosine represented only a small portion of the catabolites released from B-lymphoblasts whereas hypoxanthine and inosine were the major catabolites released from the cells. Acadesine increased the amounts of adenosine in the medium in a dose dependent fashion, but the effect was relatively small. The effect of acadesine on adenosine concentration could be explained by the inhibition of the re-uptake of released adenosine or by an elevation of adenosine release from the cells.

Effect of acadesine on adenosine uptake into cells

The effect of acadesine on adenosine uptake into human B-lymphoblasts was studied using a method of centrifugation of cells through an oil layer (14). It was found that acadesine could inhibit adenosine uptake into the cells only at sub-μmolar (0.2 μM) concentrations of adenosine. In contrast, acadesine was not able to inhibit adenosine uptake into B-lymphoblasts at μmolar (1 μM) (14) or mmolar (1 mM) adenosine concentrations (10). When these experiments were repeated in the presence of ADA or AK inhibitors the effect of acadesine on adenosine uptake was reduced. These results suggest that acadesine is a poor inhibitor of adenosine transport into the cells, and probably affects adenosine utilization rather than adenosine transport. Confirmatory detailed transport studies need to be perfomed.

Effect of acadesine on the adenosine formation from extracellular adenine nucleotides

Extracellular ATP degradation (ATP → ADP → AMP → adenosine) is a direct process of adenosine formation by cells expressing ectonucleotidase activities on their surface, but it is limited in vivo by ATP availability (31). It has been reported that human B-lymphoblasts contain highly active ecto-ATPase, ecto-ADPase and ecto-AMPase (12) and therefore these cells can efficiently degrade extracellular ATP to adenosine. To evaluate the effect of acadesine on adenosine produced extracellularly, we incubated B-lymphoblasts with ATP and with acadesine at different concentrations. We found that acadesine increased, in a dose dependent manner the accumulation of extracellularly produced adenosine (Fig. 1A). The accumulation of adenosine was accompanied by a decrease in inosine and hypoxanthine formation, as well as a decrease in labelling of the intracellular nucleotide pool (result not shown). A similar effect of acadesine on adenosine accumulation when AMP was a substrate for extracellular degradation was observed (10). These data suggest that acadesine inhibits adenosine utilization by human B-lymphoblasts.

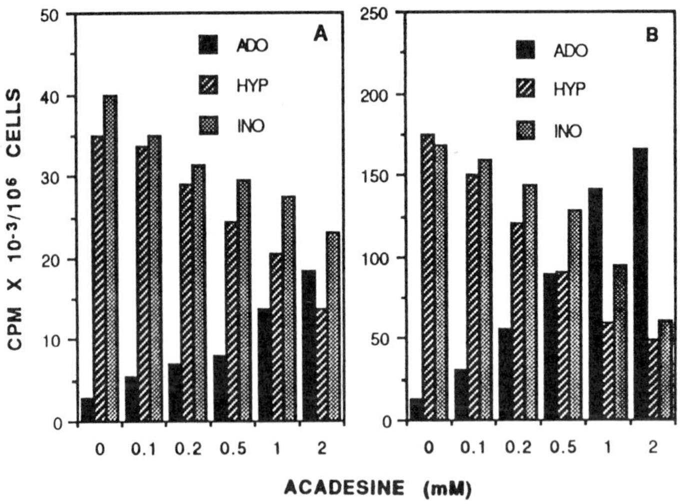

Fig. 1 Effect of acadesine on adenosine, inosine and hypoxanthine levels during extracellular ATP degradation (A) or adenosine utilization (B) in human B-lymphoblasts.

Cells (10^6/0.1 ml) were incubated for 60 min with [8-^{14}C] ATP (500μM) (A) or with [2,8-^{3}H] adenosine (1μM) (B) and different concentrations of acadesine. The amounts of radiolabelled adenosine, inosine and hypoxanthine were measured in the incubation medium after separation by TLC.

Effect of acadesine on adenosine utilization

We have recently reported that acadesine reduces adenosine utilization in B-lymphoblasts in a dose dependent fashion (Fig. 1B) (32). Preincubation of B-lymphoblasts with acadesine enhanced the inhibition of adenosine utilization, and removal of acadesine and its extracellular metabolites after incubation decreased the effect. Therefore, we suggested that in this system the presence of acadesine or its metabolites in the medium, is primarily responsible for the inhibition of adenosine utilization rather than its intracellular metabolites (10). To explain the acadesine induced inhibition of adenosine utilization in B-lymphoblasts, the effect of acadesine on the activity of enzymes involved in adenosine utilization, namely ADA, AK and S-adenosyl homocysteine hydrolase (SAHH) has been studied in cell extracts and intact B-lymphoblasts. Acadesine inhibited the activity of all three enzymes with Ki's of 2.7, 3.3, 0.12 mM respectively (10). To evaluate the effect of acadesine on activities of these enzymes in intact B-lymphoblasts, the modulation of adenosine utilization by acadesine of was studied when ADA, AK or SAHH were individually inhibited by specific inhibitors. It was found that individual inhibition of the enzymes involved in adenosine utilization had only a minor effect on the acadesine action (Fig. 2).

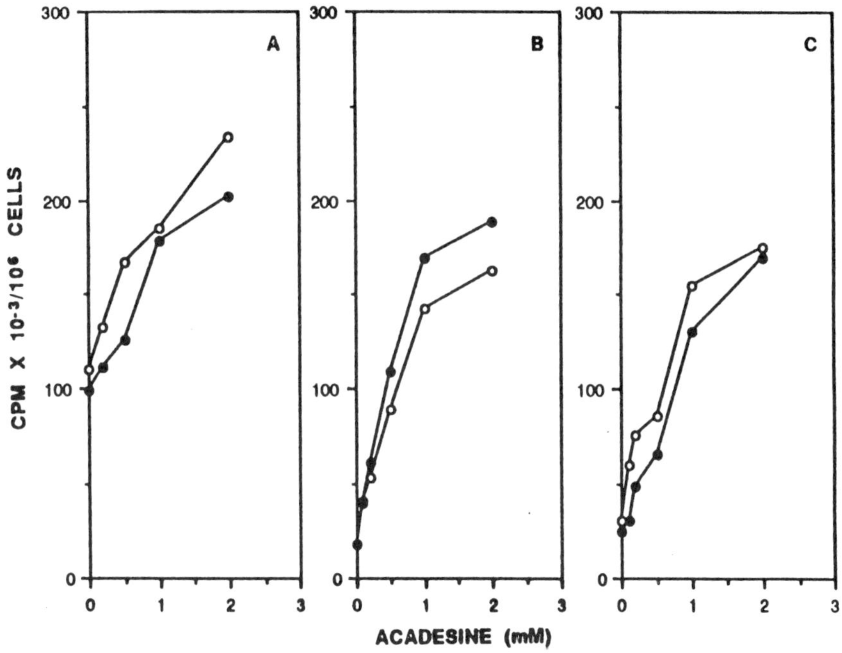

Fig. 2 Effect of acadesine on adenosine utilization when ADA, AK and SAHH were individually inhibited by specific inhibitors.

Cells ((10^6/0.1 ml) were preincubated for 60 min with media (open circle) or with inhibitor (closed circle) (A) 40μM 2′-deoxycoformyin or (B) 20μM 5-iodotubercidin or (C) 40μM adenine-9-(trans-3′-dihydroxycyclopenthyl-4-enyl) adenine to inhibit ADA, AK and SAHH respectively. Then radiolabelled [2,8-^{3}H] adenosine (1μM) and different concentrations of acadesine were added and incubation was continued for 10 min. After incubation, radiolabelled adenosine was measured in the medium after separation by TLC.

This may indicate that if only one of the major adenosine utilizing reactions (ADA or AK) is inhibited by a specific inhibitor, then the alternative reaction in adenosine utilization is inhibited by acadesine. The simultaneous inhibition of ADA and AK by 2′-deoxycoformycin and 5-iodotubercidin almost completely blocked adenosine utilization while the presence or absence of the SAHH inhibitor had little additional effect on inhibition (10). Obviously, when adenosine utilization is completely inhibited by the combination of ADA and AK inhibitors, the effect of acadesine is abolished (results not shown).

It seems that the reduction of adenosine utilization by acadesine is primarily mediated by the simultaneous inhibition of AK and ADA, but the importance of SAH hydrolase inhibition is still not clear. The inhibition of adenosine utilization by acadesine may be responsible for the observed elevation of adenosine in blood from the ischemic heart in vivo (9).

Regulation of extracellular adenosine concentrations may be of a pharmacological importance in many adenosine regulated pathophysiological functions including cardiovascular function and inflammatory responses. Even partial inhibition of adenosine utilization by acadesine should contribute to the elevation of extracellular concentrations of adenosine.

REFERENCES

1. Mullane, K.M. and Williams, B.M. in Adenosine and Adenosine Receptors. Ed M. Williams The Humana Press, Clifton, New Jersey, pp 289-333, 1990.
2. Snyder, S.H. Ann.Rev.Neurosci: 8,103-124, 1985.
3. Polmar, S.H., Birch, R.E., Mondler, R., Rudolph, S.A., pp. 249-259, Eds R.M. Berne, Roll, T.W., Rubio, R., Develop Pharmacol 2, 1983.
4. R.M. Berne, T.W. Rall, Rubio, R. (Eds.) Regulatory function of adenosine, Develop. Pharmacol. 2, Martinus Nijhoff Pub., Boston, The Hague, Dordrecht, Lancaster 1983.
5. Stefanovich, V., Rudolphi, K., Schubert, P. (Eds.) Adenosine: Receptors and modulation of cell function., IRL Press, Oxford, Washington, 1985.
6. Tritsch, G.L. (Ed.) Adenosine deaminase in disorders of purine metabolism and immune deficiency. Ann. N.Y. Acad. Sci. Vol 451, 1985.
7. Gerlach, E., Becker, B.F. (Eds.) Topics and Perspectives in adenosine research., Springer-Verlag, Berlin, Heidelberg, New York, London, Paris, Tokyo, 1987.
8. Williams, M. (Ed.) Adenosine and adenosine receptors., The Humana Press, Clifton, New York, 1990.
9. Gruber, H. E., Hoffer, M.E., McAllister, D.R., Laikind, P.K., Lane, T.A., Schmid-Schoenbein, G.W., Engler, R.L., Circulation 80, 1400 - 1411, 1989.
10. Barankiewicz, J., Jimenez, R., Uyesaka, J., McCurley, J.P., Bullough, D.A., Gruber, H.E., in preparation.
11. Barankiewicz, J., Ronlov, G., Jimenez, R., Gruber, H.E., J. Biol. Chem. 265, 15738, 1990.
12. Barankiewicz, J., Dosch, H.M., Cohen, A., J. Biol. Chem. 263, 7094, 1988.
13. Sabina, R.L., Patterson, D. and Holmes, E.W., J. Biol. Chem. 260, 6107-6114, 1985.
14. Jimenez, R., Gruber, H.E., Barankiewicz, J. (1990), Int. J.Pur. Pyr. Res. 1, 51-60, 1990.
15. Sabina, R.L., Holmes, E.W., Becker, M.A., Science 223, 1193, 1984.
16. Bochner, B.R. and Ames, B.N., Cell 29, 929, 1982.
17. Swain, J.L., Hines, J.J., Sabina, R.L. and Holmes, E.W., Circ. Res. 51, 102, 1982.
18. Mitsos, S.E., Jolly, S.R., Lucchesi, B. R., Pharmacology 31, 121, 1985.
19. Mentzer, R.M., Jr., Stephen, W.E., Lasley, R.D. and Berne, R.M., J. Thorec. Cardiovasc. Surg. 95, 286, 1980.
20. Sabina, R.L., Kernstine, K.H., Boyd, R.L., Holmes E.W. and Swain, J.L., J. Biol. Chem. 257, 10178, 1982.
21. Mauser, M., Hoffmeister, H.M., Nienaber, C., Schaper, W., Circ. Res. 56, 220, 1985.
22. Nissim, I., Yudkoff, M., Nissim I., States, B. (1989), J. Cell Physiol. 138, 536, 1989.
23. Hoffmeister, H.M., Mauser, M., Schaper, W., Basic Res. Cardiol. 80, 445, 1985.
24. Bullough, D. A., Zhoug, C. and Mullane, K.M., Int. J. Pur. Py. Res., (abstract of this symposium), 1991.
25. Metzner, E.K., Fox M.H., Potter, S., Mullane, K.M. and Bullough, D.A., Int. J. Pur. Pyr. Res. (abstracts of this symposium), 1991.
26. Marquardt, D.L. and Gruber, H.E., Biochem. Pharmacol. 35, 4415, 1986.
27. Young, M.A., Mullane, K.M., (in preparation).
28. Barankiewicz, J., Cohen, A., J.Biol.Chem. 259, 15178, 1984.
29. Van den Berghe, G., Broufman, M., Vanneste, R., Hers, M.G., Biochem. J162, 601, 1977.
30. Zoret-Shani, E., Shainberg, A., Sperling, O., Biochem, Biophys. Acto. 926, 287, 1987.
31. Gordon, J.L., Biochem.J. 233, 309, 1986.
32. Barankiewicz, J., Jimenez, R., Ronlov, G., Magill, M., Gruber, H.E., Arch. Biochem. Biophys. 282, 377, 1990.

PRESERVATION OF ADENINE NUCLEOTIDES FOLLOWING ISCHEMIA AND REPERFUSION: CORRELATION WITH FUNCTIONAL RECOVERY

Pankaj Mankad, Richard Lachno, and Magdi Yacoub

National Heart and Lung Institute
Harefield Hospital, Harefield
Middlesex. UB9 6JH U.K.

INTRODUCTION

Hypothermic storage following a single infusion of a cold hyperkalemic cardioplegic solution is currently the only clinically utilised method for ex-vivo cardiac preservation. This technique offers up to four hours of relatively safe ischemia time. A longer period of preservation may increase the donor organ pool, allow more time for tissue typing and possibly shorten the transplant waiting lists. The recently introduced University of Wisconsin (UW) preservation solution has significantly improved the solid organ preservation[1]. However, the potential role of UW solution in cardiac transplantation has not been adequately defined. The objective of this study was to perform critical evaluation of UW solution as compared with the currently used St. Thomas' Hospital cardioplegic solution (ST) for the preservation of the pig heart. The restitution of adenine nucleotide content and functional recovery following reperfusion were compared in the two groups of hearts.

MATERIALS AND METHODS

Hearts were obtained from Yucatan pigs of both sexes, weighing between 20 to 35 kg each. The animals were anethetised and ventilated. Two units of blood were collected for ex-vivo perfusion. The chest was opened and the heart excised. Diastolic arrest was achieved by infusing 300 mls of either ST or UW solution at 4°C into the aortic root (n = 6 in each). Intraventricular balloons were sutured to the tricuspid and mitral valve annuli. The aorta was clamped and a cannula inserted into the innominate artery. The heart was then mounted on a retrograde perfusion apparatus, consisting of a Watson-Marlow roller pump and a hollow fibre oxygenator with integrated heat exchanger, and placed inside an insulated glass cylinder. It was immediately perfused in a nonpulsatile fashion with enriched autologous blood (hematocrit 28-30%), containing D-ribose 1gm/L, Insulin 20IU/L and Glucose 14.25mg/min, at an aortic pressure of 80mm of Hg and perfusate temperature of 37°C. Following 20 minutes of perfusion,

Purine and Pyrimidine Metabolism in Man VII, Part A
Edited by R.A. Harkness *et al.*, Plenum Press, New York, 1991

control data regarding myocardial function were obtained. The balloons were sequentially inflated with 10, 20, 30, and 40 mls of saline. At each balloon volume, transballoon left ventricular systolic and diastolic and right ventricular systolic and diastolic pressures plus left ventricular dP/dt, were recorded.

The hearts were then rearrested with 600 mls of the same solution used previously (i.e. either ST or UW solution) and stored for 8 hours at 4°C. At the end of the ischemia period, intraventricular balloons were reinserted and the hearts perfused with the second unit of enriched autologous blood for a period of 2 hours. Repeat data for the myocardial function were collected at half hourly intervals. The composition of the two solutions was as follows (mmol/L): ST solution - NaCl 144.0, KCl 20.0, $MgCl_2$ 16.0, $CaCl_2$ 2.4, Procaine 1.0; Osmolarity 300-320 mosm/L, pH 5.5-7.5; UW solution - KH_2PO_4 25.0, $MgSO_4$ 5.0, Adenosine 5.0, Glutathione 3.0, Raffinose 30.0, Allopurinol 1.0, K Lactobionate 100.0, Pentastarch 5%, Insulin 40 IU/L, Osmolarity 320 mosm/L, pH 7.4.

Determination of Nucleotide Content

To determine control myocardial nucleotide content full-width biopsy specimens were obtained using a Tru-Cut needle from the in situ beating hearts. This was repeated on similar biopsied material taken from the hearts following 8 hours preservation period (i.e. before reperfusion) and at the end of the experiment (i.e. following reperfusion). The specimens were snap frozen and stored in liquid nitrogen until submitted to the process of extraction using 0.6M perchloric acid. The analysis for nucleotide content was performed by the reversed phase HPLC method as described in detail previously[2]. Energy charge, a quantitative estimate of the energy state of the cell, was calculated using the formula ATP + ½ ADP divided by ATP +ADP + AMP.

Expression of Results

From the values of trans-balloon systolic and diastolic pressures obtained at different balloon volumes, pressure - volume (P-V) curves were plotted for both the right and the left ventricles. The slope values of these curves were analysed for statistical significance. Differences in the pre and the post-ischemic values of dP/dt were expressed as a percentage reduction over the control values. Myocardial nucleotide data obtained at the end of ischemia (EOI) and at the end of reperfusion (EOR) were compared to each other and to the preischemic control values. Student's t test was used to assess the level of statistical significance. Significance was assumed when the P value was less than 0.05. All values are expressed as the mean ± SD.

RESULTS

The mean contents of ATP, ADP, and AMP for both groups, at control, EOI and EOR are shown in Figure 1. There was no difference in the preischemic control values of adenine nucleotides (in nM/mg protein) between the two groups (ATP - ST, 34.5 ± 2.81, UW, 35.5 ± 5.66, ADP - ST, 11.6 ± 2.23, UW,

11.3 ± 1.79, AMP - ST, 1.5 ± 0.5, UW, 1.3 ± 0.14). At the end of the ischemic period, the values were ATP - ST, 24.1 ± 3.91, UW, 27.3 ± 3.29, ADP - ST, 13.3 ± 1.78, UW, 9.4 ± 2.14, and AMP - ST, 3.6 ± 0.69, UW, 3.2 ± 0.97. The difference between the groups was not significant. However, at the end of reperfusion, the ATP levels were significantly better in the UW preserved hearts than in the ST hearts (ST, 14.8 ± 3.92, UW, 23.2 ± 1.82, P<0.01).

The mean contents of adenosine, inosine and hypoxanthine for each group, at preischemic control, EOI and EOR are compared in Figure 2. There was a marked increase in the contents of the three nucleotides over the control values at the end of ischemia in the UW group as compared with the ST group. The control values for adenosine, inosine and hypoxanthine in the two groups were respectively, ST, 0.05 ± 0.12, 0.7 ± 0.32 and 0.4 ± 0.23, UW, 0.05 ± 0.08, 0.6 ± 0.12 and 0.5 ± 0.17. On cessation of ischemia, they were ST, 0.08 ± 0.08, 1.8 ± 1.1 and 0.58 ± 0.27, UW, 1.3 ± 0.57 (P<0.01), 8.0 ± 1.17 (P<0.001) and 1.5 ± 0.43 (P<0.01).

The calculated adenine nucleotide energy charge is shown in Table 1. Energy charge remained well maintained in the UW group but in the ST group, there was a significant reduction at the end of ischemia which persisted despite reperfusion.

<u>Functional Studies</u>

In the ST group, the slope value of systolic P-V curve diminished from 4.54 ± 0.57 to 2.1 ± 0.18 (P<0.001); while in the UW group, the reduction though less, from 4.55 ± 0.49 to 3.74 ± 0.49, was still significant (P<0.01). The percentage reduction in the UW group was less than the ST group (UW, 17.7 ± 6.7%, ST, 52.4 ± 5.5%, P<0.001). Similarly, chamber stiffness, defined as a change in pressure relative to a change in volume, increased following preservation. In the ST group, the increase was from 1.13 ± 0.28 to 2.6 ±0.62 mmHg/ml (P<0.01). In the UW group, the change was from 1.33 ± 0.39 to 2.19 ± 0.49 mmHg/ml (P<0.01). The percentage increase in the UW group (65.5 ± 31.4%) was significantly less (P<0.05) than that of the ST group (125.6 ± 46.4%). Slope values of right ventricular systolic curves in both the groups were similar. However, right ventricular chamber stiffness increased by 40.2 ± 31.3% following UW preservation as compared to an increase of 112.3 ± 48.7% in the ST group (P<0.02).

Table 1 Energy charge of the preserved hearts

	Control	EOI	EOR
ST	0.850 ± 0.02	0.748 ± 0.014**	0.785 ± 0.035*
UW	0.838 ± 0.05	0.80 ± 0.02	0.838 ± 0.04

ST - St. Thomas' Hospital cardioplegic solution, UW - University of Wisconsin solution, EOI - end of ischemia, EOR - end of reperfusion, *P<0.01, **P<0.001.

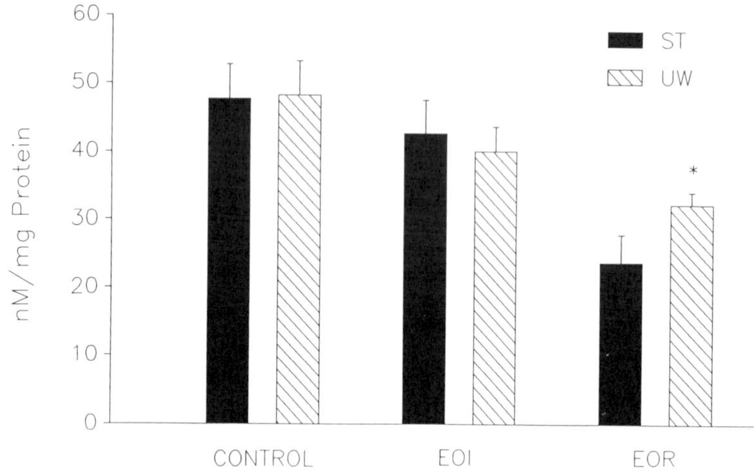

Fig. 1. Myocardial content of ATP + ADP + AMP at control, end of ischemia (EOI), and end of reperfusion (EOR) in the isolated pig heart subjected to 8 hours of preservation using St. Thomas' Hospital solution (ST) or University of Wisconsin (UW) solution. n = 6 in each. *$p<0.001$.

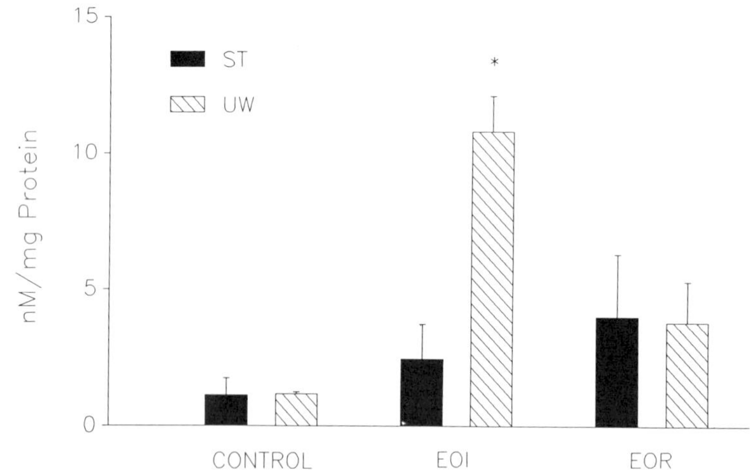

Fig. 2. Myocardial content of Adenosine, inosine plus hypoxanthine at control, end of ischemia (EOI), and end of reperfusion (EOR) in the isolated pig heart subjected to 8 hours of preservation using St. Thomas' Hospital solution (ST) or University of Wisconsin (UW) solution. n = 6 in each. *$p<0.001$.

There was a reduction in dP/dt after 8 hours of preservation in both the groups. The UW hearts showed significantly better recovery of dP/dt than the ST hearts ($P<0.001$). The UW hearts showed reduction in dP/dt from 1906.7 ± 81.1 mm Hg/sec to 1560.8 ± 87.4 mm Hg/sec ($P<0.001$), while in the ST hearts the reduction was from 2070 ± 211.7 to 1250 ± 160.5 mm Hg/sec ($P<0.001$). The percentage reduction in UW hearts (18.1 ± 4.6%) was significantly less than in the ST group (39.3 ± 8.1%, $p<0.001$).

DISCUSSION

This study has shown that a single infusion of the University of Wisconsin solution provides significantly better preservation of ATP and myocardial function upon refusion of the isolated pig heart subjected to 8 hours of hypothermic ischemic arrest at 4°C compared with St. Thomas' Hospital cardioplegic solution.

UW solution contains 5.0 mmol/L of adenosine. Superior preservation of myocardial ATP by UW solution can be attributed to its adenosine content. The extremely high concentration of inosine and elevated levels of adenosine and hypoxanthine at the end of ischemia in the UW preserved hearts suggest that adenosine from UW solution is taken up and metabolised by myocytes during the period of preservation. Inosine and hypoxanthine are known to stimulate the formation of adenine nucleotides via the "salvage pathway" through the generation of inosine monophosphate[3]. Adenosine is also a substrate for AMP and ATP synthesis directly under the influence of adenosine kinase. This could account for the better preservation of ATP in the UW group. Restitution of the myocardial ATP pool following ischemia has been shown to be associated with improved heart function[4]. Adenosine and phosphate in the UW solution has been shown to stimulate ATP synthesis in the hypothermically perfused canine kidney[5].

Adenosine, in addition, has several pharmacological properties which render it potentially useful for the preservation of ischemic myocardium. It is a powerful coronary vasodilator and is also involved in regulation of coronary flow[6]. Through its action on neutrophils, adenosine has the potential to protect the coronary endothelium from neutrophil-mediated reperfusion injury[7]. In the isolated rat heart, addition of adenosine during reperfusion has been shown to improve recovery of myocardial function[8].

Apart from adenosine, the composition of UW solution differs considerably from that of ST solution. UW solution is an intracellular type of solution with a high concentration of potassium (125 mmol/L) and a relatively low concentration of sodium (20 mmol/L). It also contains impermeants in the form of raffinose, a non-metabolizable trisaccharide, and lactobionate, an anion that substitutes for the freely permeable chloride ion. These components provide an extracellular osmotic force that has been shown to limit hypothermic cell swelling in the kidney, pancreas and liver[9]. In addition, pentastarch, a type of hydroxyethyl starch, provides a colloidal oncotic pressure that will limit the expansion of the interstitial space. It is conceivable that the presence anti-oxidants namely allopurinol and glutathione in UW solution may also have contributed to its superior qualities as addition of anti-oxidants (superoxide dismutase plus catalase, allopurinol and reduced glutathione) to cardioplegic solutions has been shown to be associated with improved recovery of postischemic function of the rat heart[10-11].

In conclusion, this study has shown that UW solution, when compared with currently used myocardial preservation solution, has the potential to improve the functional recovery and prolong the ischemic period in clinical cardiac transplantation. Which of its many components are salient to its superiority and their optimum concentrations for myocardial preservation merit further research. However, the present study suggests potentially useful role of adenosine in UW solution.

REFERENCES

1. Todo s, Nery J, Yanaga K, Podesta L, Gordon R, Starzl T. Extended preservation of human liver grafts with UW solution. JAMA 1989;261:711-714.
2. Smolenski RT, Lachno DR, Ledingham SJM, Yacoub MH. Determination of sixteen nucleotides, nucleosides and bases using high-performance liquid chromatography and its application to the study of purine metabolism in hearts for transplantation. J Chromatogr 1990;527:414-420.
3. Manfredi JP, Holmes EW. Purine salvage pathways in myocardium. Annu Rev Physiol 1985;47:691-705.
4. Hearse DJ, Stewart DA, Braimbridge MV. Cellular protection during myocardial ischemia. Circulation 1976;54:193-202.
5. Southard JH, Lutz MF, Ametani MS, Belzer FO. stimulation of ATP synthesis in hypothermically perfused dog kidneys by adenosine and PO_4. Cryobiology 1984;21:13-19.
6. Berne RM. The role of adenosine in the regulation of coronary blood flow. Circ Res 1980;47:807-13.
7. Forman MB, Puett DW and Virmani R. Endothelial and myocardial injury during ischemia and reperfusion: pathogenesis and therapeutic implications. J Am Coll Cardiol 1989;13:450-9.
8. Ledingham S, Katayama O, Lachno D, Patel N, Yacoub M. Beneficial effects of adenosine during reperfusion following prolonged cardiac arrest. Cardiovasc Res 1990,24:247-253.
9. Wahlberg JA, Southard JH, Belzer FO. Development of a cold storage solution for pancreas preservation. Cryobiology 1986;23:477-482.
10. Shlafer M, Kane PF, Kirsh MM. Superoxide dismutase plus catalase enhances the efficacy of hypothermic cardioplegia to protect the globally ischemic, reperfused heart. J Thorac Cardiovasc Surg 1982;83:830-839.
11. Chambers DJ, Astras G, Takahashi A, Manning AS, Braimbridge MV, Hearse DJ. Free radicals and cardioplegia: organic anti-oxidants as additives to the St Thomas' hospital cardioplegic solution. Cardiovasc Res 1989;23:351-358.

AN ALTERNATIVE PATHWAY OF ADENYLATE AND ATP SYNTHESIS

Celia Montero*, Ryszard T. Smolenski**, John A. Duley, and H. Anne Simmonds

Purine Research Laboratory, Clinical Science Laboratories, UMDS Guy's Hospital, London, U.K. *Instituto de Investigaciones Biomedicas del CSIC, Madrid, Spain. **Dpt. Biochemistry, Academic Medical School Gdansk, Poland

INTRODUCTION

The human erythrocyte is unable to sustain its ATP levels in the same way as nucleated cells because it lacks key enzymes essential for the <u>de novo</u> synthetic and salvage pathways. The nucleoside adenosine (Ado) has long been considered the normal ATP precursor. However, the Ado analogue, S-adenosylmethionine (SAM) has also been found to elevate erythrocyte ATP by an unidentified mechanism which appear to exclude adenosine kinase (AK), the enzyme involved in Ado anabolism (1). Recently we reported ATP formation from 2'-deoxyadenosine (dAdo) in human erythrocytes (2) which, because of its unique metabolism, could not have occurred by any known route. A novel route, involving binding of dAdo to S-adenosylhomocysteine (SAH) hydrolase with subsequent release of adenine, which was further metabolised to ATP via adenine phosphoribosyltransferase (APRT), was proposed (2).

The present study was instigated to determine the mechanism of ATP elevation mediated by SAM and whether it involves the same route as dAdo in the erythrocyte. The existence of this route in the isolated cardiomyocytes was also investigated. The results provide evidence that in both cases this effect is mediated by adenine formation.

MATERIALS AND METHODS

Fresh heparinised venous blood collected from healthy subjects and patients with complete APRT defficiency was used immediately. Cells were prepared and incubated as described (3). The final concentration of adenine, dAdo, Ado and SAM was 1.25 mM. Inhibitors of adenosine deaminase (ADA) and adenosine kinase (AK) - 2'-deoxycoformycin (dCf) and 5'-iodotubercidin (Itu), were used at 10 µM and 20 µM respectively. Erythrocyte nucleotides were separated and quantified by high performance liquid chromatography (HPLC), (3). Isolated cardiomyocytes were also prepared and incubated as described (4) with adenine, SAM and adenosine (1.25 mM) in the presence of 5 µM erythro-9(2-hydroxy-3-nonyl)adenine (EHNA). D-ribose (1 mM) was added where indicated.

RESULTS

Studies in erythrocytes from healthy controls. Incubations with Ado alone in control cells produced a slight increment in ATP concentration (Fig. 1A), which further increased in cells incubated with Ado plus dCF to inhibit ADA. This increment was reduced, but not completely inhibited, with dCf plus Itu. No further reduction was produced with 80μM ITu. Incubations with SAM, produced a mean increment in ATP levels of approximately 40% which was not altered by incubation with dCf, or dCf plus Itu. Adenine caused a similar effect on ATP concentration to SAM (about 50%) which was not affected by inhibitors. dAdo also caused an elevation of ATP during incubation with dCf and Itu.

Studies with APRT deficient erythrocytes. The results of similar studies with SAM and Ado carried out in APRT-defficient erythrocytes are presented in Fig 1B. Incubations with adenine confirmed the complete deficiency of APRT; no increment in ATP concentration being noted under any

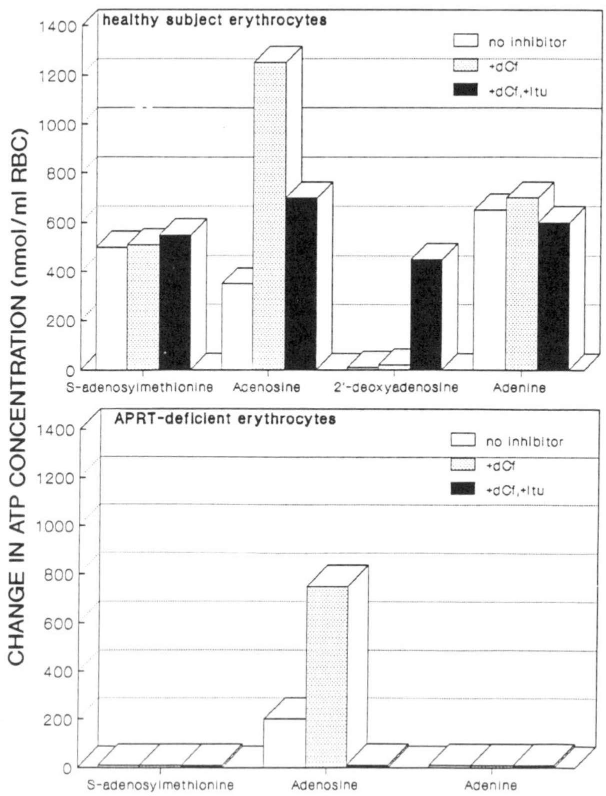

Figure 1.

Histograms showing the mean increment in ATP produced by incubating erythrocytes from healthy controls and APRT deficient subjects for 3 hours with SAM, Ado or adenine alone and with 10 μM dCF or with dCF plus 20 μM ITu. Initial ATP concentration was 1370 and 1273 μmol/l packed cells in healthy controls and in APRT deficient erythrocytes respectively and did not change during incubation without adenine compounds.

conditions. Elevation in ATP was observed only with Ado. Incubations with Ado were the only studies in these APRT deficient erythrocytes to show any increment in ATP levels. However, when using Ado together with Itu and dCf, no measurable increment in ATP was noted. Incubations with SAM likewise showed a completely different pattern. SAM, either alone or with inhibitors, produced no measurable increment in ATP levels under any conditions demonstrating conclusively the essential involvement of APRT in the SAM-induced increment in ATP levels in control cells. The virtual absence of any increment in ATP in the APRT deficient erythrocytes when using Ado with 20μM ITu plus 10μM dCF, was equally interesting. Thus some ATP must have been produced from Ado by the same APRT-dependent route demonstrated for SAM.

<u>Studies with isolated cardiomyocytes.</u> A similar increment in the adenine nucleotide pool was observed in isolated rat cardiomyocytes incubated in the presence of SAM, adenine or ado (Fig. 2), but in contrast to erythrocytes, a supplement of ribose was necessary to achieve visible changes in ATP in the case of SAM and adenine, but not with ado. This indicates clearly involvement of APRT and phosphoribose pyrophosphate in the effect observed with SAM and adenine. Initial nucleotide concentration was equal to 21, 4, and 0.7 nmol/mg cell protein for ATP, ADP and AMP respectively.

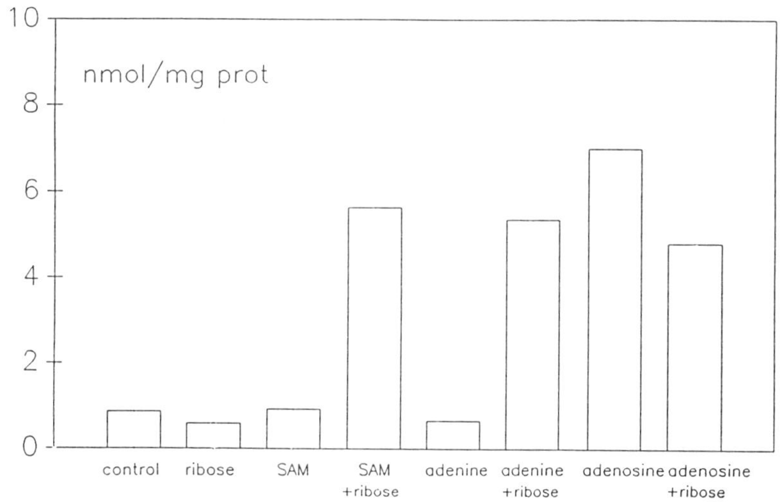

Figure 2.
Change of adenylate pool in isolated rat cardiomyocytes after 90 min of incubation in the presence of adenine derivatives. Values represent the mean from 3-7 different preparations.

DISCUSSION

SAM is a ubiquitous compound found in all living cells, participating in numerous transmethylation reactions, sulphur metabolism and polyamine synthesis (5). The only metabolic role hitherto attributed to SAM in the mature human erythrocyte has been its action as a substrate for the carboxyl methylation of membrane and cytosolic proteins (6). The studies in this paper not only confirm that SAM is even more effective than Ado in increasing erythrocyte ATP levels (1), they also demonstrate conclusively that this increment does not involve AK.

The significant observation in our study was that while the Ado-induced increment in ATP was virtually trebled when Ado catabolism was prevented with dCf and reduced significantly when AK was inhibited with Itu, neither inhibition of AK or ADA had any effect on the SAM-induced increment in ATP. Moreover, the studies in the APRT deficient erythrocytes demonstrated an absolute dependance of this SAM-induced increment in ATP on functional APRT activity. Therefore, to explain the SAM effect on ATP concentration, a mechanism other than that mediated by AK must be proposed. Our results confirm also that a flux of adenine moiety from dAdo to ATP demonstrated by us previously (2) could produce substantial changes in ATP concentration (Fig. 1). Lack of this increment without dCf could be explained by rapid deamination of dAdo. A possibile explanation of the SAM effect could be the conversion of SAM to SAH and subsequently via SAH hydrolase to adenosine which would be incorporated into ATP. However, our results contradict this possibility as the effect of SAM on ATP concentration was not attenuated by AK inhibition.

Alternatively, SAM could be metabolised by the polyamine pathway of which adenine is a metabolic by-product, or by a non-enzymatic route to MTAR, which could then be converted to adenine in the erythrocyte. Although no MTAR was detectable in the medium at any time, the possibility that some MTAR could have been generated during incubation and metabolised immediately to adenine cannot be discounted. However, this would not explain either the Ado nor dAdo results. The most likely possibility is that the high SAM concentrations used could favour accumulation of the unstable 3'-keto-adenosine intermediate in the normal catalytic cycle. Which is known to decompose rapidly with the release of adenine (2). Demonstration of similar increments in the adenylate pool in the isolated cardiomyocytes incubated with SAM suggest that proposed mechanism of adenine nucleotide synthesis could operate also in other types of cells. However, in contrast to the erythrocytes a supply of ribose was necessary as its endogenous synthesis is very low in the cardiomyocyte.

We conclude that SAM and other adenosine compounds could increase cellular ATP by a route independent of AK. We suggest also that this is mediated by a novel pathway involving APRT and SAH-hydrolase.

REFERENCES

1. N. Maeda, K. Kon, M. Sekiya, and T. Shiga, Increase of ATP level in human erythrocytes induced by S-adenosyl-L-methionine, Biochem Pharmacol . 35:425 (1986).
2. H.A. Simmonds, L.D. Fairbanks, J.A. Duley and G.S. Morris, ATP formation from deoxyadenosine in human erythrocytes: evidence for a hitherto unidentified route involving adenine and S-adenosylhomocysteine hydrolase. Biosc Rep . 9:75 (1989).
3. C. Montero, R.T. Smolenski, J.A. Duley, and H.A. Simmonds, S-adenosylmethionine increases erythrocyte ATP in vitro by a route independent of adenosine kinase, Biochem Pharmacol . 40:2617 (1990).
4. R.T. Smolenski, J. Schrader, H. de Groot, and A. Deussen, Oxygen partial pressure and free cytosolic adenosine of isolated cardiomyocytes, Am J Physiol . 260:C708 (1991).
5. G. Stramentinoli, Pharmacologic aspects of S-adenosylmethionine. Amer J Med . 83:35 (1987).
6. J.R. Barber, B.H. Morimoto, L.S. Brunauer, and S. Clarke, Metabolism of S-adenosyl-L-methionine in intact human erythrocytes, Biochim Biophys Acta . 887:361 (1986).

PRESERVATION OF NUCLEOTIDE POOL DURING HEART TRANSPLANTATION AND EVALUATION OF ADENYLATE CATABOLIC PATHWAYS IN THE HUMAN HEART

Ryszard T. Smolenski*, Amanda Suitters, David R. Lachno, and Magdi H. Yacoub

National Heart and Lung Institute, Harefield Hospital, Harefield, Middlesex, U.K. *On leave from the Department of Biochemistry, Academic Medical School, Gdansk, Poland

INTRODUCTION

Adenylate pool preservation in the donor heart during transplantation depends on efficiency of cardioprotection. In addition to diagnostic value of myocardial nucleotides and its catabolites released from the heart[1] any changes in the nucleotide pool could directly affect heart function and metabolism as adenylate interconversions are linked via ATP with high energy phosphate turnover[2] and via adenosine with the metabolic control of coronary flow, platelet aggregation, inflammatory response and the hormone action on the heart[3]. Experimental data suggest that exogenous purine supply or interaction with its endogenous metabolism could be beneficial for the heart. Detailed knowledge of adenylate metabolism in the human heart is essential if any pharmacological interventions would be undertaken under clinical conditions. Although presence of free radical generating enzyme xanthine oxidase in the human heart was the most widely discussed in the recent years, other processes including pathways and regulation of adenosine formation in the human myocardium seems to be equally important.

To address these topics we evaluated nucleotide and catabolite concentration in the donor human heart at various stages of heart or heart-lung transplantation. Purine catabolite efflux from the transplanted heart during reperfusion was also evaluated. Adenylate catabolic pathways were studied on adenosine deaminase inhibited human cardiomyocytes isolated from explanted diseased or normal heart.

METHODS

The studies were carried out during heart and heart-lung transplantations performed at Harefield Hospital. Donor heart was preserved by infusion of blood (heart-lung transplantation) or crystalloid (heart transplantation) cold potassium cardioplegia just after collection followed by cold (4°C) storage during the transport. Biopsy specimens were collected from the apex of the heart using "Tru-Cut" biopsy needles during the course of the transplantation: firstly before collection of donor organs, secondly before starting the implantation and thirdly 30 min after reperfusion. Coronary sinus and arterial blood was collected at various time after removal of the aortic clamp. Tissue and blood samples were deproteinized using perchloric acid and analysed by HPLC according to the method described[6]. Human cardiomyocytes were isolated from either right or left ventricle of the explanted diseased or normal heart[4]. The preparations used for experiments contained 30-50 % of rod shaped cells in the medium containing 1 mM Ca^{++}. Incubation conditions and the cell and medium extraction were described previously[5]. Adenosine deaminase inhibitor erythro-9-(2-hydroxy-3-nonyl) adenine (EHNA) was present in all incubations at 5 µM concentration. Nucleotide catabolism was stimulated by the addition of 10 mM deoxyglucose plus 5 µg/ml oligomycin. Perchloric acid extracts were subjected to HPLC analysis[6].

RESULTS

As may be seen from Fig. 1, ATP and the total adenine nucleotide pool (TAN) in the heart were well preserved during ischemic transport but significant depletion of TAN was demonstrated during the reperfusion. A slight metabolic changes observed during the ischemia (elevation of AMP and nucleotide catabolites concentration) were only partially reversed after reperfusion.

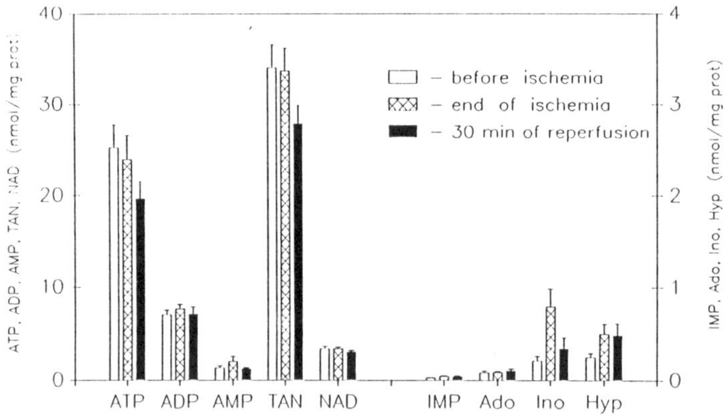

Fig.1

The influence of ischemic transport and the reperfusion after implantation on nucleotide and catabolite concentrations in the human myocardium during heart or heart-lung transplantation. Values represent the mean ±SEM of 11 series of analyses. (TAN = total adenine nucleotides).

Analysis of metabolite concentrations in coronary sinus and arterial blood indicated a massive release of inosine and hypoxanthine from the heart during reperfusion (Fig. 2). This release was significant even after 30 min of reperfusion. Transient release of adenosine was also demonstrated in the very early phase. No significant arterial-coronary sinus differences in uric acid or xanthine concentrations were observed although coronary sinus uric acid was variable in the initial phase.

The isolated human cardiomyocytes incubated under aerobic conditions in the medium containing glucose maintained initial adenine nucleotide concentration equal to 15, 5, and 0.8 nmol/mg prot for ATP, ADP and AMP respectively for 45 min of incubation. The extracellular medium contained some adenosine and inosine and a small increases were observed during the incubation (Fig. 3).

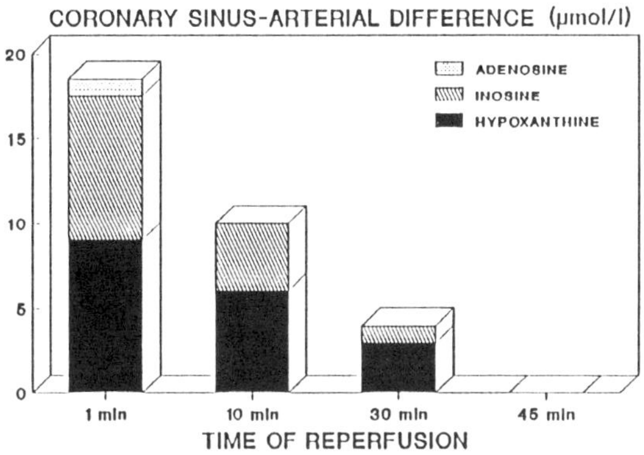

Fig.2
Adenine nucleotide catabolite release from the human heart after implantation at various time of reperfusion evaluated by measurement of arterial-coronary sinus difference. Values represent the mean of 5 series of analyses.

The presence of oligomycin and deoxyglucose caused a decrease in cellular ATP by 60% accompanied by a two-fold increase in AMP concentration after 45 min of incubation. Simultaneously, intracellular accumulation of predominantly adenosine, but also of IMP and inosine was observed (Fig. 3). Concurrently, continuous increase in adenosine and inosine concentration was observed in the medium. No accumulation of extra or intracellular hypoxanthine was demonstrated.

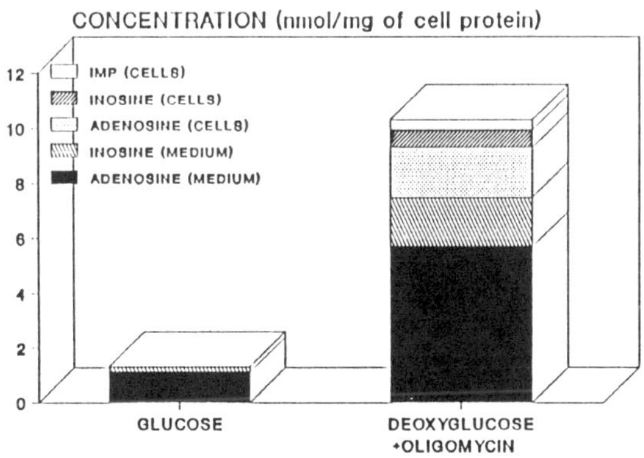

Fig.3
Intra and extracellular concentrations of adenylate catabolites in the human cardiomyocytes incubated for 45 min in the medium containing 10 mM glucose or 10 mM deoxyglucose with 5 μg/ml oligomycin. Values represent the mean from 6 preparations.

DISCUSSION

Animal studies have demonstrated nucleotide pool depletion during prolonged hypothermic ischemia imitating transport of the donor heart[7] but clinical data concerning nucleotide pool changes in the human myocardium are lacking. We have demonstrated in this study that myocardial adenine nucleotides are well preserved during ischemic transport of the donor organs but significant depletion of nucleotide pool could be observed during reperfusion (Fig. 1). Nucleotide degradation was accompanied by the release of purine catabolites from the heart (Fig.2).

Adenylate catabolism could proceed through pathway involving or bypassing formation of adenosine (via 5'-nucleotidase or AMP-deaminase). Production of this nucleoside in the human heart was demonstrated under ischemic conditions[8], (Fig. 2). However, adenosine accounted only for minimal fraction of all catabolites formed and it is impossible to estimate whether major catabolites - inosine and hypoxanthine originated from 5'nucleotidase or AMP-deaminase pathway. We addressed this question with application of isolated human cardiomyocytes incubated in the presence of adenosine deaminase inhibitor. According to the amount of adenosine accumulated in energy deprived cells (Fig. 3) we could estimate the flux through 5'nucleotidase for 70%. This is more than similar estimate in the rat cardiomyocytes[5,9]. Human myocardial cells accumulated only very little amount of IMP again contrasting rat cardiomyocytes[5,9]. Our results suggest also predominance of intracellular adenosine formation as transmembrane adenosine gradient was observed with manifold higher concentration inside the cell. An estimate of highest intracellular adenosine concentration was about 400 uM with simultaneous extracellular of about 1.3 uM. Lack of hypoxanthine accumulation suggests that as in the rat myocardial cells human cardiomyocyte activity of purine nucleoside phosphorylase is very small.

ACKNOWLEDGEMENTS

Supported in parts by the Harefield Heart Transplant Trust and the British Heart Foundation. RTS is recipient of fellowship from the Smith and Nephew Foundation.

REFERENCES

1. J.W. De Jong, Diagnosis of ischemic heart disease with AMP-catabolites, in: "Myocardial Energy Metabolism," J.W. De Jong, ed., Nijhoff Publs., Dodrecht/Boston/Lancaster (1988).
2. P.W. Achterberg: Adenine nucleotides, purine metabolism and myocardial function, in: "Myocardial Energy Metabolism," J.W. De Jong ed., Nijhoff Publs., Dodrecht/Boston/Lancaster (1988).
3. A.C. Newby, Adenosine and the concept of retaliatory metabolites, Trends in Biochem Sci. 9:42 (1984).
4. S.E. Harding, P. O'Gara, S.M. Jones, L.A. Brown, P.A. Vescovo, and P.A. Poole-Wilson, Species dependence of contraction velocity in single isolated cardiac myocytes, Cardioscience. 1:49 (1990).
5. R.T. Smolenski, J. Schrader, H. De Groot, and A. Deussen, Oxygen partial and free intracellular adenosine of isolated cardiomyocytes, Am J Physiol. 260:C708 (1991).
6. R.T. Smolenski, D.R. Lachno, S.J.M. Ledingham, and M.H. Yacoub, Determination of sixteen nucleotides, nucleosides and bases using high performance liquid chromatography and its application to the study of purine metabolism in hearts for transplantation, J Chromatogr. 527:414 (1990).
7. S.J.M. Ledingham, O. Katayama, D.R. Lachno, and M. Yacoub, Prolonged cardiac preservation: Evaluation of the University of Wisconsin preservation solution by comparison with the St. Thomas' cardioplegic solution in the rat, Circulation. 82(suppl.IV):IV.351 (1990)
8. R.T. Smolenski, A.C. Skladanowski, M. Perko, and M.M. Zydowo, Adenylate degradation product release from the human myocardium during open heart surgery, Clin Chim Acta. 182:63 (1989).
9. R.A. Altshuld, L.M. Gamelin, R.E. Kelley, M.R. Lambert, L.E. Apel, and G.P. Brierley, Degradation and resynthesis of adenine nucleotides in adult rat heart myocytes, J Biol Chem. 262:13527 (1987).

THE INFLUENCE OF ACETATE ON ADENINE NUCLEOTIDES IN RAT LIVER

Mariusz M. Zydowo, Ryszard T. Smolenski
and Julian Swierczynski

Department of Biochemistry, Academic Medical School
80-211 Gdansk, Debinki 1, Poland.

INTRODUCTION

An elevation of plasma acetate could be a consequence of the use of acetate containing fluid during hemodialysis[1] as this anion is frequently replacing bicarbonate to avoid some problems associated with pH maintenance in bicarbonate containing solutions. High plasma acetate concentration could be observed also after consumption of ethanol[2]. Metabolic conversion of the ethanol to acetate occurs in the liver with intermediate formation of acetic aldehyde.

In vivo experiments have shown that acetate is metabolised predominantly in the extrahepatic tissues[3] and therefore the influence of acetate on liver metabolism was less extensively studied. Of importance as demonstrated for instance in the heart is the fact that rapid acetate metabolism via thiokinase could lead to the depletion of ATP and the acceleration of purine catabolism[4]. This phenomenon was not studied in detail in the liver, although it was suggested that acetate increased hepatic nucleotide pool turnover[5]. The significance of this fact is that hepatic purine nucleotides could be metabolised to uric acid (in humans) and this mechanism could contribute to the hyperuricaemia associated with excessive ethanol consumption[6].

To study whether acetate could have any direct effects on hepatic nucleotides we evaluated changes in the adenine nucleotide concentration in the liver *in situ* after intraperitoneal injection of sodium acetate and in isolated rat hepatocytes incubated in acetate containing medium. We found a decrease in hepatic ATP and an increase in AMP after intraperitoneal acetate injection. Similar changes were observed in the isolated rat hepatocytes incubated in the presence of acetate, providing evidence that direct action of acetate could disturb energy equilibrium in hepatocytes which is known to be associated with accelerated nucleotide degradation.

MATERIALS AND METHODS

The experiments were performed on "Wistar" albino rats weighing 250-300 g. Sodium acetate at 1 M concentration (with pH adjusted to 7.4) was injected intraperitoneally at a dose of 20 mmol/kg body weight. Control animals received sodium chloride. After 15 min, abdominal cavity was opened under ether anaesthesia and the liver was freeze-clamped, pulverised at liquid nitrogen temperature and extracted using 0.6 M perchloric acid.

Purine and Pyrimidine Metabolism in Man VII, Part A
Edited by R.A. Harkness *et al.*, Plenum Press, New York, 1991

After protein precipitation, supernatants were neutralised and analysed for adenine nucleotide concentration using HPLC[7].

Rat hepatocytes were isolated according to the procedure of Seglen[8], involving low calcium liver perfusion followed by collagenase digestion. The incubation was started by the addition of 0.5 ml of the hepatocyte suspension to 0.5 ml of the incubation buffer containing additionally 20 mM sodium acetate, replaced by sodium chloride in controls. Incubations were terminated using 1.3 M perchloric acid and adenine nucleotide concentration was evaluated by HPLC.

RESULTS AND DISCUSSION

Plasma acetate could reach millimolar levels both after ethanol intake or during hemodialysis[1,2]. We have shown that acetate could alter energy equilibrium in hepatic cells. We have observed this *in vivo* after intraperitoneal acetate injection (Fig. 1) where a decrease in ATP and elevation in AMP concentration was demonstrated, while none of these changes were found after sodium chloride administration. No such effect was observed on NAD or NADP concentrations.

Similar effects of acetate were observed in isolated hepatocytes where a gradual decrease in ATP was associated with a sustained elevation in AMP concentration (Fig. 2). Replacement of acetate by sodium chloride abolished these changes.

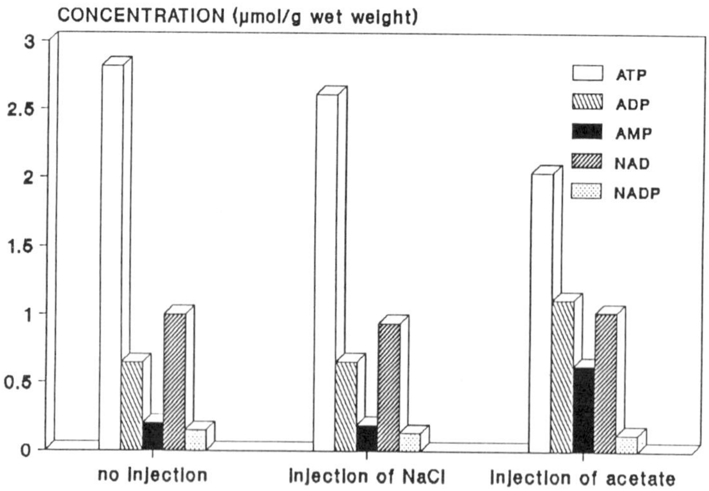

Figure 1

The influence of intraperitoneal injection of sodium acetate on hepatic adenine nucleotide concentration. Rat liver was freeze-clamped 15 min after injection of sodium acetate or sodium chloride, followed by extraction and HPLC analysis. Nucleotide concentration was also measured without injection. Values represent the mean of 4-6 determinations.

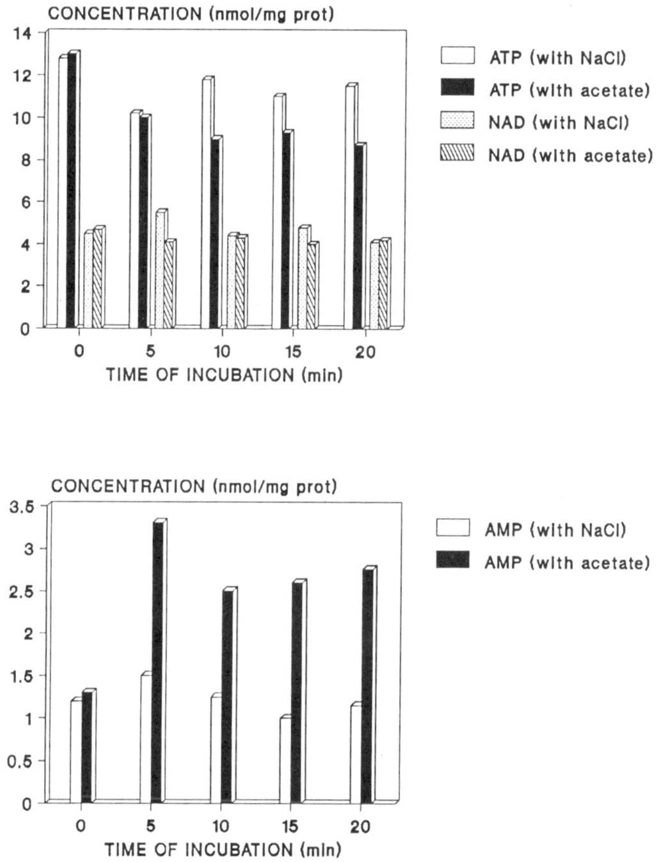

Figure 2.
Concentration of ATP, NAD and AMP in the isolated rat hepatocytes incubated at 37°C in the buffer containing additionally either 10 mM sodium acetate or sodium chloride. Values represent the mean of 5 series of incubations.

A basic aspect of this study concerns the step involving activation of acetate in the liver catalysed by thiokinase. Activity of the mitochondrial isoform of this enzyme, which is thought to be primarily involved in acetate activation, is lower in the liver than in the other tissues[10]. Also *in vivo* experiments have shown that acetate is metabolised predominantly by extrahepatic tissues[11] which again suggested that this is rather unimportant pathway in the liver. The question remained however as to whether the flux through thiokinase pathway at high acetate concentration could exceed high energy phosphate synthetic capacity in this organ causing ATP depletion and AMP elevation associated with enhanced purine nucleotide breakdown. Our results showed that the acetate could disturb the energy equilibrium in the liver, most likely via its rapid activation, and possible hepatotoxicity of acetate and involvement in uric acid overproduction must thus be taken into consideration.

ACKNOWLEDGEMENTS

These studies were supported by the Polish Committee for Scientific Research. Many helpful comments of Dr. H. Anne Simmonds are gratefully appreciated.

REFERENCES

1. G. Desh, R. Oules, C. Mion, B. Descomps, and A.C. de Paulet, Plasma acetate levels during hemodialysis, Clin Chim Acta. 85:231 (1978).
2. A. Juhlin-Dannfelt, Ethanol effects of substrate utilisation by the human brain, Scand J Clin Lab Invest. 37:443 (1977).
3. S.E. Knowles, I.G. Jarret, O.H. Fissel, and F.J. Ballard, Production and utilization of acetate in mamals, Biochem J. 142:401 (1974).
4. K.T. Kiviluoma, M. Karhunen, T. Lapinlampi, K.J. Peuhkurinen, and I.E. Hassinen, Acetate-induced changes in cardiac energy metabolism and hemodynamics in the rat, Basic Res Cardiol. 83:431 (1988).
5. J.G. Puig, and I.H. Fox, Ethanol-iduced activation of adenine nucleotide turnover. Evidence for a role of acetate.74:936 (1984).
6. J. Faller, and I.H. Fox, Ethanol-induced hyperuricemia. Evidence for increased urate production by activation of adenine nucleotide turnover, N Engl J Med. 307:1598 (1982)
7. R.T. Smolenski, A.C. Skladanowski, M. Perko, and M.M. Zydowo, Adenylate degradation product release from the human myocardium during open heart surgery, Clin Chim Acta. 182:63 (1989).
8. P.O. Seglen, Preparation of isolated rat liver cells, Meth in Cell Biol. 13:29 (1976)
9. H.R. Scholte, and P.H.E. Groot, Organ and intracellular localization of short-chain Acyl-CoA synthetases in rat and guinea-pig, Biochim Biophys Acta. 409:283 (1975).
10. C. Barth, M. Sladek, and K. Decker, The subcellular distribution of short chain fatty acyl-CoA synthetase activity in rat tissues, Biochim Biophys Acta. 248:24 (1971).

PURINE NUCLEOTIDE CATABOLISM IN RAT LIVER AFTER CASTRATION

Lucia Terzuoli, Maria Pizzichini, Anna Di Stefano,
Brunetta Porcelli, Antonella Tabucchi, Roberto Pagani

Institute of Biochemistry and Enzymology
University of Siena, Italy

INTRODUCTION

The metabolic degradation of purine nucleotides (IMP, GMP and AMP) occurs through dephosphorylation and formation of the corresponding nucleosides (inosine, guanosine and adenosine). Adenosine is further deaminated to inosine by adenosine deaminase. GMP and AMP are, respectively, reduced or deaminated to IMP which is further dephosphorylated to inosine. Inosine and guanosine are phosphorolytically split to the free bases, hypoxanthine or guanine.

Adenine and guanine - through the action of adenase and guanase - are once again trasformed, this time to hypoxanthine and xanthine: xanthine oxidase changes these bases to uric acid in man. In the rat, uric acid can be oxidized to allantoin, through the action of uricase.

Therefore, uric acid and allantoin represent the final products and are called the "key catabolic compounds of purine degradation".

Catabolism of purine nucleotide represents an alternative process for the synthesis of purine nucleotides and for regulation of their content. Purine nucleotide catabolism is usually evaluated through the determination of its "key catabolic compounds", uric acid and allantoin.

We present here a study on catabolism of purine ribonucleotides in the liver of normal and castrated rats, following the quantity of uric acid and allantoin, and their specific radioactivity, after injection of ^{14}C-formate as tracer.

In past experiments, we have demostrated that one point experiments are less precise for such evaluation, and therefore we have examined the behaviour of uric acid and allantoin killing the rats and excising their livers at different times after administration of the tracer.

MATERIALS AND METHODS

Male Albino Wistar rats, 9 weeks of age were castrated under ether, kept at 25 °C, and allowed water and normal diet ad libitum.

Control rats were kept under the same conditions and were sham-operated. After 14 days, all animals received a single dose (10 µCi/100 g b.w.) of ^{14}C-formate (54.8 mCi/mmol) intraperitoneally and were decapitated 17′-30′-60′-154′ after injection.

The livers were than rapidly excised, homogenized in 5% TCA, and centrifuged.

Purine and Pyrimidine Metabolism in Man VII, Part A
Edited by R.A. Harkness *et al.*, Plenum Press, New York, 1991

Treatment with mercuric acetate

The liver supernatant was submitted to acid hydrolysis (1 h at 100°C in 1 N H_2SO_4); after neutralization, allantoin, uric acid and free purine bases derived from nucleotides, were precipitated by addition of mercuric acetate solution (1 g of mercuric acetate and 10 g of sodium acetate in 100 ml, v/v).

After standing overnight, the flocculent precipitate was collected by centrifugation, washed with distilled water and decomposed with hydrogen sulfide (1).

After removal of mercuric sulfide, the solution was aerated and the allantoin determined by the Rimini-Schryver reaction.

Purification of allantoin and uric acid

One aliquot of the sample obtained after treatment with mercuric acetate was applied to a column (20 x 0,4 cm) of Dowex (AG 50W-X8, 200-400 mesh, cation exchange resin), equilibrated with distilled water (flow rate 12 ml/h; fractions every 10 min). Allantoin and uric acid were eluted with distilled water, purine bases with 6N HCl.

The fractions containing allantoin and uric acid were concentrated and brought to pH 2, and again submitted to chromatography on a Dowex (1-X8, 200-400 mesh, anion exchange resin) column (20 X 0,4 cm) equilibrated with distilled water; the allantoin was eluted with distilled water, uric acid with 0.01 N HCl and unidentified compounds with 0.06 N HCl (2).

The fractions containing uric acid and allantoin were checked for purity (UV spectra: constancy of specific radioactivity in the single fractions), then pooled, and examined both for total content and specific radioactivity of uric acid and allantoin.

Uric acid was determined according to Praetorius and Poulsen (3), allantoin through the Rimini-Schryver reaction (4).

The concentration of the base was determined by UV absorption: radioactivity was determined on a Nuclear Chicago Scintillation Counter, using Instagel as scintillation cocktail.

All reagents were from the highest purity commercially available, and obtained either from Merck or Fluka Company. Standard compounds were from Sigma. Resins were from Bio Rad. ^{14}C-formate (54.8 mCi/mmol) was an Amersham product.

Analysis of data were based on the concentration and specific radioactivity of uric acid and allantoin. Analysis of variance (only p values < 0.01 were considered significant) was carried out.

RESULTS AND DISCUSSION

Table 1 reports all data concerning specific radioactivity of uric acid in normal and castrated rats, at four different times after injection of ^{14}C-formate.

Table 2 reports total data concerning specific radioactivity of allantoin in normal and castrated rats, under the same conditions.

Our data should be attentively studied.

We consider the following result expecially interesting:

1) The specific radioactivity of allantoin was higher than that of uric acid, under all experimental conditions; these results cannot be explained on the basis of a different pool, since the concentration (mg/g) of uric acid and allantoin in the liver were more or less identical.

Reducing the problem to the uricase reaction and the allantoin pool in the liver is an excessive simplification: many factors and many considerations must be verified: a) uric acid and allantoin found in the liver also originate in the other organs and what we call the "pool" of the liver, represents a mixed situation; b) the turnover of both compounds

Table 1: Radioactivity content (dpm/g) and specific radioactivity (dpm/g) of uric acid in the liver of normal and castrated rats, at different times after ^{14}C-formate injection.

	NORMAL		CASTRATED	
Time	dpm/g	S.A.	dpm/g	S.A.
17´	488 ± 113	6030 ± 1034	668 ± 101	7430 ± 664
30´	638 ± 122	6118 ± 256	750 ± 74	10875 ± 2987*
60´	722 ± 121	7072 ± 1116	526 ± 69	5233 ± 735
154´	837 ± 156	9808 ± 1116	1073 ± 131	7566 ± 1246

The values represent the mean ± S.E. of four different rats.
* $p < 0.01$ referred to normal.

Table 2: Radioactivity content (dpm/g) and specific radioactivity (dpm/μmol) of allantoin in the liver of normal and castrated rats, at different times after ^{14}C-formate injection.

	NORMAL		CASTRATED	
Time	dpm/g	S.A.	dpm/g	S.A.
17´	3457 ± 212	23758 ± 2683	4688 ± 1200	30309 ± 4509
30´	4734 ± 257	28204 ± 4998	4741 ± 556	31012 ± 1416
60´	3524 ± 339	19882 ± 296	1980 ± 242*	12805 ± 1291*
154´	1235 ± 221	9789 ± 444	785 ± 125	8221 ± 868

The values represent the mean ± S.E. of four different rats.
* $p < 0.01$ referred to normal.

and their clearance or uptake for the liver could differ, c) some unknown side reactions in the sequences nucleotides --> uric acid --> allantoin, cannot be completely excluded.

Further experimentation on perfused rat liver, in culture cells, using various precursors under different conditions, is necessary: suitable scheme and mathematical expressions for the determination of the turnover and the clearance should be elaborated.

This quite unexpected result, which is difficult to interpret on the basis of actual knowledge, and which is only the first problem in any subsequent study of purine catabolism must be further investigated.

2) Uric acid concentration did not vary in castrated rats, while the concentration of allantoin tended to be lower. Such data are not reported for the sake of brevity.

Specific radioactivity of uric acid and allantoin in castrated rat liver showed a biphasic behaviour; at short times there was a slight increase, while at longer times there was a decrease. These data did not permit us to estimate accurately the rate of purine catabolism after castration but we think that this effect will be clearer when all values concerning the synthesis of purine nucleotides and nucleic acids will be analysed toghether with those relating to purine catabolism, using a mathematical model able to collect all metabolic pathways above reported. This model is almost prepared and the data will be studied using it.

REFERENCES

1) Korn E.D., in "Methods in Enzymology", Colowick S.P. and Kaplan N.O. eds, Academic Press Inc. Publishers, New York, 4, 631, 1957.
2) Di Stefano A., Pizzichini M., Resconi G., Marinello E.: Determination of allantoin in mammalian tissues. Adv. Exp. Med. Biol. 253B; 511 (1989).
3) Praetorius E., Poulsen H.: Enzymatic determination of uric acid. Scand. J. Clin. Lab. Inv. 5; 273 (1953).
4) Young E., Conway C.F.: On the estimation of allantoin by the Rimini-Schryver reaction. J. Biol. Chem. 142; 839 (1942).

ADENINE NUCLEOTIDE METABOLISM IN LIVER ISCHEMIA: EFFECT OF ALLOPURINOL

Felícitas A. Mateos, Mª. Victoria Díaz, Juan G. Puig, Teresa H. Ramos, Manuel L. Jiménez, Emilio Díaz and Emilio Burgos

Divisions of Internal Medicine, Clinical Biochemistry and Pathology
"La Paz" Hospital, Universidad Autónoma, Madrid, Spain

INTRODUCTION

Ischemia is accompanied by a decrease in ATP synthesis and an increase in ATP degradation products (1). Reperfusion provides molecular oxygen and the synthesis of purine bases and uric acid by xanthine oxidase generates free-oxygen radicals (2,3). Ischemia-reperfusion provokes severe tissue injury that has been related both to cellular ATP depletion and to the damaging effect of free radicals (3,4-6). Allopurinol, an inhibitor of xanthine oxidase, has been used to prevent tissue damage in ischemia-reperfusion situations (7-11). Depletion of high energy adenine nucleotides (ATP, ADP, AMP) cellular stores has been demonstrated in liver ischemia-reperfusion (8-10). This is followed by a rise of hypoxanthine plasma levels (11), an increase in serum alanine aminotransferase (ALAT) and aspartate aminotransferase (ASAT), and poor hepatic function (8,11). Pretreatment with allopurinol has been shown to diminish the deterioration of hepatic function caused by ischemia-reperfusion (8,9). The mechanism by which allopurinol exerts a protective effect in ischemia-reperfusion injury may be multifactorial and enhancement of ATP synthesis through hypoxanthine reutilization is a potential mechanism.

In this work we investigated the effect of allopurinol administration on intracellular liver adenine nucleotide concentrations, serum ALAT and ASAT levels, and hepatic structure, in dogs submitted to liver ischemia-reperfusion.

METHODS

Under general anesthesia a partial hepatic warm ischemia was performed on 18 dogs weighing 10-23 kg. Hepatic artery and the left branch of the portal vein were clamped for 30 minutes. During the operation and in the immediate post operative period, infusions of 0.9% saline solution were given intravenously to replace blood loss and to maintain fluid balance. Thirty minutes prior and during the ischemic period, ten dogs (control group) received a solution of 0.1N NaOH (3.3 ml/Kg weight). Eight dogs received the same solution with allopurinol (50 mg/Kg weight) (allopurinol group). Liver biopsies of ischemic tissue, blood and urine samples were taken before clamping, immediately after removing the vascular clamp and at 30, 60, 90 and 150 minutes of blood reflow. Hepatic tissue (40-60 mg) was immediately frozen (in less than 5 sec.) and stored in liquid nitrogen. The frozen samples were homogeneized in 300 μl of 3M perchloric acid and neutralized with 800 μl of 2M potassium bicarbonate. The supernatant obtained after centrifugation (1.900 g, 10 min at 4°C) was stored at -20ºC for later analysis of adenine nucleotides. Biopsy specimens were examined with an electronic microscope. ATP, ADP and AMP were assayed by high performance liquid chromatography (HPLC) (12). Total adenine nucleotides (TAN) were the sum of ATP, ADP and AMP. Serum ALAT and ASAT, urinary uric acid, and creatinine concentrations were measured in an multichannel autoanalyzer (Hitachi, 737. Boehringer Mannheim). Urinary hypoxanthine and xanthine concentrations were analyzed by HPLC (13).

RESULTS

Purine Metabolism

Table I shows modifications in liver ATP and TAN induced by ischemia-reperfusion in the control group and in the allopurinol group. A dramatic mean decrease of ATP and TAN contents (48% and 29%, respectively) at the end of the ischemic period was observed in the control group. During reperfusion, ATP an TAN gradually increased and at 150 min after the initiation of blood reflow levels were similar to baseline concentrations. The allopurinol group showed a moderate decrease of ATP following the ischemic period (mean decrease, 18% with respect to baseline values). TAN liver content in the allopurinol group did not change significantly.

Table 1. Adenosine triphosphate (ATP) and total adenine nucleotides (TAN) expressed as μg/g wet weight in hepatic tissue (mean±SD).

	ATP		TAN	
	Control	Allopurinol	Control	Allopurinol
Baseline	1.91±0.24	2.19±0.22	2.58±0.40	2.99±0.25
Ischemia	0.99±0.46[a]	1.79±0.28[a]	1.83±0.55[a]	3.02±0.41
Reperfusion				
30 min	1.48±0.30[a]	2.00±0.59	2.08±0.36[a]	2.90±0.68
60 min	1.61±0.27[a]	1.88±0.51	2.37±0.33	2.90±0.68
90 min	1.75±0.28	1.97±0.35	2.47±0.38	2.95±0.55
150 min	1.82±0.33	2.08±0.48	2.51±0.46	2.98±0.66

[a] P<0.005 vs. basaline values.

Ischemia-reperfusion elicited an increase of 3 to 9-fold in urinary hypoxanthine, xanthine and uric acid in the control group. The allopurinol group showed measurable oxypurinol plasma levels, a remarkable increase in hypoxanthine and xanthine excretion rates and a reduced urinary uric acid concentration, all of which reflects effective inhibition of xanthine oxidase by allopurinol.

Hepatic Damage

Serum ALAT and ASAT levels increased in the control group 7 and 4-fold, respectively, whereas in the allopurinol group these increases were both 3-fold ($P<0.05$). Modifications of hepatic structure are shown in Figure 1.

DISCUSION

This study shows that allopurinol administration in dogs submitted to ischemia-reperfusion is associated with: (a) a markedly lower depletion of high energy adenine nucleotides, (b) a reduced elevation of serum ASAT and ALAT, and (c) minor morphological liver changes. In addition, the simultaneous increase in urinary oxypurine excretion in the control dogs indicates an enhanced adenine nucleotide degradation which was prevented in dogs treated with allopurinol.

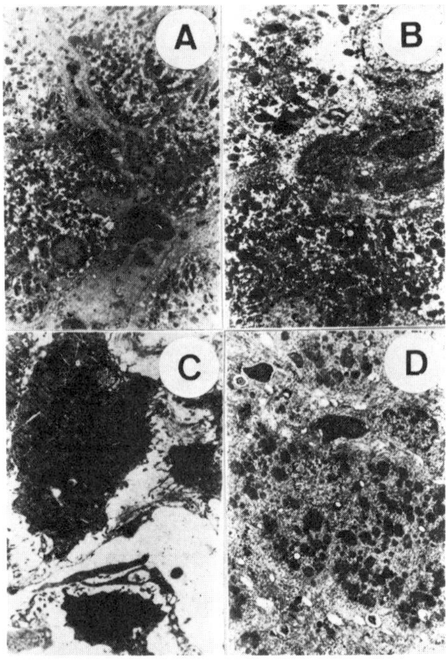

Figure 1. Hepatic structure in dogs submitted to liver ischemia-reperfusion. At the end of the ischemic period control dogs receiving saline infusion showed vacuolar degeneration, hydropic mitochondries and secondary lysosomes (x4,600) (A), whereas dogs receiving allopurinol showed no changes with respect to the baseline hepatic structure (x5,700) (B). After 150 minutes of blood reflow the hepatic structure in control dogs was almost lost with great vacuolar degeneration, edema and enlarged Disse´s spaces (x6,800) (C), whereas liver changes in dogs treated with allopurinol were minimal (x5,700) (D).

The beneficial effect of allopurinol administration in ischemia-reperfusion can be explained by a number of mechanisms. Allopurinol and oxipurinol are scavengers of the highly reactive hydroxyl radicals (14) and limit superoxide generation through xanthine oxidase inhibition (2,3). Both compounds may be converted to their corresponding ribonucleotides, and allopurinol ribonucleotide causes a 50% inhibition of 5'-nucleotidase

(15,16). This might prevent inosine and adenosine monophosphate dephosphorilitation thereby facilitating ATP resynthesis. Furthermore, inhibition of xanthine oxidase by allopurinol elicits an increased availability of hypoxanthine that may be salvaged for purine nucleotides synthesis (17). In fact, we have previously demonstrated that hypoxanthine reutilization can be enhanced by increased hypoxanthine concentrations as it occurs in hereditary xanthinuria (13). Our study showing that allopurinol attenuates adenine nucleotide depletion following experimental hepatic ischemia-reperfusion provides biochemical support to the observation that perfusion fluids containing this drug, such as the University of Winsconsin fluid, improve the results of liver transplantation.

ACKNOWLEDGEMENTS

We are indebted to Dr. Enrique de Miguel and Mrs. Mª Paz Canencia for valuable technical assistance and to Erik Lundin for help in the preparation of the manuscript. This work was supported by grants from Caja de Madrid and Fondo de Investigaciones Sanitarias de la Seguridad Social (FISS, 90/0575), Spain.

REFERENCES

1. I.H. Fox. Adenosine triphosphate degradation in specific disease. J Lab Clin Med 106:101-110 (1985).
2. J.M. McCord. Oxygen-derived free radicals in post-ischemic tissue injury. N Eng J Med 312:159-163 (1985).
3. D.N. Granger, M.E. Hölwarth and D.A. Parks. Ischemia-reperfusion injury: role of oxygen-derived free radicals. Acta Physiol Scand Suppl 548:47-63 (1986).
4. W. Kamike, F. Watanabe, F. Hashimoto et al. Changes in cellular levels of ATP and its catabolites in ischemic rat liver. J Biochem 91: 1349-1356 (1982).
5. S. Marubayashi, M Takenata, K Dohi et al. Adenine nucleotide metabolism during hepatic ischemia and subsequent blood reflow periods and its relation to organ viability. Transplantation 30:294-296 (1980).
6. C.E. Cross. Oxygen radicals and human disease. Ann Intern Med 107:526-545 (1987).
7. R.J. Nauta, E. Tsimoyiannis, M.Uribe, et al. Oxygen-derived free radicals in hepatic ischemia and reperfusion injury in the rat. Surg Gynecol Obstet 171:120-105 (1990).
8. G. Nordström, S. Torstem and P. Hasselgren. Beneficial effect of allopurinol in liver ischemia. Surgery 97:679-683 (1985).
9. W. Kamike, M. Burdelski, G.Steinhof, et al. Adenine nucleotide metabolism and its relation to organ viability in human liver transplantation. Transplantation 45:138-143 (1986).
10. W. Karwinski, A. Drange, M. Farstad, et al. Sixty min normothermic liver ischemia in rats: allopurinol improves energy status and bile flow during reperfusion. Eur Surg Res 22:27-33 (1990).
11. D. Adkinson, M.E. Höllwarth, J.N. Benoit et al. Role of free radicals in ischemia-reperfusion injury to liver. Acta Physiol Scand Suppl 548:101-107 (1986).
12. F.A. Mateos, J.G. Puig, T.H. Ramos, et al. Erythrocyte ATP (iATP) as an indicator of neonatal hypoxia. Adv Exp Med Biol 253A:345-352 (1989).
13. F.A. Mateos, J.G. Puig, M.L. Jimenez et al. Hereditary xanthinuria: Evidence for enhanced hypoxanthine salvage. J Clin Invest 79:847-852 (1987).
14. C.M. Arroyo, J.H. Kramer, B.F. Dickens et al. Identification of free radicals in myocardial ischemia-reperfusion by sppin with nitrone DMPO. FEBS Lett. 221:101-104 (1987).
15. D.A. Parks and D.N. Granger. Ischemia-induced vascular changes: role of xanthine oxidase and hydroxyl-radicals. Am J Physiol 245:G285-289 (1983).
16. M.S. Paller, J.R. Hoidal and T.F. Ferris. Oxygen free radicals in ischemic acute renal failure in the rat. J Clin Invest 74:1156-1164 (1984).
17. S. Akizuki, S.Yoshida, D.E. Chambers, et al. Infarct size limitation by the xanthine oxidase inhibitor, allopurinol, in closed-chest dogs with small infarcts. Cardiovasc Res 19:686-692 (1985).

EFFECTS OF PULSED ELECTROMAGNETIC FIELDS ON THE ADENINE NUCLEOTIDE POOL AND ENERGY CHARGE IN CELLS IN CULTURE.

Conte A., ° Petrini M., *Zaniol P., Ronca G.

Institute of Biological Chemistry, Pisa, °Clinica Medica I, Pisa, *Cattedra di Fisica Medica, Modena, Italy

INTRODUCTION

Agreement on the effects of low frequency pulsed electromagnetic fields (PEMFs) on different biological targets has not been reached suggesting that different basic parameters (waveform, frequency, intensity of magnetic and induced electric fields) may have different or even opposite actions. The interest for this kind of energy has greatly increased in recent years in view of its therapeutic and diagnostic applications and its possible adverse effects.

We have investigated the effects of PEMFs with different parameters on human lymphocytes stimulated and not stimulated by lectins. Contradictory results on lymphocyte blastogenesis have in fact been reported (1,2). The main site of PEMFs action is considered to be the cell plasma membrane and in particular at the level of the transduction of signals. We have investigated the activity of different PEMFs on energy charge (E.C.) and adenine nucleotide pool during anaerobic stress with the aim to see if they have metabolic targets inside the cells.

MATERIALS AND METHODS

Two PEMF generating apparatuses have been utilized. One experimental apparatus gives a magnetic field with an almost symmetric triangular shape, at three different frequencies of 50, 100 and 500 Hz, with magnetic strengths of 100, 40 and 10 Gauss respectively. The resulting induced electric fields for the three frequencies have an almost symmetric rectangular shape, with peak values of about +2 mV and -2 mV. Furthermore at 50 Hz the apparatus gives symmetric triangular magnetic fields with three different intensities: 10, 40 and 100 Gauss. The induced electric field intensities are ±0.25, ±1.0 and ±2.0 mV respectively. Oscilloscopic traces of the magnetic and the induced electric fields are reported in fig.1.

The second apparatus gives an asymmetrical triangular magnetic field pulsing with a frequency of 75 Hz. The strength is 25 Gauss. The duty cycle considering only the rising stage of the pulsed magnetic field is about 10%. The induced electric field has a short (1.3 msec) positive peak of 4 mV corresponding to the rising stage of the magnetic field and a longer negative peak of about 9 msec with a lower intensity (fig.1).

The magnetic and induced electric fields were determined with a Hall probe and a coil probe made with 50 turns of copper wire respectively.

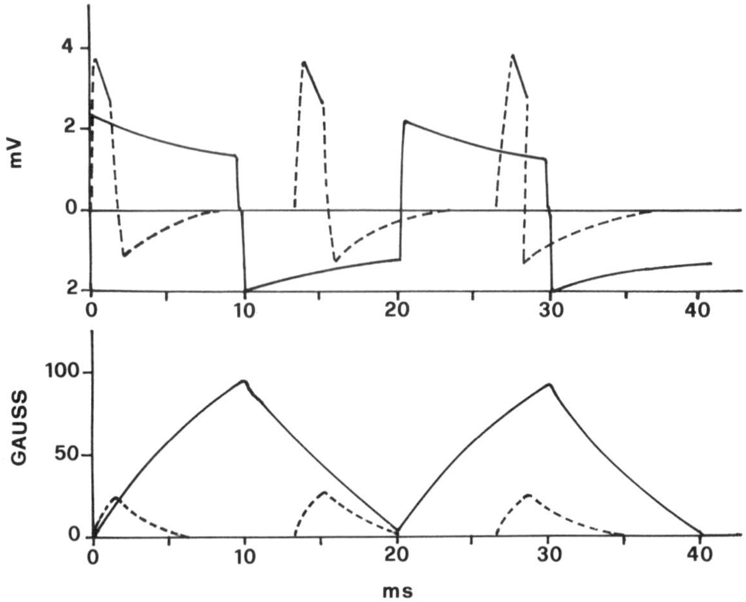

Figure 1. Traces of the pulsed fields generated by the apparatus. Continuous lines: 50 Hz 100 Gauss, symmetrical field; discontinuous lines: 75Hz 25 Gauss, asymmetrical field. Upper traces: induced electric fields; lower traces: magnetic fields.

Purified human lymphocytes were incubated for 48 h at 37°C in 5% CO_2 in microwell plates with or without an optimal concentration of phytohemagglutinin (PHA) in the presence and in the absence of PEMFs. ^{3}H-thymidine (2µci) was added and after 24 h of incubation the incorporated radioactivity was determined. Cells of line U937 grown in RPMI 1640 medium with 20% fetal calf serum and 2 mM glutamine were washed with PBS containing 1 mM glucose. The cells were then centrifuged to obtain a pellet of about $5x10^6$ cells and incubated in air in the presence and in absence of PEMFs (3). At zero time and after incubation the medium was removed and the pellet was extracted as reported by Au et al.(4). Adenine nucleotides were determined by HPLC according to Sellevold et al. (5).

RESULTS

In table 1 the effect of PEMFs on the spontaneous proliferation of lymphocytes is reported. The 50 Hz 100 Gauss field significantly inhibits cell proliferation. The other magnetic fields tested with the same waveform and frequency but with lower induced electric fields (50 Hz 10 Gauss and 50 HZ 40 Gauss) or with the same induced electric field but with higher frequencies (100 Hz 40 Gauss and 500 Hz 10 Gauss) were without effect. With PHA stimulated lymphocytes (table 2) the 50 Hz 40 Gauss and 1mV induced electric field is also effective in inhibiting cell proliferation. However its effect is lower than that obtained with the stronger field (50 Hz 100 Gauss 2mV induced electric field).
From these data it appears that the frequencies and the intensities of the magnetic and induced electric fields are important parameters in determining the antiproliferative effect.

Table 1. Effect of PEMFs on ^{3}H-thymidine incorporation in human lymphocytes

Magnetic field shape	Strength (Gauss)	Frequency (Hz) 0	50	100	500	75
		dpm (±SD)x10^2				
Triangular symmetric	0	18±10(a)				
	10		21±10	-	20±9	-
	40		19± 8	22±11	-	-
	100		6± 4*	-	-	-
Triangular asymmetric	25		-	-	-	21±10

n=6 (in triplicate) * p<.05 vs (a)

Table 2. Effect of PEMFs on ^{3}H-thymidine incorporation in PHA stimulated lymphocytes

Magnetic field shape	Strength (Gauss)	Frequency (Hz) 0	50	100	500	75
		dpm (±SD)x10^3				
Triangular symmetric	0	84±33(a)				
	10		81±28	-	82±25	-
	40		37±19*	78±32	-	-
	100		18± 1°	-	-	-
Triangular asymmetric	25		-	-	-	126±45*

n=6 (in triplicate) * p<.05 ° p<.01 vs (a)

The 75 Hz 25 Gauss field shows a low but significant proliferative activity on PHA stimulated lymphocytes (table 2). The main differences between the fields produced by the two apparatuses are the symmetry of the magnetic field and of the induced electric field and the duty cycle; these two differences probably have the major responsibility in determining the opposite actions of the fields of the two apparatuses. In fact the other parameters (intensity and frequency) are comparable.
The viability of cells is completely maintained with the various fields. We have measured the expression of interleukine 2 receptors and we have observed that with the 50 Hz 100 Gauss symmetric magnetic field (that we call SM field) the expression is strongly decreased (4±2% cells with expressed receptors) while with the 75 Hz 25 Gauss asymmetric magnetic field (that we call AM field) it is increased (65±11% cells) compared to untreated cells (53±8% cells). Since opposite effects of the field have been observed in PHA stimulated cells in which an external signal is transduced we have investigated the effect of SM and AM fields in cells under anaerobic conditions in which intracellular regulations of energy and adenine nucleotide metabolism are involved (3). Since in the U937 cell pellets anaerobic conditions developed slowly ATP concentration, adenine nucleotide pool and E.C. slowly decrease (table 3). In the SM field the cells maintain higher ATP, adenine nucleotide pool and E.C. values while in the AM field significantly lower values are observed. In this experimental model also the two fields showed opposite effects.

Table 3. Adenine nucleotide pool and E.C. variations during anaerobic stress

	Incubation time (h)				
	0	1	2	4	6
[ATP] (a)	3.1±.6	2.5±.4	2.0±.4	1.3±.3	1.1±.2*
Σ[AXP] (a)	3.8±.7	2.9±.6	2.6±.5	1.7±.4	1.5±.3*
E.C.	.91±.01	.90±.01	.87±.01	.83±.02	.78±.03*

(a) nmol x 10^6 cells ± SD n=8 * $p<.001$ vs 0 time
The other significant values are not reported

Table 4. Adenine nucleotide pool and E.C. variations during anaerobic stress in the presence of PEMFs

	No field		SM field		AM field	
Incubation time (h)	1	6	1	6	1	6
[ATP](a)	2.5±.4	1.1±.2	2.6±.5	1.3±.1*	2.4±.3	.8±.2*°
Σ[AXP](a)	2.9±.6	1.5±.3	2.9±.4	1.7±.3	2.8±.5	1.4±.2°
E.C.	.90±.01	.78±.03	.91±.01	.82±.03*	.89±.02	.71±.02*°

(a) nmol x 10^6 cells ± SD n=8 * $p<.05$ vs No field °$p<.05$ vs SM field

Although the molecular cell targets of PEMFs remain to be identified our results show that PEMFs have opposite effects on the same experimental models depending on waveform and duty cycle while the field strength determines the intensity of the biological effect. The observation that the two fields have an opposite effect also on intracelluar regulated metabolic processes suggests that cell plasma membrane is not the unique target of PEMFs. We have two well defined PEMFs with opposite effects to investigate the structural and molecular sites of action of this kind of energy.

REFERENCES

1) A.Cossarizza, D.Monti, F.Bersani, M.Cantini, R.Cadossi, A.Sacchi, and C.Franceschi, Extremely low frequency pulsed electromagnetic fields increase cell proliferation in lymphocytes from young and aged subjects, B.B.R.C.,160:692-698 (1989)
2) P.Conti, G.E.Gigante, M.G.Cifone, E.Alesse, G.Ianni, M.Reale, and P.U. Angeletti, Reduced mitogenic stimulation of human lymphocytes by extre mely low frequency electromagnetic fields, FEBS, 162:156-160 (1983)
3) S.S.Matsumoto, K.O.Raivio, and J.E.Seegmiller, Adenine nucleotide degradation during energy depletion in human lymphoblasts, J.Biol.Chem., 254:8956-8962 (1979)
4) J.L.Au, M.Su, and M.G.Wientjens, Extraction of intracellular nucleo-sides and nucleotides with acetonitrile, Clin.Chem., 35:48-51 (1989)
5) O.F.M.Sellevold, P.Jynge, and K.Aarstad, High performance liquid chromatography: a rapid isocratic method for determination of creatine compounds and adenine nucleotides in myocardial tissue, J.Mol.Cell Cardiol., 18:517-527 (1986)

MECHANISMS ACCOUNTING FOR CHANGES OF ADENINE NUCLEOTIDE CONTENT IN MITOCHONDRIA AT ISCHEMIA

Wolfgang Henke
Mathias Ziegler*
Wolfgang Dubiel*
Klaus Jung

Urological Clinic
Institute of
Biochemistry*
Humboldt-University
Berlin, Germany

CHANGES OF ADENINE NUCLEOTIDES AT ISCHEMIA

Renal mitochondria of rat contain about 30-40% of total cellular content of adenine nucleotides (AdN). Ischemia decreases oxidative phosphorylation and AdN content of renal mitochondria[1]. Restoration of blood flow after limiting ischemia of 60 min leads to a recovery of AdN content in the course of 20 days (Fig. 1).

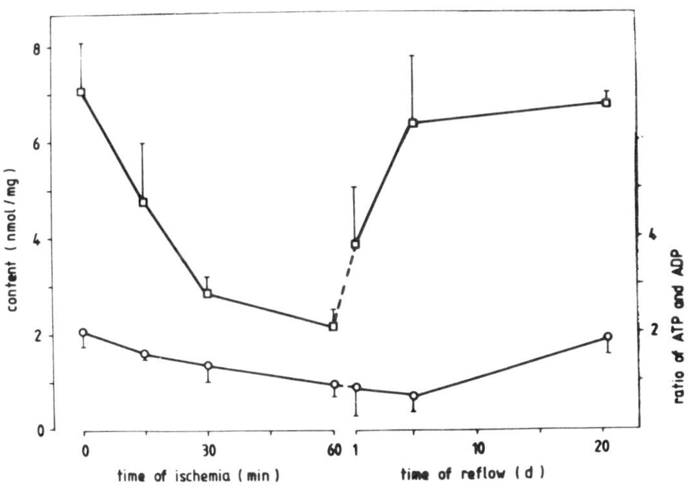

Fig. 1. Ischemic decrease of AdN content in renal mitochondria and its recovery caused by blood reflow after 60 min (by permission of BIOCHIMICA BIOPHYSICA ACTA).

Purine and Pyrimidine Metabolism in Man VII, Part A
Edited by R.A. Harkness *et al.*, Plenum Press, New York, 1991

MECHANISMS REGULATING MATRIX ADENINE NUCLEOTIDE CONTENT

De-novo Synthesis and Salvage Pathway

IMP involved in the cytosolic synthesis of AdN has not yet been detected in mitochondria. Therefore, an intramitochondrial de-novo synthesis of AdN has been ruled out[2]. A salvage pathway of precursors has not been found[3].

Catabolism of Matrix Adenine Nucleotide

The ischemia-induced decrease of AdN has been ascribed to the intramitochondrial degradation of AMP[4]. But adenosine formation by rat heart mitochondria[5] has been attributed to an efflux of AdN from mitochondria followed by a subsequent dephosphorylation of AMP in the extramitochndrial space[6,7].

Liver mitochondria form both adenosine and inosine from endogenous AdN. If mitochondria were separated from the medium, adenosine was mainly accumulated in mitochondria whereas inosine was preferentially localized

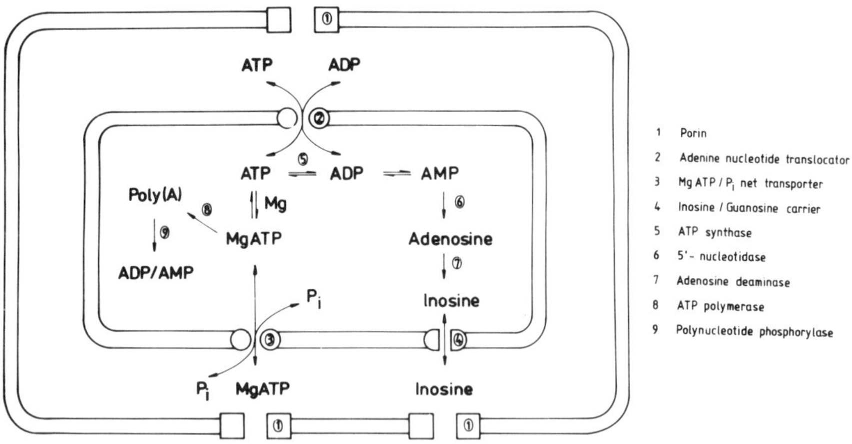

Fig. 2. Mechanisms suggested to account for changes of mitochondrial adenine nucleotides.

in medium[8]. A concentration gradient between intra- and extramitochondrial adenosine of two orders of magnitude has been estimated, a value similar to that observed in heart mitochondria[9]. It has been postulated that adenosine, formed from endogenous AdN flown out and then dephosphorylated by external ATPase and 5'-nucleotidase activities, is transported into mitochondria and accumulated there by an unknown adenosine-binding protein[9]. However, there are data showing the inability of heart mitochondria to take up adenosine[7]. Liver mitochondria are unable to form inosine from external adenosine. Therefore, it has been concluded that intramitochondrial adenosine is deaminated to inosine which may permeate through the mitochondrial membrane[8]. Inosine and guanosine, respectively, are taken up by liver mitochondria[10]. For liver mitochondria, an intramitochondrial catabolism of AdN, comprising a 5'-nucleotidase and adenosine deaminase, as well as a purine nucleoside carrier has been suggested[8] (Fig. 2).

Net Transport of Adenine Nucleotides

Adenine Nucleotide Translocator. The AdN translocator performs the 1:1 exchange of Mg-free ADP/ATP between cytosol and mitochondria. It has been shown for heart mitochondria that phosphate (P_i) stimulates an carboxyatractyloside-sensitive efflux of AdN. As P_i accumulates during ischemia, it has been proposed that the P_i-stimulated efflux of AdN via the AdN translocator may cause the decrease of matrix AdN in ischemic heart mitochondria[11].

Mitochondria lose AdN by incubating with pyrophosphate. Pyrophosphate exchanges with ADP/ATP via the AdN translocator substituting AdN[12]. Because Ca^{2+} increases the matrix content of pyrophosphate it has been proposed that this mechanism is responsible for a Ca^{2+}-regulated elevation of AdN content[13]. However, this proposal has been questionated since the pyrophosphate affinity seems to be to low[14].

MgATP/phosphate Carrier. This carrier is well established in liver mitochondria[15]. The translocation of the substrates, MgATP and MgADP, is dependent on P_i, stimulated by Ca^{2+}, modified by the membrane potential and carboxyatractyloside-insensitve[16]. By using heart mitochondria, net uptake of AMP, ADP and ATP, respectively, could not been detected[17]. Recently, it was shown that renal mitochondria also perform a net uptake of AdN[18]. The increase of AdN content observed after restoration of blood flow (Fig. 1) may be explained by this mechanism.

Storage Form of Adenine Nucleotides

Several years ago, Mowbray's group has postulated a storage form of AdN in rat heart[19]. Recently, it has been reported that rat heart mitochondria contain an acid-insoluble compound, persumedly a polyribonucleotide. This poly(A) might function as a storage form of AdN and contributes to the regulation of the matrix AdN by a cyclic process of synthesis and degradation, as supposed by an ATP polymerase and polynucleotide phosphorylase[20]. Using kidney mitochondria[21] or slow-twitch red muscle[22], an imbalance between AdN degrated and catabolites formed has been observed. As suggested, this imbalance might also be indicative for a storage form of AdN.

REFERENCES

1. W. Henke and K. Jung, Ischemia decreases the content of the adenine nucleotide translocator in mitochondria of rat kidney, *Biochim. Biophys. Acta* 1056:71 (1991).
2. H.-D. Söling, Species dependent compartmentation of guanine nucleotide metabolism in the vertebrate liver cell, *in*: "Metabolic compartmentation", p. 123, H. Sies, ed., Academic Press, London (1982).
3. M. Ziegler, W. Dubiel, W. Henke, G. Gerber, A. M. Pimenov, Yu. V. Tikhonov, R. T. Toguzov, Application of ion-pair high-performance liquid chromatography with radioisotope detection to in vitro studies of nucleoside metabolism in mitochondria, *J. Chromatogr.* 563:172 (1991).
4. K. Watanabe, W. Kamiike, T. Nishimura, T. Hashimoto and T. Tagawa, Decrease in mitochondrial levels of adenine nucleotides and concomitant mitochondrial dysfunction in ischemic rat liver, *J. Biochem.* 94:493 (1983).
5. R. D. Bukoski, H. V. Sparks, Jr. and L. M. Mela, Rat heart release adenosine, *Biochem. Biophys. Res. Commun.* 113:990 (1983).
6. G. K. Asimakis, D. E. Wilson and V. R. Conti, Release of AMP and

adenosine from rat heart mitochondria, *Life Sci.* 37:2373 (1985).
7. R. D. Bukoski, H. V. Sparks, Jr. and L. M. Mela-Riker, Mechanism of adenosine production by isolated rat heart mitochondria, *Biochim. Biophys. Acta* 884:25 (1986).
8. M. Ziegler, W. Dubiel, A. M. Pimenov, Yu. V. Tikhonov, R. T. Toguzov, W. Henke and G. Gerber, The catabolism of endogenous adenine nucleotides in rat liver mitochondria, *Molec. Cell. Biochem.* 93:7 (1990).
9. K. T. Kiviluoma, K. J. Peuhkurinen and I. E. Hassinen, Adenine nucleotide transport and adenosine production in isolated rat heart mitochondria during acetate metabolism, *Biochim. Biophys. Acta* 974:274 (1989).
10. M. Ziegler, W. Dubiel, A. M. Pimenov, Yu. V. Tikhonov, R. T. Togusov, W. Henke and G. Gerber, Mitochondrial metabolism of guanine nucleotides: possible role of guanosine, *FEBS Lett.* 248:182 (1989).
11. D. E. Wilson and G. K. Asimakis, Phosphate-induced efflux of adenine nucleotides of the roles of the phosphatehydroxyl exchanger and the dicarboxylate carrier, *Biochim. Biophys. Acta* 893:470 (1987).
12. G. K. Asimakis and J. R. Aprille, In vitro alteration of the size of the liver mitochondrial adenine nucleotide pool: correlation with respiratory functions, *Archiv. Biochem. Biophys.* 203: 307 (1980).
13. A. M. Davidson and A. P. Halestrap, Liver mitochondrial pyrophosphate concentration is increased by Ca^{2+} and regulates the intramitochondrial volume and adenine nucleotide content, *Biochem. J.* 246:715 (1987).
14. R. Krämer, Characterization of pyrophosphate exchange by the reconstituted adenine nucleotide translocator from mitochondria, *Biochem. Biophys. Res. Commun.* 127:129 (1985).
15. J. R. Aprille, Regulation of the mitochondrial adenine nucleotide pool size in liver: mechanisms and metabolic role, *FASEB J.*, 2:2547 (1988).
16. M. T. Nosek, D. T. Dransfield and J. R. Aprille, Calcium stimulates ATP-Mg/P_i carrier activity in rat liver mitochondria, *J. Biol. Chem. 265:8444 (1990).*
17. G. K. Asimakis, G. S. Sandu and V. R. Conti, Intermittent ischemia produces a cumulative depletion of mitochondrial adenine nucleotides in the isolated perfused rat heart, *Circ. Res.* 66:302 (1990).
18. W. Henke and E. Nickel, The contribution of adenine nucleotide loss to the ischemic impairment of oxidative phosphorylation from rat kidney cortex, *Biochim. Biophys. Acta* in press (1991).
19. D. J. Bates, D. Perett and J. Mowbray, Systematic variations in the content of the purine nucleotides in the steady state perfused rat heart, *Biochem. J.* 176:485 (1978).
20. P. S. Fitt, N. Sharma, J. Attia and B. Korecky, Studies of adenosine incorporation in Langendorff rat heart and rat heart mitochondria, *Molec. Cell. Biochem.* 78:37 (1987).
21. W. Henke, M. Ziegler, W. Dubiel and K. Jung, Adenosine formation by isolated rat kidney mitochondria, *FEBS Lett.* 254:5 (1989).
22. P. C. Tulson, D. M. Whitlock and R. L. Terjung, Adenine nucleotide degradation in slow-twich red muscle, *Am. J. Physiol.* 258:C258 (1990).

NEW ASPECTS OF PURINE NUCLEOTIDE METABOLISM: FORMATION OF ADP FROM INORGANIC PHOSPHATE DURING DEGRADATION OF AMP.

Daniela Vannoni, Roberto Leoncini, Maria Pizzichini, Lucia Terzuoli, Enrico Marinello and Roberto Pagani

Istitute of Biochemistry and Enzymology
University of Siena, Italy

INTRODUCTION

In recent experiments (1-3), we followed the metabolism of AMP-succinate in rat liver supernatants.

AMP-succinate (AMP-S) was incubated at 37 °C for 60 min. in mixtures containing Tris, K-phosphate or pyrophosphate buffer (pH 7.5), 2.5 mM magnesium chloride, 2.5 mM potassium fluoride and liver supernatant (obtained by centrifuging the homogenate at 45,000 rpm for 45 min), which had been treated with Norite A, to eliminate endogenous nucleotides . The deproteinized samples were analyzed by HPLC, to determine the products of AMP succinate metabolism. These were identified as AMP and fumaric acid, along with other products of purine catabolism such as IMP, nucleosides, bases and uric acid.

In the same chromatograms, we observed a peak, which differed from the known mononucleotides (IMP, AMP, GMP) and from fumaric acid. The absorption spectra and the release of adenine and adenylic acid (identified by HPLC) by the unknown compound on hydrolysis suggested that the compound was a derivative of adenylic acid. Its retention time was the same as that of ADP. The main objection to the identification of the compound as ADP was no explanation of its formation under these conditions. An identical product was obtained when AMP-succinate or AMP were added to the incubation mixtures.

In the present study, we reviewed this problem and confirm the identification of the unknown compound as ADP. Some additional information on its formation is also reported.

METHODS

1- Male albino Wistar rats (250 gr b.w. - 9 weeks age) were killed by decapitation. The livers were rapidly excised, homogenized (25%) in isotonic KCl added with EDTA 0.1 mM and DTT 0.01 mM and centrifuged at 60,000 rpm for 60 min. The supernatant was treated with Norite A for 15 min. and after centrifugation at 3,000 rpm for 10 min., dialyzed against the medium of the homogenate. The supernatant also underwent $(NH_4)_2SO_4$ fractionation, with solid ammonium-sulphate, 30-60% saturation, and the protein precipitate was dissolved in isotonic KCl (P 60).

2- The assay mixtures contained 2.5 mM AMP-S or AMP, 60 mM Tris (pH 6), 2.5 mM KF, 2.5 mM $MgCl_2$ and protein extracts containing 4 mg protein. Incubation was carried out for 60 min. at 37°C, deproteinized with PCA

Purine and Pyrimidine Metabolism in Man VII, Part A
Edited by R.A. Harkness *et al.*, Plenum Press, New York, 1991

and neutralized with 5N KOH; 20 µl of clear supernatant were separated by HPLC.
3- For HPLC separation we used a Varian Instruments Vista mod. 5500 equipped with a variable-wavelength mod.2550 UV detector and a mod. 4290 electronic integrator. A ready-to-use prepacked column (Partisil 10 SAX 250 x 4.6 mm), washed with 0.005 M ammonium phosphate (pH 2.8) completed the analytical system. Elution was carried out using a linear gradient from 0 to 80% in 24 min, with 0,5 M ammonium phosphate buffer (pH 4.8): the flow rate was 1 ml/min.
4- For ADP identification, an aliquot of the effluent containing the unknown derivative of AMP (retention time 21.5 min.) adjusted to pH 7.5, was added to incubation mixtures of the following composition: 0.4 mM PEP, 5 nM $MgCl_2$, 70 nM KCl and pyruvate Kinase (10 µl)(4). The mixtures were deproteinized with PCA, neutralized and separated by HPLC, using the same gradient for the separation of ADP and ATP.
5- Experiments performed with labelled phosphate (^{32}P) (37,000 KBQ/mmol) contained 37 KBQ/ml assay mixture. 200 µl of the mixture stopped at time 0' and 60' were injected; the fractions were manually collected and analyzed in a liquid scintillator (Packard mod.2000CA TRI-CARB).

All reagents were of the highest commercially available purity, from Merck or Fluka Company. Standard compounds and nucleotides (IMP, AMP, ADP and ATP) were obtained from Sigma. ^{32}P was purchased from Du Pont (NEN).

RESULTS

The identification of the unknown compound as ADP, was tested by checking if it was the substrate of pyruvate kinase, the enzyme which catalyzes the reaction: PEP + ADP ---> pyruvate + ATP (in the presence of magnesium ions). This reaction was performed as described in point 4 of Materials and Methods and led to the complete disappearance of the presumed ADP peak, and its substitution by a peak containing ATP.

The retention time, the coelution with internal standard, the behavior on hydrolysis and the effect of pyruvate kinase all indicated that the unidentified peak was ADP.

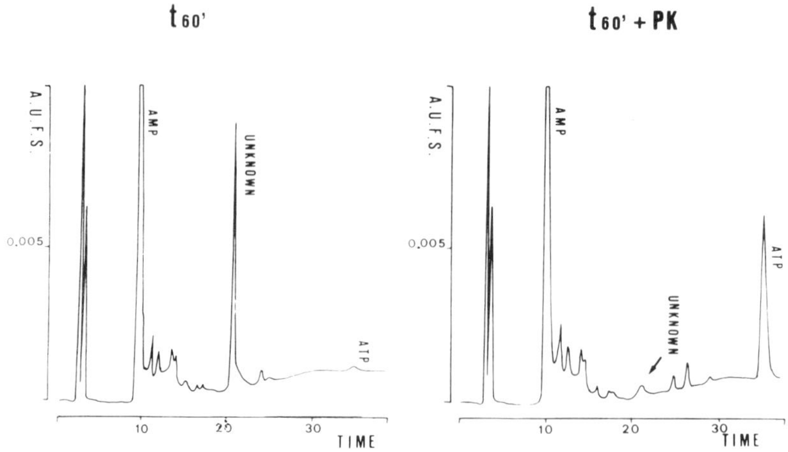

Figure 1. Chromatogram of the incubation mixture stopped at 60' before and after treatment with pyruvate kinase (PK).

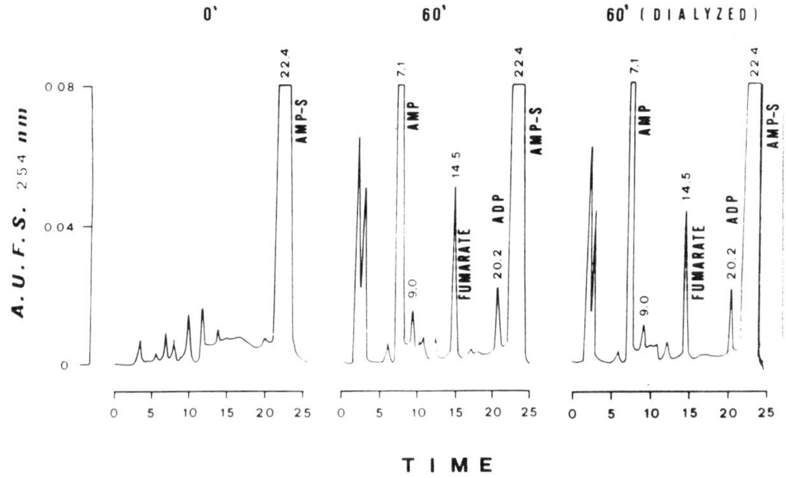

Figure 2. Incubation of AMP-S with dialyzed supernatant and fraction P6Ø.

The formation of ADP was also evident when AMP or AMP-S were incubated with rat liver supernatant, dialyzed supernatant and the fraction obtained by ammonium sulfate saturation (P 6Ø) (Fig. 2).

The formation of ADP was optimal at pH 6, slightly enhanced when K-phosphate or K-pyrophosphate were added to the incubation mixtures, and strongly enhanced after the addition of PRPP (Fig 3). When ^{32}P was added to the incubation mixture, we noted a peak of radioactivity in the fractions corresponding to ADP peak only at time 6Ø'.

DISCUSSION

Although there is still no clear explanation of these results, a number of possibilities emerge from the present study. ADP formation may depend on the presence of even a very small quantity of PRPP in the rat liver supernatant: it is not known if PRPP is freely dialyzable. However

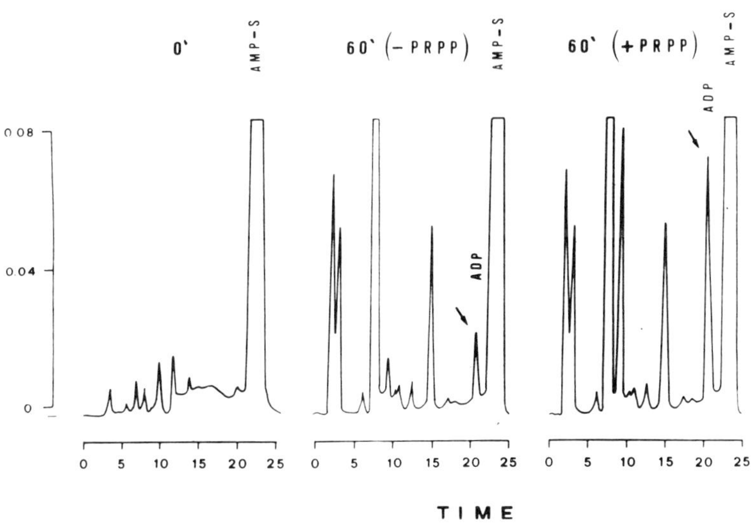

Figure 3. Incubation of AMP-S with PRPP 5mM

the existence of a PRPP---> AMP phosphotrasferase has never been demonstrated, nor has the reversibility of the PRPP synthetase reaction, ATP + R5P <--> PRPP + AMP in rat liver (5). The pH of the incubation mixtures is not in line with the optimal pH of PRPP synthetase (6) and we do not find any accumulation of ATP which could be the source of our ADP. Moreover these hypothesis do not agree with the incorporation of ^{32}P into ADP. The results may depend on the reversibility of PRPP synthetase or the presence of a new pathway, but these possibilities require further investigations.

REFERENCES

1. E. Marinello and D. Vannoni, Metabolism of AMP-S succinate in rat liver, It. J. Biochem. 38: 386 (1989)
2. E. Marinello, D. Vannoni, B. Porcelli, A. Tabucchi, G. Cinci, R. Pagani, Metabolismo dei nucleotidi: studi sull'acido AMP-S, Atti II Giornata di Facoltà. Siena, 3 Nov. 1990
3. D. Vannoni, R. Leoncini, A. Tabucchi, M. Molinelli, P. Valerio, E. Marinello, R. Pagani, Formation of ADP in liver and in lymphocyte extracts, in press on It. J.Biochem.
4. D.G. Roth, C. White and T.F. Devel, Ribosephosphate Pyrophosphokinase (rat liver) in: Methods in Enzymol., P.A. Hoffe and M.E. Jones eds., Academic press, New York, 51: 12 (1978)
5. T. Bucher and G. Pflleiderer, Pyruvate Kinase from muscle, in: Methods in Enzymol., S.P. Colowick and N.O. Kaplan eds.,Academic press, New York, 1: 435 (1955)
6. D.G. Roth, E. Shelton and T.F.Deuel, Purification and Properties of phosphoribosyl pyrophosphate from rat liver, J. Biol. Chem. 249: 291 (1974)

PHOSPHORYLATION OF ADENOSINE BY AN EXCHANGE REACTION BETWEEN AMP AND ADENOSINE IN ANOXIC HEPATOCYTES

F. Bontemps, M. Mimouni and G. Van den Berghe

Laboratory of Physiological Chemistry, International Institute of Cellular and Molecular Pathology, UCL 7539, B-1200 Brussels, Belgium

INTRODUCTION

We have previously demonstrated (1) that in normoxic isolated rat hepatocytes, adenosine (Ado) is continuously produced from AMP and immediately reconverted into AMP by adenosine kinase, without being degraded into allantoin. The present work was initially aimed at evaluating the operation of a futile cycle between AMP and Ado in anoxic hepatocytes. It has led to the unexpected observation that Ado can be incorporated into AMP in the absence of ATP, by an exchange reaction.

METHODS

Hepatocytes were isolated from overnight fasted male Wistar rats as in (2), and incubated in KRB buffer with 10 mM glucose and 1% bovine serum albumin. Cells were first preincubated during 15 min in O_2/CO_2 with, when indicated, 1 µM [U-^{14}C]adenine to label the adenine nucleotide (AN) pool, and/or 0.05 µM deoxycoformycin (dCF) to inhibit Ado deaminase. Anoxia was induced by replacing O_2/CO_2 by N_2/CO_2. Concentration and radioactivity of AN and Ado were determined as given in (3).

Cytosolic fractions of rat liver were prepared by homogeneization in 2 vol. of 25 mM Hepes (pH 7.1), 20 mM KCl, 1 mM dithiothreitol (buffer A), followed by centrifugation, first for 15 min at 8,000 x g, thereafter for 45 min at 100,000 x g. The high speed supernatant (HSS) was dialyzed overnight against 100 vol. of buffer A and filtered on Sephadex G-25 fine to remove all ATP. DEAE-Sepharose chromatography was performed on 30 % polyethylene-glycol-6000 pellets dissolved in 50 ml of buffer A without KCl and applied to a 2.5 x 25 cm column equilibrated with buffer A containing 10 mM KCl. The column was washed with 150 ml of the same buffer and eluted with a 10-600 mM gradient of KCl in buffer A, at a flow rate of 60 ml/h.

The exchange reaction was measured by incubating 20 µl of filtered HSS or fraction with 20 µM [8-^{14}C]Ado (50 µCi/mmol), 1 mM AMP, 50 mM

Hepes (pH 7.2), 10 mM Pi, 10 mM $MgCl_2$ and 0.05 μM dCF in a total vol. of 100 μl at 37°C. The reaction was terminated by spotting 10 μl of the incubate on cellulose t.l.c. plates, on which 50 nmol of AMP and Ado had been applied. After migration of Ado in water, the spot corresponding to AMP, remaining at the origin, was cut out and counted. Ado kinase was measured after (4) and cytosolic 5'-nucleotidase as in (5), with 1 mM substrate.

RESULTS AND DISCUSSION

Experiments in isolated hepatocytes

Effect of anoxia on concentration and production of Ado. Anoxia induced a near-total disappearance of hepatocyte ATP, a 75 % decrease in ADP, and a 20-fold increase in AMP. Concomittantly, Ado increased from ~0.2 to ~10 μM in the cell suspension (Fig. 1). Addition of 5-iodotubercidin (ITu), a potent inhibitor of Ado kinase (6), increased both Ado and the production of allantoin in normoxia but not in anoxia. [In the latter condition, uric acid becomes the main AN nucleotide catabolite]. This suggested that anoxia suppresses the activity of Ado kinase. Measurements of rates of production of Ado in cells incubated with ITu and dCF [to inhibit both rephosphorylation and deamination of Ado], showed that it accumulated almost linearly at a rate of 21.1 ± 2.0 nmol/min per g of cells in anoxia as compared to 11.7 ± 1.9 in normoxia. This indicates that the 50-fold elevation of Ado in anoxia results to a major extent from a decrease of its rephosphorylation by Ado kinase, rather than from an increase of its production.

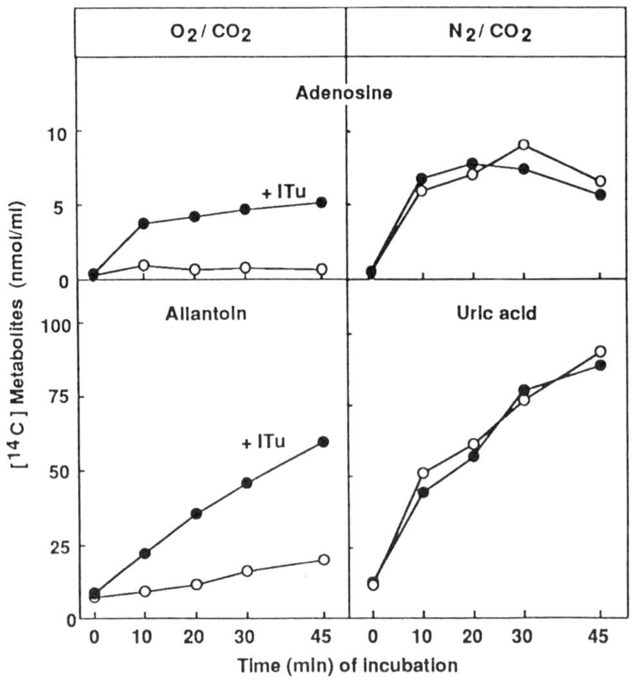

Figure 1. Influence of anoxia and ITu on Ado and purine catabolites.

Effect of anoxia on conversion of exogenous Ado into AN. To assess more directly the activity of Ado kinase, hepatocytes were incubated with 50 µM [8-^{14}C]Ado. Five min after its addition to normoxic cells, 75 % of Ado was incorporated into the AN. Surprisingly, 5 min after its addition to hypoxic cells, 33 % of [8-^{14}C]Ado was still recovered in the AN. Under both conditions, Ado incorporation was suppressed by ITu. This suggested, in contradiction with the results depicted in Fig. 1, that recycling of Ado remained possible in anoxia. The rate of incorporation of Ado under anoxic conditions, although difficult to calculate owing to dilution of [8-^{14}C]Ado by endogenous unlabelled Ado, reached at least 60 nmol/min per g. It would thus be sufficient to rephosphorylate all Ado produced in anoxia. However, whereas in accordance with (1), addition of 1 mM unlabelled Ado to normoxic cells increased their AN nucleotide pool, no increase was recorded upon addition of 1 mM Ado to anoxic cells. Taken together, the results suggested an exchange reaction between Ado and AMP.

Experiments in cell-free systems

Demonstration of an exchange reaction between Ado and AMP in rat liver cytosol. When a filtered HSS was incubated in the presence of 20 µM [8-^{14}C]Ado, 10 mM Pi and 10 mM $MgCl_2$, no formation of [8-^{14}C]AMP could be detected. However, upon addition of 1 mM AMP the nucleotide became labelled, indicating an exchange reaction between free [8-^{14}C]Ado and the un-

Additions	Activity (%)
none	100
2,3-bisphosphoglycerate (3 mM)	490
diadenosine tetraphosphate (1 mM)	32
5'-deoxy-5'-isobutylthioadenosine (5 mM)	21
5'-deoxy-5'-aminoadenosine (1 mM)	11
6-methylmercaptopurine riboside (1 mM)	48

Table 1. Effect of various compounds on the Ado-AMP exchange reaction. Activities are expressed as % of the radioactivity incorporated from Ado into AMP in the absence of additions. Results are means of duplicate assays.

labelled Ado moiety of AMP. The rate of the reaction was difficult to quantify owing to degradation of AMP into Ado in the crude preparation but reached at least 700 nmol/min per g of protein. The exchange reaction was dependent on the presence of $MgCl_2$, independent of that of Pi, and suppressed by 100 µM ITu (not shown). It was stimulated (Table 1) by 2,3-bisphosphoglycerate (2,3-BPG) a potent stimulator of cytosolic IMP-GMP 5'-nucleotidase (5,7). Diadenosine tetraphosphate, also a stimulator of cytosolic 5'-nucleotidase (8) but an inhibitor of Ado kinase (9), 5'-deoxy-5'-isobutylthioadenosine, an inhibitor of cytosolic 5'-nucleotidase (10), and 5'-deoxy-5'-aminoadenosine and 6-methylmercaptopurine riboside, both inhibitors of Ado kinase, were also inhibitory.

DEAE-Sepharose chromatography of rat liver cytosol. The results given above suggested that the incorporation of [8-^{14}C]Ado into AMP could be effected by either adenosine kinase or cytosolic 5'-nucleotidase. The latter has been shown to catalyze the phosphorylation of inosine by an exchange reaction (11). Cytosolic fractions of rat liver were therefore chromatographed on DEAE-Sepharose. The Ado-AMP exchange activity was eluted with the washing buffer, together with the activity of Ado kinase, before the start of the linear gradient of KCl. As already observed in the crude HSS, the exchange activity was dependent on the presence of AMP, supressed by ITu, and strikingly stimulated by 2,3-BPG. A first peak of 5'-nucleotidase activity eluted at 75 mM, and a second at 350 mM KCl. The activity of the second peak was strongly stimulated by 2,3-BPG, indicating that it corresponded to the cytosolic IMP-GMP 5'-nucleotidase (5,7). These results clearly show that the Ado-AMP exchange reaction is catalyzed by adenosine kinase rather than by cytosolic 5'-nucleotidase(s). Our observations suggest that liver Ado kinase acts by way of a two-site ping-pong mechanism, with a phosphorylated enzyme intermediate. For Ado kinase, this mechanism has hitherto only been proposed for the enzyme from murine leukemia L1210 cells (12).

ACKNOWLEDGEMENTS

The authors wish to express their appreciation to Mrs. Anne Delacauw for expert technical assistance. This work was supported by grant 3.4539.87 of the Fund for Medical Scientific Research (Belgium) and by the Belgian State - Prime Minister's Office for Science Policy Programming. G. Van den Berghe is Director of Research of the Belgian National Fund for Scientific Research.

REFERENCES

1. F. Bontemps, G. Van den Berghe & H.G. Hers, Evidence for a substrate cycle between AMP and adenosine in isolated hepatocytes, *Proc. Natl. Acad. Sci. USA.* **80** : 2829-33, 1983
2. G. Van den Berghe, F. Bontemps & H.G. Hers, Purine catabolism in isolated rat hepatocytes. Influence of coformycin, *Biochem. J.* **188** : 913-20, 1980
3. F. Bontemps & G. Van den Berghe, Mechanism of adenosine triphosphate catabolism induced by deoxyadenosine and by nucleoside analogues in adenosine deaminase-inhibited human erythrocytes, *Cancer. Res.* **49** : 4983-9, 1989
4. C.F. Hawkins & A.S. Bagnara, Adenosine kinase from human erythrocytes: kinetic studies and characterization of adenosine binding sites, *Biochemistry* **26** : 1982-7, 1987
5. F. Bontemps, G. Van den Berghe & H.G. Hers, 5'-Nucleotidase activities in human erythrocytes. Identification of a purine 5'-nucleotidase stimulated by ATP and glycerate 2,3-bisphosphate, *Biochem. J.* : 687-96, 1988
6. J.F. Henderson, A.R.P. Paterson, I.C. Caldwell, B. Paul, M.C. Chan & K. Lau, Inhibitors of nucleoside and nucleotide metabolism, *Cancer Chemother Rep Part 2,* **3** : 71-85, 1972
7. F. Bontemps, M.F. Vincent, F. Van den Bergh, G. van Waeg & G. Van den Berghe, Stimulation by glycerate 2,3-bisphosphate: a common property of cytosolic IMP-GMP 5'-nucleotidase in rat and human tissues, *Biochim. Biophys. Acta* **997** : 131-4, 1989
8. R.M. Pinto, J. Canales, M.A. Günther-Sillero & A. Sillero, Diadenosine tetraphosphate activates cytosol 5'-nucleotidase, *Biochem. Biophys. Res. Commun.* **138** : 261-7, 1986
9. P. Rotllan & M.T. Miras Portugal, Adenosine kinase from bovine adrenal medulla, *Eur. J. Biochem.* **151** : 365-71, 1985
10. A.C. Skladanowsky, G.B. Sala & A.C. Newby, Inhibition of IMP-specific cytosolic 5'-nucleotidase and adenosine formation in rat polymorphonuclear leucocytes by 5'-deoxy-5'-isobutylthio derivatives of adenosine and inosine, *Biochem. J.* **262** : 203-8, 1989
11. Y. Worku & A.C. Newby, Nucleoside exchange catalysed by the cytoplasmic 5'-nucleotidase, *Biochem. J.* **205** : 503-10, 1982
12. C.H. Chang, S. Cha, R.W. Brockman & L.L. Bennett, Jr., Kinetic studies of adenosine kinase from L1210 cells: a model enzyme with a two-site ping-pong mechanism, *Biochemistry* **22**: 600-11, 1983

HYPOXANTHINE AND INOSINE METABOLISM IN CULTURED RAT CARDIOMYOCYTES

Esther Zoref-Shani, Yael Bromberg, Chaim Shirin, Yechezkel Sidi, and Oded Sperling

Department of Chemical Pathology, Sackler School of Medicine, Tel Aviv University, Tel Aviv and Departments of Clinical Biochemistry and of Medicine D, Beilinson Medical Center, Petah-Tikva, Israel

INTRODUCTION

Myocardial ischemia is associated with fast degradation of ATP to nucleosides and bases, some of which are potential precursors for salvage nucleotide synthesis (1,2,4,6). Nevertheless, since these purines efflux from the cardiomyocytes, the regeneration of ATP, if reoxygenation occurs, may depend relatively more on de novo AMP synthesis, which is very costly in energy consumption. Exogenous administration of some of the lost precursors has been considered in order to overcome this problem. Hypoxanthine is a probable substrate for salvage nucleotide synthesis (8), but its metabolic fate in the heart is not yet fully clarified (19). Inosine was also suggested as a precursor for nucleotide synthesis, apparently following its degradation to hypoxanthine (3,7,8,16,17). In this connection, we used cultured cardiomyocytes for the study of the metabolic fate of hypoxanthine and inosine in the heart. The following questions were investigated: The proportion of hypoxanthine incorporated into nucleotides in comparison to that degraded to xanthine and uric acid; the pathway by which inosine is converted to IMP; and the fate of the IMP produced. 8-aminoguanosine (8-AGuo), a potent inhibitor of purine nucleoside phosphorylase (PNP) (15), was used for the last two purposes.

MATERIALS AND METHODS

Cardiomyocyte cultures were prepared as described before (10,19). For the study of the metabolic fate of hypoxanthine and inosine, (8-^{14}C)hypoxanthine, or (8-^{14}C)inosine (1-4 μM) were added to fresh culture medium (Eagle's minimal essential medium with glutamine). In some experiments, 8-AGuo was added (4 μM) in order to inhibit PNP (15). Following incubation, the medium was separated from the cells and cold perchloric acid was added to both (final concentration of 1 M). The various purine derivatives in the deproteinized cell lysates and the acidified medium were separated by bidimentional chromatography on thin layer cellulose plates (19).

RESULTS AND DISCUSSION

The rate of incorporation of hypoxanthine and inosine into nucleotides was similar (Figs. 1 & 2). 8-AGuo did not affect the rate

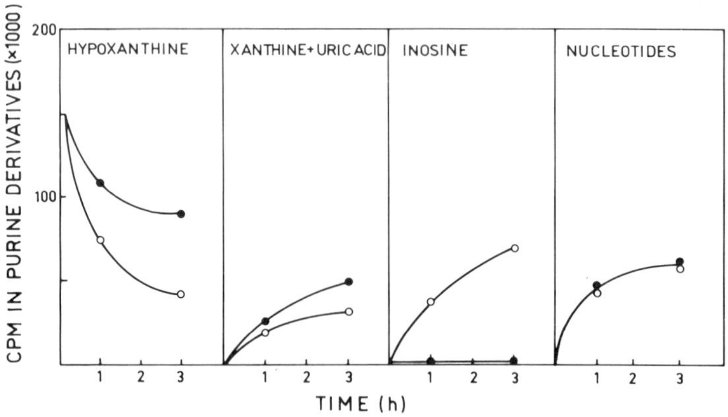

Fig 1. Distribution of label from (8-^{14}C)hypoxanthine (4μM) among purine derivatives (total counts in cellular and extracellular purines); ●---● control, o----o with 8-aminoguanosine (4μM).

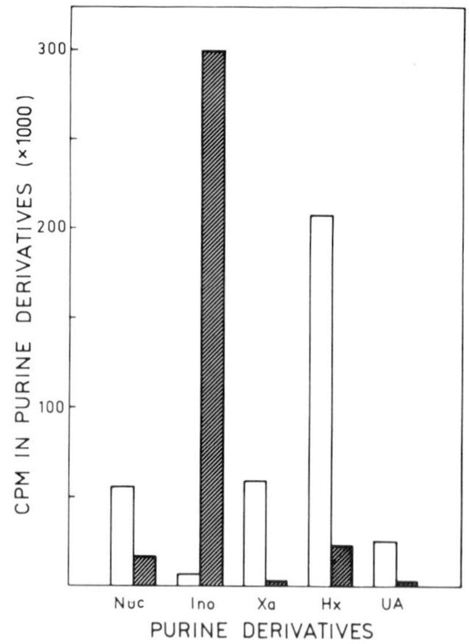

Fig 2. Distribution of label from (8-^{14}C)inosine (4μM) among purine derivatives (total counts in cellular and extracellular purines) following incubation for 3 h.
▭ - control; ▨ with 8-aminoguanosine (4μM);
Nuc - nucleotides; Ino - inosine; xa - xanthine;
Hx - hypoxanthine; UA - uric acid

Table 1. Metabolic fate of IMP* produced from hypoxanthine

Hypoxanthine concentration (μM)	Distribution of Label (% of total)			
	Adenine nucleotides	IMP	Guanine nucleotides	Inosine
1	51.9	1.2	8.6	38.3
4	36.5	1.4	7.2	54.9

of incorporation of hypoxanthine (Fig. 1), but reduced the rate of inosine incorporation by 75% (Fig. 2). This finding indicates conclusively that the pathway by which inosine is converted to nucleotides proceeds through its degradation by PNP to hypoxanthine.

Incubation was carried with (8-^{14}C)hypoxanthine as substrate, for 3 h, in presence of 8-AGuo (4 μM). Following incubation, the various purinederivatives in the cells and medium, were separated as described in Materials and Methods. *Total IMP produced (calculated by adding the radioactivity in all nucleotides + inosine).

This was expected in view of the lack of inosine kinase in mammalian tissues (17).

Inhibition of PNP by the addition of 8-AGuo to both labeled hypoxanthine or inosine as substrates, resulted in manyfold increase in the labeling of the inosine fraction, associated with a marked reduction in the labeling of the inosine degradation products hypoxanthine, xanthine and uric acid (Figs. 1 & 2).
In the experiments in which hypoxanthine was utilized as substrate, these results demonstrates a high rate of IMP degradation. Activity of IMP 5'-nucleotidase was demonstrated in extracts from cultured cardiomyocytes in our laboratory before (19), but in the presence of intact cells its activity could not be detected, unless the activity of PNP was inhibited, since the inosine formed was rapidly degraded to hypoxanthine. Calculating the fluxes from IMP to adenine nucleotides, guanine nucleotides and inosine, revealed that the IMP 5'-nucleotidase, degrading IMP to inosine, consumed about 40 to 50 % of the total IMP produced (Table 1).

This high rate of IMP degradation to inosine probably reflects the increased generation of IMP in the cardiomyocytes, due to the artificially induced salvage of hypoxanthine to IMP. Indeed, as can be seen from the Table, the degree of IMP degradation to inosine was greater at 4 μM than at 1 μM hypoxanthine concentration. High activity of IMP 5'-nucleotidase was reported recently in human red blood cells (14), in which due to the lack of adenylosuccinate synthetase (11), most of the IMP produced was found to be rapidly degraded to inosine and further on to hypoxanthine. Therefore, the real magnitude of the rate of hypoxanthine incorporation into red cell nucleotides, could be assessed only in the presence of 8-AGuo. Also in the cardiomyocytes, most of the inosine formed was efficiently degraded by PNP to hypoxanthine, which was partly further degraded to xanthine and uric acid and partly reutilized for IMP synthesis.

These results indicate the activity in the cardiomyocytes of a futile cycle hypoxanthine-->IMP-->inosine-->hypoxanthine (Fig. 3). The cycle is consuming energy in the form of ATP, which is required for the

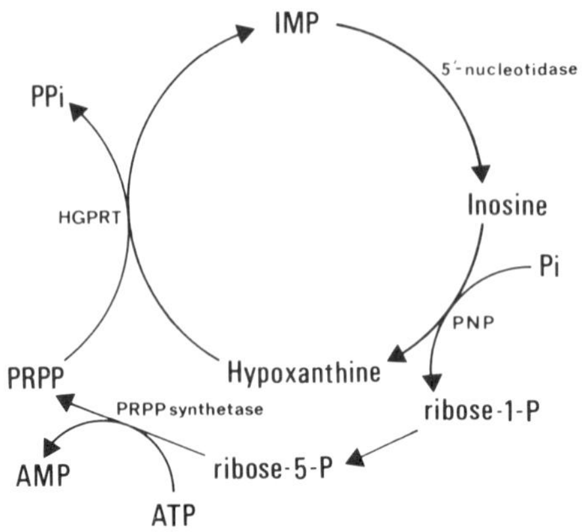

Fig 3. The futile cycle connecting hypoxanthine, IMP and inosine.

regeneration of PRPP, consumed for the HGPRT-catalyzed salvage of hypoxanthine to IMP. Apparently, in the experimental system used, only about 50 % of the energy spent in association with the salvage of hypoxanthine to IMP is of anabolic value, contributing to the synthesis of adenine and guanine nucleotides (Table 1). The physiological rational for the activity of this futile cycle is not clear, but when considering administration of hypoxanthine or inosine as precursors for salvage nucleotide synthesis in the heart, one should keep in mind that the salvage of these precursors to adenine nucleotides may be associated with relatively high wastage of energy.

In view of the presence in cardiomyocytes of xanthine oxidase (13), the degradative pathway of hypoxanthine to xanthine and uric acid was also investigated. It was found that the activity of xanthine oxidase, in comparison to that of HGPRT was relatively low, consuming at 1 μM hypoxanthine about 8% of the total hypoxanthine metabolized and at 4 μM about 28% of the total hypoxanthine metabolized. However, the activity of xanthine oxidase in the cardiomyocytes may be damaging to the cells due to the formation of free oxygen radicals (13).

In conclusion, hypoxanthine and inosine may be utilyzed for salvage purine nucleotide synthesis in the cardiomyocyte, but this is associated with futile energy expenditure, due to the marked activity of IMP 5'-nucleotidase, and with the formation of harmful free oxygen radicals, due to the activity of xanthine oxidase.

REFERENCES

1. Achterberg PW, Harmsen E, de Jong, JW (1985). Adenoine deaminase inhibition and myocardial purine release during normoxia and ischemia. Cardiovasc Res 19: 593-598.
2. Achterberg PW, Stroeve RJ, de Jong JW (1986). Myocardial adenosine cycling rates during normoxia and under conditions of stimulated purine release. Biochem J 235: 13-17.
3. Aussedat J, Verdys M, Rossi A (1985). Adenine nucleotide synthesis from inosine during normoxia and after ischaemia in the isolated perfused rat heart. Can J Physiol Pharmacol 63: 1159-1164.
4. Deuticke B, Gerlach E (1966). Abbau freier Nucleotide in Herz, Skeletmuskel, Gehirn, und Leber der Ratte bei Sauerstoffmangel. Pflugers Arch 292: 239-254.

5. Geisbuhler T, Altschuld RA, Trewyn RW, Ansel AZ, Lamka K, Brierley GP (1984). Adenine nucleotide metabolism and compartmentalization in isolated adult rat heart cells. Circ Res 54: 536-546.
6. Gerlach E, Deuticke B, Dreisbach RH (1963). Der Nucleotid-Abbau im Herzmuskel bei Sauerstoffmangel und seine mogliche Bedeutung fur die Coronardurchblutung. Naturwiss 50: 228-229.
7. Goldhaber SZ, Pohost GM, Kloner RA, Andrews E, Newell JB, Ingwall JS (1982). Inosine, a protective agent in an organ culture model of myocardial ischemia. Circ Res 51: 181-188.
8. Harmsen E, De Tombe PP, de Jong JW, Achterberg PW (1984). Enhanced ATP and GTP synthesis from hypoxanthine or inosine after myocardial ischemia. Am J Physiol 246: H37-H43.
9. Itoh R, Oka J, Ozasa H (1986). Regulation of rat heart cytosol 5'-nucleotidase by adenylate energy charge. Biochem J 235: 847-851.
10. Kessler-Icekson G, Sperling O, Rotem O, Wasserman C (1984). Cardiomyocytes cultured in serum free medium. Growth and creatine kinase activity. Exp Cell Res 155: 113-120.
11. Lowy BA, Dorfman BZ (1970). Adenylosuccinase activity in human and rat erythrocyte lysates. J Biol Chem 245: 3043-3046.
12. Meghji P, Holmquist GA, Newby AC (1985). Adenosine formation and release from neonatal-rat heart cells in culture. Biochem J. 229: 799-805.
13. Schoutsen B, de Jong JW, Harmsen E, De Tombe PP, Achterberg PW (1983). Myocardial xanthine oxidase/dehydrogenase. Biochem Biophys Acta 762: 519-524.
14. Sidi Y, Gelvan I, Brosh S, Pinkhas J, Sperling O (1989). Guanine ribonucleotide metabolism in human red blood cells. Evidence for a high rate of GMP dephosphorylation. Biochem Med and Metabol Biol 41, 149-154, (1989).
15. Stoeckler JD, Cambor C, Khuns V, Chu SH, Parks RE Jr. (1982) Inhibitors of purine nucleoside phosphorylase. C(8') and C(5') substitutions. Biochem Pharmacol 31: 163-171.
16. Thomas JX Jr, Jones CE (1979). Effect of inosine on contractile force and high-energy phosphates in ischemic hearts. Proc Soc Exp Biol Med 161:468-472.
17. Wiedmeir T, Rubio R, Berne RM (1972). Inosine incorporation into myocardial nucleotides. J Mol Cell Cardiol 4: 445-452.
18. Zoref-Shani E, Kessler-Icekson G, Wasserman L, Sperling O (1984). Characterization of purine nucleotide metabolism in primary rat cardiomycyte cultures. Biochim Biophys Acta 804: 161-168.
19. Zoref-Shani E, Kessler-Icekson G, Sperling O (1988). Pathways of adenine nucleotide catabolism in primary rat cardiomyocyte cultures. J Mol Cell Cardiol 20: 23-33.

THE MECHANISM OF CONVERSION OF XANTHINE DEHYDROGENASE TO OXIDASE AND THE ROLE OF THE ENZYME IN REPERFUSION INJURY

Takeshi Nishino and Isao Tamura

Department of Biochemistry and Department of Surgery
Yokohama City University School of Medicine
Fukuura 3-9, Kanazawa-ku
Yokohama 236, Japan

THE MECHANISM OF CONVERSION FROM XANTHINE DEHYDROGENASE TO OXIDASE

In 1968 Della Corte and Stirpe reported that mammalian xanthine oxidase was originally a dehydrogenase, and was converted to the oxidase by modification of the enzyme protein during extraction or purification (Della Corte and Stirpe, 1968, 1972). This conversion occurs either reversibly by sulfhydryl oxidation, or irreversibly by proteolysis (Della Corte and Stirpe, 1968, 1972; Stirpe and Della Corte, 1969; Waud and Rajagopalan,1976; Nakamura and Yamazaki, 1982; Ikegami and Nishino, 1986; Saito and Nishino, 1989). Although the mammalian enzyme is easily converted from the dehydrogenase to the oxidase, the enzyme can be purified in a reversible form by careful purification procedures (Waud and Rajagopalan,1976; Nakamura and Yamazaki, 1982; Ikegami and Nishino, 1986; Saito and Nishino, 1989). The dehydrogenase form has very low oxidase activity (3-4% of dehydrogenase activity) when enzyme activity is determined by the standard assay system, but detailed kinetic analysis shows that even the dehydrogenase has an appreciable amount of oxidase activity. Table 1 shows the kinetic parameters obtained from rat liver xanthine dehydrogenase and oxidase (Saito and Nishino, 1989). For the dehydrogenase, the Vmax value for dehydrogenase activity has about four times higher than the value for oxidase activity. The Km value for oxygen of the dehydrogenase is five times higher than that of the oxidase. The oxidase activity of the dehydrogenase is almost completely inhibited by NAD. While dehydrogenase activity of the oxidase type is almost negligible, the Vmax value for oxidase activity of this form is similar to that of the dehydrogenase form for dehydrogenase activity, indicating the same rate limiting step of urate release from the enzyme. It should be noted that the dehydrogenase produces more ratio of O_2^- formation than the oxidase if NAD is absent. However, O_2^- is almost completely inhibited in the presence of NAD. In contrast to the mammalian enzyme, conversion of the dehydrogenase form to the oxidase form has never been reported for the avian enzyme, but the reactivity of chicken liver xanthine dehydrogenase with oxygen is similar to that of rat liver xanthine dehydrogenase.

The cDNA for the rat liver enzyme has been cloned and the amino acid sequence has been determined in our laboratory. The enzyme consists of 1319 amino acids (Amaya et al, 1990). When this enzyme was treated with trypsin, it was converted irreversibly to an oxidase with the concomitant formation of nicks at residues 184-185 and 539-540 but without dissociation of any peptide fragments. When this cleaved enzyme was applied to SDS-PAGE, three fragments of molecular mass 20KD, 40KD and 85KD were separated. When the native enzyme was treated with fluorodinitrobenzene, a nucleophilic reagent, the enzyme was converted reversibly to the oxidase form with modification of cys-525, cys-980 and cys-1313 (Nishino et al, unpublished).

Table 1. Kinetic parameters obtained from purified rat liver xanthine oxidizing enzymes.

substrate	NAD		O_2		
	Vmax (mol/mol/min)	Km (μM)	Vmax (mol/mol/min)	Km (μM)	O_2^- /urate (mol/mol)
Dehydrogenase type	810	8.5	270	260	0.68
Oxidase type			1030	46	0.30

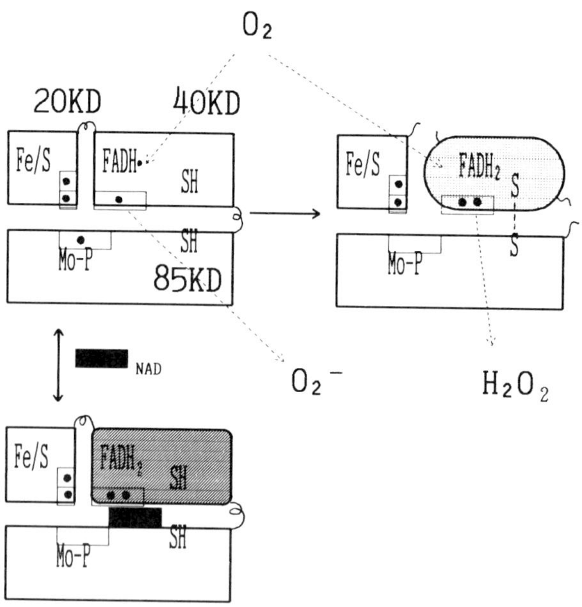

Fig. 1. Conformational change by proteolysis, sulfhydryl oxidation or NAD binding and electron distribution of 4-electron reduced state.

From detailed experimental analysis we assume that formation of disulfide bridge between cys-525 and cys-980 is responsible for the reversible conversion of the dehydrogenase to the oxidase. The modification of cysteine residues or the formation of proteolytic nicks results in a conformational change, paticularly in the FAD environments (Saito et al, 1989).

Pronounced differences in protein structure surrounding the bound flavin were indicated by the spectral properties of the two enzymes replaced with flavins containing ionizable -OH or -SH substituents at the flavin 6- or 8- position (Massey et al. 1989; Saito et al., 1989). The oxidase form either displays no preference for binding of a neutral or anionic flavin (8-OH-FAD) or a slight preference for the anionic form of the flavin (6-OH-FAD, 6-SH-FAD and possibly 8-OH-FAD). On the other hand the dehydrogenase type had a dramatic preference for binding the neutral forms of all four flavins, perturbing the pK of the ionizable substituent more than 4 pH units. This implies the existence of a strong negative charge in the flavin binding site of the dehydrogenase form, which is absent in the oxidase form. Such a change in FAD enviroments results in a change of redox potential of FAD couples as well as a loss of NAD binding site. Although no information about the three-dimensional structure from X-ray crystallography is available, Fig. 1 shows the models of 4-electrons reduced states and the localization of redox centers suggested by analysis of the amino acid sequence and the chemical modification studies (Nishino and Nishino, 1989; Amaya et al, 1990). Two iron sulfur centers are associated with 20KD fragment, the NAD and FAD binding sites are associated with 40KD fragment, and the molybdenum is associated with 85 KD fragment. Although these cofactors are not directly associated with each other, the distance between them might not be so great since intramolecular electron transfer is very rapid (Olson et al, 1974). The conformation of the native dehydrogenase is such that FAD is stabilized mostly in the semiquinone state, which reacts slowly with oxygen to form primarily superoxide anion (Nishino et al, 1989; Saito and Nishino, 1989). When the enzyme is modified by proteolysis or sulfhydryl-oxidation, the protein conformation has been changed particulary around FAD. In this new conformational state FAD prefers the fully reduced state, which reacts rapidly with oxygen to form hydrogen peroxide. If NAD is present, a conformational change also occurs favoring the fully reduced state of FAD, which reacts rapidly with NAD (Nishino and Nishino, 1989; Saito and Nishino, 1989).

Fig. 2 shows another scheme of the reaction mechanism which illustrates more acculately the steady state turnover between the 4-electron and 2-electron reduced levels, which is a main shuttle during steady state turnover of the dehydrogenase type (Nishino et al, 1989a). The 2-electron-reduced enzyme is reduced by xanthine, the enzyme substrate, to form the 4-electron-reduced level that has several species in equilibrium. In this figure only the main two forms are shown. The B form, a FAD semiquinone state, is the dominant form in the dehydrogenase conformation; the A form is dominant in the oxidase conformation. The B form reacts with oxygen by sequential steps to form superoxide. A minor amount of the A form reacts relatively rapidly with oxygen to form hydrogen peroxide. If NAD is present, the A form is the dominant form, which reacts very rapidly with NAD to form NADH. On the other hands, in the oxidase form of the enzyme, as enzyme reacts much more rapidly with oxygen, the reaction proceeds further 1-electron reduced and fully oxidized levels (Olson et al, 1974). At the 2-electron reduced level, two electrons are divided between FAD and an iron sulfur center. The FAD semiquinone reacts with oxygen to form superoxide anion. The 1-electron state is an obligatory semiquinone state which reacts with oxygen as one electron transfer reaction. Thus, the oxidase form can produce hydrogen peroxide and superoxide anion.

THE ROLE OF XANTHINE DEHYDROGENASE/OXIDASE IN REPERFUSION INJURY

It has been proposed that the oxyradical species derived from xanthine oxidase are responsible for recirculation injury (McCord, 1985). According to the hypothetical mechanism, hypoxanthine, a substrate of xanthine oxidase, accumulates during ischemia due to catabolism of adenine nucleotides. In the mean time, xanthine

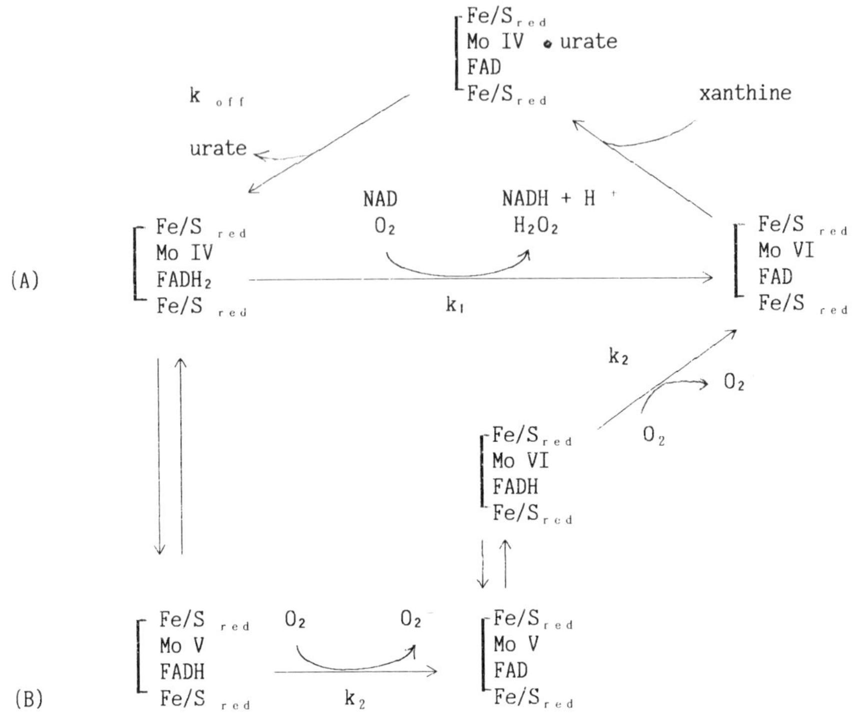

Fig. 2. Mechanism of reaction of xanthine dehydrogenase during steady state.

dehydrogenase is converted to the oxidase by proteolytic cleavage, probably by a Ca-dependent proteinase (Roy and McCord, 1982). With reoxyganation, this hypoxanthine is oxidized to uric acid with formation of superoxide and hydrogen peroxide, which cause tissue damage.

In this hypothesis, the perticipation of xanthine oxidase is suported by the following points:

1. The enzyme exists in endotherial cells.
2. The substrates hypoxanthine and xanthine accumulate during ischemia.
3. Xanthine dehydrogenase is converted to the oxidase during ischemia.
4. Inhibitors of xanthine oxidase protect tissue from damage to some extent.
5. There have been some reports that superoxide or hydoxyradical were detected after reoxigenation using the spin-trap method (Zweier et al., 1988) or chemiluminescence method.

The first point seems to be true in some species (Jarasch et al., 1981). But the two observations about the localization of the enzyme must be taken into account. One point is that the enzyme exist not only in endotherial cells but also in other parenchymal cells such as hepatocytes. However, there are reports that the cells of some organs such as human cardiac cells lack this enzyme activity (Downey et al., 1987; Eddy et al.,1987; de Jong et al., 1990), although the possibility of an inactive form (such as desulfo form: Bray, 1974; Ikegami and Nishino, 1986) is not excluded. The second observation is about the subcellar localization of this enzyme. The enzyme is not a peroxisomal enzyme but an exclusively cytosolic.

Many repports support the second point (Saugstad et al., 1988), and our data also agree with this. However, there were conflicting reports regarding the fifth point, as to whether radicals were detected. At present it is difficult to reach a definite conclusion.

The third and fourth point are deserving of detailed discussion. The problem is whether xanthine dehydrogenase is actually converted to an oxidase during ischemia. There are reports that in intestinal cells, conversion of the dehydrogenase to the oxidase occurrs very rapidly during ischemia, and that this conversion is blocked by previous treatment with protease inhibitor(Roy and McCord, 1982). On the other hand, there are conflicting reports that this conversion takes very long time. For example, Harkness reported data which were obtained from human organs post-mortem (Wajner and Harkness, 1989). Although these organs from autopsy samples seemed to be preserved for a ralatively long time under ischemic conditions, xanthine oxidase activity was still lower than xanthine dehydrogenase activity. In our own experiments, we observed only slight conversion (normal condition about 20%; after ischemia 30%). Dog lung in which edema is presumably caused by recirculation injury has been used for experimental model. In this model, the arteries and bronchi are clamped momentarily, and with unclamping recirculation injury occurs. Although edema does occur, xanthine dehydrogenase was reversibly converted to the oxidase, and was only 30% at most. Such slight conversion of the dehydrogenase to the oxidase has been noted by many other experimental reports (Gerber, 1989). It is possible that reports of large-scale conversion are due to secondary phenomenon, such as protease likage. For example, intestine may be easily contaminated with protease or the damaged cells of even other organs may be also contaminated with protease due to lysozomal rupture.

Neverseless, there are some reports showing that inhibitors of xanthine oxidase are effective for preventing recirculation injury (Akizuki et al., 1985; Chamber et al., 1985; Charlat et al., 1987). However, the effects of inhibitors of xanthine oxidase are not simply described. Allopurinol and oxypurinol are reported to be direct scavengers of oxyradical species (Moorhouse et al., 1987; Das et al., 1987; Gerber et al., 1989), although conflicting reports hold that the direct-scavenger role of this inhbitors is not an important one (Terada et al., 1991). It is also possible that inhition of xanthine oxidase affects the amount of hypoxanthine available for salvage to nucleotides (Downey et al., 1987).

An attempt should be made to estimate the degree of the increase in superoxide anion after recirculation. As mentioned previously, conversion of the dehydrogenase to the oxidase is only slight, and the increase in the content of xanthine oxidase is at most about 1.5 times normal level(from 20% to 30%). Another factor which must be considered is the inhibition of xanthine dehydrogenase specifically by NADH, which may have accumulated during anoxia. If NADH strongly inhibits xanthine dehydrogenase then the oxidation of the accumulated substrates hypoxanthine and xanthine may be catalized mainly by xanthine oxidase, which would result in the production of oxyradical species. NADH is a good inhibitor of xanthine dehydrogenase (Della Corte, 1970; Kaminski and Jezewska,1981). In our experiments inhibition is competitive with NAD, and the Ki value is in the micro molar range- a rather strong inhibitor.

Fig. 3 a) shows the effect of NADH on the formation of superoxide from the enzyme which has dehydrogenase/oxidase activity ratio 5, which is considered to be the standard content of the oxidase form in normal cells, where the content of oxidase is about 20%. Formation of superoxide is followed by reduction of cytchrome c. As this reduction is completely inhibited by superoxide dismutase, reduction is due to superoxide anion. When 10 µM xanthine was added to the enzyme-cytochrome c mixture, reduction proceeded. But when xanthine became depleted the rate of the reduction slowed. This second phase is due to NADH oxidation. First phase is short, but is prolonged by the addition of NADH. This effect of NADH is due to inhibition of xanthine dehydrogenase. But since xanthine oxidase is not inhibited, cytochrome c reduction is unaffected. Fig. 3 b) shows the effect of NADH on the velocity of xanthine oxidation as well as on the velocity of cytochrome c reduction. The rate of cytochrome c is not affected, but urate formation and particullary the rate of NADH formation are decreased. The inset shows the ratio of superoxide to urate formation with increasing NADH. However, the degree of the increase of NADH during anoxia is not known. The only available method for the estimation of NAD/NADH in cytosol is from pyruvate/ lactate ratio (Williamson et al.,1967) The isolated hepatocytes NAD/NADH ratio was estimated from the ratio of pyruvate/lactate after making the cells anoxic. The ratio changed from 700 to 30 after 2 hours.

From the previous plot, this ratio would cause only about 20% inhibition. However, estimation of the NAD/NADH ratio from the pyruvate/lactate ratio is based on the assumption that the pyruvate/lactate system is rapidly equilibrated. Thus, an accurate estimate of the inhibition is difficult, but it should be less than 50%.

In any case, the increase of superoxide can be roughly estimate as follow:

$$\underbrace{2 \times 1.5}_{\text{(velocity increase)}} \times \underbrace{1.2 \times 3}_{\text{(substrate pool)}} = 10$$

The velocity increase, a factor of two, is due to substrate saturation; the normall concentration of hypoxanthine and xanthine are near the Km value of approximately 2 μM (Tamura and Nishino, unpublished; Saito and Nishino, 1989). Conversion of the dehydrogenase to the oxidase would account for at most a 1.5-fold increase in the amount of oxidase, and thus the overall increase in velocity would

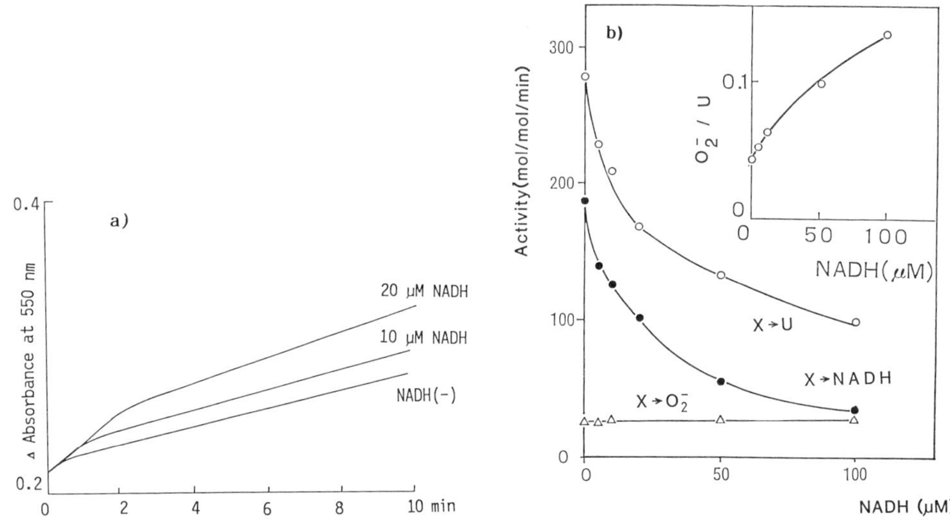

Fig. 3. Formation of superoxide anion from xanthine dehydrogenase/oxidase system. a) 10 uM xanthine was added to the mixture of xanthine dehydrogenase/oxidase, 100 uM NAD and 50 uM cytochrome c in the presence and absence of different concentration of NADH.
b) replot of the velocity of urate formation and cytochrome c reduction versus NADH concentration.

be at most three hold. The inhibition of the dehydrogenase by NADH would contribute slightly (1.2-2 times) to the substrate pool, and the increase in substrates from catabolism would be around 3 depending on the conditions (Tamura and Nishino, unpublished). Thus, the total increase is around ten times of normall concentration without considering NADH oxidation which is much slower reaction.

However, this argument does not consider the scavenger systems of the organ. Thus the evidence for the perticipation of oxyradical species derived from xanthine oxidase reaction in tissue injury is not convincing, because the participation of oxyradical species in tissue injury is not a simple problem. Under normal conditions, oxyradical species might be balanced in the cell between formation and scavenging. Oxyradicals can come not only from xanthine oxidase system but other cellular systems as well. Furthermore, sacvenging systems consists not only of enzymes such as superoxide dismutase and catalase but also of low molecular weight compounds such as uric acid. If oxyradical species increase more than the capacity of scavenging system, they may overflow and damage the tissue. Neither the rate of oxyradical production nor the capacity of scavenging system is known with cetainty, so it is difficult to predict whether a only three-fold increase in verocity or a ten-fold increase in total amount would be harmfull for the tissue.

Acknowledgements

We thank to Drs. Y. Amaya, Tomoko Nishino and T. Saito for thier helpful discussion and to Dr. T. page, the University of California San Diego, for his critical reading of manuscript. This work was supported in part by Grant-in-aid 2044123 from the Japanese Ministry of Education, Science and Culture and in part by a research grant for intractabledisease from Japanese Ministry of Health and Welfare.

REFERENCES

Akizuki,S., S. Yoshida, D.E. Chamber, L.J. Eddy, L.F. Parmley, D.M. Yellon and J.M. Downey, 1985, Cardiovasc. Res. 19: 686-692

Amaya, Y., K.Yamazaki, M.Sato, K.Noda, T.Nishino and T.Nishino, 1990, J. Biol. Chem. 265: 14170-14175

Bray, R.C., 1975, in "The Enzyme", 3rd ed., Vol.12, part B, pp299-419, Academic Press, New York

Charlat, M.L., P.G. O'Neill, J.M. Egan, D.R. Abernethy, L.M. Michael, M.L. Myers, R. Roberts and R. Bolli, 1987, Am. J. Physiol. 252:H566-H577

Das, D.K. , R.M. Engelmann, R. Clement, H. Otani, M.R.Prasad and P.S.Rao, 1987, Biochem. Biophys. Res. Commun.,148:314-319

Della Corte, E. and F. Stirpe, 1968, Biochem. J. 108:349-351

Della Corte, E. and F. Stirpe, 1970, Biochem. J. 117:97-100

Della Corte, E. and F. Stirpe, 1972, Biochem. J. 126:739-745

Downey,J.M., T.Miura, L.J.Eddy, D.E.Chambers, T.Mellert, D.J.Hearse and D.M.Yellon 1987,J. Mol. Cell. Cardiol., 19:1053-1060

Eddy,L.J., J.R. Stewart, H.P. Jones, T.D. Engerson, J.M. McCord and J.M. Downey, 1987, Am. J. Physiol. 253:H709-H711

Gerber,G., W.Siems, A.Werner, R.Stoesser, Z.Foeldes-Papp, J.Kowalewski, W.Schneider and T.Grune, 1989, in"Purine pyrimidine metabolism in man VI" Part B, 497-504

Ikegami, T. and T. Nishino, 1986, Arch. Biochem. Biophys. 247:254-260

Jarasch,E.D., C.Grund, G.Bruder, H.W.Heid, T.W.Keenan and W.W.Franke, 1981, Cell, 25:67-82

KaminskiZ.W. and M.M. Jezewska, 1981, Biochem. J. 200:597-603

J.W. de Jong, P. van der Meer, A.S. Nieukoop, T. Huizer, R.J. Stroeve and E. Bos, Circulation Res., 67:770-773

Massey,V., L.M.Schopfer, T.Nishino and T.Nishino, 1989, J.Biol. Chem. 264:10567-10573

McCord,J.M. 1985, N. Engl. J. Med. 312:159-163

Moorhouse,P.C., M. Grootveld, B. Halliwell, J.G. Quinlan and j.M.C. Gutteridge, 1987, FEBS-letters, 213:23-28

Nakamura, M. and I. Yamazaki, 1982, J. Biochem. 92: 1279-1286

Nishino, T. and T. Nishino, 1987, Biochemistry, 26:3068-3072

Nishino, T. and T. Nishino, 1989, J. Biol. Chem. 264:5468-5473

Nishino,T., T.Nishino, L.M.Schopfer and V.Massey, 1989, J. Biol. Chem. 264:2518-2527

Olson,J. S., D.P. Ballou, G. Palmer and V. Massey, 1974, J. Biol. Chem. 249:4363-4382

Roy, R.S. and J.M. McCord, 1982, Can. J. Phys. and Pharm.,60:1346-1352

Saito, T. and T. Nishino, 1989, J. Biol. Chem. 264:10015-10022

Saito,T., T. Nishino and V. Massey,1989, J. Biol. Chem. 264:15930-15935

Saugstad, O.D. 1988, Pediatr.Res., 23:143-150

Stirpe, F. and R.E. Della Corte, 1969. J. Biol. Chem. 244:3855-3863

Terada,L.S., J.D. Rubinstein, E.J. Lesnefsky, L.D. Horwitz, J.A. Leff and J.E. Repine Am. J. Physiol. 260: H805-H810

Wajner, M. and R.A. Harkness, 1989, Biochem. Biophys. Acta, 991:79-84

Waud, W.R. and K.V. Rajagopalan, 1976, Arch. Biochem. Biophys. 172: 354-376

Williamson, D.H., P. Lung and H.A. Krebs, 1967, Biochem. J. 103: 514-527

Zweier,J.L., P.Kuppusamy and G.A.Lutty, 1988, Proc.Natl.Acad.Sci.USA, 85:4046-4050

PURIFICATION OF XANTHINE OXIDASE FROM HUMAN MILK

Roger Harrison, Shahla Abadeh and Mustapha Benboubetra

Department of Biochemistry
University of Bath
Bath, BA2 7AY, U.K.

Following evidence (Granger *et al.*, 1981) that xanthine oxidase could be a source of superoxide free radicals, responsible for ischaemia-reperfusion damage in the gut, a similar role for this enzyme has been proposed in many other tissues. Thus, xanthine oxidase has become a focus of interest as a source of free radicals also in myocardium (Thompson-Gorman and Zweier, 1991), joints (Blake *et al.*, 1989), and liver (Brass *et al.*, 1991). The proposed pathogenic mechanism involves ischaemia-induced proteolytic conversion of the naturally-occurring dehydrogenase form of the enzyme to its oxidase form. Concurrently, ischaemia also leads to degradation of nucleotides in the affected tissue and to accumulation of hypoxanthine. Subsequent reperfusion then provides oxygen, which, in the presence of hypoxanthine and xanthine oxidase, is reduced to hydrogen peroxide and superoxide anion. These reactive species, together with hydroxyl radicals derived from them are the proposed agents of tissue damage.

The above hypothesis is largely based on the known properties of xanthine oxidase from rat or chicken liver and particularly of that from bovine milk, which has been intensively studied (Bray, 1975). The proposed involvement of xanthine oxidase in ischaemia-reperfusion injury to human myocardium aroused special interest, which has, however, recently waned somewhat, because of the failure of some groups to detect appreciable levels of xanthine oxidase in human heart (Eddy *et al.*, 1987; Muxfeldt and Shaper, 1987; Grum *et al.*, 1989) in apparent contradiction to immunohistochemical evidence for its presence (Jarasch *et al.*, 1981). In fact, despite the many publications devoted to the proposed pathogenic role of xanthine oxidase, there is surprisingly little information concerning the human enzyme from any tissue. Krenitsky *et al.*, (1986) described a preparation, from post-mortem human liver, which had a protein to flavin ratio (A_{280}/A_{450}) of 5.2. The human liver enzyme showed a molecular weight of approximately 300,000 on size-exclusion chromatography, but was nevertheless, considerably proteolysed, as judged by SDS-PAGE. More recently, Zeise and Zikakis (1987) described a heterogeneous preparation of xanthine oxidase, from human colostrum, that showed multiple bands in both gel-filtration and SDS-PAGE, while Graham *et al.*, (1989) briefly reported a preparation, with apparent subunit size of 122,000D from human milk that was used to generate rabbit polyclonal antibodies. We now outline a purification and partial characterisation of xanthine oxidase, from human milk, that is homogeneous in both gel filtration and SDS-PAGE.

Frozen human breast milk (250ml) was thawed, extracted with Triton X-100, and subjected to ammonium sulphate fractionation, before freeze-thawing and centrifugation to give 'crude xanthine oxidase', essentially as described by Zikakis *et al.*, (1983) for bovine milk.

'Crude xanthine oxidase' was dialysed overnight against 0.2M Na_2HPO_4 (pH6.0), containing 1mM EDTA and 1.25mM sodium salicylate, and applied to a column (15cm x 1.5cm) of calcium phosphate gel-cellulose, previously equilibrated in the same buffer. The column was washed with this buffer until no more protein (A_{280}) was eluted. Bound xanthine

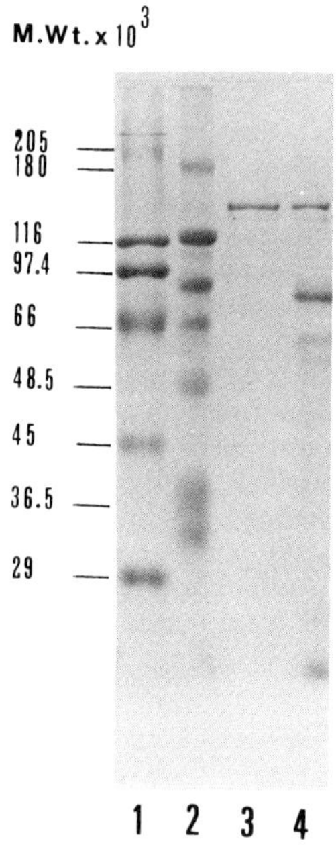

Fig. 1. SDS-PAGE of purified human xanthine oxidase (lane 3) and 'crude' human xanthine oxidase (lane 4). MWt markers are shown in lanes 1,2.

oxidase was then eluted from the column, by using the above buffer, containing, additionally, 5% (w/v) $(NH_4)_2SO_4$. Protein-containing fractions were combined and dialysed overnight against suitable buffer to give 6-10mg (n = 8) purified xanthine oxidase. These yields are comparable with those of xanthine oxidase obtained by similar methods from bovine milk (Benboubetra, 1989).

The purified enzyme showed a single symmetrical peak, in gel filtration FPLC, using either Superose 6 or Superdex 200 columns, corresponding to Mr 288,500. SDS-Polyacrylamide gel electrophoresis patterns of 'crude' and purified human xanthine oxidase are shown in Fig 1, in which it can be seen that purified enzyme gives essentially a single band. This corresponds to Mr 145,000.

In common with bovine milk enzyme, purified, as here, in the absence of thiol reagents, (Batelli *et al.*, 1973), the human enzyme showed little xanthine dehydrogenase activity. The xanthine oxidase activity, as assayed spectrophotometrically, by urate production (Nakamura and Yamazaki, 1982), was, however, very low (10-20mU/mg) compared with values (2000-5000 mU/mg) commonly obtained with bovine milk enzyme (Bray, 1975). In view of the essential homogeneity of the purified human enzyme protein, its low activity would be consistent with a high (greater than 95%) content of demolybdo- or desulpho-forms of the enzyme, in which the xanthine binding sites (molybdenum sites) are inactive (Bray, 1975). This suggestion is supported by the NADH-ferricyanide oxidoreductase activity of the purified human enzyme (1200mU/mg), which is similar to that (900mU/mg) shown by the bovine milk enzyme (Biozyme Laboratories, Blaenavon Wales); NADH-ferricyanide oxidoreductase activity is known (Komai *et al.*, 1969) not to require a functional molybdenum site. That the low human xanthine oxidase activity is attributable to desulpho-, rather than to demolybdo-enzyme is shown by its molybdenum content of approximately 1.0 atoms per 145,000 subunit. Further support for the presence of the desulpho-form in the enzyme purified from human milk is given by the observation that its xanthine oxidase activity could be increased, approximately five fold, by incubation with sulphide and dithionite under anaerobic conditions (Wahl and Rajagopalan, 1982).

The presence of 20-40% desulpho-enzyme in xanthine oxidase purified from bovine milk has long been known and was thought to arise as a result of the purification procedure or storage (Bray, 1975). Ikegami and Nishino (1986), however, have, more recently shown that 30-40% desulpho-form is present, not only in purified rat liver xanthine dehydrogenase, but also in the fresh liver homogenates. We find that xanthine oxidase activity of human milk is some 100 times lower than that of bovine milk, despite the similar yields, in terms of protein, of enzyme purified from the two sources. Similarly low activities in bovine milk have been previously noted (Zikakis *et al.*, 1976). This suggests strongly that desulpho-enzyme (if such is the 'inactive' form) makes up the overwhelming majority of the enzyme both *in vivo* and in the purified enzyme. The apparent proportion of desulpho-enzyme in human milk xanthine oxidase is much higher than that reported from other mammalian sources and its significance is unclear, although regulatory sulphuration/desulphuration systems have been proposed for several enzymes (Coughlan, 1981). It is worth noting that similar predominance of 'inactive' xanthine oxidase in human heart would serve to explain its detection by immunohistochemistry but not by conventional enzyme assay.

REFERENCES

Batelli, M.G., Lorenzoni, E and Stirpe, F., 1973, Milk xanthine oxidase Type D (Dehydrogenase) and Type O (oxidase), Biochem J., 131: 191.

Benboubetra, M., 1989, Characterization and significance of anti-(bovine MFGM) antibodies, Ph.D. Thesis, University of Bath.

Blake, D.R., Unsworth, J., Outhwaite, J.M., Morris, C.J., Merry, P., Kidd, B.L., Ballard, R. and Gray, L., 1989, Hypoxic-reperfusion injury in the inflamed human joint, Lancet, i:289.

Brass, C.A., Narciso, J. and Gollan, J.L., 1991, Enhanced activity of the free radical producing enzyme, xanthine oxidase, in hypoxic rat liver:regulation and pathophysiologic significance, J.Clin. Invest., 87:424.

Bray, R.C., 1975, Molybdenum iron-sulfur flavin hydroxylases and related enzymes, in: "The Enzymes", Vol XII, 3rd ed., P.D. Boyer, ed., Academic Press, New York.

Coughlan, M.P., 1981, Is protein function regulated by the reversible incorporation of sulphur? Biochem. Soc. Trans., 9, 307-308.
Eddy, L.J., Stewart, J.R., Jones, H.P., Engerson, T.D., McCord, J.M. and Downey, J.M., 1987, Free radical-producing enzyme, xanthine oxidase, is undetectable in human hearts, Am. J. Physiol., 253: H709.
Graham, K., Fleming, J.E., Young, R and Bensch, K.G., 1989, Preparation of antibodies against xanthine oxidase from human milk, Int. J. Biochem., 21: 715
Granger, D.N., Rutili, G. and McCord, J.M., 1981, Superoxide radicals in feline intestinal ischemia, Gastroenterology, 78: 474.
Grum, C.M., Gallagher, K.P., Kirsh, M.M. and Shlafer, M., 1989, Absence of detectable xanthine oxidase in human myocardium, J. Mol. Cell. Cardiol., 21: 263.
Hart, L.I., McGartoll, M.A., Chapman, H.R. and Bray, R.C., 1970, The composition of milk xanthine oxidase, Biochem. J., 116: 851
Ikegami, T., and Nishino, T., 1986, The presence of desulfo xanthine dehydrogenase in purified and crude enzyme preparations from rat liver, Arch. Biochem, Biophys., 247, 254-260.
Jarasch, E-D., Grund, C., Bruder, G., Heid, H.W., Keenan, T.W. and Franke, W.W., 1981, Localization of xanthine oxidase in mammary gland epithelium and capillary endothelium, Cell, 25: 67.
Komai, H., Massey, V. and Palmer, G., 1969, The preparation and properties of deflavo xanthine oxidase, J. Biol. Chem., 244: 1692.
Krenitsky, T.A., Spector, T. and Hall, W.W., 1986, Xanthine oxidase from human liver: purification and characterization. Arch. Biochem. Biophys., 247: 108.
Muxfeldt, M. and Schaper, W., 1987, The activity of xanthine oxidase in heart of pigs, guinea pigs, rabbits, rats and humans, Basic Res. Cardiol., 82: 486.
Nakamura, M. and Yamazaki, I., 1982, Preparation of bovine milk xanthine oxidase as dehydrogenase form, J. Biochem., 92: 1279.
Thompson-Gorman, S.L. and Zweier, J.L., 1990, Evaluation of the role of xanthine oxidase in myocardial reperfusion injury, J. Biol. Chem., 265: 6656.
Wahl, R.C. and Rajagopalan, 1982, Evidence for the inorganic nature of the cyanolyzable sulfur of molybdenum hydroxylases, J. Biol. Chem., 257: 1354.
Zeise, L. and Zikakis, J.P., 1987, Characterization of human colostral xanthine oxidase, J. Agric. Food. Chem., 35: 942.
Zikakis, J.P., Dougherty, T.M. and Biasotto, N.O., 1976, The presence and some properties of xanthine oxidase in human milk and colostrum, J. Food. Sci., 41: 1408.
Zikakis, J.P., Dreisel, M.A. and Silver, M.R., 1983, Bovine, caprine and human milk xanthine oxidases: isolation, purification and characterisation, in: "Instrumental analysis of foods", G. Charalambous, ed., Academic Press, Orlando, FL.

MECHANISMS OF PROTECTION OF HEPATOCYTES AT ANOXIA AND REOXYGENATION BY THE XANTHINE OXIDASE INHIBITOR OXYPURINOL

Wolfram Schneider, Werner G. Siems, Tilman Grune, Christoph Schneider, Gerhard Gerber

Institute of Biochemistry, Medical School Charité
Humboldt University, Berlin O-1040, Germany

INTRODUCTION

Oxygen-free radicals generated during postischemic reperfusion are thougt to be responsible for the damage which occurs on reoxygenation. However, the participation of free oxygen radicals on reperfusion injury in liver is still controversial,[1,2] including the effects of xanthine oxidase inhibitors allopurinol and oxipurinol. The enzyme release which occurs upon reperfusion in the ischemic perfused rat liver was decreased by allopurinol.[3,4] In contrast, allopurinol had no appreciable effect on recovery of the ATP level in ischemic perfused rat liver during reoxygenation.[5] The discussion was stimulated since a radical scavinging activity of allopurinol and oxipurinol was prescribed.[6,7]

We carried out experiments with isolated hepatocytes (i) to observe changes of the energy and of the oxygen radical metabolism as two important components of the cell damage and (ii) to test the influence of different concentrations of allopurinol and oxipurinol on nucleotide metabolism and free radical formation.

MATERIALS AND METHODS

Hepatocytes were prepared from male Wistar H-strain rats (180-250 g) with collagenase perfusion technique. Only starved (24 hours) animals were used and no glucose was added to the incubation medium (37°C). Cells were incubated in Krebs-Henseleit-bicarbonate buffer with 1% albumin. Anoxia was induced by replacement of the O_2/CO_2 gas phase by N_2/CO_2 (95:5 v/v). The final cytocrit was 2-3% while 1% cytocrit corresponded to $1.68 \pm 0.11 \times 10^6$ cells/ml. Trypan blue uptake was used to estimate cell damage.

Samples of cell suspension (1 ml) for assessing reduced and oxidized glutathione (GSH, GSSG) and HPLC separation of nucleotides, nucleosides and nucleobases were deproteinized in 6% ice-cold $HClO_4$ and centrifuged (6 min, 1200 g). For

HPLC an aliquot of supernatant was neutralized with triethanolamine-K_2CO_3, filtrated and frozen at -20°C. Adenine and guanine nucleotides and their catabolites were analysed by ion-pair reversed-phase HPLC.[6,7] Oxidative stress was estimated by means of GSH and GSSG concentrations.[10,11]

RESULTS AND DISCUSSION

Incubation of hepatocytes from starved rats under anoxic conditions caused a rapid loss of adenine nucleotides. The restoration of ATP during reoxygenation strongly depends on the preceding time of anoxia. While after 30 minutes of anoxia a fast and complete ATP recovery could be observed, it reached only 43 % of initial value after 60 minutes of anoxia.

The degradation of ATP under anoxic conditions is accompanied by a transitory increase of purine catabolites: maximal concentrations were measured at 10 min for ADP, at 20 min for AMP and at 30 min of anoxia for IMP and adenosine. IMP showed the most pronounced rise reaching maximal values of nearly 1.5 mmoles/l cells at 30 min. Because of no further accumulation of IMP, adenosine, guanosine and hypoxanthine after 30 min of anoxia, it can be suggested that the purine nucleotide pool was exhausted at this time. This might be one reason for the incomplete ATP restoration after 60 minutes of anoxia.

Table 1. Concentrations of and purine metabolites in isolated hepatocytes subjected to 30 minutes of anoxia followed by 30 minutes of reoxygenation. Values expressed as mmoles/l cells (mean values ± S.E.M., n=3)
Oxy = Oxypurinol 1 mM, HYX = Hypoxanthine, n = no (without Oxy), y = yes (with Oxy)

	Oxy	Normoxia	Anoxia 30 minutes	Reoxygenation 30 minutes
ATP	n	2.16 ± 0.17	0.21 ± 0.12	2.35 ± 0.23
	y		0.25 ± 0.16	2.38 ± 0.51
ADP	n	0.43 ± 0.04	0.25 ± 0.08	0.22 ± 0.05
	y		0,28 ± 0,09	0,20 ± 0,11
AMP	n	0,27 ± 0,03	0,34 ± 0,06	0,16 ± 0,04
	y		0,36 ± 0,06	0,20 ± 0,07
IMP	n	0,31 ± 0,05	1,45 ± 0,09	0,47 ± 0,08
	y		1,56 ± 0,05	1,25 ± 0,10
Ado	n	0,12 ± 0,02	0,62 ± 0,42	0,26 ± 0,04
	y		0,67 ± 0,06	0,31 ± 0,06
HYX	n	< 0,01	0,60 ± 0,16	0,11 ± 0,04
	y		1,52 ± 0,18	1,89 ± 0,10

It was suggested that the inhibition of xanthine-oxidoreductase (XOR) should preserve purine metabolites for the salvage pathways. Thus, an improved ATP restoration during reoxygenation could be expected. We used the inhibitor oxypurinol with a final concentration of 1 mM to guarantee the complete inhibition of XOR. However, oxypurinol did not improve ATP recovery during reoxygenation neither after 10, 30 nor 60 minutes of anoxia. Only a marked preservation of hypoxanthine occured, whereas other metabolites like IMP, inosine and adenosine sustained not significantly higher compared to the group without application of the inhibitor.

Xanthine and hypoxanthine are the endproducts of nucleotide degradation before conversion to uric acid by XOR. The oxidase form of this enzyme has been suggested to be a main source of free radicals in hypoxia and reoxygenation. To

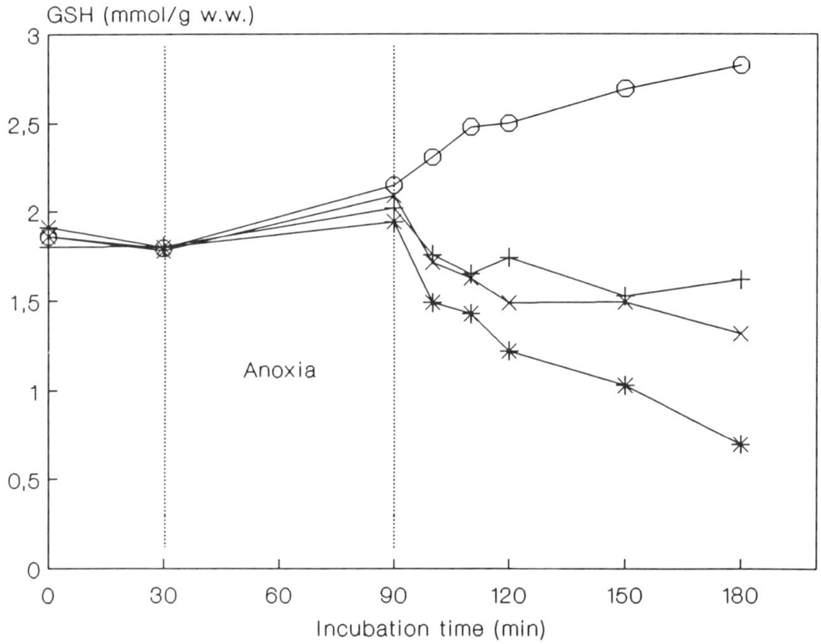

Figure 1: Concentration of reduced glutatione (GSH) in isolated hepatocytes during anoxia and reoxygenation.
(Groups: o = Control/Normoxia, * = Anoxia for 60 minutes, + = Anoxia with oxypurinol 1 mM, x = Anoxia with oxypurinol 0.02 mM)

The loss of GSH was partially prevented by oxypurinol, but there was no difference between the effects of 20 μM and demonstrate oxidative stress in hepatocytes GSH concentrations were measured. Figure 1 shows the changes of GSH level during and after 60 minutes of anoxia. Within 30 minutes of reoxygenation GSH concentrations decreased to 61 % of initial values, indicating an oxidative stress in the initial phase of reoxygenation.

1 mM levels of the inhibitor. It was reported that already 17 μM oxypurinol inhibits sufficiently the xanthine oxidase catalyzed production of superoxide.[7] On the other hand, a scavenger mechanism for oxypurinol was reported.[6] In this case the concentration of 1 mM oxypurinol should exert an additional protective effect against oxidative stress, which could not be established in our experiments.

It can be concluded, that oxypurinol reduced the cell damage of hepatocytes during reoxygenation mainly by preventing free radical generation via xanthine oxidase, whereas the preservation of purine metabolites for ATP restoration by accelerated salvage pathways seems to be in vain. Rather the ATP restoration depends on the function of mitochondria. This is confirmed by the diminished oxygen consumption (-37%) of the cells in the early reoxygenation period.

REFERENCES

1. H. J. Jaeschke, C. V. Smith, and J. R. Mitchell, Reactive oxygen species during ischemia-reflow injury in isolated perfused rat liver, J. Clin. Invest. 81:1240 (1988).
2. D. N. Gallagher, A. J. Buda, D. Pace, R. A. Geren, and M. Shlafer, Failure of superoxide dismutase and catalase to alter size of infarction in conscious dogs after 3 hours of occlusion followed by reperfusion, Circulation 73:1065 (1987).
3. M. Marotto, R. G. Thurman, and J. J. Lemasters, Early midzonal cell death during low-flow hypoxia in the isolated, perfused rat liver: Protection by allopurinol, Hepatology 8:585 (1988).
4. M. Younes and O. Strubelt, The involvement of reactive oxygen species in hypoxic injury to rat liver, Res. Comm. Chem. Pathol. Pharmacol. 59:369 (1988).
5. W. Kamiike, F. Watanabe, T. Hashimoto, K. Tagawa, Y. Ikeda, K. Nakao, and Y. Kawashima, Changes in cellular levels of ATP and its catabolites in ischemic rat liver, J. Biochem. 91:1349 (1982).
6. D. K. Das, R. M. Engelman, R. Clement, H. Otani, M. R. Prasad, and P. S. Rao, Role of xanthine oxidase inhibitor as free radical scavenger: A novel mechanism of action of allopurinol and oxipurinol in myocardial salvage, Biochem. Biophys. Res. Commun. 148:314 (1987).
7. T. Spector, Oxipurinol as an inhibitor of xanthine oxidase catalyzed production of superoxide, Biochem. Pharmacol. 37:349 (1987).
8. A. Werner, W. Schneider, W. Siems, T. Grune, and C. Schreiter, Ion-pair reversed phase HPLC determination of nucleotides, nucleosides and nucleobases - application to nucleotide metabolism in hepatocytes, Chromatographia 27:639 (1989).
9. W. Schneider, T. Grune, W. Siems, A. Werner, and G. Gerber, Nucleotide concentrations in hepatocytes during anoxia and reoxygenation in presence of allopurinol and oxypurinol, Biomed. Biochim. Acta 48:40 (1989).
10. E. Beutler, O. Duron, and B. M. Kelly, Improved method for the determination of blood glutathione, J.Lab.Clin.Med. 61:882-888 (1963).

OXYPURINOL REDUCES ISCHEMIC BRAIN INJURY IN THE GERBIL AND RAT

J.W. Phillis and Y. Lin

Department of Physiology
Wayne State University School of Medicine
Detroit, Michigan, U.S.A.

The pathophysiology of brain ischemia is characterized by a complex sequence of events, which include biochemical, hemodynamic and electrophysiological processes. Decreases in cerebral blood flow (CBF) below a critical threshold result in energy failure, tissue acidosis, disturbed ion homeostasis, with cellular Na^+ and Ca^{++} influx and K^+ efflux, membrane depolarization, and cytotoxic edema. There is a massive release of excitatory amino acid neurotransmitters (glutamate and aspartate) which triggers further membrane depolarization and an additional accumulation of intracellular Ca^{++}. This accumulation of free cytosolic intracellular Ca^{++} appears to play a key role in the progression of events towards irreversible neuronal damage, in that it leads to an activation of a series of neurotoxic processes including lipid peroxidation, free radical generation, activation of proteolytic enzymes and pathological gene activation. The formation of vasogenic tissue edema is another important consequence of brain ischemia, which frequently complicates the clinical situation.

The catabolism of adenosine 5'-triphosphate (ATP) during ischemia results in an efflux of adenosine and its metabolites, inosine, hypoxanthine, xanthine and uric acid into the interstitial spaces of the brain. The final stages of purine metabolism, hypoxanthine to xanthine, and xanthine to uric acid are catalyzed by the enzyme xanthine oxidase, which exists in two forms, xanthine dehydrogenase and xanthine oxidase. Under ischemic conditions, xanthine oxidase is converted from the dehydrogenase into the oxidase form, which utilizes molecular oxygen rather than NAD^+ as the electron acceptor. In this reaction, the degradation of hypoxanthine to xanthine and uric acid results in the release of superoxide radicals. In the presence of iron or other transition metals, superoxide can be converted to highly toxic hydroxyl radicals, readily able to cross cell membranes. Free radicals have been implicated in the pathogenesis of ischemic injury to brain.[1] The lipid peroxides, which have been identified in the brain following cerebral ischemia, may be a consequence of membrane breakdown by free radicals.

Catabolism of hypoxanthine to uric acid by xanthine oxidase represents a potential source of free radicals in the ischemic/reperfused brain. Xanthine oxidase is preferentially localized in capillary endothelial cells,[2] where the free radical formation resulting from activation of this enzyme could critically affect the properties of the blood-brain barrier, contributing amongst other effects to vasogenic brain edema.

The involvement of xanthine oxidase in ischemic tissue injury is supported by findings that allopurinol, a selective xanthine oxidase inhibitor, protects against reperfusion injury in intact rat and gerbil brains.[3] Allopurinol is itself an excellent substrate for xanthine oxidase, with superoxide radical generation occurring as it is oxidized to oxypurinol. As an alternative substrate for the enzyme, allopurinol competes with the physiological substrates, hypoxanthine and xanthine for binding and subsequent oxidation. It inhibits xanthine oxidase - catalyzed production of uric acid only by virtue of its own substrate activity. By contrast, oxypurinol is a true dead-end inhibitor of xanthine oxidase.[4] Furthermore, because hypoxanthine and xanthine compete with allopurinol as substrates for binding to xanthine oxidase, the concentrations of allopurinol required to inhibit the enzyme must be increased in direct proportion to any increases in the concentrations of competing substrates (hypoxanthine and xanthine) - the conditions which pertain during ischemia. Hence oxypurinol offers the advantages of being a more effective inhibitor of xanthine oxidase in the ischemia/reperfused brain, as well as of not providing an additional source for free radical generation. For these reasons oxypurinol, rather than allopurinol, is the preferred inhibitor of xanthine oxidase for use in experiments on cerebral ischemia.

Purine Release from the Ischemia/Reperfused Cerebral Cortex

Release of purines from the normoxic and hypoxic/ischemic rat cerebral cortex was studied with the cortical cup technique. Oxypurinol (20 mg/kg, i.v.) and allopurinol (100 mg/kg, i.v.) had comparable effects on ischemia-evoked purine release. Both agents reduced uric acid release to less than 10% of predrug levels, whereas the releases of adenosine, inosine, hypoxanthine and xanthine were elevated.[5] The failure of the xanthine oxidase inhibitors to depress xanthine levels was unexpected and may be a result of the formation of xanthine by alternative pathways such as the catabolism of guanosine to xanthine by guanine deaminase, as well as of the accumulation of any now non-metabolizable xanthine present. By preventing the catabolism of hypoxanthine, oxypurinol and allopurinol would exert a "purine-sparing" effect since hypoxanthine can be reconverted to inosine 5'-monophosphate (IMP) by the enzyme hypoxanthine phosphoribosyltransferase and thus reincorporated into the adenine nucleotide pool.

Purine release from the cerebral cortex into cortical cups was studied in rats subjected to 20 min of four vessel occlusion cerebral ischemia.[6] Adenosine and inosine release rose rapidly during the period of ischemia, peaked in the first (10 min) post-ischemic collection period and then declined to basal levels within 60 min of the onset of reperfusion. Hypoxanthine and xanthine levels rose more gradually, peaking 20-40 min after the onset of reperfusion. Xanthine levels were still significantly elevated 2 hr after the onset of reperfusion.

Cerebroprotective Actions of Oxypurinol

Oxypurinol was tested in two models of cerebral ischemia. The Mongolian gerbil has been extensively used as a model of transient forebrain ischemia since, as a result of an incomplete Circle of Willis in this animal, global forebrain ischemia can be induced by ligation of the common carotid arteries. Circulation to the hindbrain through the basilar artery is not affected by this procedure. Cerebral ischemic injury following a brief (5 min) period of forebrain ischemia can be quantitated as a readily measurable hypermotility or histologically by measuring the extent of damage to the CA1 hippocampal pyramidal cell layer 5 days later. The degree of hippocampal damage is positively correlated with the increase in locomotor activity.[7] Oxypurinol (40 mg/kg), administered either 20 min prior to or 30 min post-ischemia, significantly reduced both the hypermotility and the extent of hippocampal CA1 pyramidal cell damage observed in control saline injected animals subjected to cerebral ischemia.[8,9] The cerebroprotective actions of oxypurinol were not secondary to reductions in body temperature, which was measured with telemetry modules inserted into the peritoneal cavity prior to the ischemic episode.

The gerbil model of forebrain ischemia can be regarded as replicating a brief cardiac arrest. In order to utilize a model which more closely parallels a stroke in humans, experiments were conducted on rats with permanent occlusion of the middle cerebral artery (MCA) in tandem with occlusion of the ipsilateral common carotid artery. Oxypurinol (40 mg/kg, i.p.) administered either 30 min prior to or 60 min post-ischemia significantly reduced the ischemic cerebral infarct (Fig. 1) (measured by incubating brain slices in triphenyltetrazolium chloride solution), reduced the extent of brain swelling and attenuated the degree of neurological deficit observed in non-treated MCA lesioned rats. Protection against ischemic damage was not observed when oxypurinol administration was delayed for 5 hr post-stroke.[10] Body temperature measurements with telemetry capsules confirmed that the cerebroprotective effects of oxypurinol were not a result of hypothermia.

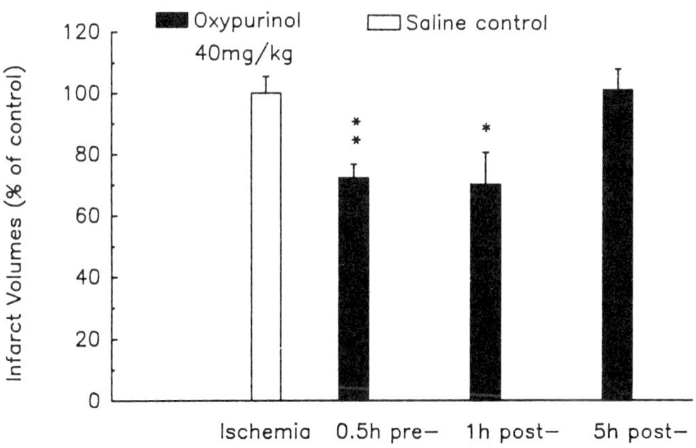

Fig. 1. Percentage reductions in cerebral hemisphere infarct volumes in rats treated with oxypurinol (40 mg/kg) 0.5 h prior to or 1 hr post (but not a 5 h post) middle cerebral artery occlusion.

In conclusion, these observations demonstrate that oxypurinol has cerebroprotective activity against ischemic injury. Various potential mechanisms can be proposed to account for oxypurinol's beneficial effects, and several of these may operate simultaneously. They include: (a) enhancement of extracellular levels of adenosine. Adenosine dilates cerebral blood vessels; depresses excitatory amino acid release from presynaptic nerve terminals; inhibits membrane Ca^{++} fluxes; and inhibits platelet and neutrophil aggregation, limiting clot formation and free radical generation by neutrophils; (b) promotion of purine salvage by conserving hypoxanthine for resynthesis into IMP and thus replenishing ATP stores; (c) prevention of toxic free radical generation by xanthine oxidase during the conversion of hypoxanthine to xanthine and uric acid. It may be significant that xanthine oxidase is localized in capillary endothelial cells, as these will inevitably be an important site of free radical generation. The resultant destruction of endothelial cells could account for the failure of the blood-brain-barrier and the vasogenic edema which frequently occur in stroked brains. The protective actions of the enzymes superoxide dismutase and catalase, which would not be expected to cross the blood-brain-barrier, against ischemic injury,[11] is likely to be a consequence of their inactivation of superoxide and hydroxyl radicals generated by endothelial cells lining cerebral capillaries. The fact that allopurinol has been extensively employed as a treatment for gout and uric acid nephropathy indicates that allopurinol and oxypurinol can be safely tolerated by humans over an extended time period. Oxypurinol promises, therefore, to be a useful therapeutic agent for the prevention and treatment of ischemic neuronal injuries such as those occurring during cardiac arrest and stroke.

REFERENCE

1. C. N. Oliver, P. Stark-Reed, E. R. Stadtman, G.L. Liu, J. M. Carney and R. A. Floyd, Ischemia/reperfusion induced oxidative damage to proteins in gerbil brain, Proc. Natl. Acad. Sci. U.S.A. 87:5144 (1990).
2. Betz, A. L., Identification of hypoxanthine transport and xanthine oxidase activity in brain capillaries, J. Neurochem. 44:574 (1985).
3. D. Martz, G. Rayos, G.P. Schielke and A. L. Betz, Allopurinol and dimethylthiourea reduce brain infarction following middle cerebral artery occlusion in rats, Stroke 20:488, 1989.
4. T. Spector, Oxypurinol as an inhibitor of xanthine oxidase-catalyzed production of superoxide radical, Biochem. Pharmacol. 37:349 (1988).
5. M. H. O'Regan, J. W. Phillis and G. A. Walter, The effects of the xanthine oxidase inhibitors, allopurinol and oxypurinol on the pattern of purine release from hypoxic rat cerebral cortex, Neurochem. Int. 14:91 (1989).
6. J. W. Phillis, G. A. Walter and R. E. Simpson, Brain adenosine and transmitter amino acid release from the ischemic rat cerebral cortex: Effects of adenosine deaminase inhibitor deoxycoformycin, J. Neurochem. 56:644 (1991).
7. S. C. Gerhardt and C. A. Boast, Motor activity changes following cerebral ischemia in gerbils are correlated with the degree of neuronal degeneration in the hippocampus, Behav. Neurosci. 102:301 (1988).
8. J. W. Phillis, Oxypurinol attenuates ischemia-induced hippocampal damage in the gerbil, Brain Res. Bull. 23:467 (1989).
9. J. W. Phillis and C. Clough-Helfman, Oxypurinol, but not deoxycoformycin, administered post-ischemia, protects against CA1 hippocampal damage in the gerbil, Int. J. Purine Pyrimid. Res. 1:31 (1990).
10. Y. Lin and J. W. Phillis, Oxypurinol reduces focal ischemic brain injury in the rat, Neurosci. Lett. 126:187 (1991).
11. T. H. Liu, J. S. Beckman, B. A. Freeman, E. L. Hogan and C.Y. Hsu, Polyethylene glycol-conjugated superoxide dismutase and catalase reduce ischemic brain injury, Am. J. Physiol. 256:H589 (1989).

ACADESINE (AICA RIBOSIDE) ATTENUATES REPERFUSION INJURY AND OXIDANT-INDUCED DAMAGE OF THE HEART

Metzner E. K., Fox M H, Potter S., Mullane K M, and Bullough D A.

Gensia Pharmaceuticals, Inc.
San Diego, CA 92121, USA.

Acadesine (AICA riboside; 5-amino-1-β-D-ribofuranosyl imidazole-4-carboxamide) has beneficial effects in reducing ischemic damage which are mediated, at least in part, by its ability to increase local adenosine concentration (Gruber et al. Circulation, 80:1400, 1989). Adenosine also protects the heart from the deleterious effects of post-ischemic reperfusion. Therefore in this study acadesine was administered solely at reperfusion to isolated guinea pig hearts perfused in a Langendorff mode and subjected to ischemia for 45 min followed by reperfusion for 30 min. Addition of acadesine (10-50μM) to the perfusion buffer at the onset of reperfusion improved the recovery of left ventricular developed pressure (LVDP) at 30 min from 68 ± 3% in controls (n=12) to e.g. 82 ± 3% at 20μM ($P < 0.01$, n=6) implying prevention of reperfusion injury. Acadesine also attenuated myocardial damage induced by electrolysis of the perfusion buffer (1mA for 1 min) to produce a mixture of free radicals and oxidants that have been implicated in reperfusion injury. The 38 ± 4% (n=8) decrease of LVDP after 10 min in untreated hearts subjected to electrolysis was completely prevented by 20μM acadesine. Hypochlorous acid (HOCl), implicated in the electrolysis-induced damage and a major oxidant formed by neutrophils, was reduced by acadesine to H_2O and Cl^- in vitro suggesting that its beneficial effects are at least partly related to its antioxidant activity. Other antioxidants tested reduced HOCl and attenuated electrolysis-induced damage but had no beneficial effect in the ischemia/reperfusion model. This suggests that the antioxidant activity may not be sufficient to account for the protection from reperfusion injury and that acadesine has additional protective properties.

Purine and Pyrimidine Metabolism in Man VII, Part A
Edited by R.A. Harkness *et al.*, Plenum Press, New York, 1991

EFFECTS OF H_2O_2 AND CHPO ON ATP-, CP-CONTENT AND 14C-HX-UPTAKE BY HUVECS

A. Griesmacher, G. Weigel, A. Windischbauer, and M.M. Müller

2nd Dept. of Surgery, Div. of Clinical Biochemistry
Spitalgasse 23, 1090 Vienna, Austria

INTRODUCTION

The endothelium of the vascular wall deserves particular attention, since it is responsible for an actively controlled metabolites' transport to the underlying strata of the blood vessels. Furthermore the endothelial cells show high metabolic activity and thus influence on a great number of physiological processes (1).

Reactive oxygen species, generated during reperfusion after oxygen deficiency and inflammatory situations, are reported to have a damaging effect on vascular endothelium. Such an injury of the endothelium is thought to be responsible for the initiation of atherosclerotic processes. We and others could show that the cellular metabolism as well as the cellular function (increase of lipidperoxides, decrease of the intracellular glutathion, stimulation of the eicosanoid-synthesis) are effected by reactive oxygen species generating substances.

Recently, we could show that in human endothelial cells purine salvage pathway is predominating over the de novo synthesis (2). In this study the effects of reactive oxygen generating agents on purine metabolism were investigated. Hydrogen hydroperoxide (H_2O_2) and cumene hydroperoxide (CHPO) served as model substances. Intracellular ATP as well as CP were measured after exposure to H_2O_2 or CHPO. In order to investigate the relationship between purine salvage and de novo purine synthesis in presence of reactive oxygen species the uptake of labelled 14C-hypoxanthine and 3H-glycine was examined.

MATERIAL AND METHODS

Cell Culture

Endothelial cells were prepared using human umbilical veins. Cells were isolated and cultured according to a modified standard procedure reported by Jaffe (3). Briefly, fresh human umbilical veins were filled with 0.1% collagenase solution (Sigma) and incubated at 37°C for 5 min. Thereafter

Purine and Pyrimidine Metabolism in Man VII, Part A
Edited by R.A. Harkness *et al.*, Plenum Press, New York, 1991

the veins were perfused with M 199 (Gibco) medium containing 20 % pooled human serum (pH 7.4). Cells were collected from the perfusate by centrifugation at 800 g for 5 min and seeded into culture T-75 flasks precoated with human fibronectin (Biomedica). Cells were cultured in M 199 medium (pH 7.4) containing 20% human serum, 100.000 U/l penicillin (Gibco) and 100.000 µg/l streptomycin (Gibco), 100.000 U/l low molecular weight heparin (Sigma) and 30 mg/l bovine hypothalamic growth factor (Biomedica). The confluent primary monolayers (approx. 8.000.000 cells/flask) were washed and trypsinized. The cell suspensions were transferred into each well of a 24 well culture plate and cultivated for four days. Only cells from these first subcultures were used for the experiments described below.

Incubation with Reactive Oxygen Species

After two washings the confluent monolayers were incubated with phosphate buffered saline (PBS) plus 5 mmol/l glucose containing increasing concentrations (0.01-1 mmol/l) of H_2O_2 or CHPO for 60 min at 37°C. ATP and CP contents were determined after 60 min of incubation in controls and oxidative stressed cells. Purine salvage and de novo synthesis were measured over a period of 60 min under the conditions described below.

Measurement of ATP and CP

ATP and CP were determined in neutralised perchloric acid extracts by means of bioluminescence as described earlier (4). CP was converted to ATP by adding creatine kinase (Sigma) and ADP (Sigma). The measurements were performed immediately after the incubation.

Uptake of 14C-hypoxanthine and 3H-glycine

In another series of experiments the endothelial cells were incubated with PBS containing 5.2 µmol/l (8-14C)-labelled hypoxanthine (Amersham, 1.86 GBq/mmol) or (2-3H)-labelled glycine (Amersham, 403 GBq/mmol) in addition to the reactive oxygen species for 60 min at 37°C. Thereafter the supernatants were withdrawn and the cell layers were washed with ice-cold 0.9% NaCl and lysed with a 0.5 mol/l perchloric acid. After centrifugation the supernatants were neutralised with 2 mol/l potassium hydroxide. Neutralised cell extracts as well as cell supernatants were stored at -70°C. After addition of unlabelled tracers the purine nucleotides were separated by thin layer chromatography on TLC aluminium silica gel 60 F_{254} precoated sheets (MERCK) using n-propanol, methanol, ammonia (33%), H_2O (45:15:30:10) (5). The zones corresponding to standards were located under UV-light, cut out and their radioactivities measured.

RESULTS

Table 1 and 2 shows the effects of H_2O_2 and of CHPO on intracellular ATP and CP content as well as on 14C-hypoxanthine uptake and incorporation into purine nucleotides. Significant increases in ATP- and CP-content were found after incubation with 0.01 mmol/l H_2O_2 (ATP: +51,3%, CP: +18,2%) or CHPO (ATP: +77.5%, CP: +66.5%) in comparison to controls. Although in presence of 0.1 mmol/l H_2O_2 and CHPO (data not shown) a remarkable transient increase in intracellular ATP between 5 and 30 min could be observed, both ATP- nor CP-contents were found to be significantly effected after 60 min of incubation. The incubation of the endothelail cells with 1 mmol/l H_2O_2 or CHPO led to a decrease of ATP- or CP-levels close to zero.

Table 1. H_2O_2-Induced Effects on Purine Metabolism.

H_2O_2 mmol/l	ATP nmol/mill cells	CP nmol/mill cells	14C-HX-Salvage nmol/mill cells	Triphosphates %
CONTROL	11.5±0.7	21.2±1.0	1.9±0.2	52.1
0.01	17.4±1.5*	24.3±2.8*	2.6±0.4*	62.0
0.1	12.7±2.4	21.5±2.4	2.1±0.3	55.7
1.0	0.6±0.2*	0.2±0.2*	not tested	not tested

14C-hypoxanthine into triphosphates is expressed as percentage of label incorporated into purine nucleotides (=100%). * = Statistical Significance ($p<0.01$)

Table 2. CHPO-Induced Effects on Purine Metabolism.

CHPO mmol/l	ATP nmol/mill cells	CP nmol/mill cells	14C-HX-Salvage nmol/mill cells	Triphosphates %
CONTROL	12.0±0.5	21.0±1.2	1.9±0.3	53.3
0.01	21.3±2.5*	35.3±4.1*	2.6±0.3*	66.6
0.1	12.4±0.9	22.3±2.2*	3.2±0.5*	65.4
1.0	0.8±0.4*	0.0±0.3*	not tested	not tested

14C-hypoxanthine into triphosphates is expressed as percentage of label incorporated into purine nucleotides (=100%). * = Statistical Significance ($p<0.01$)

The total uptake of 14C-hypoxanthine into purine nucleotides was 1.9 ± 0.2 nmol/mill cells in control experiments. 0.01 mmol/l H202 increased the total 14C-hypoxanthine-uptake by 39%, 0.01 mmol/l CHPO by 38%, 0.1 mmol/l H_2O_2 by only 11% and 0.1 CHPO by 68%. Furthermore increasing amounts of reactive oxygen species increased the relative amount of 14C-label in purine triphosphates. Compared to controls the label increased from approximately 52% to 62% and 67%, when incubating cells with 0.01 mmol/l H_2O_2 and CHPO, respectively. In contrast the uptake of 3H-glycine and its incorporation into purine nucleotides (0.49±0.23 nmol/mill cells) was not influenced by the radical oxygen generating systems.

DISCUSSION

In vivo hypoxia with subsequent reoxygenation causes breakdown of cellular nucleotides followed by an accumulation of hypoxanthine. Due to lack of intracellular scavengers simultaneously generation of oxygen reactive species are generated. Recent in vitro studies showed that millimolar concentrations of H_2O_2 lead to profound DNA damage (6). Moreover, the adenine uptake is reported to be decreased 30 min after exposure to 0.5 mmol/l H_2O_2 (7), a concentration, which certainly only occurs under pathophysiological conditions concomitantly with cell death. Our results show that levels higher than 0.1 mmol/l H_2O_2 or CHPO decrease intracellular ATP and CP, leading to persisting cell functions.

In contrast, lower concentrations of radical generating systems result in a temporary increase of high energy phosphates. These elevated levels could be explained either by a metabolic disturbance based on decreased ATP-consumption due to inhibition of ATPases by oxygen radicals or by an elevated rate of ATP-synthesis. In other experiments we observed that concentrations of H_2O_2 or CHPO lower than 0.1 mmol/l inhibit the uptake of ferritin by human endothelial cells (data not shown). Although these results would confirm the hypothesis of an inhibition of nucleotides- requiring processes (DNA-synthesis, transport processes), the significantly enhanced 14C-hypoxanthine uptake and incorporation into purine nucleotides in the presence of 0.01 mmol/l H_2O_2 and CHPO support also the latter hypothesis. Therefore decreased ATP consumption and increased purine salvage should explain the temporary highly increased ATP-levels. The 3H-glycine uptake, as measure for de novo synthesis, was not influenced under the experimental conditions. This is a strong hint, that as reported in phsysiological conditions also under oxidative stress purine salvage pathway predominates de novo synthesis in endothelial cells.

REFERENCES

1. Ross R., Atherosclerosis: a problem of the biology of arterial wall cells and their interactions with blood components, Atherosclerosis 1:293 (1981).
2. A. Griesmacher, G. Weigel, A. Windischbauer and M. M. Müller, Purine metabolism in human endothelial cells, Intern. J. Purine Pyrim. Res.: in press (1991).
3. E. A. Jaffe, R. L. Nachman, C. G. Becker and C. R. Minck, Culture of human endothelial cells derived from umbilical veins, J. Clin. Invest. 52:2745 (1973).
4. G. Schopf, H. Rumpold and M. M. Müller, Alterations of purine salvage pathways during differentiation of rat heart myoblasts towards myocytes, Biochim. Biophys. Acta 884:319 (1986).
5. N. Kolassa,H. Roos and K. Pfleger, A Seperation of purine derivatives by thin-layer chromatography on silica gel plates suitable for metabolic studies, J. Chromatogr. 66:175 (1972).
6. S. P. Andreoli, Mechanisms of endothelial cell ATP depletion after oxidant injury, Pediatr. Res. 25:97 (1989).
7. A. Ager and J. L. Gordon, Differential effects of hydrogen peroxide on indices of endothelial function, J. Exp. Med. 159:592 (1984).

ABSENCE OF XANTHINE OXIDOREDUCTASE ACTIVITY IN HUMAN MYOCARDIUM

Thomas Podzuweit, Antje Müller, Hilderose Beck,
Hans H. Scheld, Wolfgang Schaper

Max-Planck-Institute, Department of Experimental Cardiology
D-6350 Bad Nauheim, F. R. Germany

INTRODUCTION

The enzyme xanthine: oxygen oxidoreductase (trivial name xanthine oxidase, EC 1.1.3.22) (Fig. 1) has been proposed to be an important source of oxygen free radicals in the ischaemic-reperfused heart [1,2]. However, more recent investigations have shown species variability of the xanthine oxidoreductase enzyme system, that raised questions concerning the applicability of the xanthine oxidase hypothesis to human hearts [3]. Published data on xanthine oxidase activity of human myocardia were sparse and conflicting results had been obtained (Table 1). Because of this shortcoming we set out in June 1986 to examine some 100 human hearts for xanthine oxidase as well as xanthine dehydrogenase activities, using myocardial specimens obtained during open-heart surgery, and highly sensitive and selective assay procedures. In addition, human papillary muscle homogenates were analyzed for the presence of pharmacological and/or endogenous inhibitors, with rat heart xanthine oxidoreductase as reference enzyme.

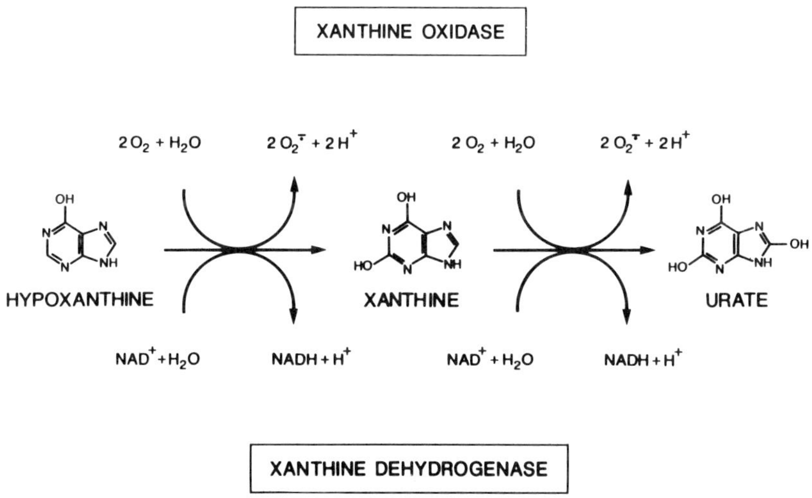

Fig. 1. Xanthine oxidoreductase reaction scheme

METHODS

Xanthine oxidoreductase activity was measured in vitro according to the depicted reaction scheme (Fig. 1), using HPLC with electrochemical and/or radiochemical detection of hypoxanthine, xanthine and urate (Figs. 2,3) [3]. Whenever possible, myocardial homogenates were incubated threefold: 1) with xanthine, 2) with xanthine and NAD^+, and 3) with hypoxanthine or radioactively labeled hypoxanthine in 100 $mmol \cdot l^{-1}$ Tris-HCl buffer pH 8.0. All incubations were aerobic with approx. 250 $\mu mol \cdot l^{-1}$ molecular oxygen present due to equilibration of the reaction mixtures with room air at ambient barometric pressure and temperature. Enzyme activity was measured at 25 °C with incubation times ranging from 10 min to 180 min. Specimens: a) isolated perfused rat myocardia and b) left ventricular needle biopsies and papillary muscles obtained during elective cardiac surgery (chiefly aortic and/or mitral valve replacement; aortocoronary bypass operation) (n=105 patients). Biopsies were frozen in liquid nitrogen and were analyzed within approx. 14 days.

RESULTS

Homogenization of human papillary muscles in Tris-HCl buffer pH 8.0 at ambient temperature caused significant (endogenous) accumulation of hypoxanthine but not xanthine or urate (Results not shown). Likewise, following incubation of crude homogenates from human myocardium with xanthine (50 or 750 $\mu mol \cdot l^{-1}$) or hypoxanthine (50 $\mu mol \cdot l^{-1}$) as electron donors, and NAD^+ (500 $\mu mol \cdot l^{-1}$) and/or O_2 (atmospheric at approx. 250 $\mu mol \cdot l^{-1}$) as electron acceptors no production of urate was detected. Similarly, during aerobic incubation of human papillary muscle homogenates with ^{14}C-hypoxanthine neither ^{14}C-xanthine nor ^{14}C-urate were formed (Fig. 3). In addition, xanthine oxidoreductase activity was undetectable in human papillary muscle extracts that were subjected to either ultrafiltration (Centricon 10 or 30) or gel filtration (Sephadex G-25) (Fig. 4).

In contrast, the rat heart was found to contain abundant xanthine oxidoreductase activity (Figs. 3,4). The rat heart enzyme was inhibited by both allopurinol and oxypurinol but remained active when mixed with human papillary muscle homogenates (Results not shown).

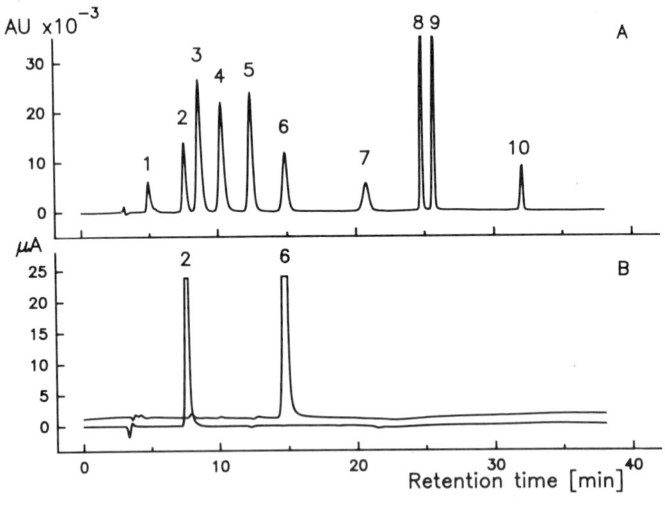

Fig. 2. HPLC separation of a mixture of 10 purine standards.
Column: LiChrospher RP-18e (250x4 mm; 5 μm packing). Sample: 20 μl water containing 0.2 nmol components 1,7,10 and 1 nmol components 2-6,8,9. Linear gradient.
Detection: absorbance at 254 nm (A), electrochemical at 0.30 V (lower trace) and 0.60 V (upper trace), resp. (B).
1=GTP, 2=urate, 3=ATP, 4=ADP, 5=hypoxanthine, 6=xanthine, 7=AMP, 8=NAD^+, 9=inosine, 10=adenosine.

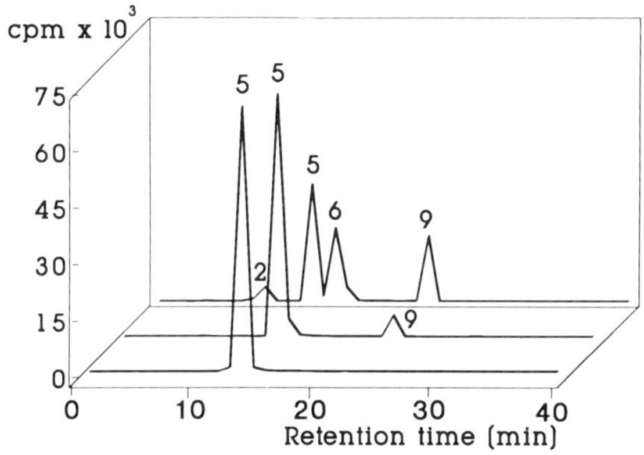

Fig. 3. Assay of xanthine oxidoreductase activity using HPLC with radiochemical detection of hypoxanthine (5), xanthine (6) and urate (2).
Substrate: [8-^{14}C] hypoxanthine (front). Degradation profile with human papillary muscle homogenate (middle) and rat heart homogenate (back); 60 min incubation. Separation conditions as in Fig. 2.

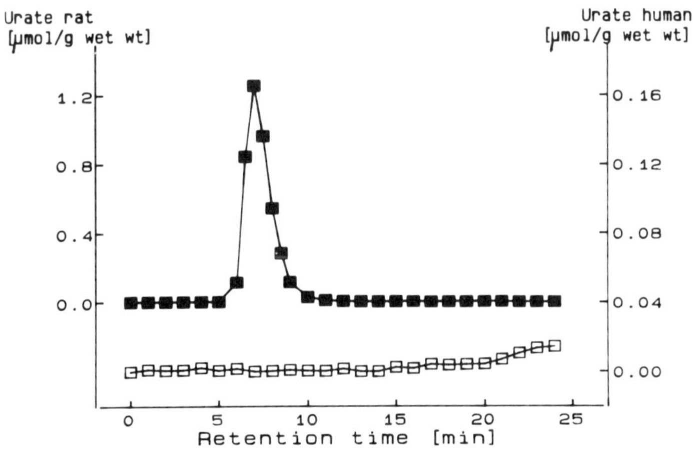

Fig. 4. Assay of xanthine oxidoreductase activity of rat heart (■) and human papillary muscle extracts (□) following gel filtration on a Sephadex G-25 medium column. Substrate: xanthine (50 µmol·l^{-1}). Assay: HPLC with electrochemical detection. Note 10fold higher magnification for human as compared to rat heart.

Table 1. Published data on xanthine: acceptor oxidoreductase activities of human myocardium

Electron acceptors: O_2	$O_2 + NAD^+$	Ferri-cyanide	Methylene blue	Number of hearts assayed	Reference	
0.009(0.005-0.013)*	-	-	-	2	Watts,	1965[4]
nd	-	-	-	1(?)	Ramboer,	1969[5]
-	-	74	-	1	Krenitsky,	1974[6]
0.013	1.30(±0.02)	-	-	2	Muxfeldt,	1987[7]
nd	nd	-	nd	4	Eddy,	1987[8]
4.6(0.0-23.8)*	42.0(26.3-69.3)*	-	-	9	Wajner,	1989[9]
nd	nd	-	-	11	Grum,	1989[10]
-	0.31(±0.10)	-	-	7	de Jong,	1990[11]

Enzyme activity in mUnits/g wet wt. Values are means (SD or range). *Recalculated data assuming a protein: wet weight ratio of 0.035, nd = not detectable.

DISCUSSION

The consistent result of the present investigation (involving 105 patients) was that the left ventricular human myocardium was lacking both xanthine oxidase and xanthine dehydrogenase activities. The lack of enzyme activity was not due to the presence of endogenous inhibitors. These findings, which are supported by the majority of the published data on human heart xanthine oxidase summarized in Table 1, indicate that xanthine oxidase is not a source of oxygen free radicals in the human myocardium. It follows that pharmacological inhibition of xanthine oxidase in man aimed at minimizing the injurious effects of myocardial ischaemia-reperfusion is bound to fail in its purpose.

REFERENCES

1. J.M. McCord, Oxygen-derived free radicals in postischemic tissue injury, N Engl J Med 312:159 (1985).
2. D.J. Hearse, A.S. Manning, J.M. Downey, and D.M. Yellon, Xanthine oxidase: a critical mediator of myocardial injury during ischemia and reperfusion?, Acta Physiol Scand 548 (Suppl.):65 (1986).
3. T. Podzuweit, W. Braun, A. Müller, and W. Schaper, Arrhythmias and infarction in the ischemic pig heart are not mediated by xanthine oxidase-derived free oxygen radicals, Basic Res Cardiol 82:493 (1987).
4. R.W.E. Watts, J.E.M. Watts, and J.E. Seegmiller, Xanthine oxidase activity in human tissues and its inhibition by allopurinol (4-hydroxypyrazolo [3,4-d] pyrimidine), J Lab Clin Med 66:688 (1965).
5. C.R.H. Ramboer, A sensitive and nonradioactive assay for serum and tissue xanthine oxidase, J Lab Clin Med 74:828 (1969).
6. T.A. Krenitsky, J.V. Tuttle, E.L. Cattau Jr, and P. Wang, A comparison of the distribution and electron acceptor specificities of xanthine oxidase and aldehyde oxidase, Comp Biochem Physiol 49B:687 (1974).
7. M. Muxfeldt and W. Schaper, The activity of xanthine oxidase in heart of pigs, guinea pigs, rabbits, rats, and humans, Basic Res Cardiol 82:486 (1987).
8. L.J. Eddy, J.R. Stewart, H.P. Jones, T.D. Engerson, J.M. McCord, and J.M. Downey, Free radical-producing enzyme, xanthine oxidase, is undetectable in human hearts, Am J Physiol 253:H709 (1987).
9. M. Wajner and R.A. Harkness, Distribution of xanthine dehydrogenase and oxidase activities in human and rabbit tissues, Biochim Biophys Acta 991:79 (1989).
10. C.M. Grum, K.P. Gallagher, M.M. Kirsh, and M. Shlafer, Absence of detectable xanthine oxidase in human myocardium, J Mol Cell Cardiol 21:263 (1989).
11. J.W. de Jong, P. van der Meer, A.S. Nieukoop, T. Huizer, R.J. Stroeve, and E. Bos, Xanthine oxidoreductase activity in perfused heart of various species, including man, Circ Res 67:770 (1990).

LOWER XANTHINE OXIDOREDUCTASE ACTIVITY IN ISOLATED PERFUSED HEARTS IF XANTHINE REPLACES HYPOXANTHINE AS SUBSTRATE

M. Janssen, J.W. de Jong, A.S. Nieukoop and
E. Keijzer

Cardiochemical Laboratory, Thoraxcenter
Erasmus University Rotterdam, The Netherlands

INTRODUCTION

In normal cardiac tissue xanthine oxidoreductase is present in the dehydrogenase form. During ischemia xanthine dehydrogenase is converted to xanthine oxidase, which generates free radicals during reperfusion. Both forms catalyze the breakdown of hypoxanthine to xanthine and xanthine to urate. Its activity can be measured with hypoxanthine or xanthine as substrate. Most widely xanthine is used as substrate because only one product is formed and therefore the enzyme activity can easily be determined. With hypoxanthine as substrate the difficult two-step reaction can raise problems if one wants to calculate the activity. Since either substrate is used in studies of xanthine oxidoreductase, this can explain the controversial data reported for various species.[1,2,3]

We decided to perfuse rat and guinea-pig hearts according to Langendorff with buffer containing hypoxanthine or xanthine, to measure xanthine and/or urate in the effluent, and to calculate xanthine oxidoreductase activities observed ex vivo.

MATERIALS AND METHODS

Rat and guinea-pig hearts were removed and arrested in ice-cold saline. Cannulation of the aorta was done within one minute of arrest. Hearts were perfused retrogradely according to Langendorff with a modified Tyrode's buffer (10 mM glucose), oxygenated with 95%O_2/5%CO_2 at 9.6 kPa and 37°C, as described previously.[4] Apex displacement was used to monitor function and heart rate. Coronary flow was measured with a flow probe. The perfusion system consisted of three reservoirs. Reservoir #1 contained medium; #2: medium and 50 μM xanthine; and #3: medium and 50 μM hypoxanthine. Both substrate concentrations were sufficient for maximal enzyme activity. The perfusion protocol is depicted in fig. 1. After a stabilization period of 30 min, the hearts were infused for 15 min with hypoxanthine or xanthine. To obtain basal levels of xanthine and urate in the effluent, this period was followed by 15 min perfusion without substrate.

Purine and Pyrimidine Metabolism in Man VII, Part A
Edited by R.A. Harkness *et al.*, Plenum Press, New York, 1991

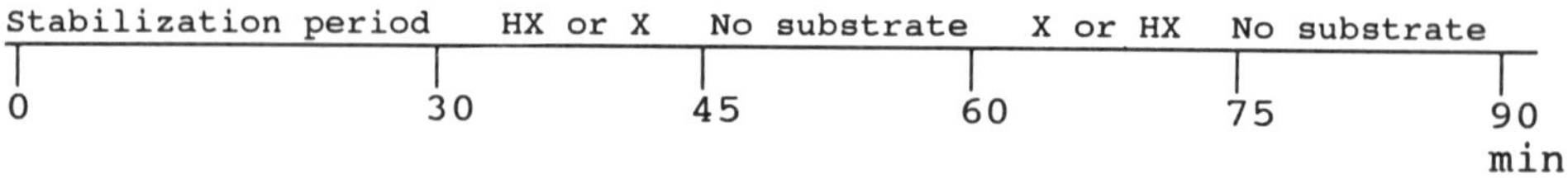

Fig. 1. Perfusion protocol. HX=hypoxanthine and X=xanthine

Then the alternative substrate was infused during 15 min. Heart weights were 1.2-1.8 g (rat) and 1.2-1.8 g (guinea pig).

For purine determination, coronary effluent samples were collected in tubes, containing dipyridamole (20 μM) and EHNA (10 μM). They were assayed by high performance liquid chromatography[4], with detection at 295 nm (urate) and 254 nm (xanthine, hypoxanthine). If hypoxanthine had been used as substrate, xanthine oxidoreductase activity was calculated as xanthine + 2x urate production. During xanthine infusion xanthine oxidoreductase activity equalled urate production.

RESULTS

Changes in heart rate and apex displacement were minimal during the experiment. Coronary flow (ml/min per g wet weight) was 10.7 ± 1.0 in rat and 10.9 ± 2.1 in guinea-pig hearts. In the course of the experiment, flow decreased insignificantly. We measured the catabolites xanthine and urate during perfusion with hypoxanthine. Effluent xanthine concentrations (μM) were 2.28 ± 0.09(rat) and 0.15 ± 0.03(guinea pig); urate concentrations (μM) were 1.2 ± 0.05 in rat and 0.81 ± 0.05 in guinea pig. During xanthine infusion urate concentrations (μM) were 2.00 ± 0.06 and 0.76 ± 0.04, respectively. The production of xanthine and urate as well as the calculated xanthine oxidoreductase activities during hypoxanthine perfusion are shown in table 1. The guinea-pig heart released little xanthine. The activity of xanthine oxidoreductase in rat heart exceeded that of guinea pig twofold.

Table 1. Xanthine (X) and urate (UA) release from isolated hearts, perfused with hypoxanthine. X and UA formed are expressed as nmol/min/g wet weight, xanthine oxidoreductase (XOD) activity is given in mU/g wet weight; means ± SEM (n=4-7).

species	X	UA	XOD
rat	19.0 ± 1.3$^{\#}$	10.5 ± 0.7	40.0 ± 2.6*
guinea pig	2.1 ± 0.3	8.0 ± 0.7	18.2 ± 1.5

$^{\#}$ P < 0.001, * P < 0.005 vs guinea pig

Table 2. Urate release from isolated hearts, perfused with xanthine. Urate formed is expressed as nmol/min/g wet weight, xanthine oxidoreductase (XOD) activity as mU/g wet weight; means ± SEM (n=4-7).

species	UA	XOD
rat	17.2 ± 1.5*	17.2 ± 1.5*
guinea pig	8.4 ± 1.5	8.4 ± 1.5

* $P < 0.005$ vs guinea pig

As can be seen from table 2, XOD activity in guinea-pig hearts with xanthine as substrate was twofold lower than that in rat hearts. This was comparable with the difference in activities between the species with hypoxanthine. There is a significant reduction of measured activity with xanthine ($P<0.002$; compare tables 1 and 2).

DISCUSSION

Guinea-pig hearts release little xanthine during hypoxanthine perfusion.[1] It seems that xanthine breakdown exceeds its formation, but then the rate of urate production with xanthine as substrate should be higher than that in rat hearts. The opposite appeared to be true; furthermore urate production in guinea-pig hearts is the same with both substrates. A different Km value of xanthine oxidoreductase in guinea-pig hearts, for the substrate xanthine, could explain the discrepancy.

The major finding in this study is the decreased xanthine oxidoreductase activity if xanthine replaces hypoxanthine as substrate. We observed this in perfused rat and guinea-pig hearts. Because the breakdown of xanthine gives only one product, urate, many investigators prefer this substrate.[2,3,5] However, the controversial data about the xanthine oxidoreductase activity in heart tissue could be due to the use of different substrates.

Conclusion

The use of hypoxanthine as substrate in xanthine oxidoreductase determinations is preferable, since a 50% lower activity is found with xanthine. Because of the intermediate product, xanthine, calculation with hypoxanthine as substrate is somewhat more complicated.

ACKNOWLEDGEMENT

We gratefully acknowledge the Netherlands Heart Foundation for financial support.

REFERENCES

1. J. W. de Jong, P. van der Meer, A. S. Nieukoop, T. Huizer, R. J. Stroeve, and E. Bos, Xanthine oxidoreductase activity in perfused hearts of various species, including humans, Circ. Res. 67:770 (1990).
2. J. M. Downey, D. J. Hearse, and D. M. Yellon, The role of xanthine oxidase during myocardial ischemia in several species, including man, J. Mol. Cell. Cardiol. 20 (suppl. II):55 (1988).
3. L. J. Eddy, J. R. Stewart, H. P. Jones, T. D. Engerson, J. M. McCord, and J. M. Downey, Free radical-producing enzyme, xanthine oxidase, is undetectable in human hearts, Am. J. Physiol. 253:709 (1987).
4. T. Huizer, J. W. de Jong, and P. W. Achterberg, Protection by bepridil against myocardial ATP-catabolism is probably due to negative inotropy, J. Cardiovasc. Pharmacol. 10:55 (1987).
5. M. Muxfeldt, and W. Schaper, The activity of xanthine oxidase in hearts of pigs, guinea pigs, rabbits, rats and humans, Basic Res. Cardiol. 82:486 (1987).

EVIDENCE FROM IN VITRO STUDIES THAT DIHYDROOROTATE DEHYDROGENASE MAY BE A SOURCE OF TOXIC OXYGEN SPECIES

G. Lakaschus, H. Krüger, D. Heese, and M. Löffler*

Institut für Physiologische Chemie
Philipps-Universität
D-3550 Marburg, F.R.G.

INTRODUCTION

The dihydroorotate dehydrogenase (dihydroorotate:ubiquinone oxidoreductase, DHO-DH [EC.1.3.3.1]) is the fourth enzyme in the *de novo* synthesis of uridine monophosphate. The enzyme activity was found to be located in the mitochondrial membrane. It is intimately associated with the electron transport chain, to which it passes electrons directly, possibly via ubiquinone [1,2,3]. Since the dehydrogenase step is likely a focal point for the anaerobic failure in pyrimidine biosynthesis, fluctuations in oxygenation level of cells and tissues could cause a shift of enzyme activity. The lack of nucleic acid precursors as a result of local oxygen deficiency has been proposed to explain the diminished growth fraction in solid tumors. This was followed from observations of cell cycle specific accumulation of cultured cells under experimental hypoxia and from restimulation of cell cycle progression and growth improvement after exogenous supply with nucleosides and deoxynucleosides [4,5,6]. Furthermore, recent work correlated the reduction of cell growth under graduate hypoxia with relation to functional pyrimidine supply [7].

Concerning normal tissues, putative alterations in enzyme characteristics after oxygen shortage and reoxygenation—as described for xanthine dehydrogenase/oxidase—have not been taken into consideration so far. The available information on oxidase activity of dihydroorotate dehydrogenase is controversial [3,8,9]. The rationale of the present work was to clarify the functional oxidase activity of mammalian DHO-DH.

METHODS

Protein Purification

The DHO-DH was purified from porcine, rodent and EAT cell mitochondria to apparent homogeneity in five steps in a modified procedure according to Hines et al [10]:

I. *Mitochondria* were isolated from liver and cell-homogenate, respectively.

*This investigation was funded by the Wilhelm Sander-Stiftung, FRG (grant ML 890091).

II. *Solubilization* of mitochondrial membranes: Mitochondria were frozen and thawed, mechanically and enzymatically disrupted. Afterwards, the membrane proteins were solubilized via detergent extraction with Triton X-100. The supernatant fluid collected after centrifugation (31.000 g) contained solubilized DHO-DH.

III. *Ammonium sulfate precipitation:* The solubilized proteins from step II were fractionated by salting-out with ammonium sulfate. The precipitate from 20–40% saturation contained solubilized DHO-DH. The salt precipitate was resuspended (in 50 mM Tris/HCL, pH 9.0) and dialysed.

IV. *Gel-filtration-chromatography:* The enzyme solution was applied on a Sephacryl S300 column and eluted with 50 mM Tris/HCL buffer.

V. *Ion-exchange-chromatography:* Active fractions were applied in appropriate ionic strength on a Q-sepharoseFastFlow column, which had previously been equilibrated with 20 mM Tris/HCL, 0.1% Triton X-100, pH 9.0. After washing, the protein was eluted with a linear salt gradient (0–300 mM KCL) in correspondent buffer. The active fractions were concentrated by ultrafiltration on a YM10 membrane.

Enzyme Assays

The *standard spectrophotometric method,* based on 2,6-dichlorophenolindophenol (DCIP) reduction, contained 20 μM DCIP, 75 μM DHO, 100 μM ubiquinone (n = 10), 0.1% Triton X-100 and enzyme in 50 mM Tris/HCL, pH 8.0. The reaction was monitored by the loss of absorbance at 600 nm. Alternatively, direct formation of orotic acid was monitored at 280 nm. The *spectrophotometric assay of the oxidase activity* was based on the hydrogen peroxide production coupled to the oxidation of leuco-dichlorofluorescein-diacetate (LDCF) [11]. The reaction was catalyzed by exogenous horseradish peroxidase (HRP). The sample contained 10 units HRP, 0.3 mM LDCF and enzyme in 50 mM Tris/HCL, pH 8.0. For calibration H_2O_2-standards (1–6 μM) were used. The increase of absorbance was monitored at 500 nm.

NBT method. [12]: Enzymes were separated on native PAGE and incubated at 37°C with 0.1 mM nitrobluetetrazoliumchloride (NBT), 0.1 mM ubiquinone (n = 10), 0.075 mM DHO, 0.1% Triton X-100 in 100 mM phosphate buffer, pH 8.5. A specific blue precipitate demonstrated dehydrogenase activity.

Cerium method [13,14]: The measurement of oxidase activity was based on production of H_2O_2, which forms a complex with Ce^{3+} under the reaction conditions. This complex could be stained with 4-chloro-1-naphthole in a second reaction procedure. A typical reaction mixture contained 3 mM $CeCl_3$, 1 μM FAD, 75 μM DHO, in 200 mM Tris/HCL pH 8.5. The gel was incubated 30 min at 37°C. The following reaction mixture contained 1.5 mM $CoCl_2$, 1.5 mM 4-chloro-1-naphthole in 100 mM sodium acetate buffer, pH 5.5. A specific coloured precipitate on the gel demonstrated active oxidase.

Electrophoresis

SDS-polyacrylamid gel electrophoresis (SDS-PAGE) was performed with a 5% acrylamid stacking and 12% acrylamid running gel in the discontinuous buffer system of Laemmli [15]. Native PAGE was performed with a 5% stacking and a 7.5% running gel without SDS.

Metal Content Determination

Zn^{2+} and Fe^{2+} content of the purified DHO-DH was determined using atomic absorption spectroscopy.

RESULTS

Enzyme Purification

The enrichment of DHO-DH during purification to apparent homogeneity was monitored on SDS-PAGES for each species: rat 63 kD (major), 72 kD (minor); hog 55 kD (major), 60 kD (minor); EAT 32 kD (major), 27 kD (minor).

Kinetic investigations on the purified DHO-DH gave congruous Michaelis-Menten constants (K_m = 6.1 ± 0.75 µM) for the substrate L-DHO to be seen in Table 1.

Measuring the metal content using atomic absorption spectroscopy revealed two mol Fe^{2+} per subunit for rat and porcine DHO-DH and 2–4 mol of Zn^{2+} per subunit.

Furthermore, two different activities of the enzyme are compared in Table 1, the dehydrogenase activity and the oxidase activity. These were followed using spectrophotometric methods (DCIP and LDCF) and additional activity stains on native PAGE (NBT and Ce^{3+}). The data verify the oxidase activity in rat, porcine and rodent tumor. The specific oxidase activity was calculated to be 10–20% of the specific dehydrogenase activity. Activity stains on native PAGE were in conformity with these findings. The cerium method, originally developed for light-microscopic visualization of D-aminoacid oxidase, proved to be of superiour sensitivity compared with conventional terazolium/formazan (here NBT) technique of histochemistry.

DISCUSSION

The present results give evidence that mammalian DHO-DH has additional oxidase activity. Controversial findings on oxidase activity have been reported for the enzymes from beef and rat liver so far [1,3]. In parasitic protozoans, it is also a subject of debate [8,9]. In a strict sense, it is not possible to compare the different statements on oxidase activity, as various

Table 1. Enzyme characteristics and kinetic data

	rat mitochond. particle	rat purified DHO-DH	hog mitochond. particle	hog purified DHO-DH	EAT mitochond. particle	EAT purified DHO-DH
MW [kD] SDS-PAGE		58 (major) 68 (minor)		60 (major) 66 (minor)		32 27
K_m [µM]		6.1±0.9		6.09±0.75		n.d.
Metall content [mol/subunit]		2 - 4 Zn^{2+} 2 Fe^{2+}		2 - 4 Zn^{2+} 2 Fe^{2+}		n.d. n.d.
spec. activity [U/mg] DCIP method	a) $6.65 \cdot 10^{-3}$	0.05	a) $3.3 \cdot 10^{-5}$	0.03	$8.7 \cdot 10^{-4}$	0.064
spec. activity [U/mg] LDCF method	$6.5 \cdot 10^{-7}$	$3.3 \cdot 10^{-3}$	+	$2.7 \cdot 10^{-3}$	+	0.011
NBT method native PAGE		+		+		+
Ce^{3+} method native PAGE		+		+		+

a) direct formation of orotic acid was monitored λ = 280 nm analytical methods were used. Therefore, the generated H_2O_2 was analysed using identical assay conditions for the proteins of the three species, and congruous results were obtained.

The production of H_2O_2 as observed here implies a putative origin of activated oxygen species *in vivo* [16]. This strengthens the possible role of the dihydroorotate dehydrogenase in hypoxia and reperfusion injuries and following pathology.

REFERENCES

1. Miller R.W., Curry J.R., *Can. J. Biochem.* (1969) **47**, 725–734.
2. Chen J.J., Jones M.E., *Arch. Biochem. Biophys.* (1976) **176**, 82–90.
3. Forman H.J., Kennedy J., *Prep. Biochem.* (1977) **7**, 345–355.
4. Löffler M., *Eur. J. Biochem.* (1980) **107**, 207–215.
5. Löffler M., Schimpff-Weiland G., Follmann H., *FEBS Lett.* (1983) **156**, 72–76.
6. Löffler M, *Exp. Cell Res.* (1989) **182**, 673–680.
7. Löffler M., *Cell Proliferation* (1991), in press.
8. Pascal R.A., Le Trang N., Cerami A., Walsh C., *Biochemistry* (1983) **22**, 171–178.
9. Kungkrai J., Cerami A., Henderson G.B., *Biochemistry* (1991) **30**, 1934–1939.
10. Hines V., Keys L.D., Johnston M., *J. Biol. Chem.* (1986) **261**, 11386–11392.
11. Small G.M., Burdett K., Connock M.J., *Biochem. J.* (1985) **227**, 205–210.
12. Heeb M.G., Gabriel O., *Methods Enzymol.* (1984) **104**, 416–439.
13. Angermüller S., Fahimi H.D., *J. Histochem. Cytochem.* (1988) **36**, 23–28.
14. Seitz J., Keppler C., Fahimi H.D., Völkl A., *Electrophoresis* (1991), in press.
15. Laemmli, *Nature* (London) (1970) **227**, 680–685.
16. Halliwell B., Gutteridge M.C., *Biochem. J.* (1984) **219**, 1–14.

INFLUENCE OF OXYPURINOL ON COLITIS INDUCED IN RAT COLON BY TRINITROBENZENE SULFONIC ACID TREATMENT

Werner Siems, Tilman Grune, Andreas Werner, Gerhard Gerber, Peter Buntrock, and Wolfram Schneider

Institute of Biochemistry and Institute of Pathology, Medical Faculty (Charité), Humboldt University, Hessische Str. 3/4, O-1040 Berlin, F.R.G.

INTRODUCTION

Chronically inflammatory intestinal diseases such as Morbus Crohn and Colitis ulcerosa are severe diseases highly impairing the condition of affected patients, not seldom leading to inability to work, and often require radical surgical measures with corresponding consequences.

A principle of a causal therapy of such inflammatory diseases is not yet available. Treatment is effected with sulfasalazine or mesalazine, but it is mostly unsatisfactory. Frequently, the application of corticoides in high doses is required. According to recent findings, reactive oxygen species and lipid peroxidation processes play an important role in the pathogenesis and in the sustaining of chronically inflammatory intestinal diseases. Therefore, the inhibition of radical generation would represent a strategy of an effective treatment.

A substantial share of superoxide radicals being formed in the intestine is generated by the xanthine oxidase-mediated uric acid formation from hypoxanthine and xanthine under conditions of an increased purine degradation rate. This study aimed at investigating the nucleotide pattern and the oxygen radical-mediated injuries in an experimental model for an inflammatory bowel disease of rats which is induced by the hapten, trinitrobenzene sulfonic acid (TNB) (Morris et al., 1984) and its treatment by the xanthine oxidase inhibitor oxypurinol.

MATERIALS AND METHODS

A colitis was triggered by rectal instillation of 25 mg of trinitrobenzene sulfonic acid (TNB) in 10 rats with a body weight of 200-250 g. The 25 mg of TNB were solved in

0.25 ml of a 30% ethanolic solution. A further group of 10 animals having been treated in the same manner received from the 3rd to the 21st day after TNB administration 5 mg of oxypurinol (sodium salt) which was dissolved in 0.5 ml of glucose solution. The oxypurinol was given per os via probang in 8 hour-intervals. A third group of 10 animals received only 0.25 ml of 30% ethanol solution via rectal instillation. A fourth group of 10 animals was the control group without any treatment.

The intestine - especially the rectum - of the animals was histologically investigated. Furthermore, in tissue samples from the animal groups characterized above at the 21st day after the rectal instillation of TNB the content of purine nucleotides and the concentration of thiobarbituric acid-reactive substances (TBA-RS) were measured. All samples were taken immediately into liquid nitrogen and kept there until homogenization. The preparation of samples for the nucleotide measurement included the deproteinization with chilled 10% perchloric acid, centrifugation for 10 min at 1200 g, neutralization with triethanolamine-potassium carbonate and filtration. After filtration, 50 µl of the supernatant were analysed by an ion-pair reversed-phase HPLC separation as described by Werner et al. (1989a and 1989b).

For the HPLC analysis of nucleotides an apparatus from Perkin-Elmer (Norwalk, CT, USA) was used, consisting of an M410 pump system, an LC 95 variable-wavelength detector, an LCI-100 integrator and a Rheodyne injector system. The columns were C18 Sil-X-5 (250mm x 4.6mm I.D.) analytical columns and C18 Sil-X-5 (25mm x 4.6mm I.D.) guard-columns. As eluents the buffers A and B were used: A was 10mM ammonium phosphate containing 2mM PIC (tetrabutyl ammonium phosphate) and buffer B was a 80:20 (v/v) mixture of buffer A and acetonitrile. The elution profile was the following: a 12min concave gradient from 100% buffer A to 80% buffer B - 20% buffer A; 25min isocratic elution with a 80:20 (v/v) mixture of buffer B and buffer A; in 2min to 100% buffer A; 5min 100% buffer A. The flow-rate was 1.3ml/min. The detection was performed at 254nm. The concentration of TBA-RS was measured according to Ohkawa et al. (1979).

RESULTS AND DISCUSSION

The intestine (rectum and colon descendens) of the animals treated with TNB showed a chronically relapsing colitis of degree II - III. In the intestine of some of the rats incomplete erosions were observed, particularly in the areas of the surface epithelium being situated above the lymphatic follicles, very dense infiltrates in the submucosa, and plenty of perivascular mast cells within the surrounding fatty tissue. The animals treated with TNB and oxypurinol showed inflammatory symptoms being clearly less distinctive (colitis of degree I) including a slight increase of lymphocytes and mast cells and a mild oedema. The morphological changes were compared with changes of the nucleotide pattern and of the TBA-RS level. The contents of adenine and guanine nucleotides in the colon of the different groups of animals are demonstrated in Table 1. It should be mentioned that the values of the group treated only with an ethanol

Table 1. Nucleotide concentrations, ATP:ADP, ATP:AMP and GTP:GDP ratios of the colon of rats following the rectal instillation of 0.25 ml 30% ethanol with 25 mg trinitrobenzene sulfonic acid (TNB). The nucleotide concentrations are given as μmoles/g w.w. and % of the control group in parentheses (mean values from 10 animals in each group). AdN = sum of adenine nucleotides.

	Control		TNB		TNB + Oxypurinol	
ATP	1.95	(100)	1.10	(56)	1.81	(92)
ADP	0.22	(100)	0.29	(132)	0.30	(136)
AMP	0.09	(100)	0.19	(211)	0.10	(111)
GTP	0.38	(100)	0.14	(37)	0.35	(92)
GDP	0.09	(100)	0.12	(133)	0.08	(89)
AdN	2.26	(100)	1.58	(70)	2.21	(98)
ATP:ADP	8.86	(100)	3.79	(42)	6.03	(68)
ATP:AMP	21.67	(100)	5.79	(26)	18.10	(83)
GTP:GDP	4.22	(100)	1.17	(27)	4.38	(104)

instillation (solvent for the TNB) were identical with those of the control group. After treatment with TNB as initiating substance for the inflammatory processes the ATP content of the tissue, the total concentration of adenine nucleotides and the GTP concentration markedly decreased. The GTP pool was more sensitive against the effects of TNB than the ATP pool. It is suggested that structure and functions of mitochondria are affected which disturbs the balance between energy-consuming and energy-producing processes in intestinal cells. Oxypurinol treatment prevented the decline of ATP and GTP concentrations by effects of TNB. The beneficial effect of oxypurinol can be demonstrated by the values of ATP:ADP, ATP:AMP and GTP:GDP ratios, too. This protective action can be explained by (i) the acceleration of salvage pathways of purine metabolism, (ii) the decrease of xanthine oxidase-mediated superoxide radical/hydrogen peroxide formation because of the inhibition of this enzyme by oxypurinol and (iii) the radical scavenging effects of oxypurinol (Das et al., 1987). The plasma and tissue concentration of oxypurinol in the experiments presented here was high enough to allow its radical scavenging, too. It could not be decided yet which component of the oxypurinol effect plays the dominant role for the maintaining of nucleotide levels and of cell and tissue structure. Nevertheless, it was clearly demonstrated that TNB at least partially acts by an increased formation of reactive oxygen species and by an accelerated rate of lipid peroxidation and that this oxidative injury of the colon is prevented by oxypurinol. Table 2 shows the accumulation of TBA-RS following the TNB treatment and the antioxidative effect of oxypurinol. There is a multitude of similarities of the functional and structural changes under the conditions of the experimentally TNB-induced colon inflammation and the changes of

Table 2. Accumulation of thiobarbituric acid-reactive substances (TBA-RS) in the colon of rats following the application of TNB. The values are given as nmoles/mg protein and % of the control in parentheses (mean values from 10 animals in each group).

	Control		TNB		TNB + Oxypurinol	
TBA-RS	0.113	(100)	0.243	(215)	0.168	(149)

nucleotide pattern, balance between pro- and antioxidative processes and morphological deteriorations observed during postischemic reperfusion of the intestine (Younes et al., 1987; Granger, 1988; Siems et al., 1989 and 1991). The similarities also include the protective effects of xanthine oxidoreductase inhibitors like allopurinol or oxypurinol.

REFERENCES

Das, D. K., Engelman, R. M., Clement, R., Otani, H., Prasad, M. R., and Rao, P. S., 1987, Role of xanthine oxidase inhibitor as free radical scavenger, Biochem. Biophys. Res. Commun., 148:314.

Granger, D. N., 1988, Role of xanthine oxidase and granulocytes in ischemia-reperfusion injury, Am. J. Physiol., 255:H1269.

Morris, G. P., Rebetro, L., Herridge, M. M., Szewezuk, M., and Depew, W., 1984, An animal model for chronic granulomatous inflammation of the stomach and the colon, Gastroenterology, 86:A1188.

Ohkawa, H., Onishi, N., and Yagi, K., 1979, Assay for lipids in animal tissues by thiobarbituric acid reaction, Anal. Biochem., 95:351.

Siems, W., Kowalewski, J., Werner, A., Schimke, I., and Gerber, G., 1989, Radical formation in the rat small intestine during and following ischemia, Free Radical Res. Comms., 7:347.

Siems, W., Kowalewski,J., David, H., Grune, T., and Bimmler, M., 1991, Discrepancy between biochemical normalization and morphological recovery of jejunal mucosa during postischemic reperfusion in presence of the xanthine oxidase inhibitor oxypurinol, Cell. Molecul. Biol., 37:in press.

Werner, A., Schneider, W., Siems, W., Grune, T., and Schreiter, C., 1989a, Ion-pair reversed phase HPLC determination of nucleotides, nucleosides and nucleobases, Chromatographia, 27:639.

Werner, A., Siems, W., Kowalewski, J., and Gerber, G., 1989b, Interrelationships between purine nucleotide degradation and radical formation during intestinal ischaemia and reperfusion, J.Chromatogr. Biomed. Appl., 491:77.

Younes, M., Mohr, A., Schoenberg, M. H., and Schildberg, F. W., 1987, Inhibition of lipid peroxidation by superoxide dismutase following regional intestinal ischemia and reperfusion, Res. Exp. Med., 187:9.

A XANTHINURIC FAMILY - THE PROPOSITA HAVING IMMUNOLOGICALLY REACTIVE XANTHINE OXIDASE BUT NO XANTHINE OXIDASE ACTIVITY

Tetsuya Yamamoto, Yuji Moriwaki, Sumio Takahashi, Toshikazu Hada, Michio Suda, Hiroyasu Imanishi, Oluyemi E Agbedana[1)], Masahisa Nanahoshi, Kazuya Higashino

Third Department of Internal Medicine, Hyogo College of Medicine, Mukogawa-cho 1-1, Nishinomiya, Hyogo 663 Japan
[1)]Department of Internal Medicine, Kinki Central Hospital of the Mutual Aid Association of Public School Teachers Kurumazuka 3-1, Itami, Hyogo 663 Japan

SUMMARY

The antibody was raised against purified human liver xanthine oxidase in a rabbit. In a xanthinuric patient, the double immuno-diffusion method demonstrated the existence of an immunologically reactive duodenal mucosa xanthine oxidase which did not possess xanthine oxidase activity. These results indicated that xanthine oxidase protein is abnormal in structure and/or amino acid sequence.

In addition, urinary excretion of pyrazinamide and its metabolites on pyrazinamide load in her mother was investigated. The result suggested that her mother was a heterozygote.

INTRODUCTION

Hereditary xanthinuria is a rare disorder of purine metabolism. Most patients with this disorder are asymptomatic but some of them suffer from stones of the urinary tract and/or myopathy due to deposits of oxypurines. More than 50 cases of xanthinuria have been reported since it was first observed in 1954 (1). However, the measurement of xanthine oxidase has not been performed immunologically in any of these cases using an antibody against xanthine oxidase, although xanthine oxidase activity was measured in some cases. Therefore, we investigated whether or not xanthine oxidase could be detected immunologically in the proband of a familial xanthinuria and also the metabolism of pyrazinamide in the proband's mother.

MATERIALS AND METHODS

The patient, the proposita of a familial xanthinuria, was reported previously (2). She oxidized neither pyrazinamide to 5-hydroxypyrazinamide nor pyrazinoic acid to 5-hydroxypyrazinoic acid in the previous study (2). In this patient, the serum uric acid was only 0.5 mg/dl and the xanthine oxidase activity of the duodenal mucosa was

Purine and Pyrimidine Metabolism in Man VII, Part A
Edited by R.A. Harkness *et al*., Plenum Press, New York, 1991

beyond the lower detection limit (2). The activity of xanthine oxidase was stained by the modified method of Holmes et al (3) using 1 mM hypoxanthine and immunological detection of xanthine oxidase performed by Ouchterlony's method (4).

The antibody against xanthine oxidase purified from normal human liver of a cadaver was raised in a rabbit. Ouchterlony's double immunodiffusion method demonstrated that a hypoxanthine-stained precipitation line of liver xanthine oxidase was consistent with the Coomassie brilliant blue R-stained one in the position of precipitation.

Xanthine oxidase was purified as described below. The partially purified xanthine oxidase, which was prepared as reported previously (5), was dialyzed against 20 mM potassium phosphate buffer (pH 7.4) and was then treated by the method of Roussos et al using a diethylaminoethyl cellulose column (6). Then, the fractions demonstrating enzyme activity were purified by the method of Nishio et al (7) using a Sepharose 4B/floate column. Oxypurines in plasma and urine, and pyrazinamide and its metabolites in urine were measured by high-performance liquid chromatography (HPLC) (3,8). To measure pyrazinamide and its metabolites in urine, 3.0 g of pyrazinamide was administered to the mother of the proposita and to 10 normal subjects orally in one dose and urine was then collected for 24 hours.

RESULTS

Ouchterlony's double immunodiffusion method demonstrated the hypoxanthine-stained precipitation line of the partially purified liver xanthine oxidase of the cadaver but did not demonstrate that of the patient's duodenal xanthine oxidase. However, Ouchterlony's double immunodiffusion demonstrated that a Coomassie brilliant blue R-stained precipitation line of the patient's duodenal xanthine oxidase fused with that of the liver xanthine oxidase of the cadaver.

The plasma concentrations and the urinary excretions of hypoxanthine and xanthine of the patient and her family are shown in Table 1. In the patient, her second elder brother and her younger sister, the plasma levels of hypoxanthine and xanthine were high, as were the 24-hour urinary excretions of hypoxanthine and xanthine when compared with those in the control subjects. In the patient, her second elder brother and her younger sister, both the plasma level of uric acid and the 24-hour urinary excretion of uric acid were low.

Table 1. Plasma concentrations and urinary excretions of hypoxanthine (Hx) and xanthine (X)

	plasma (mg/L)		urine (mg/day)	
	Hx	X	Hx	X
father	0.02	0.02	24.8	12.4
mother	0.03	0.01	6.2	6.8
elder brother (1)	0.02	0.015	5.0	5.8
elder brother (2)	0.31	0.38	64.0	270.9
proposita	0.44	0.18	87.0	205.7
younger sister	0.51	0.31	56.4	278.4
normal subjects (N=10)	< 0.03	< 0.02	< 15.4	< 11.7

Although xanthine oxidase activity was not determined in any subject other than the proposita, her second elder brother and her younger brother were also diagnosed as having xanthinuria by the measurement of oxypurines in their urine.

Table 2 shows that small amounts of 5-hydroxypyrazinamide and 5-hydroxypyrazinoic acid were excreted in the urine in the mother of the proposita on pyrazinamide load, compared with those excreted in the urine in the normal subjects.

Table 2. Urinary excretions of pyrazinamide and its metabolites in the proposita and a heterozygote of xanthinuria

	heterozygote	proposita	control subjects
5-hydroxypyrazinamide (mg/day)	82	N.D.	576±88
5-hydroxypyrazinoic acid (mg/day)	94	N.D.	505±182
pyrazinoic acid (mg/day)	765	1164	458±159
pyrazinamide (mg/day)	94	84	88±9

DISCUSSION

From the period 1954 to 1974, only 24 cases of familial xanthinuria had been reported by 1974. Thereafter, 3 cases have been reported in Japan to date. However, the detection of xanthine oxidase has not been performed immunologically in any of these cases. The present study demonstrated that the patient, the proposita of the familial xanthinuria, did not possess xanthine oxidase activity (3) like the cases previously reported (9-11) but possessed immunologically detectable xanthine oxidase. Therefore, this suggests that this patient had a genetic defect producing a nonfunctioning enzyme detected immunologically as xanthine oxidase.

Although the traits of these familial xanthinuria cases have been suggested to be autosomal recessive (12), there were no reliable methods for detecting heterozygotes because of the non-availability of definite criteria. Therefore, we administered pyrazinamide to the mother of the xanthinuric patient, a heterozygote to investigate whether or not heterozygotes could be detected by the pyrazinamide load test. The main metabolism of pyrazinamide in humans consists of a reaction sequence by which pyrazinamide is deamidated to pyrazinoic acid, followed by oxidation to 5-hydroxypyrazinoic acid by xanthine oxidase and another reaction by which pyrazinamide is oxidized to 5-hydroxypyrazinamide by xanthine oxidase. Since neither 5-hydroxypyrazinamide nor 5-hydroxypyrazinoic acid had been demonstrated to be excreted in urine on the administration of 3.0 g of pyrazinamide to the proposita who possessed no xanthine oxidase activity in a previous study (2), the results (the decreased urinary excretion of 5-hydroxymetabolites) from the pyrazinamide load test to the mother of the proposita (Table 2) suggest that the mother is a heterozygote despite the normal urinary excretion of oxypurines.

In this study, we attempted to prove immunologically the existence of a nonfunctioning xanthine oxidase and we have demonstrated for the first time the existence of immunologically detectable nonfunctioning xanthine oxidase in a xanthinuric patient. In addition, further investigation revealed that the pyrazinamide load test might be a useful test in the heterozygote of xanthinuria.

REFERENCES

1. D. E. Dent and G. R. Philpot, Xantinuria, an inborn error (or deviation) of metabolism, Lancet 1:182-185 (1954).
2. T. Yamamoto, K. Higashino, N. Kono, M. Kawachi, M. Nanahoshi, S. Takahashi, M. Suda and T. Hada, Metabolism of pyrazinamide and allopurinol in hereditary xanthine oxidase deficiency, Clin. Chim Acta.
3. R. S. Holms, L. R. Leijten, and J. A. Duley, Liver aldehyde oxidase and xanthine oxidase genetics in the mouse, Anim. Blood Groups Biochem. Genet. 12:193-199 (1983).
4. O. Ouchterlony, Antigen-antibody reactions in gels, Ark. Kemi. Mineral. Geol. 26B(14):1-9 (1949).
5. T. Yamamoto, Y, Moriwaki, S. Takahashi, T. Hada and K. Higashino, In vitro conversion of pyrazinamide into 5-hydroxypyrazinamide and that of pyrazinoic acid into 5-hydroxypyrazinoic acid by xanthine oxidase from human liver, Biochem Pharmac. 36:3517-3318 (1987).
6. G. G. Roussos, Xanthine oxidase from bovine small intestine. In: L. Grossman, K. Moldave eds. Methods in Enzymology. New York Academic Press, 1:5-16 (1966).
7. T. Nishino, T. Nishino and K. Tsushima, Purification of highly active milk xanthine oxidase by affinity chromatography on sepharose 4B/folate gel, Febs Letters. 131:369-372 (1981).
8. T. Yamamoto, H. Yokoyama, Y. Moriwaki, S. Takahashi, M. Suda, T. Hada and K. Higashino, The effect of completely purine-free diet of low sodium content on purine intermediates and end-product, Eur. J. Clin. Nutr. 44:659-664 (1990).
9. H. Yamanaka, K. Nishioka, T. Suzuki and K. Kohno, Allopurinol metabolism in a patient with xanthine oxidase deficiency, Ann. Rheum. Dis. 42:684-686 (1983).
10. R. A. Chalmers, R. W. E. Watts, C. Pallis, L. Bitensky and J. Chayen, Crystalline deposits in striped muscle in xanthinuria, Nature 221:170-171 (1969).
11. R. Frayka, I. Salti, A. Haider, U. AL-Khalidi and K. Hemady, Hereditary xanthinuria and xanthine urolithiasis: an additional 3 cases, J. Urol. 109:871-873 (1973).
12. E. W. Holmes and J. B. Wyngaarden, Hereditary xanthinuria. In: J. B. Stanbury, D. S. Fredrickson et al eds. The metabolic basis of inherited disease. McGraw-Hill Book Company 1192-1201 (1983).

HUMAN HEPATIC PEROXISOMES WITH CRYSTALLOID CORES ASSOCIATED WITH URATE OXIDASE ACTIVITY

L.A. De-Netto, P.S. Tappia, Z.A. Malik, A.J. Wood, V.M. Mann, C.J.P Jones[1], K. Burdett[2], J.P. Neoptolemos[3] and M.J. Connock

School of Applied Sciences, Wolverhampton Polytechnic, Wolverhampton, U.K; 1 & 2, Dept of Pathological Sciences and Dept of Biochemistry, University of Manchester, Manchester, U.K; 3 Dept Surgery, University of Birmingham, Birmingham, U.K.

INTRODUCTION

Mammalian hepatic peroxisomes generally contain a dense core with a distinct crystalloid structure. The major protein of this core is urate oxidase (urox) (1,2). In man and anthropoid primates hepatic peroxisomes lack cores and urox is absent (3,4). Recent DNA studies (5,6,7) indicate that a urox gene is present in man. Also there have been sporadic observations of core containing peroxisomes in pathological human livers (8-11). Here we describe subcellular fractionation studies of human liver taken from patients with colorectal cancer. In two of nine patients peroxisomes contained crystalloid cores and isolated cores exhibited urox activity.

METHODS

Tissue was homogenised in 8.5% sucrose (1mM EDTA, 3 mM imidazole pH 7.4) with 3 strokes of a Potter Elvhjem homogeniser. Post nuclear supernatant (4350 g x 2.5 min) or (patient A) resuspended "lambda" fraction (15600 g x 35 min pellet from 4250 g x 12 min supernatant) were prepared and loaded into a B XIV zonal rotor. The rotor contained: a dense cushion of fluorochemical FC 43 (1.9 g/ml), a 70 ml linear gradient of Metrizamide and sucrose (13%, 1.118 g/ml to 30%, 1.273 g/ml), and 350 ml of 12% sucrose. An overlay of 6% sucrose was added and the rotor was centrifuged at 30000 rpm for 6 hours. The gradient was fractionated into 4 ml fractions which were analysed for marker enzymes and protein by conventional procedures (12). The densest peroxisome-enriched fractions were pooled and dialysed (4°C overnight) against 1 mM EDTA 3 mM imidazole pH 7.4. Except for patient A, dialysed peroxisomes were treated with .2% Triton X-100 and centrifuged to harvest cores (13). For patient A dialysed peroxisomes were first separated out by centrifugation in a linear sucrose gradient and then treated as above. Harvested cores were suspended in 8.5% sucrose and analysed for urox activity and by SDS PAGE,

Uricase was assayed in a dual beam spectrophotometer by monitoring urate-dependent H_2O_2 generation (14). Dye-peroxidase mixtures were made as described (15) except that the buffer was 50 mM Tris pH 8. Mixtures were kept ice cold and dark and withdrawals were always made with a fresh pipette tip. Identical ref and test cuvettes were set up containing: 150

Purine and Pyrimidine Metabolism in Man VII, Part A
Edited by R.A. Harkness *et al.*, Plenum Press, New York, 1991

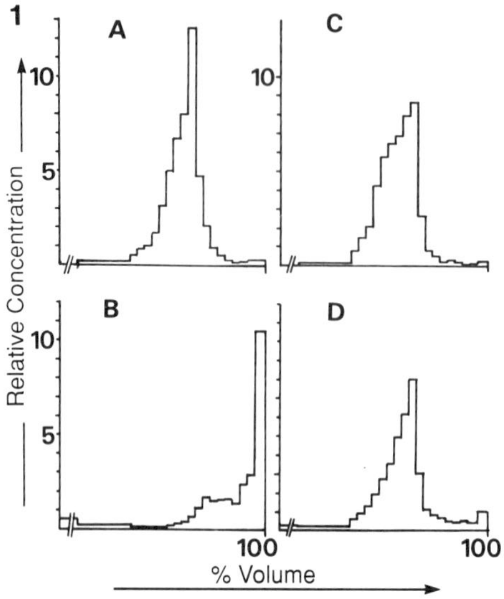

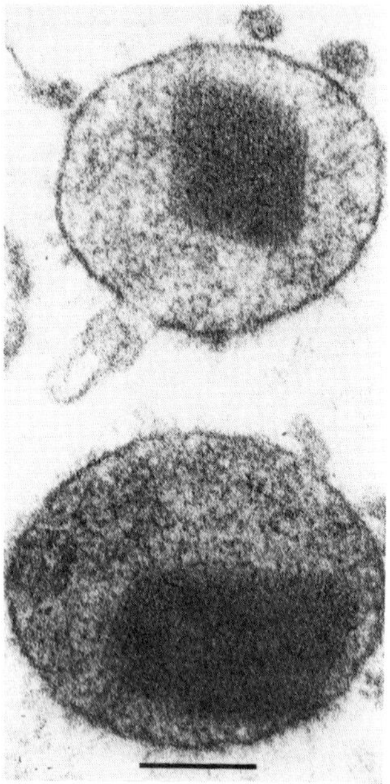

Fig 1 Separation of organelles from lambda fraction of patient A using a sucrose-Metrizamide density gradient. A, succinic dehydrogenase; B, catalase; C, Esterase; B, protein. The dense end of the gradient is to the right.

Fig 2 Electron micrograph of two peroxisomes from the sucrose-Metrizamide gradient of patient A illustrating dense core with crysalloid structure. Bar = .1 um

mM Tris pH 8, 40 mM 3-amino-1,2,4-triazole, 36 ug/ml horseradish peroxidase, 42 ug/ml dichlorofluorescin. After monitoring absorbance for 6 min potassium urate (pH 8) was added to the test (116 uM) and an equal volume of water to the ref. Absorbance was again monitored for 6 min. Control assays were done in which 8.5% sucrose was substituted for sample.

Samples of subcellular fractions and liver tissue were fixed in 4% glutaraldehyde (pH 7.4) supplemented with 8.5% sucrose or gradient solutes at the same concentrations as the fraction. They were then processed for electron microscopy by conventional procedures.

RESULTS AND DISCUSSION

Marker enzymes employed were catalase and dihydroxyacetone phosphate acyltransferase for peroxisomes, esterase for ER, succinic dehydrogenase for mitochondria and acid phosphatase for lysosomes. The lambda fraction from patient A contained 13% and 19% of the homogenate's mitochondria and ER together with 13% of the catalase. When this fraction was separated on the Metrizamide sucrose gradient markers distributed as shown (Fig 1). The densest fractions contained relatively uncontaminated peroxisomes. Electron Microscopic Examination of these fractions revealed peroxisomes containing cores with crystalloid structure (Fig 2). After dialysis and separation on a sucrose gradient these peroxisomes had a hollow appearance except for the presence of the core (Fig 3). The cores were harvested after Triton X-100 treatment and centrifugation (13). SDS PAGE analysis of cores revealed a prominent polypeptide (34 kDa) of almost identical MW to the predominant polypeptide of similarly prepared cores from rat liver (Fig 4). Cores of patient A exhibited urox activity (660 pmol/min/mg protein).

More liver tissue was available from patient A (who underwent partial liver resection) than from subsequent patients. Of 8 of these electron microscopic examination of liver tissue revealed some peroxisomes containing crystalloid cores in one individual (patient B) but not others. Density gradient fractionation was performed using post nuclear supernatant so as to maximize peroxisomal yield. Distributions for markers in gradients for these patients was similar to that shown in Fig 1, with peroxisomes separating at higher density than other organelles. Dense

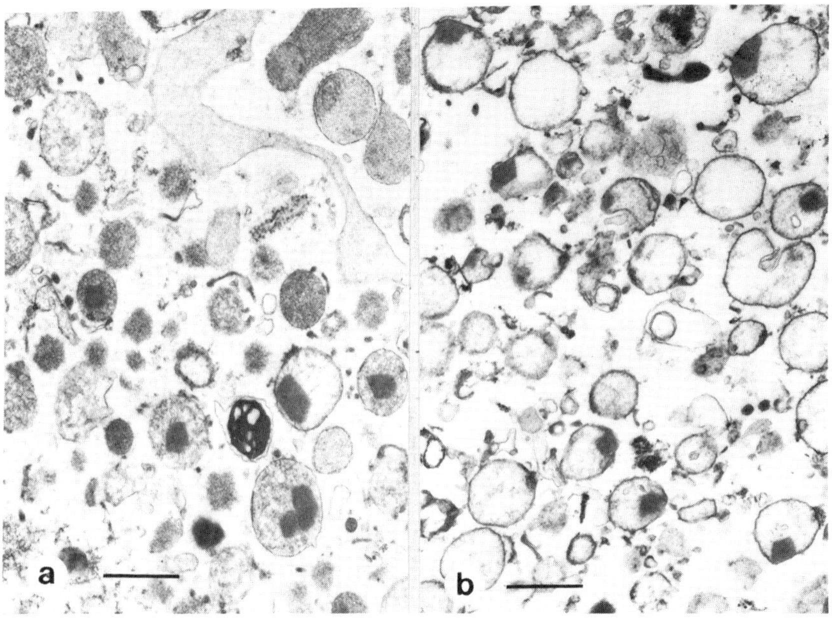

Fig 3 Electron micrographs of subcellular fractions isolated from patient A. a: catalase-enriched fractions from the dense end of the sucrose - Metrizamide gradient. b: dialysed peroxisomes after isolation in a sucrose gradient. Bar = .5 um.

peroxisome-enriched fractions from patient B were subjected to Triton X-100 treatment and pelleting to harvest cores. This preparation exhibited urate oxidase activity (86 pmol/min/mg protein) (Fig 5). Similar preparations from individuals in which no cores were observed by EM lacked detectable urox activity. The present results indicate that a human urox gene may be expressed in cases of colorectal cancer, and that the gene codes for a protein with catalytic activity and which is capable of targeting to the peroxisomes. Urox genes have recently been studied in rat, mouse, pig, baboon and man (5-7, 16,17). A human gene showing 96% homology with that in the baboon has been reported (6). This human gene contains premature chain termination codons. This would appear inconsistent with expression, however, it is interesting to note that bovine liver contains a predicted

4

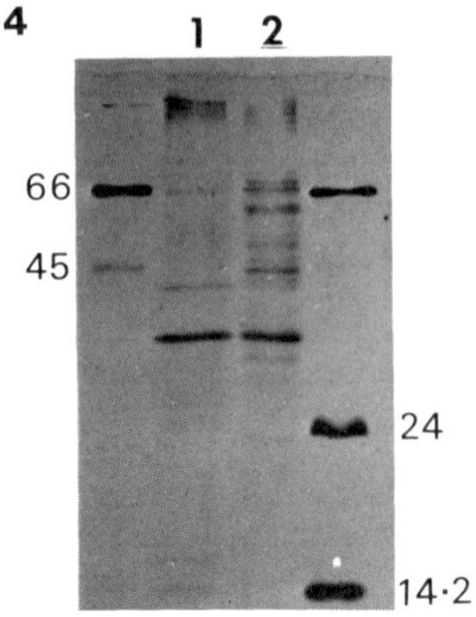

5

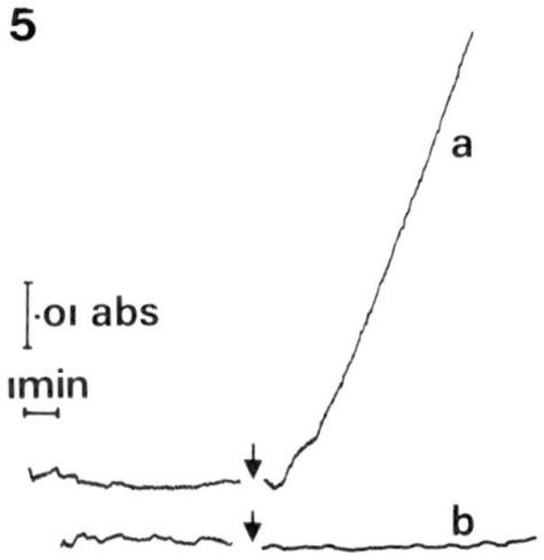

Fig 4 SDS PAGE of core preparations. Patient A track 2 and rat liver track 3. Molecular mass numbers were: bovine albumin (66 kDa), ovalbumin (45 kDa), trypsinogen (24 kDa) and lactalbumin (14.2 kDa).

Fig 5 Register of urox activity using core preparation from patient B. a: absorbance recording (502 nm) with core sample in reference and test cuvettes. b: recording in absence of core sample. The arrow indicates addition of substrate to the test and an equal volume of water to the reference.

suppressor t RNA for this codon. It is also possible that the termination codon does not occur in all human genomes, or that other urox genes may be present.

REFERENCES

1. Z. Hruban, and M.Swift, Science. 146: 1316-18 (1964).
2. S. Angermuller, and H.D. Fahimi, J.Histochem. Cytochem. 34: 159-165 (1986).
3. B.A. Afzelius, J.Cell. Biol. 26: 835-8 (1965).
4. L.B. Sorensen, 'Extrarenal disposal of uric acid', in: "Uric acid" W.N. Kelly and I.M Weiner, ed., Springer - Verlag, New York (1979).
5. C.C. Lee, X.Wu, R.A. Gibbs, R.G. Cook, D.M. Muzny, and C.T. Caskey, Science. 239: 1298-91 (1988).
6. X. Wu, C.C. Lee, D.M. Muzny and C.T. Caskey, Proc. Natl. Acad. Sci. USA. 86: 9412-16 (1989).
7. A.V. Yeldandi, X. Wang, K. Alvares, S. Kumar, M.S. Rao, and J.K. Reddy, Biochem. Biophys. Res. Commun. 171: 641-646 (1990).
8. L. Biempica, J. Cell. Biol. 29: 383-386 (1966).
9. I. Sternlieb, and N. Quintana, Lab. Invest, 36: 140-149 (1977).
10. H. Gryzbek, B. Panx, M.Petelenz, Z. Gonciarz, S. Kusmierski, and B. Kocek. Eur. J.Cell. Biol, 41: suppl 14: 16 (1986).
11. J.C. Collins, I.H. Scheinberg, D.R. Giblin, and I. Sternlieb, Gastroenterol. 97: 766-770 (1989).
12. B.N. Patel, M.I Mackness, and M.J. Connock. Biochem. J. 244: 443-448 (1987).
13. V.D. Antonenkov, and L.F. Panchenko, FEBS Lett. 88: 151-154 (1978).
14. H. Kochli, and J.P. Von Wartburg, Anal. Biochem. 84: 127-135 (1978).
15. G.M. Small, K. Burdett, and M.J. Connock, Biochem. J. 227: 205-210 (1978).
16. M. Ito, M. Suzuki and J. Takagi, Eur. J.Biochem. 173: 459-463 (1988).
17. K. Motojima, S. Kamaya, and S. Goto, J.Biol. Chem. 263: 16677-81 (1988).

STRUCTURAL ANALYSIS OF THE RAT URICASE GENE AND EVIDENCE THAT LYSINE 164 IS INVOLVED IN THE SUBSTRATE-BINDING SITE OF THE ENZYME

Masaki Ito *, Masamichi Nakamura , Seiya Kato , Hisamitsu Ogawa and Yasuyuki Takagi

*Laboratory of Biology, Saga Medical School, Saga 849, Japan
Division of Molecular Genetics, Institute for Comprehensive Medical Science
Fujita Health University, Toyoake, Aichi 470-11, Japan

Introduction

Since the degradation of purines is less complete in higher animals, comparative studies on the genes encoding the enzymes responsible for this process may provide valuable information on the process of gene and protein evolutions. In addition, the loss of uricase activity in human is assumed to prevent cancer and other diseases (1) and also to arouse gouty arthritis and the symptomes of Lesch-Nyhan syndromes (2,3). Some significant hypotheses which explain the relationship between the exon-organization and functional domains of enzymes have been proposed (4,5). Uricase catalizes a conversion of uric acid to allantoin in the purine metabolism. . Therefore, we analyzed the structure of the gene encoding rat uricase and identified the functional site, especially the site involved in the substrate-binding of the enzyme.

Materials and Methods

Cloning and characterization of the rat uricase gene. The gene was screened by using a Charon 4A genomic DNA library (Clonetech), an EMBL genomic DNA library (Clonetech) and a cosmid pHC79 genomic DNA library (a gift from Dr. Oda, T., Hamamatsu Medical School, Japan) Positive clones were selected by a hybridization technique with a rat uricase cDNA, reffered to as rURC-1 (6). Sequences were determined by the dideoxy chain termination method with a 370A sequencer (applied Biosystems Inc.).

Construction and expression of a cDNA encoding rat uricase in Escherichia coli. A cDNA encoding rat mature uricase was ligated into a expression vector Bluescript SK (+) with the *Eco* Ri site. The constructed cDNA produce a fusion protein containing 39 amino acids derived from the linker portion of the vector under the control of *lac* promotor. The recombinant uricase produced behaves as an inclusion body in *E. coli* and exhibits similar specific activity as the native enzyme.

Measurements of uricase activity and substrate-binding activity. The uricase activity was measured by a reduction of an absorption at 292 nm, derived from a conversionn of uric acid to allantoin in 1 ml reaction containing 50 mM sodium borate buffer (pH9.0), 50 uM uric acid and enzyme. 1 unit of the activity was determined as the reduction of 1 nmol uric acid/min in 1 ml reaction. The substrate-binding activity was measured in a mixture

Purine and Pyrimidine Metabolism in Man VII, Part A
Edited by R.A. Harkness *et al.*, Plenum Press, New York, 1991

containing 10 nmol uricase and 2 nmol of C-xanthine. The binding between uricase and the competitive inhibitor xanthine was assayed by a gel filtration column, NAP-10 (Pharmacia Co.).

Result

Structural analysis of rat uricase gene. Three genomic clones (GU-1, GU-2, GU-3) were isolated from a total 5x10 plaques in a Charon 4A library, one clone (GU-5) from a total of 5x10 plaques in a EMBL library and one clone (GU-4) from a total of 1x10 colonies in a cosmid library. Restriction maps and Southern blot analyses suggested that these isolated clone covered the full lenght of the cDNA sequence. Sequence determination revealed that the gene spans 40 kb and consists of 8 exons (Fig. 1). It also indicated tha t all the boundary sequences around the exon-intron junction conform to the canonical GT/AG rule, and that the exogenic sequences of the gene are substantially identical to the cDNA sequence reported previously (7).

Suggestion of the substrate-binding site by a computer search. In order to evaluate the amino acid sequences confering the enzyme function, we aligned amino acid sequences between rat (7), fly (8) and soybean (9) uricases and three conserved sequences (amino acid positions of rat uricase 62-75, 156-168 and 231-239) were found out. A computer search showed that the sequence of rat uricase, Leu 160 to Lys 164 is also conserved in hypoxanthine-guanine phosphoribosyltransferase (HPRT, EC.2.4.2.8) of human (10) and xanthine-guanine phosphoribosyltransferase (XGPRT, EC.2.4.2.22) of *E. coli* (11). Their enzymes use a purine as the substrate for the catalytic reaction. This search suggests that the sequence could be involved in the substrate-binding site of rat uricase.

Constructions of cDNAs encoding rat uricase and the mutant proteins. In order to determinethe site suggested above, we tried to construct the systems for the expression of rat uricase and the mutant proteins in *E. coli.* A cDNA encoding mature rat uricase was ligated intoBluescript SK (+). The recombinant uricase was induced in the presence of isopropyl -D-thiogalactopyranoside (IPTG) and occupied 5 % of the total cell protein. The uricase was concentrated in the insoluble fraction as an inclusion body, washed with

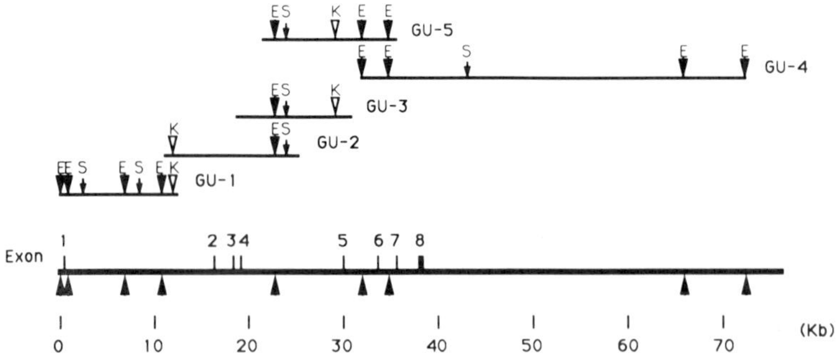

Fig. 1. Restriction map of the cloned rat uricase gene. The five upper lines represent the overlapping genomic clones, spanning more than 70 kb. Restriction sites indicate: E, *EcoRI*; K, *KpnI*; S, *SmaI*. Several *KpnI* sites in the cosmid clone GU-4 are omitted because the exact locations have not been determined. The bold line shows the location of 8 exons, indicated by vertical lines with exon numbers. *EcoRI* sites in the gene are indicated by arrowheads.

Table I. Construction of mutant uricase by oligonucleotide-directed mutagenesis. Closed circles indicate the mutated base and underline indicate the altered amino acids.

Mutant	Sequence used for the construction of mutant uricase
WILD	5' AAG GTC TTG AAA ACA ACC CAG 3' LYS VAL LEU LYS THR THR GLN
K164E	AAG GTC TTG GAA ACA ACC CAG LYS VAL LEU GLU THR THR GLN
K164I	AAG GTC TTG ATC ACA ACC CAG LYS VAL LEU ILE THR THR GLN

0.1 M sodium borate buffer (pH10.0) and solubilized in 0.1 M sodium carbonate buffer (pH11.0). The uricase preparation showed a single polypeptide band in SDS-polyacrylamide gel electrophoresis (data not shown).

We tried to alter the amino acid in the conserved sequence (Leu 160 to Lys 164) by oligonucleotide-directed mutagenesis according to a modification of Kunkel (12). In this report, we show results obtained from the substitution of Lys 164 to Glu (K164E) or Lys 164 to Ile (K164I). The oligonucleotide sequences which were utilized to direct these mutations are presented in Table 1. The desired alterations in the mutated cDNA were confirmed by sequencing about 200 nucleotides around the mutation site.

Disappearence of the enzyme activity and the substrate-binding activity in mutant proteins- The mutant proteins (K164E and K164I) were purified by the same procedure as the wild type protein. The enzyme activities of both mutant proteins were found to have decreased extremely compared to the wild type protein (Table 2). This observation suggests that Lys 164 is involved in the site crucial to the catalytic reaction. Further experiments were carried out on the substrate-binding activity. A competitive inhibitor, xanthine was used for the experiments. The binding of xanthine to uricase was detected by using a gel filtration column. Xanthine exhibits a specific binding to wild type uricase in the experimental condition (Fig. 2 a). However, the binding with K164E or K164I mutant protein was reduced, consistent with the disappearence of the enzyme activity (Fig. 2 b or c). These observations indicate that Lys164 is involved in the substrate-binding site of the enzyme.

Table 2. Specific activity of purified uricase. Each uricase was purified from 1 liter of IPTG-induced culture and showed a single band in SDS-polyacrylamide gel. The uricase activity was measured as described in Materials and Methods.

	total activity (nmol/min)	total protein (mg)	specific activity (nmol/min/mg)
wild type	5,800	1.4	4,143
K164E	< 10	0.6	< 10
K164I	< 10	1.1	< 10

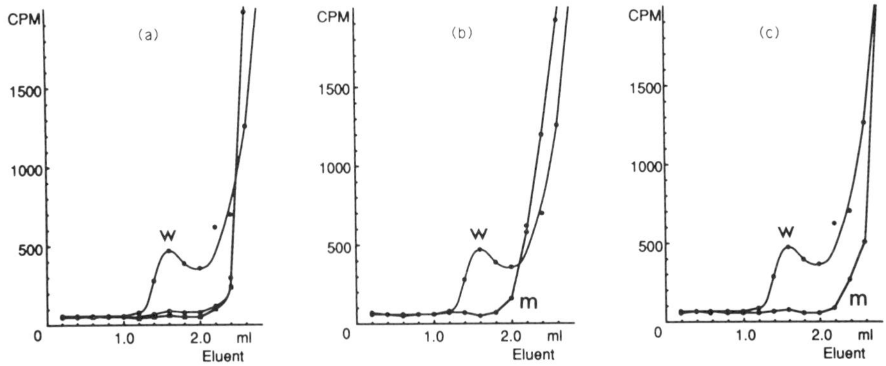

Fig. 2. Measurement of the substrate-binding activity in mutant uricase. 10 nmol of uricase was incubated with 2 nmol of the competitive inhibitor, ^{14}C-xanthine and the gel filtration assay of the mixture showed that ^{14}C-xanthine-bound uricase appeared in the void fraction. (a) Control experiment: closed circule, wild type uricase and ^{14}C-xanthine; open circule, competition with 50 fold excess of cold xanthine; open rectangle, ^{14}C-xanthin alone. (b) closed circle, wild type uricase and ^{14}C-xanthine; open circle, K164Emutant uricase and ^{14}C-xanthine. (c) closed circle, wild type and ^{14}C-xanthine; open circle, K164I mutant uricase and ^{14}C-xanthine.

REFERENCES

1. B.N. Ames, R. Cathcart, E. Schwiers and P. Hochstein, Uric acid provides an antioxidant defense in humans against oxidant- and radical- caused aging and cancer. Proc. Natl. Acad. Sci. USA. 78: 6858 (1981)
2. M. Hooper, K. Hardy, A. Handyside, S. Hunter and M. Monk, HPRT-deficient (Lesch-Nyhan) mouse embryos derived from germ line colonization by cultured cells. Nature. 326: 292 (1987)
3. M.R. Kuehn, A. Bradley, E.J. Robertson and M.J. Evans, A potential animal model for Lesch-Nyhan syndrome through introduction of HPRT mutations into mice. Nature. 326: 295 (1987)
4. W. Gilbert, Why genes in pieces? Nature. 271: 501 (1978)
5. M. Go and M. Nosaka, Protein architecture and the origin of introns. Cold Spring Harb. Symp. Quant. Biol. 52: 915 (1987)
6. M. Ito, M. Suzuki and Y. Takagi, Nucleotide sequence of cDNA and predicted amino acid sequence of rat liver uricase. Eur. J. Biochem. 173: 459 (1988)
7. K. Alvares, M.R. Nemali, P.G. Reddy, X. Wang, M.S. Rao and J.K. Reddy, The nucleotide sequence of a full length cDNA clone encoding rat liver urate oxidase. Biochem. Biophys. Res. Comm. 158: 991 (1989)
8. L.L. Wallrath, J.B. Burnett and T.B. Friedman, Molecular characterization of the *Drosophila melanogaster* urate oxidase gene, an ecdysone-repressible gene expressed only in the malpighian tubles. Mol. Cell. Biol. 10: 5114 (1990)

9. T. Nguyen, M. Zelechowska, V. Foster, H. Bergmann and D.P.S. Verma, Primary structure of the soybean nodulin-35 gene encoding uricase II. Proc. Natl. Acad. Sci. USA. 82: 5040 (1985)
10. A. Edward, H. Voss, P. Rice, A. Civitello, J. Stegemann, C. Schwager, J. Zimmermann, H. Erfle, C.T. Caskey and W. Ansorge, Automated DNA sequecing of the human HPRT locus. Genomics. 6: 593 (1990)
11. J. Nuesch and D. Schumperli, Structural and functional organization of the *gpt gene region* of *Escherichia coli.* Gene. 32: 243 (1984)
12. T.A. Kunkel, J.D. Robert and R.A. Zakour, Rapid and efficient site-directed mutagenesis without phenotypic selection.. Methods in Enzymology. 154: 367 (1987)

STUDY OF DNA REPAIR IN THE PERIPHERAL BLOOD LYMPHOCYTES OF PATIENTS WITH RHEUMATIC DISEASES UNDER GOING TREATMENT

A. Lukšienė, R. Filipavičiūtė

Research Institute of Experimental
and Clinical Medicine
Vilnius, Lithuania

INTRODUCTION

The DNA damage is a rather frequent occurrence, with its developing even if no such powerful influence as ionizing radiation. It also develops in normal cells, although being subject to elimination by a repair system which is controlling the reliability of the genetic apparatus. A constant break/repair ratio in the DNA molecule, characterizing molecular stability and its ability to secure internal homeostasis and optimal cellular metabolism when exposed to physical and chemical influences under normal and pathologic conditions it occeurs both in resting and proliferating lymphocytes[1]. But the bounds for the repair possibilities of the cell are extensively study and remain unclear. Apparently, the efficacy of the DNA repair system is limited.

In oncology, antineoplastic drugs are widely examined for their effect on the stability of the DNA molecule to establish their immunosuppressive action[2]. As applied to rheumatology, the mechanism of action of immunosuppressants in rheumatic diseases has also been not fully ascertained and it is not known how are affecting the DNA molecule stability. This work is aimed to study the stability of the DNA molecule of the peripheral blood lymphocytes (PBL) and its changes under the influence of the treatment with immunosuppressants in the patients with rheumatoid arthritis (RA) and systemic lupus erythematosus (SLE).

MATERIALS AND METHODS OF THE STUDY

A total of 69 patients; 21 had SLE, 48 RA, and group of 20 healthy persons as control were investigated. The group of SLE consisted of female patients aged 16 to 59 years and with a disease duration of 2 to 14 years, all indicating the Degree II activity of the disease. The group of RA was comprised of 35 female and 13 male patients aged 17 to 66 years, with a disease duration of 1 to 25 years, 18 patients of this groups were found to have the Degree I and 30 the Degree II and III activity of the inflammatory process. RA was established to be seronegative or seropositive in 28 and 20 patients, respectively. The control group was made up to the blood donors, 14 female and 6 male, aged 20 to 54 years.

The patients were receiving treatment with the immunosuppressants; methotrexate (10-25 mg weekly), prospidin (100 mg every other day) or azathioprine (100-150 mg daily). The treatment course lasted 3-4 weeks. Prior to their administration and after conclusion of the treatment, the patients underwent a thorough clinical and laboratory examination as well as a determination of the stability of DNA of the PBL.

To isolate the PBL for the DNA moleculer stability assay, the technique of C.Birboim was used[3]. From the suspension of PBL prepared, samples were being taken for calculating the concentration and evaluating the purity degree of the preparation obtained. The DNA moleculer stability was determined with the Hitachi MP - 2 A spectrofluorometer, according to the method of N.I.Ryabchenko[4]. The figures obtained were processed using techniques of variation statistics.

RESULTS OF THE STUDY

Our data shows that the stability of the DNA molecule of PBL was ascertained to be 67.8% for healthy persons and lower for both the RA and the SLE patients, by 13% and 15.9% respectively. The analysis of the DNA molecule stability depending on age and sex of the patients with RA and SLE failed to reveal any distinct fluctuations of it. But there was a definite dependency of the DNA stability changes in RA patients on the degree of activity of the process and on the form of the disease. So, the DNA stability in RA was almost normal with a minimal activity of the process, whereas with a pronounced activity of it (of the Degree II and III the molecule of DNA in PBL was established to be strongly destabilized as compared to the normal one 44.6% and 62.3%, respectively). The DNA molecule stability was less in a seropositive disease (43.2%)than in a seronegative one (53%).

The changes in the DNA stability depending on the immunosuppressant administered and the therapeutic effect achieved are analysed in the Table.

Table 1

The group studied	DNA molecule stability, %		
	before treatment	after treatment, with its effect being	
		good (n 6-13)	satisfactory or absent (n 3-6)
RA united, n=37	56.2± 9.3	68.2± 8.4	-
n=11	55.1±14.6	-	54.8±13.5
Methotrexate, n=18	58.8± 12.3	63.8± 13.4	57.2± 12.4
Prospidin, n=16	53.5± 14.2	68.5± 12.6	54.8± 12.6
Azathioprin, n=14	54.2± 13.0	70.3± 14.3	53.7± 13.8
SLE united, n=12	42.7± 13.9	56.7± 13.5	-
n=8	46.0± 15.3	-	51.3± 16.5
Methotrexate, n=11	44.5± 10.9	57.5± 12.6	46.9± 17.3
Azathioprine, n=9	47.4± 11.9	61.5± 13.4	45.8± 16.3
Healthy persons n=20	67.8± 9.2	-	-

From the analysis, the following regularity can be ascertained: with good therapeutic results obtained, irrespective of the disease, the DNA stability of PBL increases, by 12% for RA, with its returning to the norm, and by 14% for SLE, with its remaining, though having been increased, lower by 11.1% than in the control.

Regarding the efficacy of separate preparations, we established that methotrexate activates considerably the stabilization of the DNA molecule (quite by 13%) for the SLE patients only, the RA patients indicate a slight increase (by 5%) in the DNA stability. In RA, the elevated DNA molecule stability is maintained by prospidin and azathioprine. In cases of good therapeutic results, these preparations increase the DNA molecule stability for RA patients by 15% and 16%, respectively. In SLE patients, azathioprine restores well the DNA molecule stability (14%).

Thus, our data demonstrate that the immunosuppressants azathioprine, prospidin and methotrexate influence the stability of the DNA molecule of PBL in the patients with rheumatic diseases. It is possible that this may mediate their therapeutic activity.

REFERENCES

1. T.J.Schrader and I.G.Walter, Low levels of DNA excision repair in undamaged bovine lymphocytes, Mutation Research, 194: 65 (1988).
2. A.Begleiter, L.Pugh, Enhaced cytotoxicity and inhibition of DNA damage repair in irradiated murine L/51784 lymphoblasts and human chronic lymphocytic leukemia cells treated with 2'-deoxycoformycin and deoxyadenosine in vitro, Cancer Research, 48: July 15 (1988).
3. C.Birboim, J.Jevcak, Fluorometric breaks in human white blood cells produced by low doses of radiations, Cancer, 41: 1889 (1981).
4. N.I.Ryabchenko, S.I.Proskuriakov, Fluorometric study breaks of DNA in the cells produced by radiations, Radio biology, 24: 154 (1984).

NUCLEOSIDE TRANSPORT IN LYMPHOCYTES FROM ALCOHOLICS AND NON-ALCOHOLICS

Adrienne S. Gordon, Sharon W. Krauss, Laura Nagy and Ivan Diamond
Ernest Gallo Clinic and Research Center and the Department of Neurology
University of California, San Francisco 94110

Our laboratory is studying the biochemical mechanisms underlying tolerance and dependence on ethanol. Another aspect of our work is to identify genes that may predispose individuals to alcoholism. Our initial studies centered on the effects of ethanol on the cAMP signal transduction system. Using model cell culture systems, we found that acute exposure to ethanol causes increased receptor-stimulated cAMP production. After chronic exposure to ethanol, there is an "adaptive" response resulting in heterologous desensitization of receptor-stimulated cAMP production (1). We examined circulating lymphocytes obtained from actively drinking alcoholics and found they had a 76% decrease in receptor-dependent cAMP production when compared to lymphocytes from non-alcoholics (2).

Adenosine mediates the effects of ethanol on cAMP signal transduction. The heterologous desensitization produced by chronic ethanol exposure is similar to that caused by many agonists. In the continued presence of a stimulatory agonist, cells undergo an adaptive response, resulting in attenuated cAMP levels. Therefore, it is possible that ethanol causes the extracellular accumulation of an agonist that is coupled via its receptor to cAMP production. We focused on adenosine for several reasons. First, most cells continuously release and take up adenosine. Second, recent evidence suggests that adenosine may mediate acute ethanol-induced motor disturbances in rats and sedation in mice.

To investigate the role of adenosine in ethanol-dependent changes in cAMP signal transduction, we used the S49 lymphoma cell line. Mutants in both cAMP signal transduction and adenosine transport have been isolated from this cell line. We found that in wild type S49 cells, as well as in other cell lines and cultured human lymphocytes, acute ethanol exposure causes increased extracellular adenosine accumulation (3). When wild type S49 cells were co-incubated with ethanol and adenosine deaminase, there was no heterologous desensitization. Moreover, in S49 mutant cells unable to transport nucleosides and therefore unable to accumulate extracellular adenosine, ethanol did not cause heterologous desensitization (3).

Ethanol inhibits adenosine uptake. These data suggest that the nucleoside transporter may be a primary target for ethanol. Ethanol-induced increases in extracellular adenosine concentration could be due to decreased adenosine uptake. Indeed, a 30-40% inhibition of adenosine uptake by 200 mM ethanol was detected at 5 seconds and maintained for at least 1 minute (4). The rate of accumulation of extracellular adenosine in wild type S49 cells treated acutely with ethanol can be completely accounted for by a 35% inhibition of adenosine uptake. Adenosine uptake is inhibited with increasing concentrations of ethanol from 50-200 mM, reaching a maximum at 200 mM; the apparent K_i was 74 ± 8 mM. Kinetic analysis showed that acute ethanol was a noncompetitive inhibitor of adenosine uptake in S49 cells. V_{max} for uptake was decreased by 50% in ethanol-treated cells, with only a small change in affinity. Ethanol also inhibited the uptake of uridine, another nucleoside transported by the nucleoside transporter. Inhibition of adenosine uptake by alcohols increased with increasing chain length, indicating that inhibition of uptake by ethanol was not due to a change in osmolality. We have also found that adenosine efflux and metabolism are not altered by ethanol (4). Taken together, our data demonstrate that acute ethanol exposure results in increased extracellular accumulation of adenosine by inhibiting its re-uptake.

Chronic effects of ethanol on adenosine uptake and extracellular adenosine accumulation. Chronic exposure of several different cell lines to 100 mM ethanol for 48 hours caused a reduction in adenosine uptake when measured in the absence of ethanol. In addition, on rechallenge with ethanol, adenosine uptake was no longer inhibited and extracellular adenosine was not increased (4).

It is possible that the decreased cAMP levels observed after chronic ethanol exposure may be responsible for the loss of sensitivity of adenosine transport to ethanol. To determine whether the cAMP-mediated signal transduction pathway regulates inhibition of adenosine uptake by ethanol, we studied adenosine uptake in S49 cell lines that have altered cAMP signal transduction. Acute ethanol inhibits adenosine uptake in S49 wild type cells but does not inhibit uptake in *kin*$^-$ cells, which lack cAMP-dependent protein kinase activity, or in *unc* cells, where G_s is present but receptor activation is not coupled to adenylyl cyclase (Nagy et al., submitted). These results suggest that an active cAMP-dependent phosphorylation pathway is required for ethanol to inhibit adenosine uptake.

Adenosine transport in cells from alcoholics. We recently found that acute ethanol inhibited adenosine uptake by 71% in circulating lymphocytes from non-alcoholics, while adenosine uptake in cells from alcoholics was insensitive to ethanol inhibition (Fig. 1).

Moreover, adenosine uptake in the absence of ethanol was reduced by 59% in cells from alcoholics (Fig. 1). These changes could be due to long-term effects of ethanol exposure in vivo. In addition, genetic factors could play a role in the development of many cases of alcoholism. It is possible that a major target for ethanol, such as the adenosine transport system, might be regulated differently in these alcoholics. Therefore, we examined nucleoside transport and its sensitivity to ethanol in human lymphocytes cultured in the absence of ethanol for 7 days. We found that the magnitude of adenosine uptake in cells from alcoholics was the same as in cells from non-alcoholics. However, adenosine uptake in cells from alcoholics remained insensitive to ethanol inhibition (Fig. 2). The uptake of

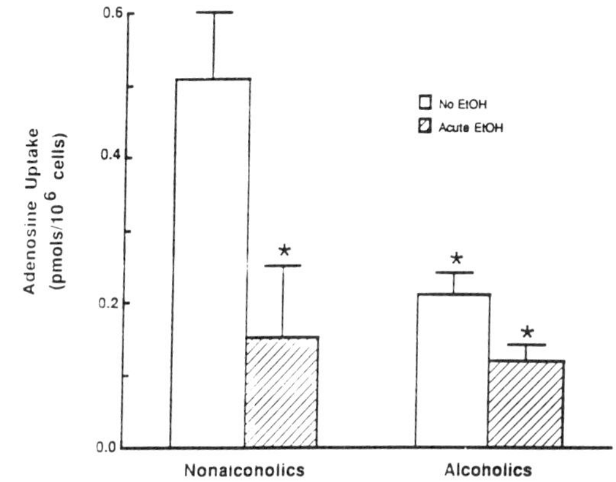

Figure 1

Ethanol inhibition of adenosine uptake in freshly isolated lymphocytes from alcoholics and nonalcoholics. Uptake of 0.3 μM [^{3}H]-adenosine was measured in the presence or absence of 200 mM ethanol in lymphocytes from nonalcoholics (n=9) and alcoholics (n=7). Values represent mean ± SEM, * $p<0.001$ compared to uptake in lymphocytes from nonalcoholics in the absence of ethanol.

[^{3}H]-formycin B, a non-metabolized nucleoside, was also insensitive to ethanol in cells from alcoholics, indicating that nucleoside transport rather than metabolism is altered in lymphocytes from alcoholics. The persistence of this altered phenotype in lymphocytes from alcoholics after 4-6 generations of culturing without ethanol suggests a genetic difference underlying altered nucleoside transport. This may be a useful marker for genetic analysis of the heritability of alcoholism. To establish whether this phenotype can be inherited, it will be necessary to determine whether insensitivity of nucleoside uptake to ethanol occurs in cells cultured from non-drinking offspring of alcoholics.

In summary, we have shown that acute ethanol treatment inhibits nucleoside uptake. After chronic ethanol exposure, nucleoside uptake becomes insensitive to rechallenge with ethanol. In addition, we have identified a cellular response that distinguishes cultured lymphocytes of alcoholics from those of non-alcoholics: nucleoside

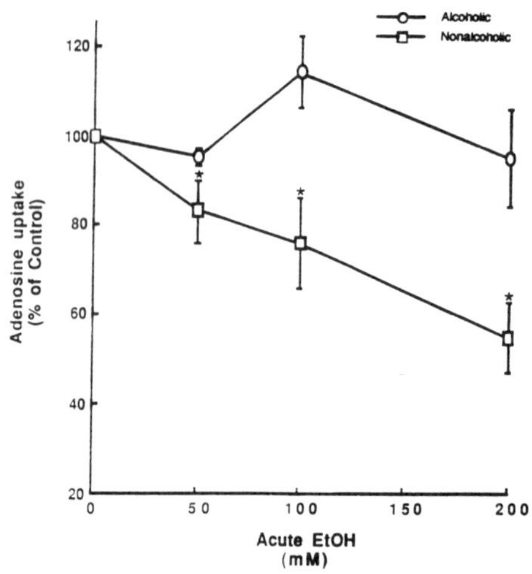

Figure 2

[^{3}H]-adenosine uptake in lymphocytes after culture for 7 d in the absence of ethanol. Uptake is expressed as a percentage of uptake in the absence of ethanol. Values represent mean ± SEM, n=12 for nonalcoholics, n=6 for alcoholics. * $p<0.03$ compared to uptake in the absence of ethanol.

uptake in cultured lymphocytes from alcoholics is insensitive to acute ethanol exposure. Our results suggest that there may be genetic differences in alcoholics underlying ethanol insensitivity of nucleoside transport which may prove useful in developing biological markers for susceptibility to alcoholism. Since adenosine has been implicated in mediating many physiological effects of ethanol, abnormal regulation of nucleoside tranport may influence the development of alcohol dependence in populations at risk for this disease.

References

1. A. S. Gordon, K. Collier, and I. Diamond, Ethanol regulation of adenosine receptor-stimulated cAMP levels in a clonal neural cell line: An in vitro model of cellular tolerance to ethanol, Proc. Natl. Acad. Sci. USA 83:2105-2108 (1986).
2. I. Diamond, B. Wrubel, W. Estrin, and A. Gordon, Basal and adenosine receptor-stimulated levels of cAMP are reduced in lymphocytes from alcoholic patients, Proc. Natl. Acad. Sci. USA 84:1413-1416 (1987).
3. L. E. Nagy, I. Diamond, K. Collier, L. Lopez, B. Ullman, and A. S. Gordon, Adenosine is required for ethanol-induced heterologous desensitization, Mol. Pharmacol. 36:744-748 (1989).
4. L. E. Nagy, I. Diamond, D. J. Casso, C. Franklin, and A. S. Gordon, Ethanol increases extracellular adenosine by inhibiting adenosine uptake via the nucleoside transporter, J. Biol. Chem. 265:1946-1951 (1990).

INITIAL RATE MEASUREMENTS OF CELLULAR NUCLEOSIDE TRANSPORT: TIME COURSE, INITIAL CURVATURE OF TIME CURVES AND METABOLISM

Hans P. Baer, Patrick O. J. Ogbunude, Azadali Moorji and Vincent Serignese

Department of Biological and Medical Research
King Faisal Specialist Hospital and Research
Centre, MBC 03, Riyadh 11211, Saudi Arabia

INTRODUCTION

Transport of physiological nucleosides including adenosine and inosine usually is followed by rapid intracellular phosphorylation in most mammalian or protozoal cells. Aspects of nucleoside transport measurements including the consequences of concomitant metabolism of permeants have been discussed extensively in the literature[1-4] and it has been stated that rapid metabolism of permeants complicates measurement of initial transport rates[2]. We are discussing here the point that in fact the presence of rapid cellular metabolism of nucleosides to non-permeant nucleotides facilitates rather than complicates the measurement of initial transport rates by maintaining the initial rate-determining concentration gradient.

Theoretical Considerations

The rate of transport of a permeant across a membrane is dependent on the driving force, i.e. the concentration gradient, and a resistance term R:

$$\text{Rate} = R \times \Delta c \quad (1)$$

In case of simple diffusion, R is a constant determined by the physical chemical properties of the permeant and the membrane. In case of facilitated, carrier-mediated diffusion, R is a function of the permeant concentration c reflecting the characteristics of the permeant-transporter interaction (usually well described by a Michaelis-Menten type equation) as well as the mobility M of the carrier within the membrane:

$$R = f(c) + M$$

Although R thus varies with the permeant concentration in a nonlinear fashion, it can be considered constant within limited concentration ranges so that the concentration gradient again remains the determinant of the transport rate (equation 1). Accordingly, a discussion on whether or not in a given situation initial rates are being measured should focus on whether or not the concentration gradient is maintained during assay.

Practical Considerations for Initial Rate Measurements

Experimentally, the external permeant concentration is fixed and considered constant, a condition which may need verification in some circumstances. The concentration gradient becomes a function of the intracellular permeant concentration, if its magnitude and change are significant relative to the external concentration. These factors indeed become significant in the absence of intracellular metabolism and in this case a curvature in transport-time curves is evident within seconds of incubation. From these true initial rates have to be obtained by graphical or mathematical analysis (applying integrated rate equations[2-4]) and fitting a tangent to the time function at zero time. In some cases special procedures not applicable to all cells may be required to sample over extremely short time intervals[4,5]. If metabolism occurs intracellularly, this would serve to reduce the rate at which the intracellular permeant concentration increases, thus delaying a decrease in the concentration gradient and extending the time period during which initial data points approximate the slope of the zero time tangent. If metabolism is extensive, matching the rate at which permeant enters the cell, the intracellular concentration of permeant will be held at near-zero and constant; thus, the initial concentration gradient will be essentially maintained and the transport-time curve is linear. Theoretical transport-time functions, representing conditions with no intracellular metabolism of permeant, slow metabolism and rapid metabolism for both passive diffusion and carrier-mediated diffusion are summarized in a graphic illustration shown in Fig. 1.

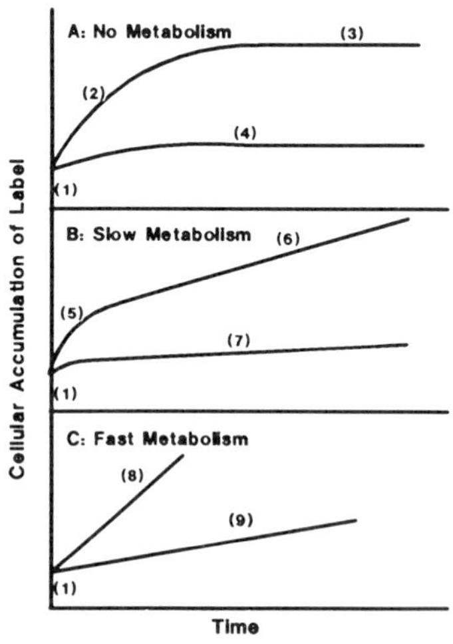

Figure 1. Schematic representation of transport-time curves. In all cases there is a zero time blank value (1) to which curves or lines extrapolate; the difference between upper and lower curves in each panel represents facilitated diffusion. A - No Metabolism: Data points lie on a curve until reaching equilibrium (conc. gradient = zero) for both total (3) and passive (4) diffusion. B - Slow Metabolism: The initial portion of the function (5) is steeper and closer to linearity, while the subsequent linear portion reflects the capacity of metabolism, both in case of total (6) and passive (7) diffusion. C - Fast Metabolism: The initial steep and linear section of total (8) and passive (9) diffusion, reflecting initial rate of transport at a given and constant concentration gradient, is extended.

A range of artificial situations will arise when studying nucleoside transport in isolated cells with nonphysiological substrates[1-4]. Such studies, of course, are necessary for the purpose of describing experimentally some of the features of transporters. For example, the basic question of whether a given transport process is concentrative or nonconcentrative can be studied most directly with nonmetabolizable permeants. In this case, analysis of permeant distribution at equilibrium and steady state is of interest rather then that of initial rates. Yet in case of physiologically meaningful permeants such as adenosine or inosine, and within their physiologically relevant concentration range of 1 uM or so, intracellular phosphorylation will always facilitate initial rate measurements, whereby the time interval within which transport-time curves remain sufficiently linear may vary from cell to cell.

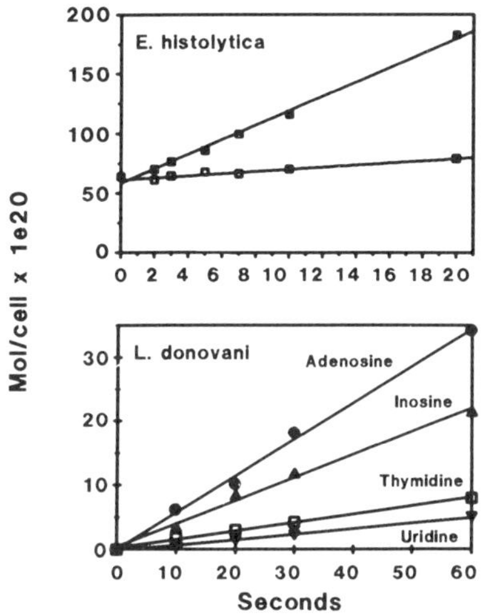

Figure 2. Transport-time curves obtained with Entamoeba histolytica, using [3H]adenosine as permeant and Leishmania donovani promastigotes, using different permeants as indicated. Permeant concentrations were 1 μM, and reactions were stopped by dilution with high concentrations of unlabeled permeants and centrifugation of cells through oil layers. The upper panel (E. histolytica) shows control values (reactions stopped at zero time, lower line) and adenosine transport (upper line), both linear fits through data points showing a common zero-time intercept. In the lower panel (L. donovani) respective control values have been subtracted from data points in case of each permeant. No initial curvatures were observed in these transport-time functions, and analysis of intracellular label revealed full or extensive phosphorylation of permeants (unpublished data).

Recently we studied parasitic protozoa, analyzing aspects of their nucleoside transport systems. Transport-time curves with adenosine or inosine as permeants remained linear over extended time periods (10 to 60 s or longer) while rapid intracellular phosphorylation occurred without evidence of any significant levels of free permeant intracellularly (cf. Fig 2). In case of such parasitic organisms, a rapid overall salvage system appears to be physiologically meaningful because these organisms generally are auxotrophic for purine compounds. In these and other cells where nucleoside transport is of physiological significance, one should expect the rate of phosphorylation of permeants to match or exceed that of transport, otherwise transport activity would be wasted.

The fact that mammalian cells indeed are capable of rapid phosphorylation of nucleosides is illustrated by recent studies with mouse macrophages which have a concentrative, sodium-dependent nucleoside transporter of high capacity: Despite rapid transport, permeants such as adenosine are phosphorylated intracellularly at rates matching or exceeding that of transport[6].

In this brief discussion, not all practical situations can be adequately described. Metabolism of permeants by nonphosphorylating routes, for example deamination of adenosine to inosine, may create a set of problems requiring special attention. In case of adenosine deaminase this interference can be eliminated by employing potent inhibitors of the enzyme, channeling metabolism towards phosphorylated, nonpermeable products which again should facilitate initial rate measurements.

Particular attention has to be paid to effective reaction stop procedures and the measurement of a valid zero-time control value to which transport-time curves must be extrapolated (cf. Fig. 2). In case of inhibitable, facilitated diffusion, the use of transport inhibitors such as 6-[(4-nitrobenzyl)-mercapto]purine ribonucleoside or dilazep allows not only effective stopping of reactions but also measuring selectively the activity of the facilitated transport system[7]. Procedures such as cold medium stop, dilution of reaction volumes and addition of high concentration of unlabeled permeant do not discriminate between different transport routes.

REFERENCES

1. A. R. P. Paterson, N. Kolassa, and C. E. Cass, Transport of nucleoside drugs in animal cells, Pharmac. Ther. 12:515-536 (1981).
2. P. G. W. Plagemann, R. M. Wohlhueter, and C. Woffendin, Nucleoside and nucleobase transport in animal cells, Biochim. Biophys. Acta 947:405-433 (1988).
3. P. G. W. Plagemann and R. M. Wohlhueter, Permeation of nucleosides, nucleic acid bases, and nucleotides in animal cells, Curr. Top. Membr. Transp. 14:225-330 (1980).
4. R. H. Wohlhueter and P. G. W. Plagemann, Measurement of transport versus metabolism in cultured cells, Methods in Enzymology 173:714-732 (1989).
5 A. R. P. Paterson, E. R. Harley, and C. E. Cass, Inward fluxes of adenosine in erythrocytes and cultured cells measured by a quenched flow method, Biochem. J. 24:1001-1008 (1984).
6. H. P. Baer and A. Moorji, Sodium dependent and inhibitor insensitive uptake of adenosine in mouse peritoneal exudate cells, Biochim. Biophys. Acta 1026:241-247 (1990).
7. A. R. P. Paterson, E. R. Harley, and C. E. Cass, Measurement and inhibition of membrane transport of adenosine, _in_ Methods in pharmacology, D. M. Paton, ed., Plenum Publishing Corporation, New York (1985).

RELATIONSHIPS BETWEEN METABOLIC ENZYMES AND THE NUCLEOSIDE TRANSPORT

R. Franco, C. Lluis, E.I. Canela, J. Mallol,
J.J. Centelles, J.M. Arán, J. Blanco and J. Sayós

Department of Biochemistry and Physiology
School of Chemistry. University of Barcelona
Diagonal 647. Barcelona 08028. Spain

INTRODUCTION: CARRIERS AND ENZYMES

The understanding of nucleoside incorporation into cells requires the integration of all factors related directly or indirectly with the transport phenomenon. This is implicitely assumed when using the term Uptake whose definition is: the total intracellular accumulation of radioactivity from exogenous labelled substrate regardless of metabolic conversions.

Since long time ago there has been controversy as to whether the nucleoside carrier molecule is physically linked to some metabolic enzyme. In this case, a molecule is, at the same time, transported and derivatized or processed. This seems to be true in the case of inosine uptake by some cells which involves a membrane-localized purine nucleoside phoshorylase acting in a group translocation reaction (Cohen and Martin Jr 1977, Dowd and Quinlan 1977). In the case of the "classical" nucleoside transporter which mainly deals with adenosine and uridine. Agarwal and Parks (1975) working with erythrocytes suggested an interaction between intracellular adenosine deaminase and the nucleoside transport system. In subsequent years the hypothesis was forgotten and the unexpected results explained by excess substrate inhibition of adenosine kinase.

In experiments performed in our laboratory we have studied adenosine uptake in brush-border membrane vesicles (BBM) from proximal tubule of rat kidney. The time course of uptake shows a saturation at 15 minutes. In the presence of metabolic inhibitors such as 2'-deoxycoformycin (strong inhibitor of adenosine deaminase) and iodotubercidin (strong inhibitor of adenosine kinase), the uptake diminishes and no saturation was found at up to 30 minutes (Sayós et al 1991). These results fit well with those expected. What was also expected was that, in presence of the inhibitors, accumulation of adenosine inside the vesicles does occur. But as indicated in figure 1, adenosine does not accumulate. It is obvious that if the enzymes which metabolize adenosine are blocked, a rise of adenosine level should occur. If this does not occur, only two possibilities exist:

- One (or both) of the inhibitors (deoxicoformycin or iodotubercidin) also inhibits the transporter
- There is an interaction of the transporter and one (or both) metabolic enzyme(s)

The first possibility seems difficult since it would require some similarity between the active site of the metabolic enzyme and the "active site" of the transporter molecule. The second possibility is simply difficult to demonstrate. Another interesting finding was the high degree of phosphorylated compounds inside the vesicles despite the low ATP content (figure 1). In our laboratory we are addressing these questions by using a number of different experimental strategies.

The importance of adenosine deaminase as a key enzyme of purine metabolism was evident after the demonstration that 13 out of 22 patients of severe combined immunodeficiency had a congenital defect of the enzyme (Meuwissen et al 1975). Thus adenosine deaminase is crucial for the development of immune system and for maduration of immunocompetent cells.

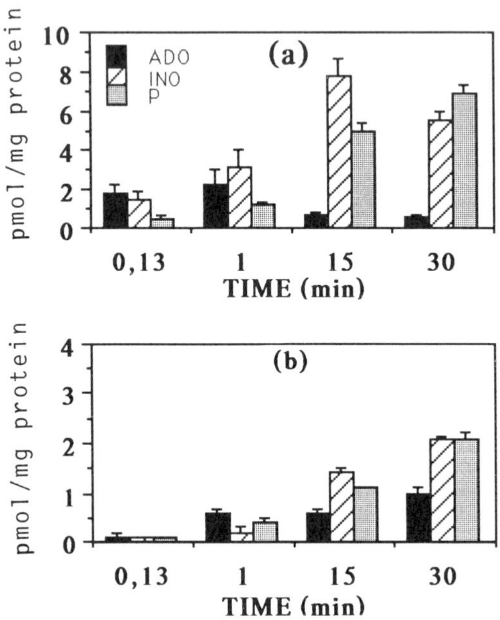

Figure 1. IDENTIFICATION OF THE INTRACELLULAR RADIOLABELLED COMPOUNDS. BBM were incubated with 2.5 µM [^{3}H]Ado at 25ºC. Values are mean±SD of four independent experiments with quadruplicates. **Ado**: Adenosine; **Ino**:Inosine; **P**:AMP+ADP+ATP+IMP. **a**) Control membranes **b**) Membranes with 1µM **DCF** and 10 µM **Itu.** (From Sayós et al 1991).

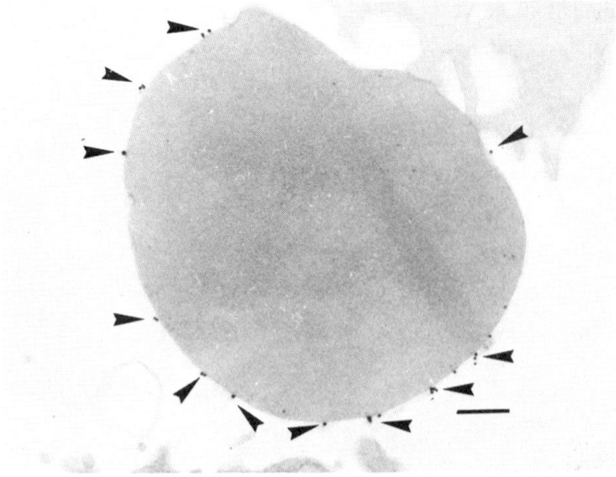

Figure 2. ELECTRON MICROGRAPH OF RED BLOOD CELL GOLD-LABELLED WITH ANTI-ADENOSINE DEAMINASE RABBIT ANTIBODY. Bar: 0.5 µm. From Franco et al 1990.

By immunocytochemical techniques, adenosine deaminase was found to be associated with the plasma membrane of erythrocytes (figure 2) and platelets (Franco et al 1990). Similar results, using a cytochemical technique, were found by Byelat and Tritsch (1989) who indirectly demonstrated that the enzyme can be considered an ectoenzyme. In recent studies (Aran et al 1991) we have reported a great variation in the expression of ecto-adenosine deaminase in the plasma membrane of mononuclear blood cells. Thus up to 90 % of monocytes and B lymphocytes present adenosine deaminase in their surface whereas the percentage is relatively low 10-20 for T lymphocytes. These results may indicate that the ecto-enzyme is also important, perhaps even more important than the intracellular enzyme, for the maduration of lymphocytes.

In the past, when one spoke of metabolic enzymes, everybody knew that they were intracellular enzymes. Now the situation has changed. Since the discovery of ecto-adenosine deaminase, adenosine can be converted into inosine within the extracellular compartment.

Usually, deoxicoformycin is used in transport experiments and then it does not matter where it is localized the enzyme. However, when comparisons are made from experiments performed in absence or in presence of deoxicoformycin (or EHNA, another strong inhibitor) care must be taken in the interpretation. Perhaps deoxicoformycin is inhibiting ecto-adenosine deaminase instead of (or further to) intracellular adenosine deaminase, mainly if there is not a preincubation period with the inhibitor. But irrespective of whether how accurately we perform the assays and interpret the results the philosophy should be to think that ecto-adenosine deaminase is there for some reason.

The connection between ecto-adenosine deaminase and the transporter may be obvious. If only ecto-adenosine deaminase and the Na^+-independent, dipyridamole sensitive carrier molecule is taken into account, the link is evident: ecto-adenosine deaminase produces extracellular inosine which can be incorporated by the same transport system than adenosine. Thus, in presence of ecto-adenosine deaminase, the incorporation/uptake of adenosine plus inosine is the process which is physiologically relevant.

SOME ABOUT FLUX CONTROL

We have simulated the whole process of extracellular degradation of adenine nucleotides towards adenosine and inosine and the subsequent transport and metabolism of the nucleosides (data submitted for publication). In our opinion the results have been very important because of their predictive power. First of all, in a variety of simulation conditions it appeared that the intracellular anabolic flux towards adenine nucleotides is controlled by ecto-5´-nucleotidase and the carrier. In any case, adenosine kinase was a less important enzyme in controlling that flux. On the other hand, the occurence of both, adenosine and inosine, in the intracellular space, may lead to strange kinetics of uptake. For instance, there exist conditions, in terms of extracellular versus intracellular balance, which may lead to transport against chemical gradients. This leads to an exquisite balance of the uptake depending upon the chemical gradients of adenosine and inosine.

The conditions of the simulation have been taken from the literature. This may limit the validity of the results because it is not possible to get homogeneous data corresponding to the same system. Besides, the intra and extracellular concentration of purine compounds vary from tissue to tissue and from technique to technique, i.e. from lab to lab. The predictive power of this theoretical tool will increase enormously when the concentration of purine compounds is accurately determined in a variety of systems and of metabolic situations.

CONCLUDING REMARKS

Nucleoside uptake must be a well regulated phenomenon. Recently the group of Miras-Portugal has provided evidence of regulation by covalent modification (more details are given in other chapter of this book). Thus, if indirect regulation via second messengers is demonstrated, we may be pretty sure that there must be direct ways of modulation of the nucleoside transporter. In essence, modulation must be exerted according to a very practical rule, i.e. the cells must incorporate adenosine when it requires it and not otherwise. Moreover, in some circumstances, the cells may prefer to incorporate inosine than adenosine.

To elucidate which are the direct modulators of the transport will not be an easy task. However there are some approaches that may be tried by the groups working in this field. In our opinion, some of the questions that should be addressed are:

- Whether or not adenosine deaminase binding protein is able to transpose adenosine deaminase accross cell membrane in one or both directions. Alternatively, search for other mechanisms of adenosine deaminase internalization or release.
- Search for dynamic changes in the number of transporter molecules in the plasma membrane in response to different functional and metabolic conditions.
- Design of specific experiments to know whether or not the transporter is physically linked to metabolic intracellular or extracellular (ecto-) enzymes.
- Analysis of the conditions which lead to the release of adenosine. We mean conditions apart from those which are obvious such as in hypoxia
- Measurement of intracellular and extracellular metabolite concentrations in conditions close to those *in vivo*. This would very much improve the design of a plausible theoretical model of the whole process
- Measurement of flux control coefficients to validate our hypothesis that intracellular anabolic fluxes may be, in some circumstances, controlled by ecto-enzymes.

"for truth is always strange, stranger than fiction" (Byron: Don Juan)

ACKNOWLEDGEMENTS

This work was supported by Grants 87/1389, 89/629 and 91/0272 from Fondo de Investigaciones Sanitarias.

REFERENCES

Agarwal R.P. and Parks Jr. R.E., 1975, A possible association between the nucleoside transport system of human erythrocytes and adenosine deaminase. Biochem. Pharmacol., 24: 547-550.

Aran J.M., Colomer D., Matutes E., Vives-Corrons J.L. and Franco R., 1991, Presence of adenosine deaminase on the cellular surface of blood mononuclear cells: An immunochemical localization by means of electron microscopy. J. Histochem. Cytochem., In the Press.

Bielat K. and Tritsch G.L., 1989, ECTO-enzyme activity of human erythrocyte adenosine deaminase. Mol. and Cell. Biochem. 86:135-142

Cohen A. and Martin Dr. D.W., 1977, Inosine uptake by cultured fibroblasts from normal and PNP-deficient humans. J. Biol. Chem., 252: 4428-4430

Dowd D.J., Quinlan D.C. and Hochstadt J., 1977, Mechanism of purine nucleoside handling and transport in isolated membrane vesicles from polyoma transformed BHK/21 cells. Biochemistry, 16: 4526-4532

Franco R., Aran J.M., Colomer D., Matutes E. and Vives-Corrons J.L., 1990, Association of adenosine deaminase with erythrocyte and platelet plasma membrane: An immunological study using light and electron microscopy. J. Histochem. Cytochem., 38: 653-658

Meuwissen, H.J., Pollara, B. and Pickering R.J., 1975, Combined immunodeficiency disease associated with adenosine deaminase deficiency J. Pediatr., 86: 169-181

Sayós J., Centelles J.J., Mallol J., Canela E.I., Lluis C. and Franco R., 1991, Adenosine (Ado) uptake in brush-border membranes from rat kidney (BBM) Biochem. Soc. Trans., 19: 323S.

ACTIVE TRANSPORT OF ADENOSINE INTO PRIMARY CULTURES OF BRAIN CELLS AND ITS METHODOLOGICAL CONSEQUENCES

Leif Hertz

Depts. of Pharmacology and Anaesthesia
University of Saskatchewan
Saskatoon, Sask., Canada S7N OWO

INTRODUCTION: WHY DO CELLS TRANSPORT NUCLEOSIDES AND HOW ARE TRANSPORT RATES DETERMINED

In spite of the ability of most mammalian cells to synthesize nucleosides, the presence of transport mechanisms for these compounds across cell membranes is well established. Their presence is of importance for several reasons: 1) some cells, e.g., erythrocytes, leucocytes, bone marrow cells, cells in the gastrointestinal tract and perhaps some brain cells are deficient in *de novo* synthesis of purines or pyrimidines and thus need to accumulate purines or pyrimidines which have either been ingested or produced in other cells; 2) the physiological role of certain cells, e.g., in the kidney or in the choroid plexes, is to transport different compounds (which may include nucleosides) across a barrier between different compartments, and 3) some cells, predominantly in the central nervous and cardiovascular systems, express receptors for nucleosides on their cell surface, and cellular uptake is an important mechanism to terminate the action of many transmitters.

Elucidation of transport mechanisms for nucleosides have to a large extent been made using erythrocytes and other blood cells (group 1 above) as well as transformed cultured cells (*cell lines*). Most of these cell types have been found to behave in a similar manner, i.e., adenosine *is accumulated by facilitated diffusion which can only be determined during ultrashort incubation periods, since this uptake ceases once the extracellular and intracellular concentrations of adenosine itself are identical.* In contrast, longer incubation times mainly reflect the subsequent metabolism which leads to continued adenosine accumulation by maintaining a concentration gradient for adenosine itself. More recently, some authors have attempted to apply these findings and concepts in studies of cells from the central nervous system (group 3 above), whereas others have studied transport in neural cells in an analogous way as transport of the amino acid transmitters, glutamate and GABA, i.e., *determined active uptake during longer incubation periods.* If it is attempted to determine this uptake during ultrashort periods, the measurement reflect initial diffusion rather than active uptake. These two different approaches have led to very different concepts and conclusions as discussed in more detail elsewhere (Hertz, 1991).

Facilitated diffusion has in many cases been studied not only by applying an ultrashort incubation period (sec or fractions of sec) but also in non-metabolizing cells, obtained by administration of metabolic poisons (e.g., cyanide, iodoacetate). When this is done, it should obviously be ascertained that the abolishment of energy metabolism has no deleterious effect on the transport mechanisms (facilitated diffusion or active transport) themselves. This has been well established, e.g., for the uptake of uridine into cultured Chinese hamster ovary cells which is known to occur by facilitated diffusion (Paterson et al., 1981) but not cells from the central nervous system (CNS).

Purine and Pyrimidine Metabolism in Man VII, Part A
Edited by R.A. Harkness *et al.*, Plenum Press, New York, 1991

The most important difference between nucleoside transport in, e.g., blood cells and in cells from the CNS is that there is an intense, easily demonstrable *active transport*, at least of the nucleoside adenosine, into cells from the central nervous system. This applies both to neurons and astrocytes (glial cells). In addition to the active uptake there is a diffusional mechanism which, at least at low extracellular concentrations, will be of negligible importance for the uptake determined after longer incubation time (e.g., 5 min) but may distort the determination of the rate of active transport, if ultrashort incubation times are used.

Evidence for the existence of an active transport mechanism for adenosine into neurons is presented in Table 1. Table 1 shows *cell contents* of adenosine and its major metabolites in primary cultures of cortical neurons during incubation in the presence of 10 μM adenosine with or without the adenosine deaminase inhibitor deoxycoformycin. At the concentration used (10 μM) this inhibitor has a relatively specific inhibitory effect on the adenosine deaminase. It can be seen that the intracellular content of adenosine itself during prolonged (60 min) incubation remains relatively constant at 0.1-0.2 nmol/mg protein in the control cultures and that the variability in adenosine content is quite pronounced (reflecting the uncertainty of determination of such low quantities). In the presence of deoxycoformycin the adenosine content is very considerably increased, whereas the inosine content is decreased. The increase in adenosine content is obvious already 5 sec after exposure to the inhibitor and it reaches a steady state before or around 15 min. From then onwards, the concentration of adenosine remains steady at about 7 nmol/mg protein. The intracellular water content of these cells is probably around 5 microliter per mg protein and certainly not above 10 microliter per mg protein . Calculation of an intracellular concentration using the latter value, which will result in the minimum estimated concentration of adenosine in the intracellular water phase, shows that the content of *adenosine itself* during the period 15-60 min has *increased* to a level of at least 7 nmol adenosine per 10 microliter intracellular water, which equals 700 nmol/ml water or an intracellular adenosine concentration of 700 μM. *This is 70 times more than the extracellular adenosine concentration and provides unequivocal proof of an active, concentrative transport of adenosine into the neurons.* In the absence of the deaminase inhibitor, the adenosine concentration does not rise significantly above 0.1 nmol/mg protein (Hertz, 1991), yielding an intracellular adenosine concentration close to the extracellular concentration of 10 μM. Thus, the reason for the high adenosine concentration in the presence of the inhibitor is not a faulty technique, causing a degradation of adenosine nucleotides before the measurements.

TABLE I

Cellular Contents (Pool Sizes) of Adenosine and Its Metabolites (Nanomoles per Milligram Protein)

	5 sec		15 min		30 min		60 min	
Compound	Contr.	Inhib.	Contr.	Inhib.	Contr.	Inhib.	Contr.	Inhib.
ADO	0.10	5.05	0.10	7.07	0.23	6.94	0.25	7.27
ATP+ADP	20.8	28.5	23.7	29.6	24.5	331.0	20.6	33.0
AMP	7.35	6.91	6.39	6.12	7.79	5.37	5.20	6.16
INO	4.40	2.66	4.03	0.53	3.65	0.55	3.10	0.52
HYP	2.97	2.73	4.42	4.14	2.58	2.55	2.48	2.62

Note: ADO; Adenosine; ATP: adenosine triphosphate; ADP: adenosine disphosphate; AMP; adenosine monophosphate; INO; inosine; HYP; hypoxanthine; Contr.: control; Inhib.:+10 μM deoxycoformycin. Modified from Hertz, 1991.

In neurons, at least 50% of the accumulated adenosine is normally metabolized to the deaminated products inosine and hypoxanthine. Interference with deaminase activity therefore abolishes a major part of adenosine metabolism. If this metabolism was of importance in "driving" the trans-membrane accumulation of adenosine (see Introduction), it would be expected that the total uptake of radioactivity in *adenosine plus all its metabolites*

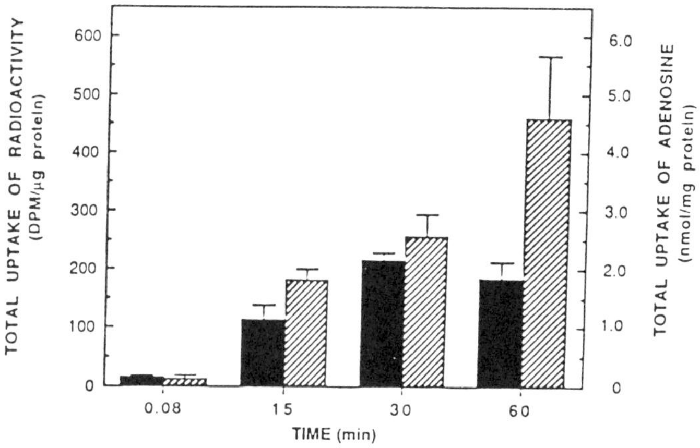

Figure 1. Total accumulation of labeled adenosine and its metabolites (nanomoles per milligram protein) into cultured neurons during incubation with (hachured bars) and without (solid bars) the adenosine deaminase inhibitor, 2'-deoxycoformycin. The adenosine concentration was 10 μM. Note that the time periods indicated are only relatively equidistant from 0.08 to 30 min (From Hertz, L. and Matz, H., Neurochem. Res., 14, 755, 1989. With permission)

would decrease in the presence of the deaminase inhibitor. That this is not the case (but that the total uptake of radioactivity rather is increased in the presence of deoxycoformycin, probably on account of incorporation into the phosphorylated compound rather than into inosine, which is readily released from the cells) can be seen from Fig. 1, showing the total uptake of radioactivity after varying times of incubation with 10 μM adenosine. This finding lends further support to the *conclusion that adenosine is accumulated into cultured neurons by a continuous, active, concentrative uptake, not by metabolism-driven facilitated diffusion;* therefore, *long- term studies* are not only permissible but *required* to determine the true continued transport rate of adenosine into neurons. However, due to the increasing efflux of radioactivity ("back-flux") when the intracellular specific activity increases, the transport should be determined only during the period of a rectilinear increase (0-30 min in Fig. 1). It can be seen that this uptake amounts to 0.10 nmol/min per mg protein at an extracellular adenosine concentration of 10 μM. The initial uptake within the first 5 sec. uptake is faster (≈ 0.15 nmol/mg protein after 5 sec., or ≈ 1.0 nmol/min per mg protein). This uptake represents the sum of diffusional uptake and active uptake; moreover, it cannot be ignored that remnants of adherent medium may contribute to the measured radioactivity. This initial uptake rate would dominate in the total accumulation, *if it were able to continue for a longer period;* however after 5 sec, the diffusional uptake has ceased and, if this value is used to calculate trans-membrane permeation of adenosine during normal *in vivo* conditions, the transport rate will be grossly overestimated. Clearance of extracellular adenosine into neurons by diffusion (which may be simple or facilitated) will only be of importance at extracellular concentrations of adenosine high enough to saturate the active uptake system (Hertz, 1991) and under these conditions, especially if the extracellular concentrations of adenosine is abruptly increased to a high value.

Adenosine is accumulated not only into neurons but also into astrocytes in primary cultures. Studies on the effect of deoxycoformycin on adenosine uptake into primary cultures of brain astrocytes, where deamination plays a relatively smaller role than in neurons (Matz and Hertz, 1990), have yielded results similar to those observed in neurons. The total concentration of adenosine in the water phase of the cells, calculated as for the neurons and under similar experimental conditions, becomes at least 100 μM, i.e., 10 times higher than the the extracellular concentration, when they are exposed to the deaminase inhibitor; again the total radioactivity of adenosine plus all its metabolites is increased rather than decreased in the presence of deoxycoformycin and, again, the initial diffusional uptake during the first few seconds is faster than the continued active uptake extracellularly.

Essentially similar time courses for accumulation of labeled adenosine into primary cultures of chick astrocytes exposed to 25 μM [^{3}H] adenosine or chick neurons exposed to 4 μM [^{3}H] adenosine were observed by Thampy and Barnes (1983). The continued accumulation in non-poisoned cells was ≈0.05-0.1 nmol/min per mg protein. Again, there was a short-lasting, much faster accumulation of adenosine which was over as soon as the adenosine concentration in the intracellular water phase, calculated as above, was similar to that in the incubation medium. A ≈10-fold lower rate of continued accumulation was found by Geiger et al. (1988) and Johnston and Geiger (1989) in "isolated brain cells", probably partly as a result of cell damage during the isolation and partly as a result of the use of a lower adenosine concentration (1 μM). Again, there was a much faster initial (first 5 sec) uptake until the intracellular concentration of labelled adenosine was approximately similar to that in the incubation medium. Unfortunately, these authors used an incubation time of 15 sec (i.e., 5 sec. *mainly* (but not exclusively) diffusional uptake followed by 10 sec. active, partly sodium-dependent uptake for subsequent investigation of transport kinetics and drug inhibition. At the concentration used (1 μM adenosine), approximately one half of the total uptake in these studies must have occurred by rapid diffusional transport and the remainder by a less rapid active transport system. Accordingly, complex kinetics, suggesting a two component system for both saturation of adenosine uptake and drug inhibition were obtained. Most of the studies by Thampy and Barnes were also done using incubation periods well below 1 min. In most of these experiments, the authors used cells poisoned with inhibitors of energy metabolism to prevent subsequent metabolism. This procedure will also abolish active transport, but in these experiments at least *some* useful information is provided because *one* process only (diffusion, which probably is facilitated (Hertz, 1991)) is measured.

CONCLUDING REMARKS

The present study has unquivocally demonstrated that there is an active transport of adenosine into primary cultures of mouse neurons and astrocytes. This indicates that 1) active transport mechanisms *do* exist for adenosine, and 2) that it is likely that similar transport mechanisms are found in neurons and astrocytes *in situ*. This is because primary cultures, in contrast to cell lines, faithfully seem to express the characteristics of their in vivo counterparts (Hertz et al., 1985). This is consistent with the concept that adenosine accumulation into *these cells* are geared towards regulating the extracellular, not the intracellular concentration.

It should be strongly emphasized that a methodological consequence of the presence of active transport mechanisms is that measurements of adenosine transport during ultrashort incubation periods is misleading and provides no pharmacological or kinetic relevant information about adenosine transport *into these cells*.

REFERENCES

Geiger, J.D., Johnston, M.E., and Yago, V., Pharmacological characterization of rapidly accumulated adenosine by dissociated brain cells from adult rat, *J. Neurochem.*, 51, 283-291, 1988.

Hertz, L. Nucleoside transport in cells: kinetics and inhibitor effects. In Adenosine and the Adenine Nucleotides as Regulators of Cellular Function (J.W. Phillis, Ed.), pp. 85-107, CRC Press, Boca Raton, 1991.

Hertz, L., Juurlink, B.H.J., and Szuchet, S. Cell cultures. In Handbook of Neurochemistry, (A. Lajtha, Ed.), 2nd Ed., Vol. 8. Plenum Press, New York, pp. 603-661, 1985.

Matz, H., and Hertz, L., Effects of an adenosine deaminase inhibitor on active uptake and metabolism of adenosine in astrocytes in primary cultures, *Brain Res.* 515, 168-172, 1990.

Johnston, M.E., and Geiger, J.D., Sodium-dependent uptake of nucleosides by dissociated brain cells from the rat, *J. Neurochem.*, 52, 75-81, 1989.

Paterson, A.R.P., Kolassa, N., and Cass, C.E., Transport of nucleoside drugs in animal cells, *Pharmac. Ther.*, 12, 515-536, 1981.

Thampy, K.G., and Barnes, E.M., Jr., Adenosine transport by cultured glial cells from chick embryo brain, *Arch. Biochem. Biophys.*, 220, 340-346, 1983.

INTESTINAL TRANSPORT AND METABOLISM OF PURINE AND PYRIMIDINE NUCLEOSIDES

Bronk, J. R., Helliwell, P. A. and Stow, Ruth A.

Department of Biology
University of York
York YO1 5DD, U.K.

INTRODUCTION

We have compared the influence of mucosal metabolism on the transmural transport of purine and pyrimidine nucleosides following their absorption from the lumen of the rat small intestine. Previous studies (Bronk & Hastewell, 1988) have shown that uridine absorbed from the lumen of isolated loops of rat small intestine is split by phosphorolysis yielding uracil on the serosal side of the epithelial layer. By contrast luminal cytidine was poorly absorbed and not split by phosphorolysis, although there was some deamination to uridine. When the fate of luminal 5-methyluridine, thymidine and deoxyuridine was investigated with the same experimental system, the free base was the major form of pyrimidine appearing on the serosal side, although at higher luminal concentrations of thymidine and deoxyuridine a small amount of the unchanged nucleoside did appear in the serosal secretions and there was some tendency for all of the pyrimidines to be converted to uracil (Bronk & Hastewell, 1989).

In the isolated loop preparation of rat small intestine the submucosal spaces are not cleared by the normal vascular flow so that solutes which are transported across the epithelium tend to accumulate both on the serosal side of the epithelial layer and in the mucosal tissue. For the transport of free pyrimidines we have found that this accumulation reduces the rate of transmural transport, since the pyrimidine transport rates obtained with the dual luminal/vascular perfusion preparation are significantly higher than those obtained with isolated loops (Bronk, Lister & Lynch, 1987), but are reduced to the isolated loop values by including the free pyrimidine in the vascular perfusate.

We have now extended the earlier studies in three ways: by studying purine nucleoside transport and metabolism with the isolated loop preparation, by using the dual luminal/vascular perfusion preparation to follow the fate of luminal purine and pyrimidine nucleosides, and by using the vascular perfusion preparation to examine

Purine and Pyrimidine Metabolism in Man VII, Part A
Edited by R.A. Harkness *et al.*, Plenum Press, New York, 1991

the metabolism of nucleosides placed in the vascular perfusate. This paper gives a preliminary report of our findings.

METHODS

Two perfusion techniques for rat small intestine were used to obtain the data described below. Isolated loops of jejunum from 250g male Wistar rats were perfused *in vitro* as described by Bronk & Hastewell, 1988. The loops were perfused for an initial 50 min control period after which the luminal perfusate was changed and the perfusion continued for a further 70 min in the presence of the transport substrate; serosal samples were collected every 10 min. The second technique was a modification of the *in situ* dual luminal/vascular perfusion preparation described by Bronk, Lister & Lynch, 1987 in which the vascular bed was perfused in a single pass with a Krebs bicarbonate Ringer containing 5mM D-glucose and 5% serum albumin at a rate of 1.5ml/min. For both preparations the lumen of the intestine was perfused with Krebs bicarbonate Ringer at a rate of 7ml/min.

The luminal disappearance and serosal or vascular appearance of purine and pyrimidine nucleosides were followed by isocratic HPLC as described previously (Bronk & Hastewell, 1988). This analytical technique also gave us an accurate picture of the extent to which the various nucleosides were metabolised during the perfusions.

At the conclusion of each perfusion the intestinal segment was removed for determination of the dry weight and the preparation of deproteinised mucosal tissue extracts for estimation of the tissue concentrations of the purine and pyrimidine nucleosides and their metabolites as described earlier (Bronk & Hastewell, 1988).

RESULTS & DISCUSSION

Purine nucleoside transport and metabolism in isolated loops of rat intestine.

When jejunal loops were perfused with 0.5mM adenosine we found that the nucleoside was rapidly deaminated to inosine within the lumen, and we could find no evidence of phosphorolysis of the nucleoside to produce adenine.

Both guanosine and inosine were absorbed from the lumen but neither nucleoside could be detected in the serosal secretions. Instead both nucleosides caused a large increase in the serosal uric acid concentration. Analysis of the enzyme activities of the mucosal tissue revealed that following absorption guanosine and inosine were split by phosphorolysis to yield the free purine which was then oxidised to uric acid by xanthine oxidase. If the rate of serosal uric acid appearance from 0.15mM luminal guanosine is compared with that from luminal guanine at the same concentration, the nucleoside rate is 50% higher, although there is no significant difference at lower luminal concentrations. With luminal inosine the rate of serosal uric acid appearance does not become significantly faster than that with hypoxanthine until the luminal concentration is increased to 0.5mM, but at this concentration the transmural transport rate from luminal nucleoside is twice that found with the free base. This difference between the two nucleosides is a reflection of the 8-fold difference in the Km of the nucleoside phosphorylase for these two purine nucleosides.

Inclusion of 0.3mM allopurinol in the luminal perfusate reduces the xanthine oxidase activity to less than 20%, but even under these conditions the total serosal purine (hypoxanthine + xanthine + uric acid) is significantly higher with the nucleoside than with the free base. Also it is interesting to note that the inclusion of allopurinol causes a larger reduction in the rate of total serosal purine appearance for hypoxanthine and inosine than it does for guanine and guanosine. This is presumably due to the fact that in the latter case inhibition of xanthine oxidase will not result in the accumulation of the immediate product of the nucleoside phosphorylase since there is an extra catabolic step, deamination of guanine to xanthine by guanase.

Influence of vascular perfusion on the transport and metabolism of nucleosides.

When we examined the effects of vascular perfusion on the transport and metabolism of purine and pyrimidine nucleosides and their constituent bases, the effects on the purines and pyrimidines were very different. For both nucleosides vascular perfusion reduced the rate of serosal appearance of the uric acid or free pyrimidine by about 20%, presumably because clearance of the submucosal spaces reduced the available substrate for the nucleoside phosphorylase. However, the most dramatic effect of the vascular perfusion was to triple the rate of transmural uracil transport from luminal uracil to give a rate that was higher than that from uridine. By contrast the transmural transport of free purines was unaffected by vascular perfusion, leaving the purine nucleosides as the more effective substrates for vascular uric acid appearance. The difference in treatment of the free purines and pyrimidines is a reflection of the extra catabolic steps for the purines which ensure that the bases do not accumulate in the tissue and submucosal spaces even in the isolated loops where the submucosal spaces are not cleared by perfusion of the vascular bed.

It might be expected that clearance of the submucosal spaces by vascular perfusion would lead to some unsplit nucleoside appearing in the vascular perfusate. We did not observe this for either type of nucleoside at a luminal concentration of 0.5mM, so we increased the luminal concentration 10-fold to 5mM. In the case of uridine about 2% of the nucleoside appeared unchanged on the vascular side, although for inosine at the same concentration the unsplit nucleoside accounted for less than 1% of the purine appearing on the vascular side.

Intestinal metabolism of nucleosides perfused through the vascular bed.

When we perfused the vascular bed with uridine or inosine at 0.5mM, approximate 30% of the nucleoside was split in a single pass to yield uracil or uric acid, respectively. This shows that the small intestine is remarkably effective in the catabolism of purine and pyrimidine nucleosides entering the tissue from the vascular side as well as of those nucleosides absorbed from the lumen.

CONCLUSIONS

1. Pyrimidine nucleosides perfused through the lumen of the rat small intestine are absorbed and metabolised at widely different rates. The rate of absorption is dependent on the rate of phosphorolysis and at concentrations in the 0.05 to 10 mM range most of the absorbed

nucleoside leaves the serosal pole of the epithelial cells as free base, mainly in the form of uracil.

2. Clearance of the submucosal spaces by use of the dual luminal/vascular perfusion technique reduced the rate of transmural transport by about 20%, but caused little increase in the proportion of uncleaved pyrimidine nucleoside appearing on the serosal side of the epithelial layer.

3. Adenosine was deaminated in the lumen of the small intestine and the other purine nucleosides were split by phosphorolysis within the mucosal cells after absorption. The free purines were converted to uric acid unless xanthine oxidase was inhibited by allopurinol.

4. Clearance of the submucosal spaces by use of the dual luminal/vascular perfusion technique reduced the rate of transmural purine transport by about 20%, but caused little increase in the proportion of unsplit nucleoside appearing on the serosal side of the epithelial layer. Virtually all of the purine absorbed from the lumen in the form of nucleosides appeared on the serosal side as uric acid.

5. When 0.5mM inosine or uridine were perfused through the vascular bed of the rat small intestine approximately 30% of the nucleoside was split by phosphorolysis and converted to uric acid and uracil, respectively.

REFERENCES

Bronk, J.R. and Hastewell, J. G. , 1988, The transport and metabolism of naturally occurring pyrimidine nucleosides by isolated rat jejunum, *Journal of Physiology*, **395**:349-361.

Bronk, J.R. and Hastewell, J. G. , 1989, The specificity of pyrimidine nucleoside transport and metabolism by rat jejunum *in vitro*, *Journal of Physiology*, **408**:405-411.

Bronk, J.R., Lister, N. and Lynch, S., 1987, Absorption of 5-fluorouracil and related pyrimidines in rat small intestine, *Clinical Science* **72**:705-716.

GUANINE UPTAKE BY HUMAN ERYTHROCYTES

Giuseppe Spoto, Cosimo Pio Quaratino, and Alessandro Giacomello

Institutes of Biochemical Sciences
and Medical Physiopathology
University of Chieti
66103 Chieti, Italy

INTRODUCTION

It has been clearly established that mature human erythrocytes cannot synthesize *de novo* purines (1); consequently the turnover of purine nucleotides in red cells must result from the continual entry and release of purines. Hypoxanthine and adenine uptake by human erythrocytes has been studied in our laboratories (2–4). Hypoxanthine uptake is mediated by its incorporation into IMP. The rate of synthesis of this nucleotide is dependent on the intracellular availability of 5-phosphorybosyl-1-pyrophosphate (PPRP) which increases by increasing the concentration of inorganic phosphate (Pi) and by decreasing the pH and/or the oxygen partial tension of the incubation mixture. Human erythrocytes incorporate exogenous adenine into the nucleotide pool even in the presence of a very low intracellular PPRP concentration, insufficient to allow an appreciable uptake of hypoxanthine as IMP. Adenine nucleotide synthesis in Pi-free media is faster at physiological pH and independent of the oxygen partial pressure. Mature human red blood cells, which mount a high rate of dephosphorylation of GMP, depend almost exclusively on guanine salvage for the maintenance of their guanine ribonucleotide pool (5). Guanine uptake by human erythrocytes has not been investigated systematically. In the present paper, the effect of varying Pi concentration on guanine uptake and hypoxanthine release by human erythrocytes has been studied.

EXPERIMENTAL PROCEDURE

Human erythrocytes were separated and prepared from freshly drawn blood in heparin and washed twice in 0.9% NaCl. The packed cells were suspended to a final hematocrit of 40% in a medium containing 0.05 M Tris-HCl (pH 7.4), 0.005 M glucose, 0–10 μM guanine, 0–10 mM Pi and the appropriate amount of NaCl to form an isotonic solution, and incubated at 37°C. At regular intervals, samples of the incubation mixture were removed. The medium, separated from the erythrocytes by centrifugation, was employed for purine metabolite analysis by the RP HPLC method described (6).

RESULTS AND DISCUSSION

When human erythrocytes are suspended in the described medium not containing Pi, hypoxanthine is the only purine base released in appreciable amounts (Fig. 1). Under the same experimental conditions, guanine, added to the incubation mixture, is taken up by the cells (Fig. 2). The extent of guanine incorporation was strikingly influenced by the Pi concentration of the incubation medium. The rate of the decrease of guanine concentration in the suspension medium was approximately a linear function of |Pi| in the range studied (Fig. 3). The rate of hypoxanthine release decreased almost linearly with the increasing |Pi| with a slope value very close to that obtained for the increase of the rate of guanine uptake shown in Fig. 3. The uptake of guanine is mediated by its incorporation into GMP and other ribonucleotides (5,7,8). The synthesis of GMP is dependent on the intracellular availability of PPRP and on the activity

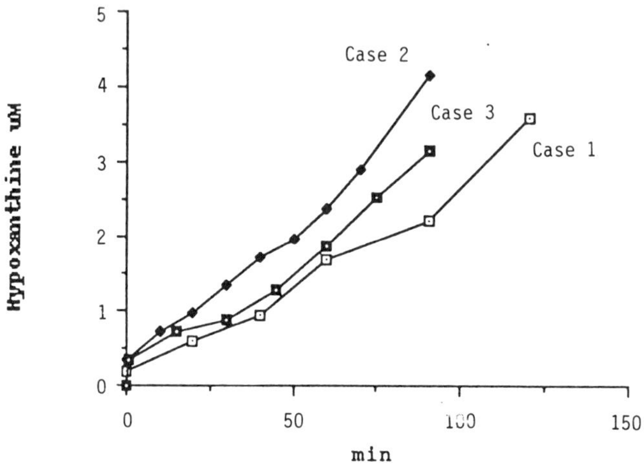

Figure 1. Hypoxanthine release by human erythrocytes obtained from 3 subjects in the absence of Pi in the incubation medium.

of hypoxanthine-guanine phosphoribosyltransferase (HGPRT). It appears reasonable to assume that the high Pi levels enhance the assimilation of guanine by favouring the formation of PPRP, a cosubstrate in the phosphoribosylation of guanine catalyzed by HGPRT. In the absence for Pi, guanine is taken up by the cells in appreciable amounts while hypoxanthine is released by the erythrocytes. When labeled hypoxanthine is added to the incubation mixture, the erythrocytes do not accumulate radioactivity from the permeant to a significantly higher concentration than that present in the medium (2,7). Guanine phosphorylation in the absence of Pi in the suspending medium could occur through two HGPRT catalyzed reactions:

1. The reaction between guanine and the very low amount of PPRP present in the cells under these experimental conditions (3,9) to form GMP and pyrophosphate.
2. The direct transfer of the phosphoribosyl moiety of IMP to guanine to form GMP and hypoxanthine (10).

Evidence has been obtained that the latter reaction could occur in human erythrocytes (9).

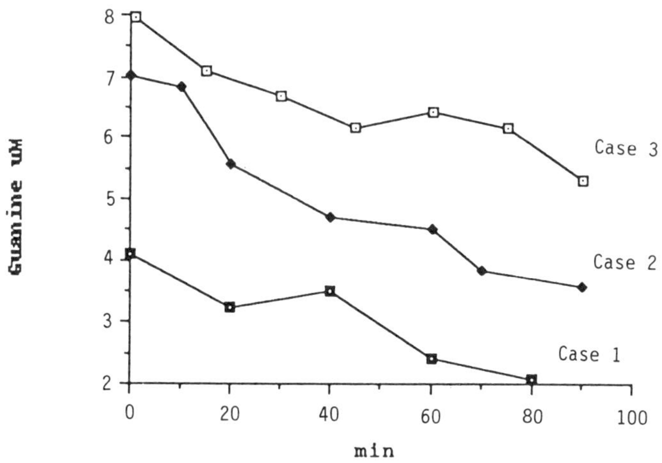

Figure 2. Guanine uptake by human erythrocytes obtained from 3 subjects in the absence of Pi in the incubation medium.

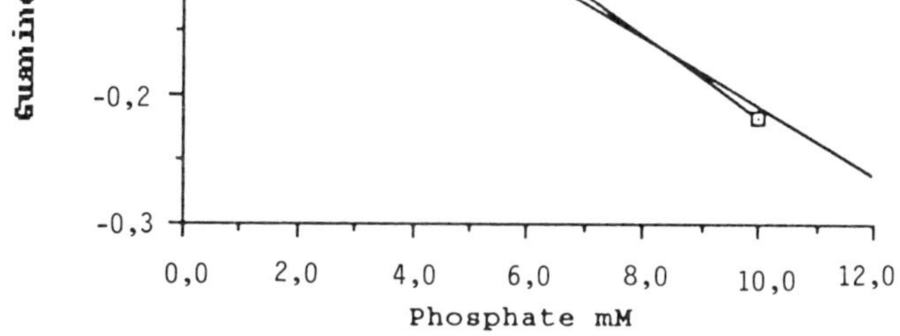

Figure 3. Rate of guanine uptake as a function of Pi in the incubation medium.

REFERENCES

1. A.W. Murray, *Ann. Rev. Biochem.* **40:**811–828 (1971).
2. A. Giacomello and C. Salerno, *FEBS Letters* **107:**203–204 (1979).
3. C. Salerno, G. Gerber, A. Giacomello, *Adv. Exp. Med. Biol.* **195B:**75–77 (1986).
4. C. Salerno, A. Werner, W. Siems, G. Gerber, A. Giacomello and C. Crifo, *Adv. Exp. Med. Biol.* **253B:**543–546 (1989).
5. Y. Sidi, I. Gelvan, S. Brosh, J. Pinkhas and O. Sperling, *Adv. Exp. Med. Biol.* **253A:**67–71 (1989).
6. C.P. Quaratino, E. Messina, P. Bertuglia, G. Spoto, G. Sciarra, G. Bucci and A. Giacomello, this volume.
7. B.A. Domin, W.B. Mahony and T.P. Zimmerman, *J.Biol. Chem.* **263:**9276–9284 (1988).
8. A. Hershko, A. Razin, Tavassah Shoshani, J. Mager, *Biochim. Biophys. Acta* **149:**59–73 (1967).
9. C. Salerno and A. Giacomello, *Biochemistry* **24:**1306–1309 (1985).
10. C. Salerno, A. Giacomello, *J. Biol Chem.* **254:**10232–10236 (1979).

RADIOLIGAND BINDING STUDIES ON THE NUCLEOSIDE TRANSPORT PROTEIN

Ad P. IJzerman, Armand Voorschuur, Marieke Kruidering, Irene M. Pirovano, Herman Van Belle* and Cor Janssen*

Center for Bio-Pharmaceutical Sciences, PO Box 9502, 2300RA Leiden, The Netherlands, and *Janssen Research Foundation, Beerse, Belgium

INTRODUCTION

The permeation of adenosine and other nucleosides across the plasma membrane of animal cells is mediated by nucleoside specific transport proteins[1]. The availability of radiolabelled transport inhibitors allows the characterisation of these carrier proteins. One of these radioligands, [^{3}H]nitrobenzylthioinosine (NBI), has been and still is extensively used for this purpose. Recently, [^{3}H]dipyridamole[2] and [^{3}H]dilazep[3] were introduced. In this report we discuss the relationship between ionization and affinity of nucleoside transport inhibitors and describe the interaction of a new radioligand with the transport protein, comparing it with data obtained for [^{3}H]NBI.

Ionization and affinity

Several compounds, such as dilazep, hexobendine and soluflazine, are capable of displacing [^{3}H]NBI binding, although their chemical structures are distinct from NBI. At physiological pH these substances are present in both unionized and monoprotonated forms. NBI, on the other hand, is an uncharged molecule at this pH, raising the question which species being active on the binding site of the transport protein.

Saturation and displacement experiments were carried out on membrane preparations from calf lung tissue[4] at pH 6.6 and pH 9.5. The affinity of [^{3}H]NBI, uncharged at both pH values, for the calf lung membranes was not largely different. Computer-assisted data analysis of the saturation isotherms at 25 °C revealed K_D values of 0.66 nM and 0.45 nM at pH 6.6 and pH 9.5, respectively. A rather marked increase in binding capacity was observed at pH 6.6 (3.8 pmol/mg protein vs 2.3 pmol/mg protein at pH 9.5 for this membrane preparation).

[^{3}H]R75231 - a new radioligand

It has been shown that the displacement of [^{3}H]NBI by lidoflazine analogues differs from reference compounds such as NBI and dipyridamole. The lidoflazine analogues are highly active on crude and solubilized calf lung membranes. The displacement curves, however, are rather steep with pseudo-Hill coefficients greater than unity for most compounds[6]. Therefore, we judged it worthwhile to further examine this anomalous behaviour. A tritiated analogue of lidoflazine, [^{3}H]R75231, was prepared. Its synthesis will be described elsewhere. The new radioligand displayed high affinity and capacity on the same calf lung membrane preparation that we used for [^{3}H]NBI. Its K_D value was 0.32 nM at 25 °C. Specific [^{3}H]R75231 binding could be displaced by both structural analogues and reference inhibitors such as dipyridamole and NBI (see also Fig. 2). The pseudo-Hill coefficients approximated unity for all compounds, clearly distinct from the [^{3}H]NBI binding data. Moreover, in a further analysis by means of 'competition kinetics[7]' it was demonstrated that the association kinetics of [^{3}H]R75231 at 0 °C in the presence of unlabelled ligands were far less influenced

Table 1. 'True' affinities of the monoprotonated ($K_{I,M}$) and uncharged species ($K_{I,U}$) species and the affinity ratios of dilazep, hexobendine, soluflazine and ST7092.

	$K_{I,M}$ (nM)	$K_{I,U}$ (nM)	$K_{I,M}$ / $K_{I,U}$
dilazep	2.35	0.58	4.0
hexobendine	10.2	0.12	81.5
soluflazine	7.79	1.64	4.7
ST7092	3.10	0.27	11.5

than in identical experiments with [^{3}H]NBI as the radioligand. Computer-assisted mathematical analysis of these time course experiments revealed that the association rate constants for the unlabelled ligands were significantly different when either [^{3}H]R75231 or [^{3}H]NBI was employed. We conclude that [^{3}H]R75231 binds to the protein in a way fundamentally distinct from [^{3}H]NBI, suggesting the existence of two different binding sites. This conclusion is corroborated by recent findings that in mouse erythrocytes soluflazine, an analogue of R75231, showed poor potency relative to NBI, whereas the same compound appeared to be more potent that NBI in the rabbit[8].

In conclusion, [^{3}H]R75231 is a new radioligand with high affinity for the nucleoside transport protein. Its characteristics are distinct from the reference radioligand [^{3}H]NBI, which render it a useful radioligand for a further biochemical characterisation of the nucleoside transport protein.

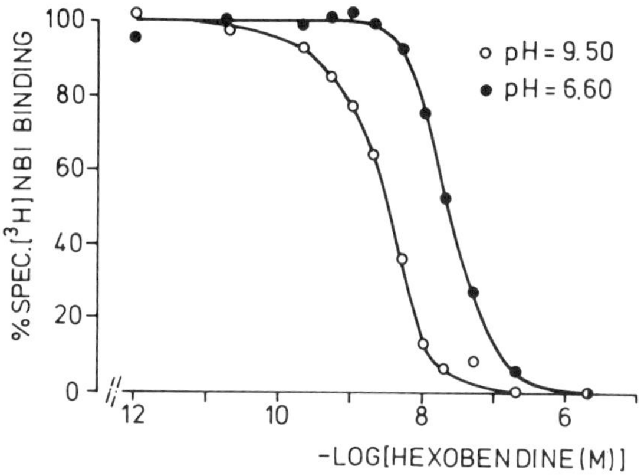

Fig. 1. The displacement of [^{3}H]NBI binding from calf lung membranes by hexobendine at pH 6.6 (•) and pH 9.5 (o). A typical experiment is shown.

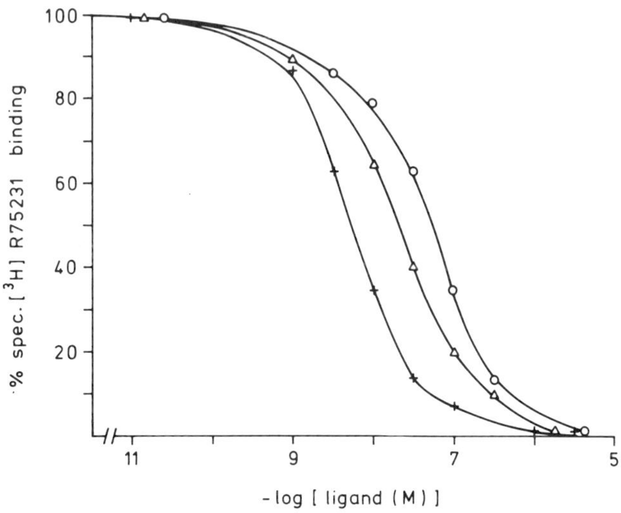

Fig. 2. Displacent of specific [^{3}H]R75231 binding from a membrane preparation of calf lung tissue by dipyridamole (o), NBI (Δ), and R75231 (+). A representative experiment is shown.

Displacement studies were carried out at 25 °C in the presence of 10 μM physostigmine in order to prevent degradation of the inhibitors, all carboxylic esters or carboxamides. The determination of the aqueous ionization constants of these transport inhibitors enabled the calculation of the fractions of both monoprotonated and uncharged species at the two pH values[5]. As an example, pK_a values for hexobendine were 4.52 and 8.25 for the two basic nitrogen atoms present in the structure, yielding 98% monocations present at pH 6.6, and 92% uncharged molecules at pH 9.5. The pK_a values for soluflazine were 5.09 and 5.74, yielding uncharged molecules almost exclusively at both pH values. Combination of all these data with the apparent IC_{50} values obtained in [^{3}H]NBI displacement studies yielded the 'true' affinity for each species. Thus, the displacement of [^{3}H]NBI by hexobendine was largely pH-dependent, the compound being more active at pH 9.5 (Fig. 1). For all compounds considered (dilazep, hexobendine, soluflazine and ST7092) it was found that the uncharged species has higher affinity than the monoprotonated form (Table 1). This phenomenon was most outspoken for hexobendine, the difference in 'true' K_I values being approximately 80-fold.

In conclusion, the NBI-binding site on the nucleoside transport protein as present on calf lung membranes preferentially recognizes uncharged molecules, although not exclusively. The nature of this interaction (competitive or non-competitive) remains to be elucidated.

REFERENCES

1. Paterson, A.R.P., Jakobs, E.S., Ng, C.Y.C., Odegard, R.D., and Adjei, A.A., Nucleoside transport inhibition in vitro and in vivo, in: "Topics and Perspectives in adenosine research," E. Gerlach and B.F. Becker, eds., Springer, Berlin, Heidelberg, New York, Tokyo (1987).
2. Young, J.D., and Jarvis, S.M., The use of ligands in the study of the nucleoside-transport complex (nitrobenzylthioinosine), in: "Methods in Pharmacology, vol. 6: methods used in adenosine research," D.M. Paton, ed., Plenum Press, New York (1985).
3. Gati, W.P., and Paterson, A.R.P., Interaction of [^{3}H]dilazep at nucleoside transporter-associated binding sites on S49 lymphoma cells, Mol. Pharmacol. 36:134 (1989).
4. IJzerman, A.P., Menkveld, G.J., and Thedinga, K.H., A refined method for the photoaffinity labelling of the nitrobenzylthioinosine-sensitive nucleoside transport protein: application to cell membranes of calf lung tissue, Biochim. Biophys. Acta 979:153 (1989).
5. IJzerman, A.P., Ionization constants of sparingly soluble substances from aqueous titration data, Int. J. Pharmaceut. 46:173 (1988).
6. IJzerman, A.P., Thedinga, K.H., Custers, A.F.C.M., Hoos, B., and Van Belle, H., Inhibition of nucleoside transport by a new series of compounds related to lidoflazine and mioflazine, Eur. J. Pharmacol. (Mol. Pharmacol. Sect.) 172:273 (1989).
7. Motulsky, H.J., and Mahan, L.C., The kinetics of competitive radioligand binding predicted by the law of mass action, Mol. Pharmacol. 25:1 (1984).
8. Baer, H.P., Haq, A., El-Soofi, A., Serignese, V., and Paterson, A.R.P., Potencies of mioflazine and its derivatives as inhibitors of adenosine transport in isolated erythrocytes from different species, J. Pharm. Pharmacol. 42:367 (1990).

NUCLEOSIDE TRANSPORT IN GUINEA-PIG MYOCARDIUM

Alan R Conant and Simon M Jarvis

Biological Laboratory
University of Kent
Canterbury, Kent, CT2 7NJ, UK

INTRODUCTION

Adenosine is an endogenously released nucleoside in mammalian heart, the levels of which rise during periods of ischaemia or hypoxia, where the oxygen supply to the myocardium is reduced. Under these conditions adenosine acts as a cardioprotective agent and enhances recovery of damaged tissue on reperfusion, whereas under normoxic conditions it has none of these functions. Adenosine's influence on the myocardium can be divided into five areas: (1) vasodilation, (2) inhibition of the effects of ß-adrenergic stimulation, (3) inhibition of impulse generation at the sinoatrial node and impulse transmission at the atrioventricular node, (4) inhibition of neutrophil adherence and superoxide anion generation, and (5) inhibition of platelet aggregation. All of these effects act to increase oxygen supply whilst reducing oxygen demand and are terminated by uptake into the surrounding tissue via nucleoside transporters (Schrader, 1990).

Nucleoside transport in mammalian cells has been investigated using a variety of inhibitors, for example dipyridamole and dilazep and the nucleoside analog nitrobenzylthioinosine (NBMPR). More recently analogs of lidoflazine have been developed as potential transport inhibitors (Ijzerman, et al., 1989). The aim of the studies described in this report was to investigate the symmetry of inhibition of nucleoside transport in guinea-pig cardiomyocytes by lidoflazine and its derivatives. Compounds that preferentially block influx as compared to efflux would be ideally suited to potentiate the beneficial effects of adenosine released during periods of ischaemia.

METHODS

Solutions

Modified Krebs Henseleit Solution (KHS). 127 mM NaCl, 4.6 mM KCl, 25 mM $NaHCO_3$, 1.2mM KH_2PO_4, 1.1 mM $MgSO_4$, 8.3 mM Glucose, 20 mM Taurine, 10 mM Creatine, 2 mM NaPyruvate, 0.5 mM Uric acid, 1mM L-Arginine, L-Glutamic acid and D-Ribose.

Assay Medium (AM). 137 mM NaCl, 2.7 mM KCl, 6.3 mM Na_2HPO_4, 1.5mM KH_2PO_4, 0.5 mM $MgCl_2$, 10.0 mM Glucose, 0.5 mM EGTA and 5% Polyvinylpyrrolidone (PVP40).

Digestion Medium (DM). 50 mls KHS plus 80 μM $CaCl_2$, 0.10 % (w/v) Collagenase 200 U/mg (Worthington Type 1), 0.012 % (w/v) Dispase (Boehringer Mannheim), 0.012 % (w/v) Trypsin 1:250 (Gibco) and 0.1 % (w/v) BSA.

Purine and Pyrimidine Metabolism in Man VII, Part A
Edited by R.A. Harkness *et al*., Plenum Press, New York, 1991

Myocyte Preparation

Cells were prepared by a combination of enzymic digestion and mechanical dissociation after Poole et al. (1990) and Dendorfer et al. (1987). Briefly, isolated guinea-pig hearts were mounted on a langendorf perfusion column at 37°C. The heart was first perfused with KHS plus 750 μM $CaCl_2$ for 4 min, then with KHS plus 100 μM EGTA for 4 min and finally with DM recirculating for 40 min. The ventricles were then cut away from the rest of the myocardial tissue, chopped coarsely and resuspended in the same protease solution, for 30 min, with gentle gassing from below to maximise dissociation. The solution was filtered through nylon mesh, 200 μm pore size, the cells pelleted at 37g for 1 min and resuspended in AM at 22°C ($\approx 10^5$ cells/ml). Cell viability was determined by the proportion of cells with rod like morphology, which was found to be equivalent to values obtained by trypan blue dye exclusion.

Adenosine influx and formycin B efflux

For influx studies, cells were preincubated with 1 ml AM ($\pm$ inhibitor) for 15 min. The cells were then gently spun out and the inhibitor solution replenished. Influx was initiated by the addition of 50 μl of 1.05mM [^{3}H]adenosine (1 μCi) to 1 ml of cells at 22°C and terminated by spinning 1 ml of the solution through silicone oil (density 1.04 g/ml). Radioactivity associated with the cell pellet was determined by liquid scintillation spectrophotometry. Transport rates were calculated after subtraction of ^{3}H-activity which became associated with the cells due to non-mediated permeation. This correction was obtained by performing parallel experiments in the presence of 20μM dilazep. In efflux studies, cells were preloaded for at least 1 h with 50 μM [^{3}H]formycin B. Stock solutions of inhibitor were then added for 15 min and efflux at 22°C was initiated by the addition of 50 μl preloaded cells to 1 ml of AM ($\pm$ inhibitor) and terminated as for influx. Non-mediated efflux was determined in the presence of 20μM dilazep. The experimental data for the inhibition curves were fitted using the computer program LIGAND.

Materials

[2,5',8-^{3}H]Adenosine (58 Ci/mmol) and [G-^{3}H]formycin B (7Ci/mmol) were purchased from Amersham, U.K. and Moravek Radiochemicals (Brea, Ca, U.S.A.). The flazine analogs (R73-335; soluflazine, R64-719; mioflazine, R51-469 and lidoflazine, R7904) and dilazep were generous gifts from Dr. H. Van Belle, Janssen Pharmaceutica (Belgium) and Hoffman-La Roche and Co. (Switzerland), respectively.

RESULTS

Adenosine influx by guinea-pig cardiomyocytes was found to be linear over the first 90 and 150 s at 20 and 200 μM adenosine, respectively. Uptake was sensitive to inhibition by NBMPR and dilazep. The concentration dependence of NBMPR-sensitive adenosine influx was saturable and gave Km values of 0.104 $\pm$ 0.033 mM and Vmax values of 3.10 $\pm$ 0.84 nmoles/10^6 viable cells/min (mean $\pm$ S.E.M.(3)). The flazine analogs were potent inhibitors of adenosine influx in cardiomyocytes (see Fig. 1) with average Ki values (nM) of 4.9 (2.4, 7.3) for R73-335, 64 $\pm$ 21 for soluflazine, 100 $\pm$ 31 for mioflazine and 240 (230, 250) for lidoflazine (mean $\pm$ S.E.M. or where $n < 3$ the average plus individual results).

In order to perform efflux studies from cells, the permeant must either be not metabolised or the cells rendered incapable of metabolising the substrate. Preliminary studies had established that formycin B, a non-metabolised nucleoside, is a permeant for the adenosine influx pathway and thus cardiomyocytes preloaded with [^{3}H]formycin B were used to investigate the effect of the inhibitors on nucleoside efflux. The initial rate of 50μM [^{3}H]formycin B efflux was linear for up to 5 min and inhibited by 20μM NBMPR. Fig. 2 compares the effect of the series of lidoflazine analogs on formycin B efflux. Inhibition constants (nM) for two separate experiments were 12 and 6.2, 19 and 14, 150 and 220, and 250 and 200 for R73-335, soluflazine, mioflazine and lidoflazine, respectively.

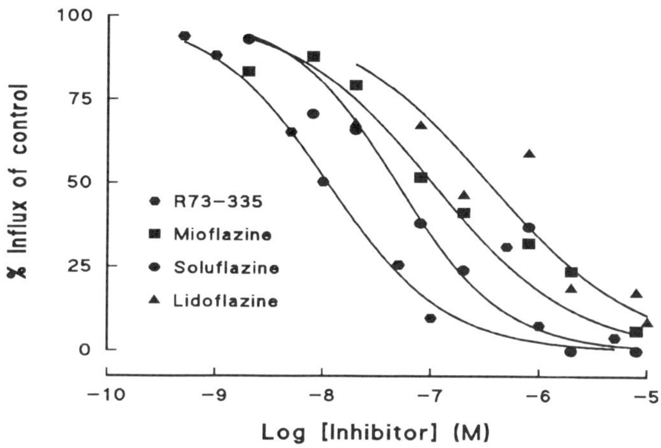

Fig.1. Effect of R73-335, mioflazine, soluflazine and lidoflazine on adenosine influx by guinea-pig cardiomyocytes. Cardiomyocytes were preincubated with inhibitor as described in Methods and the initial rates of 50μM [^{3}H]adenosine influx at 22°C (corrected for the dilazep-insensitive component) determined using a 1 min incubation interval. Results are plotted as a percentage of the control influx rate. Values are the average of triplicate estimates. Standard errors were on average 12% and for clarity are not plotted.

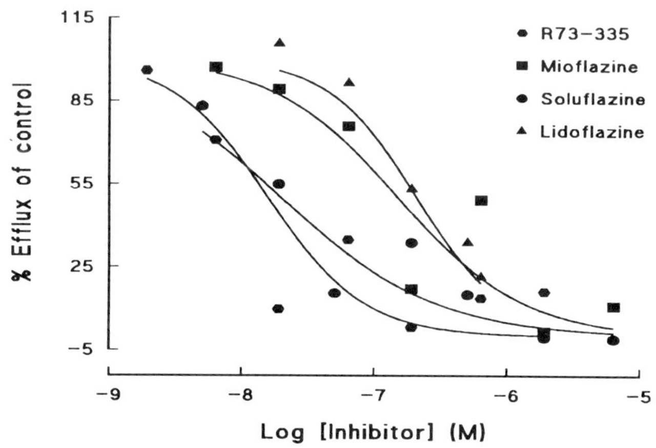

Fig. 2. Effect of R73-335, mioflazine, soluflazine and lidoflazine on formycin B efflux by guinea-pig cardiomyocytes. Initial rates of 50μM [^{3}H]formycin B efflux at 22°C were determined as described in Methods. The flux rates (corrected for the dilazep-insensitive component) are plotted as a percentage of the control efflux rate. Other details were the same as Fig.1.

DISCUSSION

In agreement with earlier studies (Heaton and Clanachan, 1987), adenosine influx in guinea-pig cardiomyocytes was shown to occur via a saturable process that is sensitive to inhibition by NBMPR. Lidoflazine and its derivative were potent inhibitors of adenosine influx with Ki values ranging from 4.9 to 240 nM. Similar inhibition constants were obtained for inhibition of formycin B efflux. These inhibition constants differ from those reported for isolated cells from other species. For example, in rabbit erythrocytes the above compounds tested are more potent with Ki values (nM) of 4.7, 8.4, 4.6 and 3.6 for lidoflazine, mioflazine, soluflazine and R73-335, respectively (Griffith, et al., 1990). These differences appear to be species related rather than cell related, as similar Ki values to those in guinea-pig cardiomyocytes have been obtained for inhibition of uridine influx by guinea-pig erythrocytes (data not shown). The present results for soluflazine and mioflazine are also in agreement with the inhibition constants estimated from quantitative autoradiographic analysis of [^{3}H]NBMPR binding to cardiac myocytes in guinea-pig heart sections (Parkinson and Clanachan, 1989). However, Parkinson and Clanachan (1989) reported that lidoflazine had an 8-fold lower potency as an inhibitor of nucleoside transport activity in myocytes compared to the results of the present study, and moreover was a selective inhibitor of [^{3}H]NBMPR binding by endothelial cells. This result raises the possibility that the present myocyte preparation may be contaminated with endothelial cells (see for example Dendorfer et al., 1987).

In conclusion, lidoflazine and the derivatives studied exhibit no selectivity for inhibition of nucleoside influx versus efflux by guinea-pig cardiomyocytes. Further studies on the possible selectivity of these compounds in inhibiting nucleoside transport in cardiac endothelial cells is planned.

ACKNOWLEDGEMENTS

ARC is in receipt of an MRC collaborative research studentship with ICI Pharmaceuticals.

REFERENCES

Dendorfer, A., Lauk, S., Schaff, A. and Nees, S., 1987, New insights into the mechanism of myocardial adenosine formation in: "Topics and Perspectives in Adenosine Research", E. Gerlach and B.F. Becker, ed, Springer Verlag, Heidelberg.

Griffith, A.D., Conant, A.R., and Jarvis, S.M., 1990, Differential inhibition of nucleoside transport systems in mammalian cells be a new series of compounds related to lidoflazine and mioflazine. Biochem. Pharmacol., 40: 2297.

Heaton, T.P., and Clanachan, A.S., 1987. Nucleoside transport in guinea-pig myocytes: comparison of affinities and transport velocities for adenosine and 2-chloroadenosine, Biochem. Pharmacol., 36:1275.

Ijzerman, A.P., Thedinga, K.H., Custers, A.F.C.M., Hoos, B., and Van Belle H., 1989, Inhibition of nucleoside transport by a new series of compounds related to lidoflazine and mioflazine, Eur. J. Pharmacol., 172:273.

Parkinson, F.E., and Clanachan, A.S., 1989, Subtypes of nucleoside transport inhibitory sites in heart: a quantitative autoradiographical analysis, Eur. J. Pharmacol., 163:69.

Poole, R.C., Halestap, A.P., Price, S.J., and Levi, A.J., 1989, The kinetics of transport of lactate and pyruvate into isolated cardiac myocytes from guinea-pig, Biochem. J. 264:409.

Schrader, J., 1990, Adenosine a homeostatic metobolite in cardiac energy metabolism. Circulation, 81:389.

COMPARISON OF THE EXISTING NUCLEOSIDE TRANSPORT INHIBITORS: *IN VITRO* AND *IN VIVO* DATA

Herman Van Belle, J. Wynants and K. Ver Donck

Janssen Research Foundation
2340-Beerse, Belgium

In order to exclude intracellular metabolism and its interference with the transport process, most studies on nucleoside transport and nucleoside transport inhibition apply either ligand binding and displacement, or rapid kinetic procedures of a few seconds (the present day programme is just another illustration). Although in these ways interesting information can be, and has been, gathered for the characterization of the transporter, the situation is far from what may happen *in vivo*. For instance, during myocardial ischemia, local concentrations of adenosine (ADO) may become very high (10^{-5}-10^{-4} M), and remain so for extended periods of time (many minutes). The role of the transporter in the endothelial cells is the removal of ADO from its site of production by carrying it into those cells where it is rapidly catabolized. The concerted action of transporter and metabolism is an absolute requirement. Applying, under these conditions, an adenosine deaminase inhibitor may leave the transporter intact but ADO will not be removed. Upon reperfusion, the transporter will facilitate the passage of ADO through the otherwise tight endothelial layer, so that it is rapidly lost by dilution into the circulation. If the ischemic area is limited, so that the circulating concentration of ADO is not substantially increased, inhibition of adenosine deaminase will hardly affect the rate of removal *during reperfusion*. Keeping that in mind, I believe that, since our goal is to prolong the concentration of local ADO whenever and wherever it is produced, our approach of looking at the effect of nucleoside transport inhibitors on the rate of disappearance of ADO in a cell suspension can be justified-provided of course that these drugs have no effect on the metabolic steps.

As a tool, the human red blood cell (RBC) is almost ideal: it can easily be obtained, is very rich in transport and metabolizing activities and, as far as the transporter is concerned, it can be regarded as representative for most other cells. Finally, it stems from man, the species we have in mind when considering potential therapeutic use. A few relatively easy techniques can be applied to estimate nucleoside transport inhibition: 1) ADO can be added to a suspension of washed RBCs and the change in optical density (ΔOD) at 265 nm can be determined in the supernatant after incubation and spinning down the cells; 2) HPLC analysis of the supernatant provides information not only on the drop in ADO but at the same time on the appearance of inosine (INO) and hypoxanthine (HYP); it can also be used when plasma is present and, instead of ADO, INO can be applied in order to exclude adenosine deaminase inhibition by the drugs; 3) as reported in more detail elsewhere[1], nucleosides, by trapping intracellular phosphate in the nucleoside phosphorylase reaction, prevent the release of phosphate from RBCs into the medium; inhibition of nucleoside uptake will reestablish phosphate efflux.

In Table 1, the IC_{50}-values are put together for dilazep (DIL), dipyridamole (DIP), nitrobenzylthioinosine (NBTI), mioflazine (MIOFL) and R 75 231 (a MIOFL analog), as obtained with the 3 methods (for more details see ref.). It is evident that all these drugs are inhibitory at 10^{-8} M, irrespective of the method applied. Using HPLC, the ΔINO is the most sensitive parameter. At least twice as much of the inhibitor is needed to affect the

ΔHYP. This difference is indicative for inhibition by these drugs of the efflux of INO whereby the intracellular catabolism is driven towards HYP. The phosphate-efflux method yields values comparable with ΔOD 265 nm or ΔHYP for DIL, MIOFL and R 75 231 but not for DIP and NBTI. We have no idea about the reason for this discrepancy. It is also interesting that the weaker the inhibitor, the more pronounced the competition is from higher substrate concentrations.

Table 1. IC_{50}-values (nM) of the existing NTI's in washed RBC's

Drug μM	ΔOD 265* 100 μM ADO	HPLC-40 μM ADO** ΔADO↓	ΔINO↑	ΔHYP↑	Release P_i** INO 10 μM	INO 40
Dilazep	9	5	4	7	15	23
Dipyridamole	33	19	12	48	160	360
NBTI	-	5	4	9	38	95
Mioflazine	39	18	12	36	40	90
R 75 231	17	11	8	15	15	20

12.5 μl packed RBCs in 1 ml MOPS-buffered saline
* 20 min room temp.
** 1 h at 37° C

Being nearly equipotent in washed RBCs, does not mean that these drugs will be equally effective *in vivo*. As illustrated in Table 2, the IC_{50}-values for DIP and MIOFL are shifted by two orders of magnitude when inhibition is measured in undiluted plasma. It follows that for efficient nucleoside transport inhibition *in vivo,* and in man (not in dogs!), doses in the micromolar range are needed. Especially for DIP it could mean that its well-documented inhibition of phosphodiesterase may come into play.

Table 2. IC_{50}-values (nM) of the existing NTIs in washed RBCs (W) *versus* in RBC's suspended in undiluted plasma (PL) (HPLC-ΔADO↑).

Drug	W	PL	PL/W
Dilazep	5	44	9
Dipyridamole	19	2300	121
NBTI	5	100	20
MIOFL	18	260	144
R 75 231	11	60	5.5

The duration of action is another important factor for efficacy of the drugs. When given intravenously, in rabbits or dogs, R 75 231 inhibits nucleoside transport, as measured *ex vivo* for much longer periods than do DIL or DIP[1]. Recent studies in volunteers revealed complete inhibition *ex vivo* 30 min after intravenous infusion of R 75 231 (10 mg given over 15 min) and still 53 % inhibition (± 11 SD; n = 7) after 4 hours. I could not find any such data for DIP or DIL, but it is well-known that the hemodynamic effects of high intravenous doses of DIP, whatever the mechanism involved, are seen for minutes but not hours. One reason for the comparatively long duration of action of R 75 231 may be its tight binding to the transporter. Evidence for this was obtained in an experiment in which we looked for transport inhibition (using the phosphate efflux) after exposure of RBCs to 2 x 10^{-7} M of the drugs and after different cycles of subsequent centrifugation and reuptake of the cells in drug-free buffer. After 2 cycles, inhibition by all drugs was lost except for R 75 231 where, even after 6 cycles, inhibition was still nearly complete as illustrated in Fig. 1.

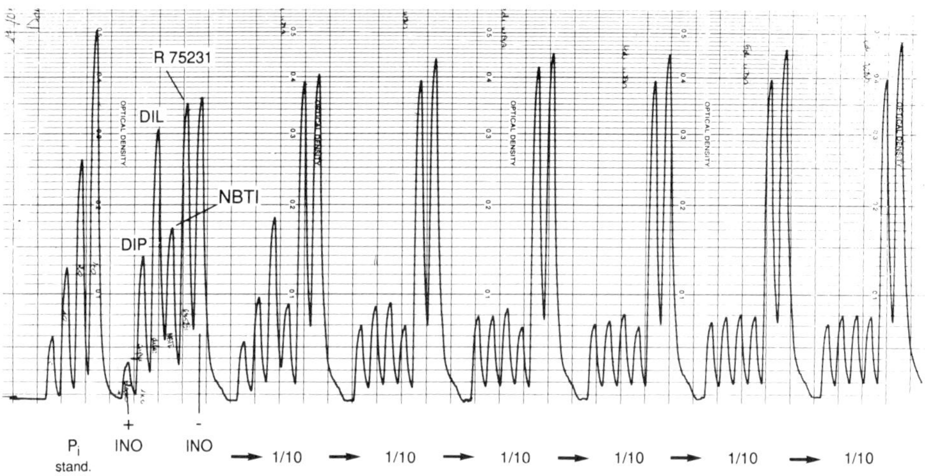

Fig. 1. Washout of inhibitory effect of NTI'rs in human RBC.
Washed RBC were added to MOPS-buffered saline (MBS) containing 2 x 10^{-7} M of the drugs (or solvent). One part of the suspension, corresponding to 12.5 µl of packed cells, was diluted to 1 ml (final volume) of 40 µM INO in MBC (= + INO) and incubated for 1 h at 37° C. After centrifugation (Eppendorf 20 sec) inorganic phosphate was determined in the supernatants (automated method-malachite-green). As a control for 100 % inhibition, INO was omitted (= - INO). Another part of drug-RBC suspension was centrifuged and the pellet resuspended in 10 volumes of drug-free MBS (= 1/10). Using this suspension the whole procedure was repeated. The data show the result of 6 repetitive cycles. The first 4 peaks are standard phosphate solutions (5, 10, 20 and 40 µM).

Remarkable differences between DIP, DIL and the mioflazine-type drugs have also been observed in studies on oral bioavailability in dogs[1]. More recent experiments with DIP and R 75 231 in pigs further indicate the absence of any oral availability for DIP (given at 1.2 mg/kg) whereas 0.2 mg/kg of R 75 231 resulted in an *ex vivo* inhibition between 40 and 70 % (n = 3) at 6 hours after administration. Studies in men with R 75 231 are now underway. It is well-known that, in men, the oral bioavailability of DIP is poor and almost nil for DIL.

Very recently the synthesis of the 2 stereoenantiomers of R 75 231 has been achieved. The l-enantiomer (R 88 021), has a potency exactly twice that of the racemate, which means not only that the d-enantiomer is inactive (which indeed it is when looked at separately) but that it also does not interfere with the transporter in any way. Taken together with excellent chemical stability (which may be questionable for DIP) we expect that the use of the enantiomers as tools may open new perspectives for the characterization of the transporter.

Much more interesting than having these molecules available as biochemical tools is their potential for therapeutic use. In several different models for cardioprotection, R 75 231 has been shown to exert remarkably beneficial effects. Most exciting is its nearly complete suppression of catecholamine-induced myocardial damage and death in rabbits even when given 1 hour after the challenge (25 deaths out of 60 control rabbits; 1 death out of 60 when treated with R 75 231 at 0.1 mg/kg i.v.). That the protective effect is related to nucleoside transport inhibition is evidenced in a study where rabbits were continuously infused with l-epinephrine (10 µg/kg/min) until death. The 50 % survival point was reached in controls (n = 16) after 2.5 hours and 90 % died within 6 hours. Bolus injection of the d-enantiomer (R 88 016, 0.1 mg/kg) at the start of the infusion did not change the survival rate when compared with controls but the same dose of the l-enantiomer was highly protective (50 % survival at 8 hours).

In conclusion, we believe that with R 75 231 and especially its l-enantiomer, R 88 021, we have now available a potent and specific nucleoside transport inhibitor with unique pharmacokinetic properties. Being devoid of any affinity for many receptors or of any effect on a variety of pharmacological tests (as we have them available at Janssen Pharmaceutica) they may evolve to interesting drugs for the management of cardiac (and possibly other) disorders involving ischemia and reperfusion - at least if studies in man will confirm what has been seen thus far in experimental animals.

REFERENCE

1. H. Van Belle, and P.A.J. Janssen, Comparative pharmacology of nucleoside transport inhibitors, Nucleosides & Nucleotides, 10: (1991), in press.

EFFECT OF MEMBRANE LIPID COMPOSITION ON THE FUNCTIONAL ACTIVITY OF A RECONSTITUTED NUCLEOSIDE TRANSPORTER DERIVED FROM EHRLICH ASCITES CELLS

James R. Hammond

Department of Pharmacology and Toxicology
University of Western Ontario
London, Ontario, Canada

Mammalian cell membranes possess a variety of nucleoside transport mechanisms that can be distinguished by their differential sensitivities to inhibitory agents such as nitrobenzylthioinosine (NBMPR) and dipyridamole (DY), and by their dependence on ion gradients[1]. These transporters play an important role in the metabolism and inactivation of bioactive endogenous nucleosides such as adenosine, and in mediating the cellular accumulation of cytotoxic nucleoside analogs used in chemotherapy[1,2]. To clearly define the functional and structural differences among these transporters it will be necessary to isolate and purify each subtype. An integral part of such work is the functional reconstitution of the isolated proteins in an environment approximating the native cell membrane. The functional integrity of a purified transport protein can only be assessed through reconstitution studies. A reconstituted preparation would also allow appraisal of the effects of membrane lipid composition on transport function and ligand binding, as well as the determination of transport kinetics in the absence of metabolic enzymes and intracellular organelles. Initial studies from this laboratory and others have reported the use of asolectin (ASO; mixture of soybean lipids) membranes as the reconstituting environment[3,4]. Although functional reconstitution was obtained in those studies, the activity of the reconstituted transporter was low compared to that of the native cell membrane (<10%). The present study was directed at improving the efficiency of liposomal membrane incorporation of transport proteins, as well as the [^{3}H]uridine translocation capacity of the reconstituted preparations, through modification of the lipid composition of the liposomal membranes.

METHODS

Plasma membranes prepared from Ehrlich ascites tumour cells were used as the source of transport protein. Ehrlich cells were grown as an intraperitoneal culture in mice (Swiss, male, ≈30 g), and plasma membranes were solubilized with a combination of 1.0% (w/v) octylglucoside (Calbiochem, USA) and 0.15% ASO (Associated Concentrates, USA), as described previously[3]. All membrane isolation and reconstitution procedures were performed at 0-6°C in a medium containing 10 mM Tris, 100 mM KCl, 0.1 mM $MgCl_2$, and 0.1 mM $CaCl_2$ (pH 7.4). The reconstitution procedure was based on a method used successfully for the functional reconstitution of the Ehrlich cell glucose transporter[5]. Solubilized membranes were supplemented with a sonicated preparation of lipids consisting of ASO, phosphatidylserine (PS, bovine brain), phosphatidylethanolamine (PE, egg), phosphatidylcholine (PC, bovine brain), and cholesterol (CH), in various proportions, plus a trace (10^5 dpm/ml) quantity of [^{14}C]cholesteryl-oleate (Amersham Canada Ltd, 0.1 mCi/ml). All the pure phospholipids and CH were obtained from Serdary Research Laboratories, London, Canada. The detergent was then removed by gel filtration (Sephadex G-50 medium; 1.5 x 28 cm column) at a flow rate of 1.6 ml/min. The void volume fractions, containing proteoliposomes (liposomes + protein), were pooled, frozen in ethanol/dry ice, and stored at -80°C for up to one month.

The specific binding of [^{3}H]NBMPR (Moravek Biochemicals Inc, USA; 1.0 mCi/ml) to proteins associated with the nucleoside transporter was assessed at room temperature (22°C), exactly as described previously[3], using 10 μM nitrobenzylthioguanosine (NBTGR; SIGMA, USA) to define the nonspecific binding. [^{3}H]Uridine (ICN Biomedicals Canada Ltd; 1.0 mCi/ml) uptake by the reconstituted preparations was measured as follows. Proteoliposome suspensions were centrifuged at 40,000 *g* for 20 minutes, resuspended in 1 ml buffer (22°C), then sonicated for 10 sec. Sonication of the final preparation led to more consistent [^{3}H]uridine uptake data, probably as a result of a more homogeneous liposome size distribution. Assays were initiated by the addition of 100 μl of a diluted (5-10 fold) proteoliposome suspension (± inhibitors) to 25 μl of [^{3}H]uridine (20 μM final) and mixed by repeated pipetting. After a defined incubation time, 100 μl of the latter mixture was layered on an ice-cold Sephadex G-50 Fine minicolumn (see below), which was then immediately centrifuged for 45 sec at ≈1000 *g*. The effluent was collected directly into pre-weighed tubes from which aliquots were removed for the determination of protein, [^{14}C], and [^{3}H] contents using established techniques. The minicolumns were prepared by filling a 1 ml syringe, fitted with a polyethylene filter, with Sephadex gel equilibrated in buffer containing 10 μM dilazep (Asta Werke, Germany) and 10 μM NBTGR; at 30-60 minutes prior to use, the gel-filled syringes were centrifuged for 45 sec at ≈1000 *g* and placed on ice. Approximately 35% of the sample material added to the minicolumns was recovered in the effluent. An estimate of the "zero-time" uptake was obtained by measuring the uptake of [^{3}H]uridine at the minimum possible incubation time (≈3 sec) using ice-cold solutions containing 10 mM adenosine (SIGMA, USA), 10 μM DY (SIGMA, USA), and 10 μM NBTGR to inhibit all transporter-mediated influx.

Results from individual experiments were normalized based on the [^{14}C]lipid concentration of the column effluent. The "zero-time" uptake of [^{3}H]uridine was subtracted from all data points, and the accumulation of [^{3}H]uridine was expressed as the number of pmol accumulated per microgram of proteoliposomal protein. Computer assisted hyperbolic curve fitting was used to obtain a estimate of the initial rate (V_i) of influx and the maximum accumulation. Mediated influx of [^{3}H]uridine was defined as the total uptake minus the uptake observed in the presence of 10 mM adenosine/10 μM DY/10 μM NBTGR (non-mediated).

RESULTS AND DISCUSSION

Octylglucoside (+ASO) solubilized 46 ± 3% of the protein and 87 ± 19% of the [^{3}H]NBMPR binding activity from the plasma membranes; 71 ± 7% of the solubilized protein and 41 ± 6% of the solubilized [^{3}H]NBMPR binding activity was incorporated into the proteoliposomes. These incorporation efficiencies were independent of the composition of the reconstituting lipid mixture. The apparent low efficiency of [^{3}H]NBMPR binding site incorporation is likely due to the instability of the solubilized binding protein; a progressive loss of [^{3}H]NBMPR binding with time has been observed at temperatures above 0°C (unpublished observations).

Table 1. Effect of lipid composition on the maximum accumulation and initial rate (V_i) of uptake of [^{3}H]uridine by proteoliposomes

Lipid Composition (%w:w)[a]		Maximum (pmol/mg)[b] Total	Mediated	V_i (pmol/mg/sec)	
ASO	(100)	280 ± 90	90 ± 10	8 ± 2	(5)[c]
ASO:CH	(90:10)	480 ± 50	260 ± 40	33 ± 10	(5)
ASO:CH	(70:30)	1290 ± 270	600 ± 120	32 ± 6	(9)
ASO:CH	(50:50)	960 ± 200	610 ± 200	33 ± 7	(3)
ASO:CH:PS	(60:30:10)	1250 ± 160	690 ± 100	51 ± 5	(15)
ASO:CH:PS:PE	(50:30:10:10)	1210 ± 240	1090 ± 490	43 ± 8	(8)
PC:CH:PS	(60:30:10)	550 ± 290	560 ± 440	5 ± 1	(3)

[a] Each reconstitution mixture contained approximately the same concentration of lipid (10 mg/ml) and protein (0.4 mg/ml).
[b] The relative contribution of non-mediated uptake to the total accumulation may be estimated from the difference between the maximum "total" uptake and the maximum "mediated" uptake values.
[c] Each value represents the mean ± SEM from the number of experiments shown in parentheses.

The use of ASO alone as the reconstituting lipid resulted in a minimal degree of proteoliposomal transport activity (Table 1). The majority (≈70%) of this influx was likely due to passive diffusion of the nucleoside through the liposomal membrane, as only 30% of the total influx was sensitive to inhibition by NBTGR, DY, and adenosine.

The addition of CH to the reconstitution mixture enhanced the accumulation of [^{3}H]uridine dramatically. Proteoliposomes prepared from a mixture of 90% ASO/10% CH (w:w) accumulated twice as much [^{3}H]uridine/mg protein, at four times the initial rate, as those prepared from ASO alone (Table 1). These results suggest that the presence of 10% CH in the lipid mixture led to an increased size or number of sealed vesicles containing the transport protein, as well as a change in the activity of the reconstituted transporter itself. Further enhancement of the maximum accumulation, but not the initial rate, was observed as the proportion of CH was increased from 10%-50% (w:w) of total lipid. The ratio of mediated to non-mediated influx also increased with increasing CH concentration. These results are compatible with the reported ability of CH to stabilize liposomal membranes by decreasing membrane fluidity and permeability to small molecules[6,7]. It is of interest to note that CH is also required for the functional reconstitution of the γ-aminobutyrate and glutamate transporters[8].

The addition of 10% PS to the ASO/CH mixture produced an additional enhancement of the proteoliposomal uptake of [^{3}H]uridine (Table 1). However, in contrast to the effect of CH, this enhancement appeared to be due solely to an increase in the activity of the transporter (increased initial rate of influx with no increase in maximum accumulation). These results are indicative of a direct interaction of PS with the reconstituted transport protein[9]. Substitution of a portion of the ASO with PE did result in a further enhancement in the maximum accumulation, but also reduced the initial rate of [^{3}H]influx slightly (Table 1). Replacement of ASO (≈50% PC) with pure PC resulted in an almost complete loss of transporter-mediated [^{3}H]uridine uptake activity, indicating that a component of ASO other than PC was essential for the functional reconstitution of nucleoside transport activity (Table 1). Experiments are in progress to identify this essential component.

Overall, the optimum starting mixture for the reconstitution of the nucleoside transporter was ASO/CH/PS (60/30/10%; w/w). Proteoliposomes of this composition incorporated an inhibitor-sensitive mechanism that mediated the accumulation of [^{3}H]uridine at an initial rate of 51 ± 5 pmol/mg/sec, with a maximum mediated accumulation of 690 ± 100 pmol/mg protein. These values are about 7-fold greater than those obtained using proteoliposomes prepared from ASO alone (Table 1). These proteoliposomes also bound [^{3}H]NBMPR specifically, and with high affinity (K_D=0.11 ± 0.01 nM; n=10), to a maximum capacity of 3.7 ± 0.5 pmol/mg protein (n=7). It can be calculated from these data that the translocation capacity of the reconstituted transporter was 13.7 molecules of uridine/NBMPR binding site/sec, which is almost identical to the translocation capacity of the transporter in its native Ehrlich cell membrane (14.2 molec/site/sec). In addition, [^{3}H]uridine influx by these proteoliposomes (20 μM, 15 sec incubation) was inhibited by adenosine, DY, dilazep, and NBMPR with IC_{50} values of 106 μM, 610 nM, 150 nM, and 6 nM, respectively (n=3). This is identical to the order of potency of these agents for inhibiting [^{3}H]uridine uptake by intact Ehrlich cells[3]. Moreover, as in Ehrlich cells, about 20% of the reconstituted DY/adenosine-sensitive influx was resistant to inhibition by dilazep and NBMPR, indicating that both subtypes of transporter (NBMPR-sensitive/resistant) were solubilized from the plasma membranes and functionally reconstituted in equal proportions. In summary, nucleoside transporters incorporated into ASO/CH/PS membranes appear fully functional in both their uridine translocation and ligand binding capabilities. This preparation should provide an excellent model for analysis of the functional integrity of purified nucleoside transport proteins.

ACKNOWLEDGEMENT

This study was supported by the Medical Research Council of Canada.

REFERENCES

1. P.G.W. Plagemann, R.M. Wohlhueter, and C. Woffendin, Nucleoside and nucleobase transport in animal cells, Biochim. Biophys. Acta 947:405 (1988).
2. T.P. Zimmerman, B.A. Domin, W.B. Mahony, and K.L. Prus, Membrane transport of nucleoside analogues in mammalian cells, Nucleosides and Nucleotides 8:765 (1989).
3. J.R. Hammond and R.M. Johnstone, Solubilization and reconstitution of a nucleoside-transport system from Ehrlich ascites-tumour cells, Biochem. J. 262:109 (1989).
4. C-M. Tse, J.A. Belt, S.M. Jarvis, A.R.P. Paterson, J-S. Wu, and J.D. Young, Reconstitution studies of the human erythrocyte nucleoside transporter, J. Biol. Chem. 260:3506 (1985).
5. J.I. McCormick and R.M. Johnstone, Asymmetric reconstitution of the glucose transporter from Ehrlich ascites cell plasma membrane: Role of alkali cations, Arch Biochem. Biophys. 248:379 (1986).
6. D. Liu and L. Huang, Role of cholesterol in the stability of pH-sensitive, large unilamellar liposomes prepared by the detergent-dialysis method, Biochim. Biophys. Acta 981:254 (1989).
7. F. Szoka Jr., Comparative properties and methods of preparation of lipid vesicles (liposomes), Ann. Rev. Biophys. Bioeng. 9:467 (1980).
8. A. Shouffani and B.I. Kanner, Cholesterol is required for the reconstitution of the sodium and chloride-coupled, γ-aminobutyric acid transport from rat brain, J. Biol. Chem. 265:6002-6008 (1990).
9 A. Levitzki, Reconstitution of membrane receptor systems, Biochim. Biophys. Acta 822:127 (1985).

ANTIBODIES AND PROTEASES AS PROBES OF THE STRUCTURES OF MAMMALIAN NUCLEOSIDE TRANSPORTERS

F.Y.P. Kwong•, H.E. Fincham•, A. Davies•, P.J.F. Henderson°, S.A. Baldwin•, and J.D. Young+

•Departments of Biochemistry & Chemistry, and of Protein & Molecular Biology, Royal Free Hospital School of Medicine (University of London), Rowland Hill Street, London NW3 2PF; °Department of Biochemistry, University of Cambridge, Tennis Court Road, Cambridge CB2 1QW; +Department of Physiology, 7-55 Medical Sciences Building, University of Alberta, Edmonton, Alberta, Canada T6G 2H7

INTRODUCTION

Permeation of nucleosides across the membranes of many mammalian cell types is mediated by a facilitated diffusion process of broad specificity, that is potently inhibited by nitrobenzylthioinosine (NBMPR)[1]. In most species, including human and pig, this transport process is also potently inhibited by the vasodilators dipyridamole and dilazep[1]. However, in rat tissues the NBMPR-sensitive nucleoside transporters are much less sensitive to inhibition by these two agents[2,3]. The transporters responsible for this NBMPR-sensitive nucleoside uptake have been identified as plasma membrane glycoproteins in a variety of tissues by photoaffinity labelling with the tritiated form of this inhibitor[4]. In human erythrocytes the labelled species migrates on SDS/polyacrylamide gels as a broad band of average apparent M_r 55000, but can be converted by deglycosylation using endoglycosidase F to a species which migrates as a sharp band of apparent M_r 45000[5]. In contrast, the photolabelled protein from pig erythrocytes migrates with apparent M_r values of 64000 and 57000 before and after deglycosylation respectively[5]. The NBMPR-sensitive nucleoside transporters of rat tissues exhibit an apparent M_r similar to that of the pig erythrocyte transporter [2-4]. However, the deglycosylated rat transporter migrates as a sharp band of apparent M_r 45000, a value identical to that for the human protein and much smaller than that for the deglycosylated pig protein[4].

The human erythrocyte glucose transporter co-migrates with the nucleoside transporter on SDS/polyacrylamide gels both before and after enzymic deglycosylation[5]. The two proteins also co-purify during DEAE-cellulose chromatography of solubilised membranes to yield a "band 4.5" protein preparation that migrates as a single broad band of average apparent M_r 55000 on SDS/polyacrylamide gels[5]. This preparation consists largely of the glucose transporter (approx. 95%) with the nucleoside transporter present as a minor component (approx. 5%). We recently reported the purification of the nucleoside transporter to near homogeneity by passage of this mixture through a column of immobilized antibodies specific for the glucose transporter[6]. The resultant preparation bound approx. 0.6 mol NBMPR/mol polypeptide and catalysed uridine uptake when reconstituted into phospholipid vesicles[6]. In the present report we describe the preparation of antibodies against this purified protein, and their use in examining the structures of, and relationships between, the nucleoside transporters of different mammalian species.

Purine and Pyrimidine Metabolism in Man VII, Part A
Edited by R.A. Harkness *et al.*, Plenum Press, New York, 1991

EXPERIMENTAL

Human erythrocyte nucleoside transporter was purified as previously described[6]. All preparations used for rabbit immunisation were shown, by their failure to bind cytochalasin B and their lack of reactivity towards anti-glucose transporter monoclonal antibodies in Western blotting experiments, to be free of apparent contamination with glucose transporters. In order to prepare antibodies, samples (100μg) of the purified protein emulsified in complete Freund's adjuvant were injected into the thigh muscle of male New Zealand White rabbits. After one month a second, subcutaneous injection (50μg) in incomplete Freund's adjuvant was made. Animals were bled after a further two weeks. Antibody titres were measured by enzyme-linked immunosorbent assay (ELISA) and additional boost injections given when appropriate. Anti-nucleoside transporter antibodies were affinity-purified from the serum by adsorption onto protein-depleted pig erythrocyte membranes (15mg membrane protein per ml serum). After extensive washing of the membranes by centrifugation, specifically-bound antibodies were eluted by incubation with 0.2M glycine-HCl, pH 2.4, followed by immediate adjustment to neutral pH. Human erythrocyte glucose transporter, free of nucleoside transporter, was prepared by expression of a human hepatoma-derived cDNA clone using the Baculovirus/insect cell system, and was the kind gift of Mr. C-K. Yi. Partially-purified pig erythrocyte nucleoside transporter was prepared by the procedure of Kwong et al.[7]. A preparation enriched in rabbit nucleoside transporters was prepared by DEAE-cellulose chromatography of octyl-glucoside-solubilized rabbit erythrocyte integral membrane proteins (Young, J.D., unpublished procedure). Tryptic digestion of the purified transporter (126μg/ml in 50mM sodium phosphate, 100mM NaCl, 1mM EDTA, pH 7.4) was performed at 25°C using 1% (w/w) trypsin. Digestion was terminated by addition of aprotinin (2%, w/w). Western blotting, ELISA, endoglycosidase F digestion, photoaffinity labelling with NBMPR, ligand binding assays and all other procedures were as previously described[5-8].

RESULTS AND DISCUSSION

Antiserum prepared from rabbits immunised with the human erythrocyte nucleoside transporter reacted strongly with the purified protein in ELISA, half-maximal responses typically being obtained with serum dilutions of greater than 1 in 3000. Pre-immune serum showed no reaction. Partially-purified rabbit and pig nucleoside transporters also reacted with the antiserum in ELISA, although the absorbances obtained using microtitre plates coated with these proteins were somewhat lower than for plates coated with the human protein. These findings suggest that the nucleoside transporters from all three species share some common structural features. Western blots of human erythrocyte membrane samples showed that the antiserum recognised a protein that migrated as a broad band of apparent M_r 55000, identical in mobility to the nucleoside transporter. The intensity of the staining was proportional to the nucleoside transporter content of the samples - for example, the purified nucleoside transporter preparation stained much more strongly than an equal amount of unfractionated human erythrocyte band 4.5 proteins. The specificity of the antibodies was also demonstrated by probing Western blots of erythrocyte membranes from other species. No protein bands were labelled in blots of cow erythrocytes, which do not exhibit detectable carrier-mediated transport of nucleosides[1], but a protein band of mobility similar to that of the human nucleoside transporter was weakly labelled on blots of rabbit erythrocytes, which transport nucleosides rapidly[1]. On blots of pig erythrocyte membranes strong labelling was obtained of a band with an apparent M_r of 64000, identical to that of the protein identified as the nucleoside transporter by photoaffinity labelling with NBMPR[5].

These findings strongly suggested that the antiserum contained antibodies capable of recognising both the native nucleoside transporters (in ELISA) and the denatured proteins (on Western blots) from several species. However, it was essential to exclude the possibility that antibodies capable of recognising the glucose transporter were also present, because the two transporters from human erythrocytes migrate with identical mobility on SDS/polyacrylamide gels. Furthermore, the nucleoside transporter used as immunogen had been purified from a large excess of glucose transporter[6]. Although apparently homogenous, it was possible that the nucleoside transporter preparations contained amounts of the glucose transporter below the levels detectable by cytochalasin B binding assays or Western blots. Reactivity of the antiserum towards the glucose transporter was therefore assessed in two ways. In the first, the antiserum was shown not to react in Western blots with glucose transport protein that had been prepared in insect cells infected with recombinant Baculovirus containing the coding region of the transporter cDNA. Western blots of this recombinant protein were in contrast strongly labelled by anti-glucose transporter antibodies (C-K. Yi, personal communication). Anti-glucose transporter activity was also tested by ELISA using plates coated with two synthetic peptides corresponding to regions of the transporter known to be strongly immunogenic (residues 231-246 and 477-492[8]). In these assays a slight response to the peptides was seen with some batches of sera, but the absorbance values were much less than for plates coated with the nucleoside transporter. Reactivity was only seen when high concentrations of sera were used, and so it may have been artefactual. Nevertheless, because it may have stemmed from the presence of low concentrations of anti-glucose transporter antibodies, attempts were made to affinity-purify nucleoside transporter-specific antibodies from the serum. Advantage was taken of the fact that the antibodies recognised the nucleoside transporter of the pig erythrocyte. This cell type has no glucose transport activity[9], and we showed by Western blotting with a range of anti-glucose transporter antibodies that it does not contain immunologically cross-reactive glucose transporter polypeptides. After adsorption to and elution from pig erythrocyte membranes, the affinity-purified anti-nucleoside transporter antibodies retained the ability to recognise pig and human nucleoside transporters in ELISA assays, but no longer gave any response with peptide-coated plates, indicating complete removal of putative anti-glucose transporter antibodies. Typical yields of affinity-purified antibody were 150μg per ml serum. Although all reactivity towards pig erythrocyte membranes was removed from the serum by adsorption with pig membranes in this fashion, substantial reactivity towards the human protein remained in the serum, indicating that only a sub-population of the anti-nucleoside transporter antibodies were cross-reactive with the pig protein.

The specificity of the affinity-purified antibodies was confirmed by immunoadsorption experiments using antibody immobilised on protein A-Sepharose CL-4B. Treatment of detergent-solubilized human erythrocyte band 4.5 membrane proteins with the immobilised antibody led to the removal of 100% of the NBMPR-binding activity (measured after reconstitution) from solution. In contrast, none of the cytochalasin B binding activity was removed. These results indicated not only that the affinity-purified antibodies were specific for the nucleoside transporter, but also that they recognised it in its native conformation. Western blotting showed that the affinity purified antibodies also retained the ability to recognise the denatured protein - bands of average apparent M_r 55000 and 64000 were strongly labelled by the antibodies on Western blots of the purified human erythrocyte nucleoside transporter and of the partially purified pig erythrocyte nucleoside transporter respectively. In order to determine whether the antibodies recognised the oligosaccharide or protein moieties of these glycoproteins, Western blotting experiments were also performed after deglycosylation of the transporters by incubation with endoglycosidase F. This treatment converted the human protein to a sharp band of apparent M_r 45000, and the pig protein to a species that migrated with an apparent M_r of 57000. Both proteins were strongly labelled by the antibodies on blots after deglycosylation, indicating that most, if not all, of the antibodies were directed against protein rather than oligosaccharide epitopes.

The observation that the antibodies recognised both the deglycosylated pig and human nucleoside transporters indicates that these proteins, despite differing in apparent M_r by approx. 12000, are related in sequence. The pig and human proteins also exhibit similarities in their sensitivity to inhibition by dipyridamole and dilazep[1]. In contrast, the transporters of rat tissues are resistant to these inhibitors but exhibit an apparent M_r similar to that of the human protein after deglycosylation[4]. It was therefore of interest to examine the reactivity of the anti-human transporter antibodies towards the nucleoside transporters of rat tissues. The antibodies were found to label a fairly broad band of average apparent M_r 60000 on Western blots of a crude membrane fraction prepared from rat liver. Control IgG showed no labelling. The M_r of the labelled band was similar to that previously reported for rat nucleoside transporters identified by photolabelling[2-4]. This finding suggests that despite its different sensitivity to inhibitors, the rat liver nucleoside transporter is related in sequence to the human and pig proteins.

The anti-nucleoside transporter antibodies were also used to investigate the structure of the human erythrocyte nucleoside transporter. Tryptic digestion of the native, membrane-bound transporter was used to cleave the protein into large fragments, which were then probed for reactivity with the antibodies on Western blots. Prolonged treatment (3h) with trypsin yielded two large fragments. One of these migrated as a broad band of apparent M_r 30000 - 40000, indicating that it bore the site(s) of transporter glycosylation. This fragment also bore the site(s) of photoaffinity labelling with [^{3}H]NBMPR. A second fragment of apparent M_r 23000 migrated as a much sharper band and so was probably not glycosylated. Neither of these bands was recognised by the antibodies, indicating that the cleavages which produced them also destroyed the relevant epitopes. However, a minor digestion product of apparent M_r 40000 was labelled by the antibodies. This fragment migrated as a sharp band and so was probably not glycosylated. Comparison of its apparent M_r with that for the intact, deglycosylated transporter (45000) suggests that the site(s) of glycosylation of the nucleoside transporter lies close to one end of the protein. The fact that the glycosylated M_r 30000 - 40000 fragment of the transporter is also photolabelled by NBMPR indicates that this site(s) of glycosylation is also in the same half of the protein as the inhibitor-binding site.

ACKNOWLEDGEMENTS

We are grateful to the Cancer Research Campaign and Medical Research Council for support, and to Mr. N. Beaumont for assistance in some of the experiments.

REFERENCES

1. J.D. Young and S.M. Jarvis, Nucleoside transport in animal cells, Biosci. Rep. 3:309-322 (1983).
2. S.M. Jarvis and J.D. Young, Nucleoside transport in rat erythrocytes: two components with differences in sensitivity to inhibition by nitrobenzylthioinosine and p-chloromercuriphenyl sulfonate, J. Membrane Biol. 93:1-10 (1986).
3. M.M. Shi, J.-S.R. Wu, C.-M. Lee, and J.D. Young, Nucleoside transport. Photoaffinity labelling of high-affinity nitrobenzylthioinosine binding sites in rat and guinea-pig lung, Biochem. Biophys. Res. Commun. 118:594-600 (1984).
4. S.M. Jarvis and J.D. Young, Photoaffinity labelling of nucleoside transporter polypeptides, Pharmacol. Ther. 32:339-359 (1987).
5. F.Y.P. Kwong, S.A. Baldwin, P.R. Scudder, S.M. Jarvis, M.Y.M. Choy, and J.D. Young, Erythrocyte nucleoside and sugar transport. Endo-β-galactosidase and endoglycosidase-F digestion of partially purified human and pig transporter proteins, Biochem. J. 240:349-356 (1986).
6. F.Y.P. Kwong, A. Davies, C.M. Tse, J.D. Young, P.J.F. Henderson, and S.A. Baldwin, Purification of the human erythrocyte nucleoside transporter by immunoaffinity chromatography, Biochem. J. 255:243-249 (1988).
7. F.Y.P. Kwong, C.-M. Tse, S.M. Jarvis, M.Y.M. Choy, and J.D. Young, Purification and reconstitution studies of the nucleoside transporter from pig erythrocytes, Biochim. Biophys. Acta 904:105-116 (1987).
8. A. Davies, T.L. Ciardelli, G.E. Lienhard, J.M. Boyle, A.D. Whetton, and S.A. Baldwin, Site-specific antibodies as probes of the topology and function of the human erythrocyte glucose transporter, Biochem. J. 266:799-808 (1990).
9. J.D. Young, A.R.P. Paterson, and J.F. Henderson, Nucleoside transport and metabolism in erythrocytes from the Yucatan miniature pig. Evidence that inosine functions as an *in vivo* energy substrate, Biochim. Biophys. Acta 842:214-224 (1985).

EXPRESSION OF SODIUM-DEPENDENT NUCLEOSIDE TRANSPORTERS IN *XENOPUS* OOCYTES

Douglas A Griffith and Simon M Jarvis

Biological Laboratory
University of Kent
Canterbury, Kent, CT2 7NJ, UK

INTRODUCTION

The identification and isolation of proteins responsible for the transport of nucleosides across membranes has been difficult due to the low abundance of these membrane proteins (estimated to be < 0.1% of the total membrane proteins). However, with the availability of specific and high-affinity inhibitors, particularly nitrobenzylthioinosine (NBMPR), much progress has been made in determining the structural identity of the NBMPR-sensitive broad-specificity facilitated diffusion nucleoside carrier[1]. For example, in mammalian erythrocytes, the NBMPR-sensitive nucleoside transporter is a band 4.5 glycoprotein that has been purified to near homogeneity[1]. As such, both antibody and synthetic oligonucleotide probes derived from partial amino-acid sequences have been obtained and are being used to screen cDNA libraries. In contrast, there is virtually no information available on the other nucleoside transport systems that exist in animal cells.

In epithelial cells, the uptake of nucleosides at the brush-border surface is catalysed by two Na^+-dependent transport systems (N1 and N2) [2,3]. Although overlapping substrate specificity has been demonstrated, the N1 system mainly accepts purine nucleosides and guanosine is a specific substrate while the N2 system has a substrate specificity for mainly pyrimidine nucleosides and thymidine is a model permeant[2]. To date, no specific inhibitors of these active nucleoside transport systems are available and thus the approaches used to identify the NBMPR-sensitive nucleoside carrier are not applicable.

A possible strategy to identify Na^+-dependent nucleoside transport systems would be their functional expression in *Xenopus laevis* oocytes. This approach has proved successful in the identification and purification of mRNA encoding the intestinal sodium-glucose cotransporter[4]. In this report, we describe the functional expression of the intestinal Na^+/nucleoside N2 transporter following the injection of exogenous poly $(A)^+$ mRNA isolated from rabbit intestinal mucosal cells.

METHODS

Preparation of Poly$(A)^+$ mRNA

Total RNA from mucosal scrapings of rabbit small intestine were obtained using guanidine thiocyanate/Cs TFA extraction and poly$(A)^+$ mRNA isolated by oligo(dT)-cellulose column chromatography and stored as an ethanol precipitate at -70°C[5].

Isolation of oocytes and microinjection of poly(A)$^+$ mRNA

Oocytes (stages 5 and 6) were isolated from *X. laevis* toads using a collagenase digestion procedure[6]. Heathy oocytes were injected with either 50nl of poly(A)$^+$ mRNA dissolved in diethyl pyrocarbonate (DEPC)-treated water or 50nl of DEPC-treated water using a semi-automatic injector (Inject + Matic-system; Gabay, Geneva, Switzerland). Injected oocytes were maintained at 18°C in calcium-containing medium (96mM NaCl, 2mM KCl, 1mM $MgCl_2$, 1.8mM $CaCl_2$, 2.5mM sodium pyruvate, 5mM Hepes pH 7.6 and 0.1 mg/ml gentamicin) for 4 days. The medium was changed daily.

Uptake measurements

Eight oocytes were used for each uptake measurement. The oocytes were incubated in Na^+-free buffer (100mM choline chloride, 2mM KCl, 1mM $MgCl_2$, 10mM Hepes/Tris, pH 7.5) and then placed in 0.2ml of either Na^+ or the Na^+-free buffer containing either radioactive nucleoside (50μCi/ml) or ^{14}C-α-methyl-D-glucopyranoside (20μCi/ml). After incubation at 22°C for 30 min, uptake was terminated by removing the incubation medium and rapidly washing the oocytes six times with 1ml aliquots of ice-cold Na^+-free medium. Blank values of uptake were determined by exposing oocytes to radioactivity at 1°C immediately followed by washing. These blank values were subtracted from measurements associated with the oocytes at 22°C to determine uptake values. Radioactivity associated with the oocytes was determined by liquid scintillation spectrometry.

Materials

[6-^{3}H]Thymidine (25 Ci/mmol), [^{14}C]-α-methyl-D-glucopyranoside (0.14 Ci/mmol) and [8-^{3}H] guanosine (11 Ci/mmol) were purchased from Amersham, UK and ICN Radiochemicals, USA, respectively. Guanidine thiocyanate, caesium trifluoroacetate and oligo (dT)-cellulose were obtained from Fluka, Pharmacia LKB and Boehringer Mannheim, respectively. Collagenase (type II), gentamicin and diethyl pyrocarbonate were from Sigma. All other chemicals were of analytical grade.

RESULTS

Uptake of Glucose and Nucleosides in Water-injected and Poly(A)$^+$ mRNA-injected Oocytes

Oocytes were injected with water or poly(A)$^+$ mRNA from rabbit intestinal mucosal cells and incubated at 18°C for 4 days. The expression of Na^+-dependent nucleoside transporters was monitored by measuring guanosine and thymidine uptake, specific permeants for the N1 and N2 systems, respectively[2]. Active glucose transport was also measured to serve as a control of the effectiveness of the expression technique. Endogenous 50μM methyl-α-D-glucopyranoside uptake in water-injected oocytes was 0.26 ± 0.05 and 0.092 ± 0.034 pmol/oocyte per h in Na^+ and Na^+-free media, respectively (mean $\pm$SEM (n=8)). This Na^+-dependent component of uptake was completely inhibited by 0.5mM phlorizin, a specific inhibitor of Na^+-dependent glucose transport[7] (see Fig 1). Following injection of rabbit intestinal poly(A)$^+$ mRNA, Na^+-dependent phlorizin-sensitive methyl-α-D-glucopyranoside uptake increased 14-fold from 0.17 to 2.44 pmol/oocyte per h. These results are similar to those observed by earlier workers[7] and demonstrate that the present batch of poly(A)$^+$ mRNA is active. Using the same batch of water-injected or poly(A)$^+$ mRNA-injected oocytes, the uptake of 50μM thymidine and guanosine was measured. Thymidine uptake in water-injected oocytes was similar in the presence and absence of sodium indicating the absence of endogenous Na^+-dependent thymidine uptake in oocytes (Fig 2). However, thymidine uptake increased to 8.0 ± 1.0 pmol/oocyte per h in poly(A)$^+$ mRNA-injected oocytes in the presence of an inwardly directed Na^+-gradient compared to 0.86 ± 0.017 pmol/oocytes per h in the presence of choline. In contrast, although 50μM guanosine uptake in poly(A)$^+$ mRNA-injected oocytes was higher than that for water-injected oocytes, no significant difference was observed between sodium and choline containing medium. Uptake of guanosine in water-injected oocytes was greater in the presence of sodium compared to choline (1.94 ± 0.34 versus 0.36 ± 0.05 pmol/oocyte per h) but this difference was not reproduced in other preparations (data not shown).

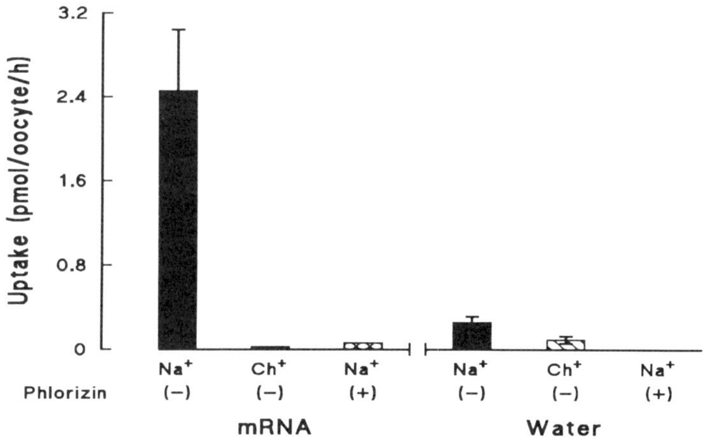

Fig. 1. Uptake of α-methyl-D-glucopyranoside in oocytes injected with either poly(A)$^+$ mRNA or water. Oocytes were injected with 50nl of water or rabbit intestinal poly(A)$^+$ mRNA (1μl/μg). After 4 days of incubation, the uptake of 50μM α-methyl-D-glucopyranoside was measured in the presence of 100mM NaCl (solid bars), 100mM choline chloride (hatched bars), and 100mM NaCl/0.5mM phlorizin (crossed bars). Uptakes are presented as the means of eight oocytes ± SEM.

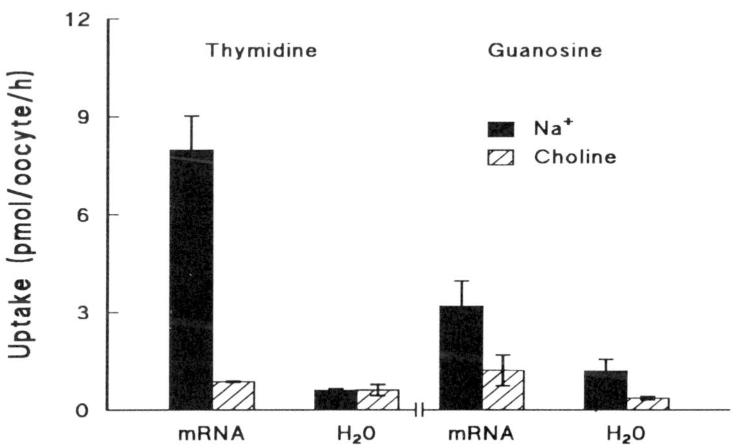

Fig. 2. Thymidine and guanosine uptake in oocytes injected with either water or poly(A)$^+$ mRNA. Oocytes from the same batch as in Fig. 1 were injected as detailed in Fig. 1 and the uptake of 50μM thymidine or guanosine measured in the presence of 100mM NaCl (solid bars) and 100mM choline chloride (hatched bars).

Specificity of the Expressed Na^+-Dependent Thymidine Transporter

In order to determine whether the expressed Na^+-dependent thymidine transport retains its substrate specificity, the abilities of nucleoside substrates for the N1 and N2 carriers to compete with radiolabelled thymidine were investigated. Na^+-dependent 50μM thymidine influx (uptake in the presence of Na^+ minus that in the presence of choline) was 2.50 ± 0.88 pmol/oocyte per h in poly(A)$^+$ mRNA injected oocytes. In the presence of 1mM uridine and cytidine, thymidine influx was inhibited (0.30 ± 0.03 and 0 pmol/oocyte per h, respectively). In contrast, 1mM inosine and guanosine had no significant effect on Na^+-dependent thymidine influx (2.00 ± 0.26 and 2.50 ± 0.28 pmol/oocyte per h).

DISCUSSION

We have demonstrated that injection of rabbit intestinal poly(A)$^+$ mRNA into *Xenopus* oocytes results in functional expression of the N2 Na^+/nucleoside cotransporter. Thus, the *Xenopus* oocyte expression system should be useful as an assay procedure for the isolation of the gene for the N2 Na^+-dependent nucleoside carrier.

ACKNOWLEDGEMENTS

This work was supported by grants from The Royal Society and The National Kidney Research Fund.

REFERENCES

1. S.M. Jarvis, Chemical and molecular probes of nucleoside transport mechanisms in mammalian tissues, in: "New Methods in the Study of Transport Across the Cell Membrane", D.L. Yudilevich, R. Deves, S. Peran and Z.I. Cabantchik, ed., Plenum, New York (1991).

2. T.C. Williams and S.M. Jarvis, Multiple sodium-dependent nucleoside transport systems in bovine renal brush-border membrane vesicles, Biochem. J. 274: 227 (1991).

3. D. Vijayalakshmi and J.A. Belt, Sodium-dependent nucleoside transport in mouse intestinal epithelial cells. Two transport systems with differing substrate specificities, J. Biol. Chem. 263: 19419 (1988).

4. M.A. Hediger, M.J. Coady, T.S. Ikeda, and E.M. Wright, Expression cloning and cDNA sequencing of the Na^+/glucose co-transporter, Nature 330: 379 (1987).

5. T. Maniatis, E.F. Fritsch, and J. Sambrook, "Molecular Cloning: A Laboratory Manual", Cold Spring Harbor Laboratory, New York (1982).

6. S.S. Tate, R. Urade, R. Micanovic, L. Gerber, and S. Udenfriend, Secreted alkaline phosphatase: an internal standard for expression of injected mRNAs in the *Xenopus* oocyte, FASEB J 4: 227 (1990).

7. M.A. Hediger, T. Ikeda, M. Coady, C.B. Gundersen, and E.M. Wright, Expression of size-selected mRNA encoding the intestinal Na/glucose cotransporter in *Xenopus laevis* oocytes, Proc. Natl. Acad. Sci. USA 84: 2634 (1987).

CONTROL OF NUCLEOSIDE TRANSPORT IN NEURAL CELLS

EFFECT OF PROTEIN KINASE C ACTIVATION

M.Teresa Miras-Portugal, Esmerilda G. Delicado, Teresa Casillas
Raquel P. Sen.

Departamento de Bioquímica, Facultad de Veterinaria
Universidad Complutense de Madrid
28040 Madrid, Spain.

INTRODUCTION

Although nucleoside transport has been extensively characterized in synaptosomal preparations, they are not suitable for regulatory studies because the cellular integrity is necessary[1,2]. Adenosine transport is a highly regulated process in cultured neural cells. Long term effects of nerve growth factor (NGF) have been described in cultured chromaffin cells, with a significant increase in the nucleoside transport[3]. Short term regulation by the action of protein kinases has also been reported. Activation of protein kinase A, directly by cAMP analogs or through membrane receptors, results in a quick inhibition of adenosine transport capacity and in a parallel decrease in the transporters number at the plasma membrane, measured by [^{3}H]nitrobenzylthioinosine binding[4,5]. There is also recent evidence of the interaction between adenosine receptors and transporters in this cellular model, where the A_2 agonist, NECA, activates the adenosine transport[6].

Due to the relevance of protein kinase C (PKC) in neural cells and its activation by secretagogue stimulation, the present study deals with the effect of PKC on the nucleoside transport in primary cultures of chromaffin cells, which present the advantage of being an homogeneous neural cell population, and non tumoral.

RESULTS AND DISCUSSION

Action of Protein Kinase C Effectors on Adenosine Transport

Treatment of cultured chromaffin cells with the protein kinase C activators, Tricaprylin and dicaproin, and also the phorbol esters (PDBu, PMA), resulted in a significant inhibition of adenosine transport (Table 1). The kinetic parameter modified was the maximal transport capacity (Vmax), but only minor changes were observed at the affinity constant (Km). The calcium ionophore, A-23187, potentiates the inhibitory actions of phorbol esters by combining the increase in cytosolic calcium with the PKC activation. On the contrary, the inhibition of PKC by H-7 resulted in a significant increase in adenosine transport (Table 3). These results indicate the relevance of PKC in adenosine transport regulation in neural cells. In this cellular model, the protein kinase C activation and calcium mobilization are associated with extracellular signals, such as secretagogues, which depolarize the plasma membrane opening the Ca^{2+}channels, or P_{2y}-receptors, coupled to inositolphospholipid metabolism[7,8,9]. Therefore, the action of both types of effectors was studied.

Table 1. Inhibition of Adenosine Transport by Protein Kinase C Activators in Cultured Chromaffin Cells.

Effector	Km(μM)	Vmax	%I
Control	1.6 ± 0.2	40 ± 5.0	--
PMA	1.4 ± 0.4	31 ± 3.0 *	22
PDBu	1.6 ± 0.2	32 ± 4.0 *	20
Tricaprylin	1.7 ± 0.3	21 ± 3.4 *	47
Dicaproin	1.7 ± 0.5	25 ± 5.0 *	37
A-23187	1.9 ± 0.3	30 ± 7.3 *	23
A-23187 + PMA	1.7 ± 0.3	15 ± 4.0 *	60

Cultured cells at a density of 250,000 cells/well were incubated at 37°C for 10 min with the indicated protein kinase C activators, PMA (100 nM), PDBu (100 nM), Tricaprylin (10 μg/ml), Dicaproin (10 μg/ml) and A-23187 (0.5 μM) for 10 min at 37°C. Adenosine transport was then measured at 37°C during the first minute which corresponded to the linear period. The transport capacity (Vmax) was expressed as pmol/10^6 cells.min. Values are the means ± SD of five experiments performed in quadruplicate. * $p < 0.001$.

Inhibition of Adenosine Transport by Secretagogues and Agonists of P_{2y}-purinergic Receptors

The secretagogues, carbachol and nicotine, significantly decreased the adenosine transport in chromaffin cells (Table 2). To assure that their action was mediated by PKC, studies were made with cells cultured in the presence of phorbol esters, in order to produce a down-regulation of PKC. As shown in Fig. 1, the effect of the nicotinic agonist, DMPP, on adenosine transport was lost after PKC depletion (time 0), however this effect was recovered at the same time as PKC activity. The inhibitory capacity of DMPP, was the same as that obtained for control cells not treated with PMA[10] 6 hours following the removal of phorbol esters.

Table 2. Inhibition of Adenosine Transport by Secretagogues and Agonists of P_{2y}-Purinergic Receptors in Cultured Chromaffin Cells.

Effector	Km(μM)	Vmax	%I
Control	1.0 ± 0.4	40 ± 1.0	--
Carbachol[a]	2.0 ± 0.3	20 ± 5.0 *	50
Nicotine[a]	1.4 ± 0.1	18 ± 4.7 *	55
ADP-ß-S	1.0 ± 0.3	26 ± 5.0 *	35
AMP-PNP	1.2 ± 0.3	32 ± 3.0 *	20

Adenosine transport was measured as described in Table 1. Cells were preincubated in the presence or absence of carbachol (50 μM) and nicotine (10 μM) for 1 min at 37°C, and with ADP-ß-S (0.1 mM) and AMP-PNP (0.1 mM) for 2 min at 37°C before starting the transport experiments. The Vmax values were expressed as pmol/10^6cells.min. Values are the means ± SD of three experiments performed in quadruplicate. * $p < 0.001$.
[a] Values from reference 5.

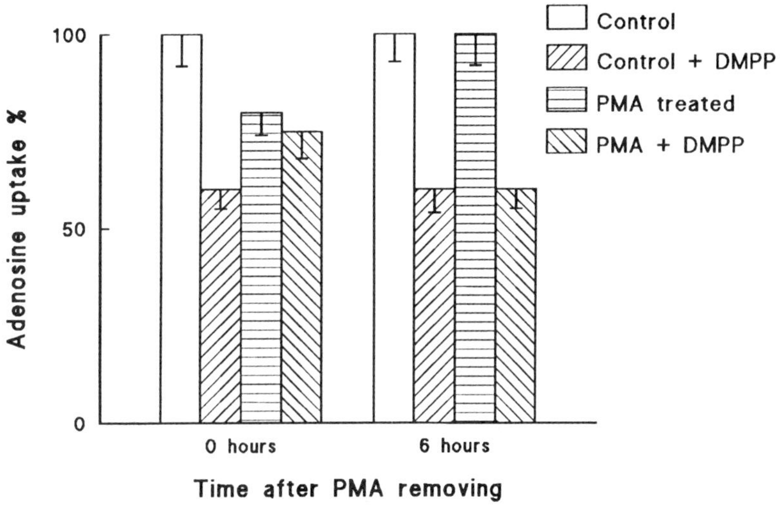

Fig. 1. Effect of the Nicotinic Agonist DMPP on Adenosine Transport after Down-regulation of PKC by Long-term Treatment with Phorbol Esters.

Cultured cells were treated with PMA (400 nM) for 24 hours . Adenosine transport was then measured at a concentration of 1 μM adenosine inmediately after the phorbol ester removing (0 hours), and after 6 hours of washing. The effect of the secretagogue, DMPP (1,1-dimethyl-4-phenyl piperazinium iodide), was studied incubating the cells for 1 min at 37°C with DMPP 10 μM.
Values are $\pm$ SD of three experiments in quadruplicate.

The presence of P_{2y}-receptors has been recently shown in chromaffin cells[8], with mobilization of calcium from internal stores[9]. Adenosine transport was inhibited by the action of the P_{2y}-agonists, ADP-ß-S and AMP-PNP as shown in Table 2, although they were less effective than secretagogues. The ATP released from the chromaffin granules therefore modulates the transport of adenosine, which is its extracellular metabolite. The

Table 3. Effects of Protein Kinase C and Protein Phosphatase Inhibition on Adenosine Transport in Cultured Chromaffin Cells.

Effector	V (pmol/10^6cells/min)	%C
Control	20.20 $\pm$ 0.5	100
H-7	24.38 $\pm$ 2.3 **	120
H-7 + Ap4A	14.80 $\pm$ 2.5 **	73
Ap4A	14.10 $\pm$ 2.9 **	70
Okadaic acid+Ap4A	13.02 $\pm$ 1.6 **	65
Okadaic acid	17.37 $\pm$ 6.5	85

Cells were incubated in the presence or absence of 1-(5-isoquinolinyl-sulfonyl)-2-methylpiperazine (H-7) (10 μM) and okadaic acid (0.1 μM) for 1 min at 37°C. Both compounds were added before diadenosine tetraphosphate (Ap_4A), and then Ap_4A was maintained for 2 min at 37°C. Adenosine transport was determined for 1 min at a concentration of 1 μM adenosine. Values are $\pm$ SD means of two experiments performed in quadruplicate. ** $p < 0.01$.

inhibitory action of a P_{2y}-agonist, the diadenosine tetraphosphate, Ap_4A, which is also stored and released from chromaffin cells[11], was decreased by combination with the PKC inhibitor H-7, and potentiated by the okadaic acid (Table 3). As this compound is an inhibitor of cellular protein phosphatases the phosphorylation by PKC cannot be reversed and, as expected, the inhibitory effect on adenosine transport was increased[12].

In cultured chromaffin cells the different manners of activation of PKC therefore result in adenosine transport inhibition. It can thus be assumed that the phosphorylation of this protein induces a conformational change that prevents the adenosine transport. Using the same cellular model, it has recently been reported that PKC activation also decreases the high affinity number of binding sites for [^{3}H]nitrobenzylthioinosine (NBTI), which is a specific ligand for adenosine transporters[5]. The effects of protein kinase C on [^{3}H]NBTI binding can also be reproduced in isolated plasma membranes with purified protein kinase C[13].

Acknowledgements: This work was supported by a Research Grant PB 89-0095 from the Spanish CICYT.

REFERENCES

1. C. W. Lee, and Jarvis, S. M., Nucleoside transport in rat cerebral-cortical synaptosomes. Evidence for two types of nucleoside transporters, Biochem. J. 249:557 (1988).
2. R. P. Shank, and W. J. Baldy, Adenosine transport by rat and guinea pig synaptosomes: basis for differential sensitivity to transport inhibitors. J. Neurochem. 55:541 (1990).
3. M. Torres, M. F. Bader, D. Aunis, and M. T. Miras-Portugal, Nerve growth factor effect on adenosine transport in cultured chromaffin cells, J. Neurochem. 48:233 (1987).
4. R. P. Sen, E. G. Delicado, and M. T. Miras-Portugal, Effect of forskolin and cyclic AMP analog on adenosine transport in cultured chromaffin cells, Neurochem. Int. 17:523 (1990).
5. M. T. Miras-Portugal, R. P. Sen, and E. G. Delicado, Nucleoside transport in neurons. Regulation by secretagogues and effectors of protein kinase C, Nucleotides and Nucleosides 10:110 (1991).
6. E. G. Delicado, A. Rodrigues, R. P. Sen, A. M. Sebastiao, J. A. Ribeiro, and M. T. Miras-Portugal, Effect of 5'-(N-ethylcarboxamido)adenosine on adenosine transport in cultured chromaffin cells, J. Neurochem. 54:1941 (1990).
7. S. A. Lee, R. W. Holz, Protein phosphorylation and secretion in digitonin-permeabilized adrenal chromaffin cells, J. Biol.Chem. 261:17089 (1986).
8. J. Pintor, M. Torres, E. Castro, and M. T. Miras-Portugal, Characterization of diadenosine tetraphosphate (Ap_4A) binding sites in cultured chromaffin cells: evidence for a P_2 site, Br. J. Pharmacol. (in press) (1991).
9. E. Castro, M. Torres, M. T. Miras-Portugal, and M. P. González, Effect of diadenosine polyphosphates on catecholamine secretion from isolated chromaffin cells, Br. J. Pharmacol. 100:360 (1990).
10. D. R. Terbush, M. A. Bittner, and R. W. Holz, Ca^{2+}influx causes rapid translocation of protein kinase C to membranes, J. Biol. Chem. 263:18873 (1988).
11. J. Pintor, M. Torres, and M. T. Miras-Portugal, Carbachol induced release of diadenosine polyphosphates - Ap_4A and Ap_5A - from perfused bovine adrenal medulla and isolated chromaffin cells, Life Sci. 48:2317 (1991).
12. P. Cohen, C. F. B. Holmes, and Y. Tsukitani, Okadaic acid: a new probe for the study of cellular regulation, TIBS. 15:98 (1990).
13. E. G. Delicado, R. P. Sen, and M. T. Miras-Portugal, Effects of phorbol esters and secretagogues on the nitrobenzylthioinosine binding to nucleoside transporters and nucleoside uptake in cultured chromaffin cells. Biochem. J. (in press) (1991).

ISOLATION OF NUCLEOSIDE PERMEASES FROM ESCHERICHIA COLI

Maurice P. Gallagher, Yianbiao Zhang and Jane Craig

Institute of cell and molecular biology
Division of Biology
University of Edinburgh
West Mains Road
Edinburgh EH9
Scotland

INTRODUCTION

Although capable of synthesizing nucleosides de novo, most living cells have the capacity to catabolize nucleosides from the surrounding environment via the salvage pathways. This involves the selective transport of such molecules across the plasma membrane and requires nucleoside-specific transport systems. Such systems provide a route for targeting chemotherapeutic agents. A molecular understanding of the mode of entry of these compounds is therefore of considerable importance for understanding their efficacy and potential action..

The low abundance of specific carrier systems in biological membranes and the lack of functional assays other than transport make such systems difficult subjects for study. Nucleoside transport systems are no exceptions and most studies have focussed on characteristics of specificity and the kinetics of uptake. Genetic analysis of transport systems in the enteric bacteria, Escherichia coli and Salmonella typhimurium, circumvent many of these limitations because of their relative abundance and the ease with which their genome can be manipulated. Furthermore, microbial transport proteins are known to exhibit homologies not only between themselves but also to membrane proteins of higher organisms (Higgins et al, 1990, Henderson & Maiden, 1990) and the comparative information gained from studying enteric bacteria may well provide valuable insights into the systems of other organisms.

Both E. coli and S.typhimurium are capable of growth on a range of natural ribo- and deoxyribonucleosides when present as sole carbon source (Doskocil, 1974; Munch-Petersen and Mygind, 1976) and studies with toxic analogues have indicated that multiple high affinity transport systems exist (Hoffmeyer & Neuhard, 1971; Beck et al, 1972). The cell envelope of these organisms consists of two distinct membranes separated by a dense peptidoglycan layer. The outer membrane is commonly regarded as a permeable layer, containing porins through which solutes with less than 600 Da can diffuse. Nucleosides can traverse this boundary readily, although E. coli also contains a minor outer membrane protein, Tsx, which enhances nucleoside transport, at low substrate concentrations via facilitated diffusion (Hantke, 1976). The cytoplasmic membrane provides a more selective barrier and, in general, specific transport systems are required for solute entry. In E. coli, two high affinity, proton motive force-driven nucleoside transport

Purine and Pyrimidine Metabolism in Man VII, Part A
Edited by R.A. Harkness *et al.*, Plenum Press, New York, 1991

systems, encoded by *nupC* and *nupG*, have been demonstrated (Munch-Petersen et al., 1979) which differ in their sensitivity to the uridine analogue, showdomycin (Komatsu and Tanaka, 1972).

Mapping studies have localised the genes for *nupG* and *nupC* to 64 minutes and 50-52 minutes on the *E. coli* chromosome, respectively. The former gene was isolated by randomly screening a chromosomal plasmid library, for the ability to restore growth on cytidine as carbon source, to transport-negative mutants (Munch-Petersen & Jensen, 1984). Further work by Westh-Hansen and colleagues (1987) led to the nucleotide sequence of this gene. Little information presently exists on the nupC locus. In the present report, we have aimed to localise the *nupG* gene to defined plasmids which contain chromosomal framents of known physical location and also, to isolate clones of the poorly characterised *nupC* locus.

MATERIALS AND METHODS

Bacterial Strains, Plasmids and Manipulations

All strains were *E. coli* K-12 derivatives. MPG100 is a derivative of wild type strain, 12033, which carries *nupC*::Tn*10*(Δ16, 17; Tc^r), *nupG*::Tn*10*(Δ16, 17; Km^r). JA200 is a multi-auxotroph (Clarke & Carbon, 1976). Plasmid pLC20-5 was obtained from the *E. coli* Genetic Stock Centre. Strains were grown in Luria Broth, containing antibiotics where appropriate (Km-50mgs/l; Tc-10mgs/l; Ap-50mgs/l) or on minimal medium (Munch-petersen & Jensen, 1984), containing cytidine (1mg/ml) for complementation studies or glucose (0.2% w/v) for toxicity studies. Conjugal transfer of pLC20-5 was carried out between JA200 and MPG100 as described by Herrero and colleagues (1990), selecting transconjugants on minimal glucose medium containing colicin E1 (0.5mgs/ml). 5'-aza-cytidine (10mgs/l; 0.01ml per disc) was used for toxicity studies. Phage lysates were amplified in NM621 and spot recombination-complementation tests were performed as previously described (Masters, 1990), using 0.004 ml of phage (10^7 p.f.u./ml) with an equal volume of recipient cells.

RESULTS

Localisation of nupG to Plasmid pLC20-5

The nupG gene was originally isolated by randomly screening the Clarke-Carbon plasmid library (1976), containing fragments of the *E. coli* chromosome, for clones which restored growth on a cytidine carbon source to a *nupC*, *nupG* double mutant (Munch-Petersen and Jensen, 1984). In the present study, we utilised the knowledge that the *speC* gene is located next to *nupG* on the chromosome (Westh-Hansen et al., 1987), to determine whether *speC* clones of known physical location from the Clarke-Carbon library might also carry *nupG*. Plasmids were introduced into MPG100, a nucleoside transport deficient strain of *E. coli* constructed in the lab., by conjugation from JA200 and colonies were screened for restoration of the ability to utilise cytidine as sole carbon source. Substantial growth was obtained with pLC20-5, indicating the presence of both *nupG* and *speC*. Complementing clones retained antibiotic resistance markers in the chromosomal *nupC* and *nupG* genes, eliminating the possibility of chromosomal reactivation.

As a further test for the presence of *nupG* on pLC20-5, complementing clones were also examined for restoration of sensitivity to the toxic nucleoside analogue, 5'-aza-cytidine, using a minimal glucose overlay. The presence of pLC20-5 in MPG100 was found to have a dramatic effect on analogue sensitivity and zones of inhibition of 31, 12 and 35 cm. (diameter) were obtained for strains 12033, MPG100 and MPG100 containing pLC20-5, respectively.

Isolation of nupC Clones

The nupC locus had previously been localised to 50-52 minutes on the E. coli chromosome (Munch-Petersen & Mygind, 1983)). We decided to screen the ordered phage Lambda library of Kohara and colleagues (1987) for clones which might carry nupC. Clones with locations between 50 and 53 minutes were prepared and phage were then introduced into MPG100, containing plasmid pUN121 which expresses the lambda cI repressor (preventing phage lysis of the host), and screened for restoration of growth on minimal cytidine media. These clones are incapable ofintegration and complementation can only occur following recombination between the chromosome and the viral insert. Two complementing clones of overlapping sequence were found (Figure 1) thereby localising the nupC

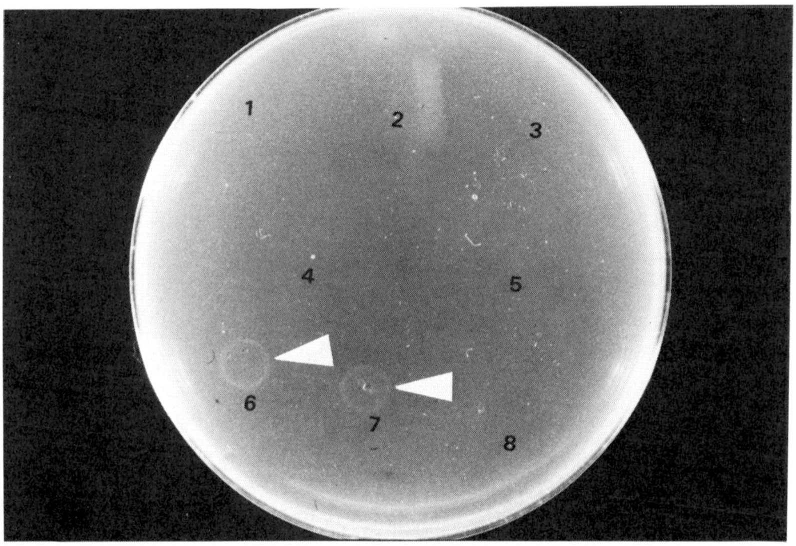

Figure 1. Cells of MPG100, carrying plasmid pUN121, were spotted on minimal cytidine plates together with phage lambda clones, reported to carry defined fragments of the E. coli chromosome (Kohara et al., 1987). Clones 6 and 7 (arrow) can be seen to restore nucleoside transport, allowing growth.

locus to a 13 Kb fragment. In each case, recombination resulted in excision of the mutated chromosomal nupC gene, as evidenced by loss of the associated antibiotic resistance marker.

DISCUSSION

In the present study, we have aimed to localise and isolate the nupG and nupC genes of E. coli. Our studies support previous data from restriction analysis (Westh-Hansen et al., 1987) that nupG is, in fact, located adjacent to the speC gene which encodes ornithine decarboxylase and we conclude that plasmid pLC20-5, from the Clarke-Carbon library, which is known to carry the speC gene, also carries nupG. Furthermore, our studies provide the first clear isolation of the nupC gene and verify earlier mapping data which located the transport function between 50 and 52 minutes of the chromosome.. Nucleic acid sequence comparisons of the nupG and nupC genes should provide valuable insights into the overlapping substrate specificity and differing analogue sensitivity of these two systems.

REFERENCES

Beck, C. F., Ingraham, J. L., Neuhard, J. and Thomassen , E. Metabolism of Pyrimidines and Purine Nucleosides by *Salmonella typhimurium*. J. Bacteriol. 110, 219-228 (1972).

Clarke, L. and Carbon, J. A Colony Bank Containing Synthetic Col E1 Hybrid Plasmids Representative of the Entire *Escherichia coli* Genome. Cell 9: 91-99 (1976).

Doskocil, J. Inducible nucleoside permease in Escherichia coli. Biochem. Biophys. Res. Commun. 56: 997-1003 (1974).

Hantke, K. Phage T6, Colicin K Receptor and Nucleoside transport in *Escherichia coli.* FEBS Letts 70: 109-112 (1976).

Henderson, P. J. F. and Maiden, M. C. J. Homologous Sugar Transport Proteins in *Escherichia coli* and their Relatives in Both Prokaryotes and Eukaryotes. Phil. Trans. R. Soc. Lond. B 326: 391-410 (1990).

Herrero, M., Lorenzo, V. and Timmins, K. N. Transposon Vectors Containing Non - Antibiotic Resistance Selection Markers for Cloning and Stable Chromosomal Insertion of Foreign Genes in Gram Negative Bacteria. J. Bacteriol. 172: 6557-6567 (1990).

Higgins, C. F., Hyde, S. C., Mimmack M. L., Gileadi, U., Gill, D. R., Gallagher M. P. Binding Protein-Dependent Transport Systems. J. Bioenerg. Biomemb. 22: 571-592 (1990).

Hoffmeyer, J. and Neuhard, J. Metabolism of Exogenous Purine Bases and Nucleosides by *Salmonella typhimurium*. J. Bacteriol. 106: 14-24 (1971).

Kohara, Y., Akiyama and K., Isono, K. The physical map of the whole *Escherichia coli* Chromosome: Application of a New Strategy for Rapid Analysis and Sorting of a Large Genomic Library. Cell 50: 495-508 (1987).

Komatsu, Y. and Tanaka, K.A. Showdomycin-resistant mutants of Escherichia coli K-12 with altered transport character. Biochim. Biophys. Acta 288: 390-403 (1972).

Masters, M. Locating genes on the Escherichia coli physical map. A request for contributions. A.S.M. News 56: 6-7 (1990).

Munch-Petersen, A. and Jensen, N. Cloning and expression of genes encoding the nupG nucleoside transport system in *Escherichia coli*. Pp.85-99 in "The cell membrane: its role in interaction with the outside world". E. Haber, ed., Plenum Press, New York (1984).

Munch-Petersen, A. and Mygind, B. Nucleoside transport systems in Escherichia coli K-12: Specificity and regulation. J. Cell Physiol. 89: 551-560 (1976).

Munch-Petersen, A. and Mygind, B. Transport of nucleic acid precursors. Pp.259-305 in "Metabolism of Nucleotides, Nucleosides and Nucleobases in Microorganisms", A. Munch-Petersen, ed., Academic Press, London (1983).

Munch-Petersen, A., Mygind, B., Nicolaisen, A. and Pihl, N, J. Nucleoside Transport in Cells and Membrane Vesicles from *Escherichia coli* K-12. J. Biol. Chem. 254: 3730-3737 (1979).

Westh-Hansen, E., Jensen, N. and Munch-Petersen, A. Studies on the Sequence and Structure of the *Escherichia coli* K-12 *nup G* Gene, Encoding a Nucleoside-Transport System. Eur. J. Biochem. 168: 385-391 (1987).

MURINE ERYTHROLEUKEMIA CELLS RESISTANT TO PERIODATE-OXIDIZED ADENOSINE HAVE LOWERED LEVELS OF NUCLEOSIDE TRANSPORTER

Jerald Hoffman

Department of Biochemistry
University of Louisville
Louisville, KY 40292 USA

INTRODUCTION

Transmethylation from S-adenosylmethionine (AdoMet) is important in the metabolism of numerous high and low molecular weight compounds. Methyltransferase reactions yield, in addition to the methylated product, the demethylated remnant of AdoMet, S-adenosylhomocysteine (AdoHcy). Since AdoHcy is a strong product inhibitor of virtually all methyltransferases, it must be removed in vivo to permit methylation reactions to continue. This is accomplished by the enzyme AdoHcy hydrolase. We have found that periodate-oxidized adenosine (Adox) is a potent, mechanism-based inhibitor of this enzyme due to its structural similarity to the enzyme-bound intermediate, 3'-ketoadenosine (1). Adox effectively inhibits transmethylation in cells and whole animals by causing accumulation of AdoHcy as shown in Fig. 1. Adox has been used in my laboratory to measure the rate of transmethylation in mouse liver (2) and to inhibit selenium methylation in mice (3). Since methylation is essential to many biosynthetic processes in the cell we decided it would be of interest to investigate whether cells could adapt to growth in Adox, and, if so, what biochemical changes were necessary for such adaptation. The results presented below show that one such adaptation is loss of nucleoside transport preventing Adox from entering the cell.

METHODS

Cell Culture: Murine erythroleukemia cells (MELC) were grown at 37°C in Dulbecco's modified Eagle's medium containing 10% fetal calf serum in an atmosphere of 5% CO_2 and 95%

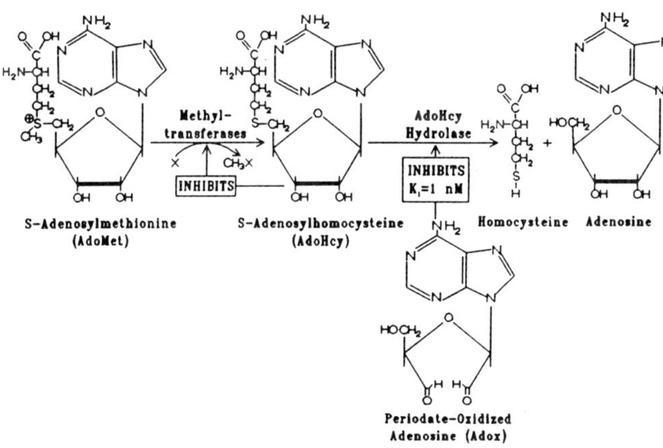

Figure 1: Adox inhibits AdoHcy hydrolase directly, and the accumulated AdoHcy inhibits transmethylation.

relative humidity. Cells were routinely reseeded at 200,000/ml every 2 days to maintain log-phase growth. MELC were adapted to growth in Adox starting at 5 μM. Once an Adox-resistant (AR) population was obtained, the Adox concentration was increased until we obtained AR MELC resistant to 1 mM Adox. Cells were enumerated in a hemacytometer using the ability to exclude trypan blue as a measure of viability.

Measurement of AdoHcy: Samples of 10^6 MELC were collected by centrifugation, extracted with 100 μl of 5% perchloric acid, and centrifuged. The resulting supernatants were analyzed for AdoHcy by cation-exchange HPLC.

Binding of S-Nitrobenzylthioinosine (NBTI): MELC were collected by centrifugation, washed with phosphate-buffered saline (PBS), and resuspended at 37°C in 0.5 ml PBS at 10^7 cells/ml. [^{3}H]NBTI was added at various concentrations, and MELC were centrifuged and washed twice with 5 ml of PBS. Final cell pellets were dissolved in 1 ml of 1% sodium dodecyl sulfate (SDS), mixed with 9 ml of liquid scintillation cocktail, and counted for ^{3}H. Nonspecific binding was assessed by addition of unlabeled NBTI at 1000-fold excess over [^{3}H]NBTI and was subtracted from total binding to give the specific binding shown in the Results section.

Adenosine Transport: MELC were collected and washed with PBS as before and then suspended in 0.4 ml of PBS at 10^7 cells/ml. [^{3}H]Adenosine was added at zero time at a final concentration of 250 μM. At 15, 30, and 45 sec samples of 100 μl were removed into 10 ml of stop solution (10 μM NBTI in PBS). Samples were centrifuged, and the MELC were washed twice with 10 ml of stop solution. Final cell pellets were dissolved in SDS and counted for ^{3}H as above.

RESULTS

Effects of Adox on Growth and AdoHcy Content of Normal and AR MELC: The results in Fig. 2 showed that normal MELC were unable to grow in 100 μM Adox, apparently due to their greatly increased content of AdoHcy which persisted throughout the incubation period. AR MELC grew equally well in the absence or presence of 100 μM Adox. In 100 μM Adox the AdoHcy content of AR MELC increased transiently but decreased after 8 hr when growth began.

Although the normal MELC did not grow in Adox, they were still viable as evidenced by their ability to exclude dye. When these normal MELC treated with Adox were analyzed by flow cytometry, they were found to be arrested at the G2/M stage of the cell cycle (data not shown). This suggests the interesting possibility that there is a critical methylation event necessary for cells to undergo mitosis.

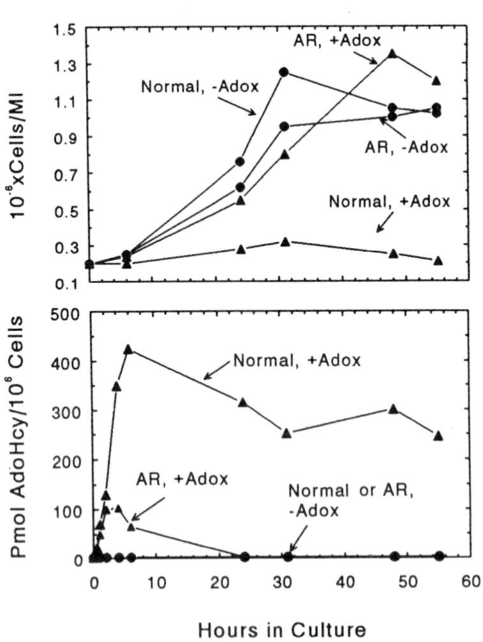

Figure 2: Effects of 100 μM Adox on growth (top) and AdoHcy content (bottom) of normal and AR MELC.

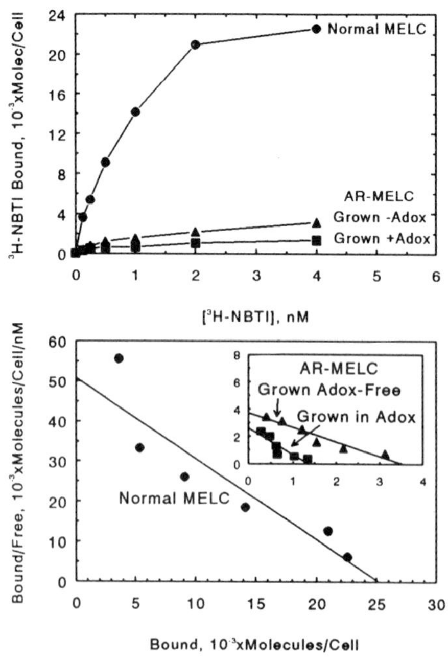

Figure 3: NBTI binding by normal and AR MELC.

Maintenance of Adox Resistance in the Absence of Selective Pressure: In order to determine if Adox resistance was inheritable, AR MELC were grown in Adox-free conditions for several weeks during which they were periodically retested for their ability to grow in 100 μM Adox. Adox resistance was maintained over the course of at least 24 generation times in Adox-free medium, indicating that such resistance is a genetic trait (data not shown).

Nucleoside Transport Properties of Normal and AR MELC: Since AdoHcy hydrolase is so sensitive to inhibition by Adox (1) it appeared that the enzyme in AR MELC was not being exposed to the inhibitor, i.e. that Adox was not being taken up by the resistant cells. Because Adox is a nucleoside analog the possibility existed that it was transported into normal MELC by the nucleoside transport system which was not functional in AR MELC. MELC are derived from erythroid precursor cells arrested at the CFU-E stage by infection with the Friend virus complex. These cells can be induced by various treatments to differentiate further along the erythroid pathway to the point where they synthesize considerable hemoglobin but do not eject their nuclei (4). It thus appeared that MELC would resemble mouse erythrocytes in their nucleoside transport properties, meaning that they would have only the NBTI-sensitive, equilibrative transporter (5). This nucleoside transporter can be quantitated by measuring the binding of the strong inhibitory ligand, NBTI.

Accordingly, normal and AR MELC were compared with respect to NBTI binding, and the results are shown in Fig. 3. The top panel compares the specific binding curves for normal MELC, AR MELC grown for 1 week in the absence of Adox, and AR MELC grown in 100 μM Adox up to the time of the binding study. Normal MELC bind much more NBTI than the AR MELC regardless if the latter are grown with or without Adox. The bottom panel shows the Scatchard transformation of the binding data which yielded the following binding constants. The normal MELC had 25,200 NBTI binding sites/cell with a K_d of 0.5 nM. AR MELC grown in Adox had 1,500 NBTI binding sites/cell with a K_d of 0.5 nM. AR MELC grown without Adox had the slightly higher number of 3,500 NBTI binding sites/cell but with a comparable K_d of 0.95 nM. These data may be compared to published values for mouse erythrocytes of 3,300 binding sites/cell (but with much less surface area/cell than MELC) and a K_d of 0.14 nM (6). These data are consistent with the adaptation procedure selecting for AR MELC expressing greatly lowered quantities of the NBTI sensitive, equilibrative transporter. If, for instance, Adox resistance was due to a qualitative change (mutation) in the nucleoside transporter one would not expect the K_d of the mutant transporter to be similar to that of the normal transporter.

While NBTI binding is generally taken as an indicator of the amount of the nucleoside transporter, the actual rate of nucleoside (adenosine) transport was compared between normal and AR MELC with the results shown in Fig. 4. Normal MELC imported adenosine from a 250 μM solution at a rate of 623 fmol/sec/10^6 cells, much higher than the rate of 95 fmol/sec/10^6 cells found for AR MELC. Adenosine uptake by normal MELC was completely inhibited by 1 μM NBTI (data not shown), indicating that these cells have only the NBTI-sensitive, equilibrative transporter. Direct measurements of Adox transport in normal and AR MELC have not been completed as yet.

NBTI Permits Growth of Normal MELC in Adox: If Adox is transported into normal MELC by the NBTI-sensitive transporter, then NBTI should block the uptake and growth-inhibitory effects of Adox. The results in Fig. 5 show that this is the case. NBTI at 1 μM had no effect on growth of normal MELC, but restored the ability to grow in 100 μM Adox.

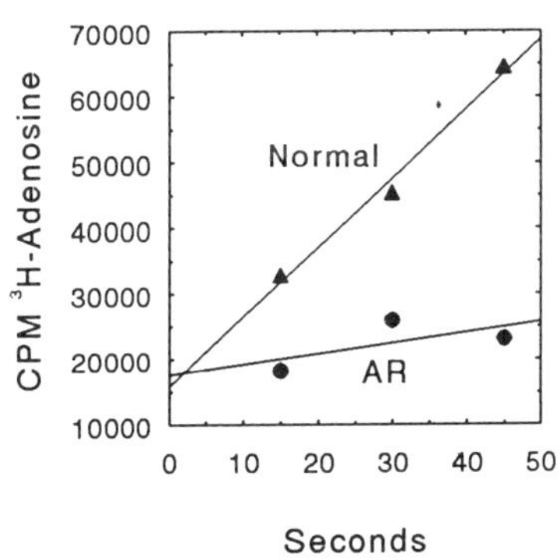

Figure 4: Adenosine transport by normal and AR MELC.

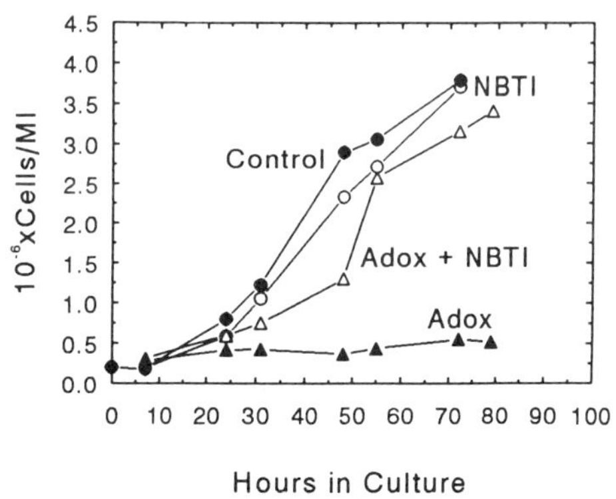

Figure 5: NBTI permits normal MELC to grow in Adox. Growth of normal MELC was measured in 1 μM NBTI, 100 μM Adox, or both.

SUMMARY AND CONCLUSIONS

The results above show that mammalian cells, as exemplified by MELC, can be selected to be resistant to Adox, and that the resistant cells have greatly decreased nucleoside transport capacity. Since no mutagen was used prior to the selection process and Adox resistance was genetically stable, it appears that within a population of normal cells there is a genetically controlled range of expression of the nucleoside transporter. On the basis of the present data we cannot determine if the low level of nucleoside transporter in AR MELC is due to an altered form of the protein or a decreased amount of the normal protein. However the similarity of the K_d for NBTI in normal and AR MELC suggests that the latter is the case. Considerable indirect evidence is presented that Adox, with its ribose converted to an acyclic dialdehyde, is a substrate for the nucleoside transporter, adding a new type of compound to this list.

ACKNOWLEDGMENTS

This research was supported by grant #RII8610671 from the National Science Foundation to the Kentucky Experimental Program to Stimulate Competitive Research. Expert technical assistance was provided by Jane Williams.

REFERENCES

1. J. L. Hoffman, Inhibition of S-Adenosyl Amino Acid Metabolism: Periodate-Oxidized Nucleosides as Potent Inhibitors of S-Adenosylhomocysteine Hydrolase, in "Transmethylation", E. Usdin, R. Borchardt, and C. Creveling, eds., pp. 181-186, Elsevier, Amsterdam (1979).

2. J. L. Hoffman, The Rate of Transmethylation in Mouse Liver as Measured by Trapping S-Adenosylhomocysteine, *Arch. Biochem. Biophys.* 205, 132 (1980).

3. J. L. Hoffman and K. P. McConnell, Periodate-Oxidized Adenosine Inhibits the Formation of Dimethyl Selenide and Trimethyl Selenonium Ion in Mice Treated with Selenite, *Arch. Biochem. Biophys.* 254, 534 (1987).

4. P. A. Marks and R. A. Rifkind, Erythroleukemic Differentiation, *Annu. Rev. Biochem.* 47, 419 (1978).

5. P. W. Plagemann, R. W. Wohlhueter, and C. Wolfenden, Nucleoside and Nucleobase Transport in Animal Cells, *Biochim. Biophys. Acta* 947, 405 (1988).

6. S. M. Jarvis, J. R. Hammond, A. R. Paterson, and A. S. Clanachan, Species Differences in NUcleoside Transport: A Study of Uridine Transport and Nitrobenzylthioinosine Binding by Mammalian Erythrocytes, *Biochem. J.* 208, 83 (1982).

EFFECT OF ISCHAEMIA ON CARDIAC ADENOSINE BINDING SITES

Simonetta Ronca-Testoni, Riccardo Zucchi, Ugo Limbruno, Paola Galbani, Giovanni Ronca, Mario Mariani

Institute of Biological Chemistry, Scuola Superiore S.Anna
Institute of Cardiology, University of Pisa, Pisa (Italy)

INTRODUCTION

Adenosine produces different effects on the heart. It markedly increases coronary blood flow and exerts negative inotropic, chronotropic and dromotropic effects on atrial myocardium. Adenosine does not appear to have any direct action on ventricular contractility but significantly reduces the positive inotropic effect of catecholamines on both atrial and ventricular myocardium.[1]

These direct and indirect effects of adenosine are considered to be mediated by the interaction with specific surface receptors which cause either inhibition (A1) or stimulation (A2) of adenylate cyclase,[2] even if cAMP-independent mechanisms cannot be excluded.[3] A1 receptors have been identified in atrial and ventricular myocardium[4,5] and are involved in the electrophysiological effects of adenosine and in the inhibition of the response to catecholamines. A2 receptors have been demonstrated in the coronary endothelium and coronary smooth muscle and appear to be responsible for the vasodilator effect of adenosine.[6]

During ischaemia adenosine is released from the myocardium but the sensitivity of the heart, and in particular of the coronary circulation, to exogenous and endogenous nucleoside is reduced.[7,8] To understand the mechanisms of this phenomenon we have examined the effect of ischaemia on A1 and A2 adenosine binding sites in the rat heart.

MATERIALS AND METHODS

^{3}H-5'-N-ethylcarboxamidoadenosine (^{3}H-NECA, 18-30 Ci/mmol), ^{3}H-N^6-cycloexyladenosine (^{3}H-CHA, 11-34 Ci/mmol), ^{3}H-8-cyclopentyl-1,3-dipropylxanthine (^{3}H-DPCPX, 120 Ci/mmol) were obtained from New England Nuclear, Florence (Italy). Unlabelled adenosine derivatives and adenosine deaminase were from Sigma Chemicals Co., St.Louis, Mo.(USA). All other chemicals were of analytical grade.

Wistar rats (250-300 g) were anaesthetized with a mixture of ether and air and after intravenous administration of 1000 IU sodium heparin the hearts were quickly excised, arrested in ice-cold saline and mounted onto the aortic cannula. The hearts were then perfused according to the working heart technique, as described elsewhere.[8] Control hearts were subjected to 30 min of aerobic perfusion, ischemic hearts to 10 min of aerobic perfusion followed by 10, 20, or 30 min of global normothermic ischaemia induced by simultaneous clamping of the aortic and atrial cannulae. Krebs-Henseleit bicarbonate buffer with 11 mM glucose was used as standard perfusion medium. The buffer was gassed with 95% O_2 and 5% CO_2, temperature was kept at 37°C and the pH was 7.4. A total of 8 hearts were perfused in the control and 20-min ischaemia groups: 3 were used for ^{3}H-DPCPX binding, 3 for ^{3}H-NECA binding, and 2 for ^{3} H-CHA binding. 3 additional hearts were perfused in the 10-min and 30-min ischaemia groups, and used for ^{3}H-NECA binding.

Immediately after removal from the perfusion system, the hearts were washed with ice-cold saline and ventricular muscle, trimmed free of atria, remnants of valves and connective tissue, was homogenized in 8 volumes of 10 mM TRIS.HCl, 1 mM EDTA and 1 mM dithiothreitol, pH 7.4. Microsomal membranes were prepared according to Bohm et al.[5]

To characterize adenosine binding sites we used ^{3}H-NECA that shows a high affinity for A2 receptors, ^{3} H-CHA and ^{3}H-DPCPX, selective A1 agonist and antagonist, respectively.[9] Membranes suspended in incubation buffer were preincubated with adenosine deaminase (0.2 IU/ml) for 40 min at 4°C to remove endogenous adenosine. Binding of ^{3}H-DPCPX was done in a final volume of 250 µl incubation buffer containing 200 µg protein and ligand concentration ranging from 0.2 to 12 nM as described by Bohm et al.[5] Binding of ^{3}H-CHA was performed according to Bruns et al.[10] with a ligand concentration ranging from 0.4 to 25 nM. Nonspecific binding was defined in the presence of 10 µM CHA. Binding of ^{3}H-NECA was carried out as described by Ronca-Testoni et al.[3] NECA is not a pure A2 ligand but can also bind to A1 receptors and to a poorly characterized "non receptor" site.[9] Therefore in additional experiments the binding of ^{3}H-NECA was also carried out as described by Bruns et al.,[9] using 50 nN N^6-cyclopentyladenosine (CPA) to eliminate the A1 component and 100 µM CPA to determine nonspecific binding. This technique allows identification of the A2 component of NECA binding.

RESULTS AND DISCUSSION

Results of ^{3}H-DPCPX binding experiments are shown in Table 1. In control hearts Kd and Bmax values of 3.6 ± 1.3 nM and 11.4 ± 4.6 fmol/mg protein, respectively, were obtained and no significant difference was observed in the ischaemic hearts. ^{3}H-CHA binding gave similar results.

Typical saturation curves for ^{3}H-NECA binding obtained in control and ischaemic hearts are reported in Fig.1. In control hearts (n=3) data

Table 1. Characteristics of ^{3}H-DPCPX binding

	Kd (nM)	Bmax(fmol/mg)
Control	3.6 $\pm$ 1.3	11.4 $\pm$ 4.6
Ischaemia (20 min)	1.8 $\pm$ 0.6	12.0 $\pm$ 6.2

Data were interpolated by a single binding site model and are mean $\pm$ SEM of 3 hearts in each group. The differences between groups are not statistically significant.

analysis by a single binding site model provided a Kd of 71 $\pm$ 16 nM and a Bmax of 398 $\pm$ 78 fmol/mg protein. In ischaemic hearts (n=3) a significantly different fit was obtained: Kd and Bmax averaged 208 $\pm$ 31 nM and 1206 $\pm$ 211 fmol/mg protein ($p < 0.02$ and $p < 0.05$, respectively). Since A1 receptors do not appear to be affected by ischaemia, such difference should be attributed either to the A2 component or to the "non receptor" component of NECA binding. In some experiments 3 nM ^{3}H-NECA binding was determined in the presence of 50 nM and 100 μM CPA: according to Bruns et al.,[9] the fraction of binding which is displaced by 100 uM CPA but is not displaced by 50 nM CPA represents A2 receptor binding. A2 NECA binding averaged 4.4 $\pm$ 0.5 fmol/mg protein in control hearts and 2.2 $\pm$ 1.1 fmol/mg protein in ischaemic hearts. This finding suggests that the changes in ^{3}H-NECA binding observed in the ischaemic group reflect, at least in part, modifications in A2 receptors. This refers in particular to the reduction in affinity, while the increase in Bmax is likely to represent increased "non receptor" binding.

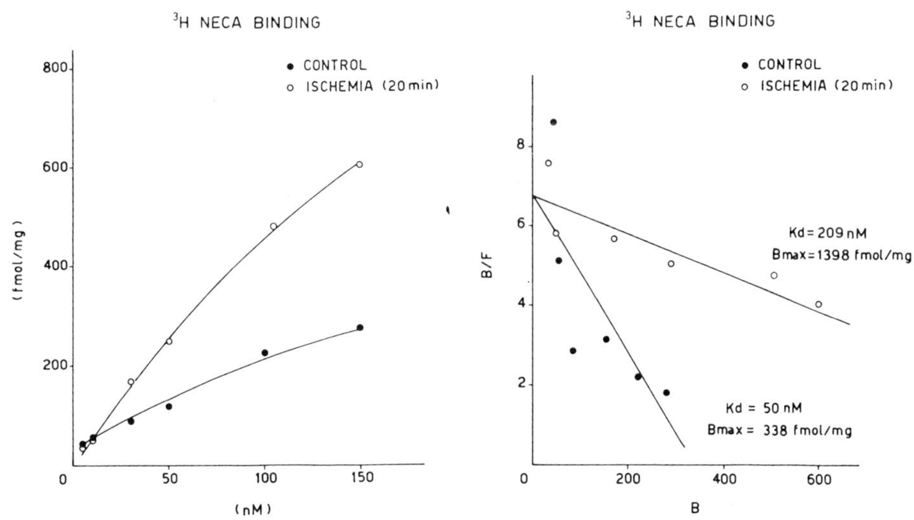

Fig. 1. Specific ^{3}H-NECA binding in a control rat heart and in a heart subjected to 20 min of ischaemia. (Left) Direct plot of a typical experiment. (Right) Scatchard plot of the data.

Table 2. Effect of ischaemia on ^{3}H-NECA binding.

	Kd (nM)	Bmax (fmol/mg)
Control	71 $\pm$ 16	398 $\pm$ 78
Ischaemia (10 min)	165 $\pm$ 30*	517 $\pm$ 11
Ischaemia (20 min)	208 $\pm$ 31**	1206 $\pm$ 211*
Ischaemia (30 min)	393 $\pm$ 48***	1732 $\pm$ 304**

Data are mean $\pm$ SEM of 3 hearts in each group.
* $p<0.05$, ** $p<0.02$, *** $p<0.01$ vs control.

As shown in Table 2 the effect on ^{3}H-NECA binding was proportional to the duration of ischaemia, and significant changes were observed even after brief ischaemia, unable to produce irreversible tissue injury.

In conclusion, ischaemia does not appear to affect A1 receptors, but determines significant changes in A2 receptors, which might contribute to the reduced coronary response to adenosine.

REFERENCES

1. Lubbe W.F., Cardiac action of adenosine, Meth. Find. Exptl. Clin. Pharmacol. 6:171 (1984).
2. Paton D.M., Classification of adenosine receptors, Meth. Exptl. Clin. Pharmacol. 6:167 (1984).
3. Ronca-Testoni S., Poddighe R., and Zucchi R., Adenosine binding sites in pig ventricular sarcolemma and interaction with calcium channels, Int. J. Tiss. Reac. 10:89 (1988).
4. Lohse M.J., Ukena D., and Schwabe U., Demonstration of Ri-type adenosine receptors in bovine myocardium by radioligand binding, Naunyn-Schmiedeberg's Arch. Pharmacol. 328:310 (1985).
5. Bohm M., Pieske B., Ungerer M., and Erdmann E., Characterization of A1 adenosine receptors in atrial and ventricular myocardium from diseased human hearts, Circ. Res. 65:1201 (1989).
6. Nees S., Herzog V., Becker B.F., Bock M., Des Rosiers C., and Gerlach E., The coronary endothelium: a highly active metabolic barrier for adenosine, Basic Res. Cardiol. 80:515 (1985).
7. Vanhaecke J., Flameng W., Borgers M., Jang I-K., Van de Werf F., and De Geest H., Evidence for decreased coronary blood flow reserve in viable postischemic myocardium, Circ. Res. 67:1201 (1990).
8. Zucchi R., Limbruno U., Poddighe R., Mariani M., and Ronca G., The adenosine hypothesis revisited: relationship between purine release and coronary flow in isolated rat heart, Cardiovasc. Res. 23:125(1989).
9. Bruns R.F., Lu G.H., and Pugsley T.A., Characterization of the A2 adenosine receptor labeled by ^{3}H-NECA in rat striatal membranes, Mol. Pharmacol. 29:331 (1986).
10. Bruns R.F., Daly J.W., and Snyder S.H., Adenosine receptors in brain membranes: binding of N^6-cycloexyl [^{3}H]adenosine and 1,3-diethyl-8-[^{3}H]phenylxanthine, Proc. Natl. Acad. Sci. USA 77:5547 (1980).

A FERRITIN-CYCLOHEXYLADENOSINE PROBE FOR ELECTRON MICROSCOPE LOCALIZATION OF A_1 ADENOSINE RECEPTOR

Claudia Martini, Umberto Montali ^, Laura Giusti, Gino Giannaccini, Alessandra Falleni*, Vittorio Gremigni*, Antonio Lucacchini

Istituto Policattedra di Discipline Biologiche, Ist. di Chimica Biologica ^, Dip. di Biomedicina Sperimentale, Infettiva e Pubblica*, University of Pisa, PISA, Italy

INTRODUCTION

Adenosine is an intermediate in the pathway of purine nucleotide degradation. Extracellularly, adenosine and its nucleotides are believed to play a regulatory function in cell to cell comunication via purinergic receptors (Burnstock, 1981). The autacoid nature of adenosine's action is thought to occur through two high affinity plasma membrane receptors. Biochemical and pharmacological studies have led to the sudvision of these receptors into the A_1 and the A_2 subtype (van Calker et al. ,1978; Londos et al., 1980). Only recently a significant progress has been made in the isolation and characterization of these receptors that are linked to adenylate cyclase (the A_1 subtype mediating an inhibition and the A_2 subtype mediating a stimulation) through the G-protein (Munshi and Linden, 1989; Nakata, 1989). Therefore the localization of A_1 adenosine receptors is important to understand the effects of purinergic stimulation. The distribution of A_1 adenosine receptors in different tissue components of various organs have been determined at the light microscope level using autoradiography (Lewis et al., 1981; Goodman and Snyder, 1982; Lee and Reddington, 1986; Palacious et al. 1987; Weber et al.,1990). This technique has not enough resolution to determine the localization of the A_1 receptor at subcellular level. The purposes of this study were to synthesize a ferritin-labeled probe that could bind to the A_1 receptor and to use this probe to visualize the A_1 adenosine receptor at the electron microscope level in synaptoneurosome preparations. Ultrastructural and biochemical studies have shown that this preparation contains both pre- and postsynaptic membranes which formed closed vescicles that retain a number of receptor mediated properties (among which accumulation of cyclic AMP in response to catecholamines, serotonin, adenosine and vasoactive intestinal peptides) and mantain an electrochemical gradient (Hollingsworth et al., 1984).

MATERIALS AND METHODS

^{3}H-CHA (^{6}N-Cyclohexyladenosine) was from New England Nuclear (specific activity 27 Ci/ mmol), ferritin was from Boehringer Mannheim, all other compounds, salts and reagents were from standard commercial sources.

Synthesis of the probe. CHA was oxidized essentially as described by Gilhman (1971) : two cold solution of 5.7 mM CHA and 5.7 mM metaperiodate in 50 mM Tris-HCl pH 7.0 were mixed and the reaction was allowed to proceed for 1 hour in the dark at 0° C . Oxidized CHA (1.1 μmoles) was incubated with ferritin (0.1 μmoles) for 40 min at 25° C in a final volume of 1.3 ml. For preparation of a stable oxidized

CHA-ferritin complex, the Schiff bases, formed by reaction of the aldheidic function of the oxidized CHA with primary amines present in ferritin, were reduced by addition of $NaBH_4$ at a final concentration of 80 mM in the incubation mixture. The reduction was allowed to proceed for 60 min. at 25°C and the reaction mixture was dyalized exhaustively against 50 mM Tris-HCl buffer at pH 7.7. The substitution (mmol of CHA/ mmol of ferritin) was determined by carrying out a parallel preparation in which labelled CHA was used.

Synaptoneurosome preparation. Synaptoneurosome were prepared from guinea pig brain as previously described by Hollingsworth et al. (1985). Briefly, cerebral cortices were dissected free from white matter and homogenized in 1:7 vol of buffer containing 20 mM HEPES-Tris ,118 mM NaCl, 4.7 mM KCl, 1.18 mM $MgSO_4$ and 2.5 mM $CaCl_2$ at pH 7.4 (T buffer) using a Potter homogenizer. After dilution the homogenate was filtered through three layers of nylon mesh (160 μm) and then filtered through a 10 μm Mitex-filter (LCWP047 Millipore). The filtered preparation was centrifuged at 1000 g for 15 min and the pellet was washed by centrifugation (1000 g for 15 min).

Binding assay. ^{3}H-CHA binding assay was performed as previously described by Martini and Lucacchini (1987). ^{3}H-CGS binding was carried out as previously described by Jarvis et al. (1989). The concentrations of CHA-ferritin that inhibit specific ^{3}H-CHA binding and ^{3}H-CGS binding by 50% (IC_{50}) were determined by log-probit analysis with six concentrations of the ferritin derivative each performed in triplicate.

Electron Microscope Localization. Synaptoneurosomes were incubated with CHA-ferritin (1 μM) without (total binding) and with (nonspecific binding) 160 μM L-phenylisopropyladenosine (PIA) in T buffer for 8 hours at 4°C. The assay mixtures were then centrifuged for 1 min in an Eppendorf centrifuge and rinsed three times in cold buffer .Pellets were cut in small pieces (1 mm^3) and fixed in 2.5 % glutaraldehyde and 1% paraformaldehyde in 0.1 M cacodylate buffer at pH 7.4 for 1 hour at 4°C. After rinsings in the buffer specimens were postfixed in 0.1M cacodylate buffered 1% osmium tetroxide at pH 7.4 for 1 hour at room temperature. Following dehydration in a gradied series of ethanols the specimens were briefly transferred in propilene oxide and embedded in Epon-Araldite. Thin sections (30-50 nm thick) were cut with a diamond knife on LKB ultratome III and placed on formvar-carbon coated grids. Control sections were stained with uranyl acetate since lead citrate obscures ferritin particles. Grids were examined with a Siemens Elmiskop 101 electron microscope. The number of ferritin particles in the synapse region was determined by analyzing photographyc fields (X 100000) for specimens treated with either CHA-ferritin and CHA-ferritin plus PIA. The neurosome membrane was analyzed with a scale Lupe 10X for the reckoning of ferritin particles. The number of ferritin-particles was referred to unit of length (1 μm) of the synaptic neurosome membrane.

RESULTS AND DISCUSSION

The aldehydic functions of the periodate-oxidized CHA may react with primary amines present in ferritin to form Schiff bases. These are reducted with $NaBH_4$ to form stable secondary amines. In our condition the substitution was approximately 1 mol of CHA/ 6 mol of ferritin.

Ultrastructural observations of control sections revealed the presence of numerous membrane-bound sacs measuring 0.4-0.8 μm in diameter. Some of them containing small translucent vescicles were often attached to apparently empty sacs (fig.1). These two types of sacs correspond to the so-called synaptoneurosome consisting of a presynaptic (synaptosome) and of a postsynaptic sac (neurosome) as described by Hollingsworth et al. (1985)in a preparation from guinea pig cerebral cortex. The CHA-ferritin was tested for its ability to bind to the adenosine receptor. The CHA-ferritin was assayed at various concentrations, and it inhibited ^{3}H-CHA binding with an IC_{50} of 17 ±0.9 nM (mean ± SD), compared with an IC_{50} for CHA of 3.0 ±0.3 nM. Moreover, the A_1 adenosine receptor selectivity of CHA-ferritin was evaluated by its displacing activity of the ^{3}H-CGS 21680 binding , a selective agonist to A_2 receptors

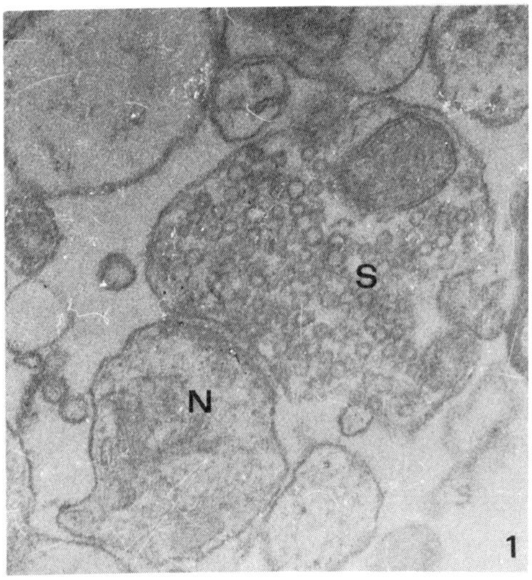

Fig.1. The presynaptic sac (S) with small translucent vescicles is attached to the postsynaptic sac (N). x 82,500

CHA-ferritin gave an $IC_{50} > 1000$ compared with an IC_{50} for CHA of 695 ± 80 indicating an increased selectivity for A_1 receptors of the probe. After incubation of CHA-ferritin solution with synaptoneurosome preparations either with or without L-PIA, the samples were examined by electron microscope. Specimens treated with CHA-ferritin showed ferritin paticles distributed on the neurosome membrane. Ferritin particles appeared more concentrated in the area where neurosome was attached to the synaptosome (fig.2).

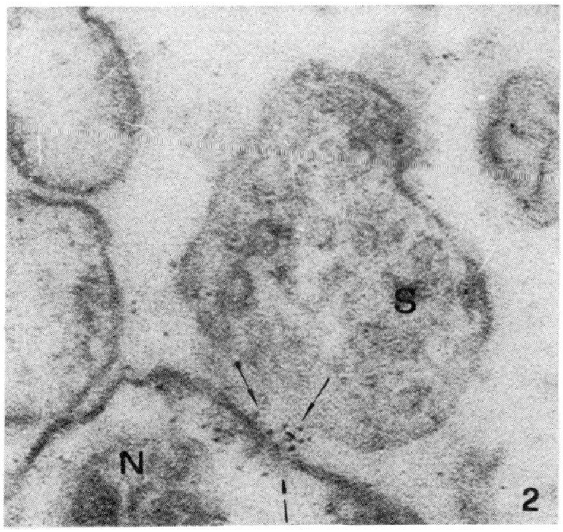

Fig.2.Ferritin particles are visible on the postsynaptic neurosomial membranes (arrows) x115,00

The examination of 26 synaptoneurosomes from mCHA-ferritin treated specimens showed 166 grains of ferritin on a total of 4.66 μm of postsynaptic membrane with an average of 39.5 ± 16.3 particles of ferritin per μm of membrane (total binding). The examination of 26 synaptoneurosome from CHA-ferritin plus L-PIA treated specimens revealed 94 grains of ferritin on a total of 5.71 μm of postsynaptic membrane with an average of 17.0 ± 12.0 particles of ferritin per μm of membrane (non specific binding). There was a significant difference between total and nonspecific binding ($p < 0.001$). These data suggest that the oxidized CHA bound to ferritin is an useful tool for the localization of the A_1 receptors.

ACKNOWLEDGMENT

This work was supported in part with a grant of M.U.R.S.T. (40%)

REFERENCES

Burnstock, G., and Brown C.M., 1981, An introduction to purinergic receptors, in "Purinergic receptors" Burnstock G. ed. p.1, Chapmann and Hall, London.

Gilham P.T., 1971, The covalent binding of nucleotides, polinucleotides and nucleic acid to cellulose, Methods Enzymol., 21 :191.

Goodman, R.R. and Snyder, S.H., 1982, Autoradiographic localization of adenosine receptors in rat brain using [^{3}H] cyclohexyladenosine, J.Neurosci., 2: 1230.

Hollingsworth E.B., Creveling C.R., Mc Neal E.T. and Daly J.W., 1984, A filtered neurosome enriched preparation from brain: morphology, receptors, enzymes, and cyclic AMP responses. Fed. Proc. 43: 1093.

Hollingsworth E.B., Mc Neal E.T., Burton J.L., Williams R.J., Daly J.W., and Creveling C.R., 1985, Biochemical Characterization of a filtered synaptoneurosome preparation from guinea pig cerebral cortex: cyclic adenosine 3':5' -monophosphate- generating systems, receptors and enzymes. J. Neurosci. 5: 2240.

Lee K.S. and Reddington, M., 1986, Autoradiographyc evidence for multiple CNS binding sites foe adenosine derivatives, Neuroscience, 19: 535.

Lewis, M.E., Patel, J., Edley,S.M., and Marangos P.J.,1981, Autoradiographic visualization of rat brain adenosine receptors using N^6-cyclohexyl[^{3}H]adenosine, Eur.J. Pharmacol. 73: 109.

Londos C., Cooper D. M. F. and Wolff J., 1980, Subclasses of external adenosine receptors, Proc. Natl. Acad. Sci. USA 77: 2551.

Martini C. and Lucacchini A., 1987, Affinity labeling of adenosine A_1 binding sites, J. Neurochem., 49: 681.

Munshi, R. and Linden, J., 1989, Co-purification of A_1 adenosine receptors and guanine nucleotide-binding proteins from bovine brain , J.Biol.Chem., 264, 14853.

Nakata,H., 1989, Purification of A_1 adenosine receptors from rat brain membranes, J. Biol. Chem., 264: 16545.

Palacios, J.M. and Pazos A., 1986, "Axonal transport of neuro transmitter receptors studied byquantitative autoradiography" in Quantitative receptor autoradiography, Boast:, Snowhill E. and Altar C.A. eds, pp. 173, Alan R. Liss, New York.

Jarvis, M.F., Schulz R., Hutchinson A.J., D o U.H., Sills M.A. and Williams M., 1989, ^{3}H- CGS 21680 a selective A_2 adenosine receptor agonist directly labels A2 receptors in rat brain, J. Pharmacol.Exper. Pharmacol.,251: 888.

H.Van Calker, D. Muller M., and Hamprecht, B., 1978, Adenosine inhibits the accumulation of cyclic AMP in cultured brain cells, Nature, 276: 839.

Weber R.G., Jones C.R., Lohse M.J., Palacios J.M., 1989, Autoradiographyc visualization of A_1 adenosine receptors in rat brain with [^{3}H]8-cyclopentyl--1,3-dipropylxanthine, J.Neurochem., 54: 1344.

EFFECT OF ADENOSINE ON THE MEMBRANE POTENTIAL CHANGES OF HUMAN NEUTROPHILS TRIGGERED BY THE TRIPEPTIDE F-MET-LEU-PHE

E. Capuozzo, W. Siems, C. Salerno, and C. Crifò

Departments of Biochemical Sciences and Human Biopathology, University of Rome La Sapienza, Italy

INTRODUCTION

The binding of adenosine to specific A_2-receptors on the surface of activated human granulocytes causes inhibition of superoxide anion generation, promotes chemotaxis but exerts little or no effect on either degranulation or aggregation [1,2]. Since the alteration of the granulocyte response appears to be restricted to Ca^{2+}-dependent stimuli (the calcium ionophore A23187, the chemotactic peptide f-met-leu-phe), it has been suggested [3] that adenosine can affect cellular functions by interfering with intracellular calcium distribution. It has been shown [3,4] that adenosine reduces calcium accumulation in A23187-activated granulocytes. As far as the activation by f-met-leu-phe is concerned, experiments with fluorescent Ca^{2+} probes gave rise to rather contradictory results [4-6].

It has been recently reported [7] that granulocyte calcium influx is strongly accelerated by voltage-gated calcium channels. Changes in transmembrane potential and in cytoplasmic Ca^{2+} concentration can be observed during the early stages of cell activation. However, the potential events are not always temporally coincident with the alterations in Ca^{2+} level [8].

In this paper, we analyse the effect of adenosine on the transmembrane potential changes triggered by f-met-leu-phe. Moreover, we report data obtained with an improved Ca^{2+} indicator not suffering from the severe and acknowledged limitations of other probes [9,10].

MATERIALS AND METHODS

The experiments were performed in isotonic phosphate-buffered saline, pH 7.4 at 37°C, unless otherwise stated. Human leukocyte preparations containing 90-98% of polymorphonuclear cells and apparently free of contaminating erythrocytes were obtained as previously described [11]. Membrane potential was evaluated by measuring fluorescence intensity of the probe 3,3′-dipentyloxacarbocyanine iodide (0.1 μM) with excitation and

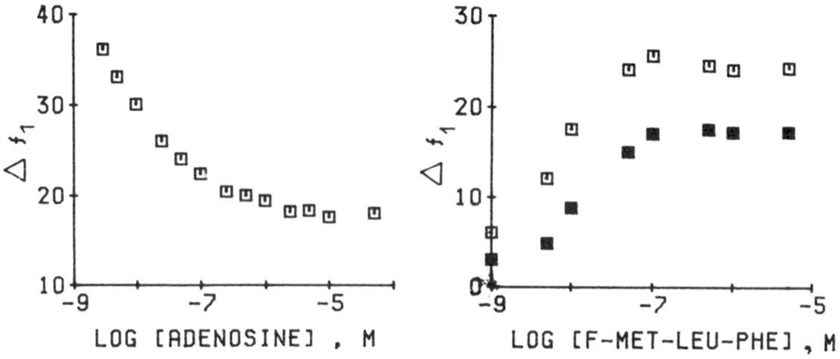

Fig. 1. *Left:* Δf_1 as a function of adenosine concentration. F-met-leu-phe concentration was 0.1 µM. *Right:* Δf_1 as a function of f-met-leu-phe concentration. Closed symbol refer to experiments in the presence of 250 µM adenosine.

Table I. Rank order of potencies for inhibition of Δf_1 signal

	Affinity constant	Extent of inhibition arbitrary units
Adenosine	22 nM	18
PIA(-)	35 nM	21
PIA(+)	65 nM	24
AMP	> 50 µM	n.d.

emission wavelengths of 460 and 510 nm, respectively [12]. Intracellular calcium was fluorimetrically monitored by loading the cells with the acetoxymethyl ester of Indo I (Molecular Probes) as described in reference [10].

RESULTS

Transmembrane potential changes

Incubation of granulocytes with micromolar concentrations of f-met-leu-phe resulted in a biphasic response [11] consisting of a rapid decrease in the potential probe signal, measured quantitatively as Δf_1, followed by an increase of the fluorescence to a maximum value (Δf_2). The duration of the response was within 5-10 min depending on the particular cell donor and the concentration of peptide used.

Addition of adenosine in the suspending medium decreased Δf_1 (Fig. 1) but did not appreciably affect Δf_2. The effect on Δf_1, still evident at nanomolar concentrations, was saturated by micromolar concentrations of adenosine. The response obtained in the presence of adenosine was unaffected by increasing f-met-

Table II. Effect of granulocyte preincubation at 37°C

Treatment	Δf_1 2 min	Δf_1 20 min	Δf_2 2 min	Δf_2 20 min
		arbitrary units		
None	37.2	26.8	96.5	94.6
50 nM Adenosine	23.9	16.9	97.0	96.9
10 μ M EHNA	31.7	26.1	93.3	89.2
1 U/ml Ado deaminase		30.0		93.26

leu-phe concentration (0.1-10 μM) and/or Ca^{2+} concentration (0-2 mM) in the suspending medium. N^6-Phenylisopropyl derivatives of adenosine (PIA) were slightly less potent than adenosine, as would be expected for A_2 receptors (Table I) [13]. AMP, at relatively high concentrations (10-500 μM), decreased Δf_1 but this effect disappeared after addition of 1U/ml adenosine deaminase. No remarkable effect was observed in the presence of micromolar concentrations of IMP, inosine, hypoxanthine, and adenine.

To determine whether endogenously generated adenosine was of significance for decreasing Δf_1, granulocytes were preincubated at 37°C before stimulation with f-met-leu-phe. We observed that Δf_1 decreased by increasing the preincubation time. Addition of adenosine deaminase (1 U/ml) in the suspending medium prevented the fall of the Δf_1 signal. Opposite effect (though less pronounced) was exerted by 10 μM EHNA (Table II).

Intracellular calcium concentration changes

Addition of $CaCl_2$ to granulocyte suspensions caused an increase of the Indo I signal (Fig. 2). After reaching a peak value, fluorescence decreased to a new steady level. F-met-leu-phe induced a burst of the fluorescence signal, which slowly fell towards basal levels. The signal amplitude increased with $CaCl_2$ concentration, reaching a maximum value in the presence of an excess of salt. None of these Indo I signals were appreciably affected by adenosine.

CONCLUSIONS

Our data indicate that adenosine does not interfere with the intracellular calcium distribution during the activation process. This conclusion is in line with the observation that adenosine inhibits only the depolarization step (Δf_1) of the membrane potential response triggered by f-met-leu-phe. In fact, membrane depolarization and changes in cytoplasmic Ca^{2+} concentration have been shown [10] to represent separate events in the granulocyte activation process.

It has been reported [14] that, in contrast to the hyperpolarization signal, Δf_1 is strictly related to the oxidative response of the activated cells [14]. Inhibition Δf_1 phase could be account for the specific impairment of the respiratory burst that has been observed in the presence of adenosine.

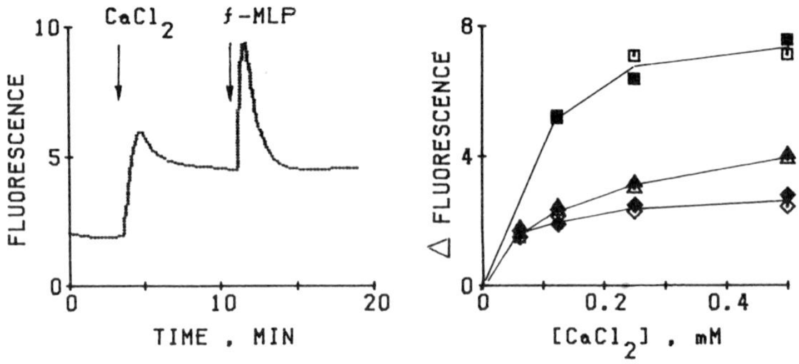

Fig. 2. *Left:* Time course of fluorescence signal of Indo I. Arrows indicate the addition of 0.5 mM $CaCl_2$ and of 0.1 μM f-met-leu-phe. *Right:* Amplitude of the fluorescence signal as a function of $CaCl_2$ concentration. ▲ : peak of the $CaCl_2$ signal; ◆: steady level of the $CaCl_2$ signal; ◘ : peak of the f-met-leu-phe signal.

REFERENCES

1. B.N.Cronstein, S.B.Kramer, G.Weissmann and R.Hirschhorn, J. Exp. Med. 158:1160-1177 (1983)
2. F.R.Rose, R.Hirschhorn, G.Weissmann and B.N.Cronstein, J. Exp. Med. 167:1186-1194 (1988)
3. M.Thiel and H.Bardenheuer, Ann. Nutr. Metab. 33:228-229 (1989)
4. F.L.Pasini, P.L.Capecchi, A.L.Pasqui, L.Ceccatelli, T. DiPerri, G.Valensin and E.Gaggelli E., Immunopharmacol. Immunotoxicol. 12:77-91 (1990)
5. K.M.Skubitz, N.W.Wickham and D.E.Hammerschmidt, Blood 72:29-33 (1988)
6. S.Tsuruta, S.Ito and H.Mikawa, FEBS letters 268:241-244 (1990)
7. A.Majander and M.Wikström, Biochim. Biophys. Acta 980:139-145 (1989)
8. K.G.Lazzari, P.Proto and E.R.Simons, J. Biol. Chem. 265: 10959-10967 (1990)
9. G.Grynkiewicz, M.Poenie and R.Y.Tsien, J. Biol. Chem. 260: 3440-3450 (1985)
10. K.Lazzari, P.J.Proto and E.R.Simons, J. Biol. Chem. 261: 9710-9713 (1986)
11. S.Jullien, E.Capuozzo, C.Salerno and C.Crifò, Biochem. Pharmacol. 41:2037-2040 (1991)
12. S.Jullien, E.Capuozzo, C.Salerno and C.Crifò, Biochem. Int. (1991) *in press*
13. J.W.Daly, J. Med. Chem. 25:197-207 (1982)
14. R.Sullivan, D.A.Melnick, H.L.Malech, T.Meshulam, E.R. Simons, K.G.Lazzari, P.J.Proto, A.S.Gadenne, J.L.Leavitt and J.D.Griffin, J. Biol. Chem. 262:1274-1281 (1987)

A_2 ADENOSINE RECEPTORS IN NEUTROPHILS FROM HEALTH VOLUNTEERS AND PATIENTS WITH RHEUMATIC DISEASE

Claudia Martini, Silvia Di Sacco, Paolo Tacchi, Laura Bazzichi*, Annalia Soletti*, Franca Bondi*, Maria Laura Ciompi*, Antonio Lucacchini

Istituto Policattedra di Discipline Biologiche, *Istituto Patologia Speciale Medica, University of Pisa, PISA (Italy)

INTRODUCTION

Inflammation is the primary response to tissue injury or microbial invasion and is characterized by the local accumulation of neutrophils. While neutrophils are essential for limiting the spread of infection, stimulated neutrophyls are capable of damaging injured tissues while en route to sites of infection or inflammation. It has been recently discovered that release of adenosine is one mechanism by which normal cells may protect themselves from activated neutrophils (Cronstein et al.,1986). Extracellular adenosine and its analogues diminish generation of toxic oxygen product (e.g. O_{2-} or H_2O_2) by activated neutrophils, yet promote neutropil chemiotaxis (Cronstein et al., 1985; Roberts et al., 1985; Rose et al., 1988). There are two types of cell surface adenosine receptors, A_1 and A_2 which are differentiated by their affinity for adenosine, the order of potency of adenosine analogues and opposing effects on cellular cAMP metabolism (van Calker et al., 1979). Recently it has been possible to label adenosine receptors by adenosine analogues such as ^{3}H-N^6-cyclohexyl adenosine (CHA) for A_1 adenosine receptors and ^{3}H-5'N-ethylcarboxamido adenosine (NECA) for A_2 receptors in cells from synovial fluid from rheumatic patients (Martini et al, 1989). In the present report we described the presence of A_2 adenosine receptors in neutrophils from health volunteers and from rheumatic patients. Initially, the sites were labeled by ^{3}H-NECA but the binding properties of this ligand did not agree with the pharmacology of A_2 receptors and indicated the presence of A_2 -like binding protein. Then, the characterization of A_2 adenosinc binding sites in human neutrophils was carried out using ^{3}H- (2 [p- (2- carboxyethyl)-phen ethylamido]-5' -N- ethylcarboxamido adenosine (CGS 21680), a novel high affinity and selectivity ligand for striatal A_2 receptors (Jarvis et al., 1989).

MATERIALS AND METHODS

^{3}H-NECA (specific activity 18 Ci/mmol), ^{3}H-CGS (specific activity 48.6 Ci/mmol) were from New England Nuclear. NECA, R-PIA (N^6- (R) phenylisopropyladenosine), 2-CADO (2-cloroadenosine) were from Boehringer Mannheim. All other chemicals were reagent grade and were from standard suppliers.Human neutrophils were isolated from heparinized whole blood after centrifugation through Hypaque-ficoll gradients sedimentation through dextran (6% wt/vol) followed by hypotonic lysis of erythrocytes. The study groups (25) of patients with different rheumatic disease. None of the patients was receiving corticosteroids at the time their PMN were studied. This procedure allows the study of populations of cells that are 98 ±2% neutrophils with few contaminating erytrocytes or platelets

(Boyun, 1968). The cells were washed with physiological solution, centrifuged at 50000Xg for 40 min at 4° C and homogenized in 20 vol of 20 mM Tris-HCl at pH 7.4 containing protease inhibitors as previously described (Martini et al., 1989) then centrifuged at 50000Xg for 40 min at 4° C. The resulting pellet was treated with adenosine deaminase (1 U/ml) at 37° C for 60 min to remove endogenous adenosine, centrifuged at 50000Xg for 40 min and used in the binding assays. ^{3}H-NECA binding assay was performed as previously described (Martini et al., 1989). Non specific binding was defined in the presence of 100 μM NECA or 100 μM R- PIA . Binding of ^{3}H-CGS to human neutrophil membranes was measured incubating membranes (0.2 mg proteins) with approximately 7 nM ^{3}H-CGS in a final volume of 0.5 ml of 50 mM Tris-HCl/10 mM $MgCl_2$ buffer at pH 7.5 for 90 min at 25° C. The samples were filtered into Whatman GF/C glass fiber filters under vacuum. After washing 3X4 ml of Tris-buffer, the radioactivity was counted in 8 ml of Ready-Protein Beckman scintillation cocktail. The specific binding was obtained in the presence of 100 μM NECA or 100 μM R-PIA.

RESULTS AND DISCUSSION

^{3}H-NECA was able to bind to human neutrophil mebranes but the binding properties of this ligand did not agree with the pharmacology of A_2 receptors. In fact while the specific binding obtained in the presence of 100 μM NECA amounting to approximately 85% of total binding, only 20% of ^{3}H-NECA bound to human neutrophil membranes could be displced by N^6-substituted adenosine analogue such as R-PIA. Competition experiments for ^{3}H-NECA binding were performed with different adenosine ligands in order to characterize the pharmacological profile of the binding site. The results are shown in the table I.

Table I

3H-NECA binding specificity

Compounds	IC $_{50}$ (nM)
NECA	224
R-PIA	> 100,000
2-CADO	1,476
Theophylline	518,000

Most adenosine receptor ligands did not compete significanty for ^{3}H-NECA binding apart from NECA itself and CADO. In addition, resistence to the addition of guanine nucleotides, a high density of binding site, and the esistence of rapid, simple one-site kinetics all suggest labeling of a site distinct from the adenosine A_2 receptor. These sites have been already demonstrated in other tissue (Hutchison and Fox, 1989). The relatively high-affinity of NECA and CADO for this site suggest the presence of a purine binding domaine, which does, however, recognize neither N^6-substituted adenosine derivative nor xanthines. Then, we using ^{3}H-CGS 21680 to label A_2 adenosine receptor in human neutrophil membranes. The specificity of ^{3}H-CGS binding is shown in the table II. The IC_{50} values of the compounds in competing for the binding of ^{3}H-CGS agree well with their affinity for the A_2 receptor. Binding characteristic studies showed that the binding of ^{3}H-CGS increased linearly with increasing protein concentration (0.1-1.5 mg proteins). The amount of specifically bound ^{3}H-CGS was temperature dependent and the highest value was reached at 25° C. The incubation time selected for the binding assay was 90 min. since at that time equilibrium had been reached between the ligand and binding sites. In

Table II

^{3}H-CGS 21680 binding specificity

Compounds	IC 50 (nM)
NECA	20
R-PIA	530
2CADO	160
Theophylline	27,100

order to test the saturability of ^{3}H-CGS specific binding sites membrane preparation were incubated with increasing concentrations of ^{3}H-CGS (1-60 nM) .

Specific ^{3}H-CGS 21680 was saturable; in contrast non specific ^{3}H-CGS 21680 binding increased linearly. Plotting the data of specific ^{3}H-CGS 21680 binding according to Scatchard method a single straight line was obtained indicating that a single population of specific binding sites and a Kd of 11.5 nM (see figure).

The ^{3}H-CGS 21680 binding satisfies essential criteria for the identification of membrane bound receptors since it was found to be specific, saturable, rapid, rapid reversible and dependent on protein concentration and temperature.
In table III the binding of 3H-CGS 21680 to membrane preparation of neutrophils from patients with rheumatic diseases is reported. Variations in the number of binding sites associated with diseases were observed. Therefore, it is likely that the activation of A2 adenosine receptors on neutrophils play an a complex role in the control of inflammatory reactions.

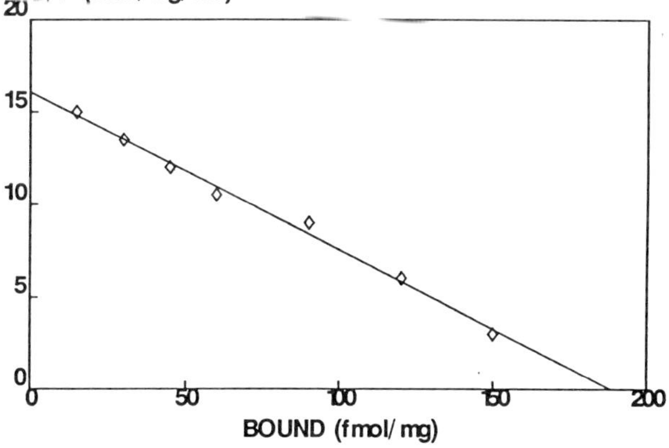

figure 1 Scatchard plot of 3H-CGS 21680 binding to neutrophil membrane

Table III

Bmax values for A2 receptors in some rheumatic disease.

Disease	fmol/mg protein
Rheumatoid arthritis(6)	142.9 ±20
Psoriatic arthritis (6)	59.2 ±12
Ostheoartrosis (6)	81.2±15
Serum negative arthritis(2)	147.7 ±50
Sarcoidosis (3)	76.8 ±15
Sjogren (2)	63.2 ±18
Healthy volunteers (14)	180.0 ±40

REFERENCES

Boyum A., 1968, Isolation of mononuclear cells and granulocytes from human blood. Isolation of mononuclear cells by centrifuga tion and sedimentation at 1 X g, Scand. J. Clin. Lab. Invest., 21:77.

Cronstein B.N., Levin R.I., Belanoff J., Weissmann G. and Hirsch horn R., 1986, Adenosine: an endogenous inhibitor of neutrophil- mediated injury to endothelial cells, J. Clin. Invest., 78: 760.

Cronstein B.N., Rosestein E.D., Kramer S.B., Weissmann G. and Hirschhorn R., 1985, Adenosine: a physiologic modulator of super oxide anion generation by human neutrophils. Adenosine acts via an A_2 receptor on human neutrophils, J. Immunol., 135: 1366.

Hutchison K.A. and Fox I.H., 1989, Purification and characterization of the adenosine A_2-like binding site from human placental membranes, J.Biol.Chem., 264, 19898.

Jarvis M.F., Schulz R., Hutchison A.J., Do U.H., Sills M.A., and Williams M., 1989, ^{3}H-CGS 21680, a selective A_2 adenosine recep tor agonist directly labels A_2 receptors in rat brain, J. Pharm. Exper. Ther., 251, 888. o7 3

Martini C., Giannaccini G., Lucacchini A., Bazzichi L., Soletti A. and Ciompi M.L., 1989, A_1 and A_2 adenosine recep tors in synovial cells from patients with rheumatic diseases, in Purine and Pyrimidine metabolism in man VI, Mikanagi K., Nishioka . and Kelley W.N. eds, p. 429, Plenum Publishing Corp. New York.

Roberts P.A., Newby A.C., Hallet M.B., and Campbell A.K., 1985, Inhibition by adenosine of reactive oxygen metabolite production by human polymorphonuclear leucocytes, Biochem. J., 277, 669.

Rose F.R., Hirschhorn R., Weissmann G. and Cronstein B.N., 1988, Adenosine promote neutrophils chemiotaxis, J. Exp. Med., 167, 1186.

van Calker D.M., Muller M. and Hamprecht B., 1979, Adenosine regulates via two different types of receptors, the accunulation of cyclic AMP in cultured brain cells, J. Neurochem., 33, 999.

INDEX